Microbial Metabolomics

Sukhminderjit Kaur
Sunny Dhiman • Manikant Tripathi
Editors

Microbial Metabolomics

Recent Developments, Challenges and Future Opportunities

Springer

Editors
Sukhminderjit Kaur
Department of Biotechnology
Chandigarh University
Mohali, Punjab, India

Sunny Dhiman
Department of Biotechnology
Chandigarh University
Mohali, Punjab, India

Manikant Tripathi
Biotechnology Program
Dr. Rammanohar Lohia Avadh University
Ayodhya, Uttar Pradesh, India

ISBN 978-981-96-4823-8 ISBN 978-981-96-4824-5 (eBook)
https://doi.org/10.1007/978-981-96-4824-5

This Springer imprint is published by the registered company Springer Nature Singapore Pte Ltd.
The registered company address is: 152 Beach Road, #21-01/04 Gateway East, Singapore 189721, Singapore

Preface

In recent years, there has been a significant surge in interest and research surrounding microbial metabolomics and its potential applications in various fields, from biotechnology and environmental science to healthcare and agriculture. Our book titled "Microbial Metabolomics: Recent Developments, Challenges, and Future Opportunities" delves into the dynamic world of microbial metabolomics, offering a comprehensive exploration of its latest advancements, persistent challenges, and the promising avenues it unveils for the future. With this book, the authors embark on a journey through the intricate landscape of microbial metabolomics, illuminating the profound insights it offers into the metabolic activities of diverse microorganisms. From bacteria to fungi, and beyond, the book unveils the intricate biochemical processes that govern microbial life, shedding light on their profound implications for fields ranging from healthcare, food, agriculture to environmental science.

The chapters unveil the recent breakthroughs in microbial metabolomics, showcasing current technologies and methodologies that have revolutionized our understanding of microbial metabolism. The book also confronts the multifaceted challenges that beset the field. Whether navigating the complexities of data integration or grappling with the intricacies of metabolite identification, the book offers invaluable insights into the obstacles that must be overcome to fully realize the potential of microbial metabolomics. The chapters also outline the future directions of microbial metabolomics, envisioning a landscape where personalized medicine, sustainable biotechnology, and ecological stewardship converge through the lens of metabolic insights. "Microbial Metabolomics: Recent Developments, Challenges, and Future Opportunities" seeks to capture the latest advancements in microbial metabolomics while also addressing the current challenges and envisioning future directions for this exciting field. This book is poised to become a go-to reference in the field, fostering innovation and knowledge dissemination in the exciting world of microbial metabolomics. The book will serve as a comprehensive resource on microbial metabolomics, with a focus on its applications and practical relevance.

Each chapter in this book, authored by experts, offers insights into foundational concepts, cutting-edge techniques, and forward-looking perspectives. We hope this resource inspires researchers, students, and professionals to advance the field of microbial metabolomics, fostering discoveries that reshape industries and improve lives.

Mohali, Punjab, India Sukhminderjit Kaur
Mohali, Punjab, India Sunny Dhiman
Ayodhya, Uttar Pradesh, India Manikant Tripathi

Contents

Editors and Contributors

About the Editors

Sukhminderjit Kaur is a Professor in the Department of Biotechnology at Chandigarh University, India. She earned her Ph.D. from Post Graduate Institute of Medical Education and Research (PGIMER), Chandigarh, in 2008 and has over 20 years of academic and research experience. She has published 58+ research and review articles in high-impact journals and contributed to several book chapters. She holds two patents for biofertilizer formulations and has edited the book "Futuristic Approaches in Biofertilizer Industry" (Springer Nature, 2 volumes). She has guided four Ph.D. students and received the Young Scientist Award in 2011 at the National Conference on Impact of Science and Technology on Societal Development. She is a member of the Indian Science Congress, the Association of Microbiologists of India, and the Gastrointestinal Society of India. Her research focuses on biofertilizers, probiotics, and bioremediation.

Sunny Dhiman an Associate Professor at the Department of Biotechnology, Chandigarh University, India, has been actively engaged in research, education, and scientific writing within his field for over 14 years. Dr. Dhiman has published more than 45 peer-reviewed articles, including 16 book chapters, 2 patents, and 3 international books. His research focuses on microbial biorefinery technologies, microbial metabolic capabilities for waste valorization, production, purification, and optimization of microbial enzymes and pigments. Beyond his research endeavors, Dr. Dhiman actively participates as a guest editor and reviewer for numerous international journals, reflecting his commitment to advancing scientific discourse and knowledge dissemination within his field.

Manikant Tripathi is an Assistant Professor in Biotechnology Program, Dr. Rammanohar Lohia Avadh University, Ayodhya, Uttar Pradesh, India. He has more than 10 years of teaching and research experience. He holds M.Sc. and Ph.D. in Microbiology. He has published 85 peer-reviewed articles, including 13 book chapters, and edited 5 international books. He is the recipient of many awards, including Young Scientist Award and Excellence in Research Award. He has guided postgraduate students on their dissertation work. He is also an Editorial Board Member of many international journals. Moreover, he has participated in several national and international conferences, seminars, and workshops. His current research work focuses on microbial biotechnology, biomass valorization, environmental and agricultural biotechnology, nanobiotechnology, and bioactive applications.

Contributors

Dickson Adom Department of Educational Innovations in Science and Technology, Kwame Nkrumah University of Science and Technology, Kumasi, Ghana

Ram Ajore Department of Hematogenomics, Lund University, Lund, Sweden

Beatriz Adriana Andrade-Espinoza Departamento de Ciencias Biomédicas, Centro Universitario de Tonalá, Tonalá, Jalisco, Mexico

Jaya Angom Department of Aquaculture, College of Fisheries, Central Agricultural University (Imphal), Lembucherra, Tripura, India

S. R. Arpitha Division of Biochemistry, ICAR-Indian Agricultural Research Institute, New Delhi, India

Jenny Arratia-Quijada Departamento de Ciencias Biomédicas, Centro Universitario de Tonalá, Tonalá, Jalisco, Mexico

Arundadhi Muthukumar Department of Microbiology-VRDL, Government Theni Medical College, Theni, Tamil Nadu, India

J. H. Ashwini Division of Plant Pathology, ICAR-IARI, New Delhi, India

K. Aswini Division of Microbiology, ICAR-Indian Agricultural Research Institute, New Delhi, India

C. V. Babu Department of Plant Biotechnology, University of Agricultural Sciences Bangalore, GKVK, Bengaluru, Karnataka, India

Farhat Bano College of Veterinary Medicine, Northeast Agricultural University, Harbin, People's Republic of China

Mukul Machhindra Barwant Department of Botany, Sanjivani Arts Commerce and Science College Kopargoan, Ahmednagar, Maharashtra, India

Prakash Bharti Department of Biological Sciences, Mahatma Gandhi Chitrakoot Gramodaya Vishwavidyalaya Chitrakoot, Chitrakoot, Madhya Pradesh, India

Sutapa Bhowmik School of Applied Sciences, Suresh Gyan Vihar University, Gyan Vihar Marg, Jaipur, India

K. K. Chetan Division of Plant Pathology, ICAR-IARI, New Delhi, India

Lucía Flores Contreras Departamento de Ciencias Biomédicas, Centro Universitario de Tonalá, Tonalá, Jalisco, Mexico

Anil Dahuja Division of Biochemistry, ICAR-Indian Agricultural Research Institute, New Delhi, India

Muhammad Ahsan Ullah Danish Department of Allied Health Sciences, The University of Agriculture, Dera Ismail Khan, Khyber Pakhtunkhwa, Pakistan

Siddhartha Das Department of Plant Pathology, Centurion University of Technology and Management, M.S. Swaminathan School of Agriculture, Paralakhemundi, Odisha, India

Sourabh Debbarma Fisheries College and Research Institute, Tamil Nadu Dr. J. Jayalalithaa Fisheries University, Thoothukudi, Tamil Nadu, India

Shubhankhi Dey University Institute of Biotechnology, Chandigarh University, Mohali, Punjab, India

Sunny Dhiman Department of Biotechnology, Chandigarh University, Mohali, Punjab, India

B. N. Ganavi Department of Plant Biotechnology, University of Agricultural Sciences Bangalore, GKVK, Bengaluru, Karnataka, India

Merugu Shashank Goud Division of Microbiology, ICAR-IARI, New Delhi, India

Namrata Gupta Department of Biotechnology, University Institute of Biotechnology, Chandigarh University, Mohali, Punjab, India

Kumar Nupur Hrishikeshan Division of Molecular Biology and Biotechnology, ICAR- Indian Agricultural Research Institute, New Delhi, Delhi, India

Harsimran Kaur Department of Biotechnology, University Institute of Biotechnology, Chandigarh University, Mohali, Punjab, India

Rajinder Kaur Department of Plant Sciences, University of Idaho Moscow, Moscow, ID, USA

Sukhminderjit Kaur Department of Biotechnology, Chandigarh University, Mohali, Punjab, India

T. Kavya Division of Microbiology, ICAR-IARI, New Delhi, Delhi, India

Kiran Kharb Department of Biotechnology, Chandigarh University, Mohali, Punjab, India

Venugopal Krishnan HIFU Lab, Department of Radiology, UT Southwestern Medical Center, Dallas, TX, USA

Suresh Babu Naidu Krishna Faculty of Health Sciences, Durban University of Technology, Durban, South Africa

Ashwani Kumar Department of Biotechnology, University Institute of Biotechnology, Chandigarh University, Mohali, Punjab, India

Anil Kumar School of Chemical and Metallurgical Engineering, University of the Witwatersrand, Johannesburg, South Africa

C. J. Tharun Kumar Division of Molecular Biology and Biotechnology, ICAR-Indian Agricultural Research Institute, New Delhi, Delhi, India

Deepak Kumar Department of Biotechnology, Chandigarh University Gharuan, Mohali, India

Parvind Kumar Department of Aquaculture, College of Fisheries, Central Agricultural University (Imphal), Lembucherra, Tripura, India

Ritik Kumar Department of Biotechnology, Chandigarh University Gharuan, Mohali, India

Shravan Kumar Department of Biochemical Engineering, Harcourt Butler Technology University, Kanpur, Uttar Pradesh, India

Rosalin Laishram Division of Biochemistry, ICAR-Indian Agricultural Research Institute, New Delhi, India

Sweety Maity Department of Molecular Biology and Biotechnology, Odisha University of Agriculture and Technology, Bhubaneswar, Odisha, India

Mallanna Malagatti Division of Seed Science and Technology, ICAR-Indian Agricultural Research Institute, New Delhi, India

Debdut Manna Division of Biochemistry, ICAR-Indian Agricultural Research Institute, New Delhi, India

Mohit Department of Prosthodontics, Faculty of Dental Science, King George Medical University, Lucknow, UP, India

Center for Advance Research, Faculty of Medicine, King George Medical University, Lucknow, UP, India

Hippolyte Tene Mouafo Institute of Medical Research and Medicinal Plant Studies, Yaounde, Cameroon

Gaurav Mudgal University Institute of Biotechnology, Chandigarh University, Mohali, Punjab, India

Karla Nuño Departament of Psychology, Education and Health, Nutrition, ITESO Jesuit University of Guadalajara, Guadalajara, Mexico

Arun Bhai Patel Department of Aquaculture, College of Fisheries, Central Agricultural University (Imphal), Lembucherra, Tripura, India

Chirag N. Patel Biotechnology Research Center, Technology Innovation Institute, Abu Dhabi, United Arab Emirates

Drug Design & Development Section, Translational Gerontology Branch, Intramural Research Program National Institute on Aging, NIH, Baltimore, MD, USA

Varsha S. Patil Department of Botany, Sanjivani Arts, Commerce and Science College Rahata, Ahmednagar, Maharashtra, India

Praveen S. Patted ICAR-CPRI, Regional Station, Ooty, Tamil Nadu, India

Muhammad Farhan Qadir School of Life Science, Shanxi University, Taiyuan, People's Republic of China

Monika Rai Division of Biochemistry, ICAR-Indian Agricultural Research Institute, New Delhi, India

Alok Raj Department of Chemical Engineering, Rashtrakavi Ramdhari Singh Dinkar College of Engineering, Begusarai, Bihar Engineering University, Patna, Bihar, India

Preeti Rajesh Department of Biotechnology, School of Biotechnology and Biosciences, Brainware University, Kolkata, India

Gopinath Ramalingam Department of Microbiology-VRDL, Government Theni Medical College, Theni, Tamil Nadu, India

B. N. Ramesh ICAR-AICRP on Post-Harvest Engineering and Technology, University of Agricultural Sciences, GKVK, Bengaluru, Karnataka, India

Rashmitha Division of Seed Science and Technology, ICAR-IARI, New Delhi, India

K. S. Rishika ICAR-National Institute for Plant Biotechnology (NIPB) LBS Centre, Pusa Campus, New Delhi, India

Víctor Ruíz-Santoyo Laboratorio de Investigación Interdisciplinaria, Área de Nanoestructuras y Biomateriales, Escuela Nacional de Estudios Superiores, Unidad León-UNAM, León Guanajuato, CP, Mexico

Sakshi Department of Biotechnology, Chandigarh University Gharuan, Mohali, India

Muhammad Asmat Ullah Saleem College of Veterinary Medicine, Northeast Agricultural University, Harbin, People's Republic of China

Reena Saxena School of Applied Sciences, Suresh Gyan Vihar University, Gyan Vihar Marg, Jaipur, India

Shalagha Sharma School of Biological Science, Shobhit Deemed to be University, Meerut, Uttar Pradesh, India

Anshul Singh School of Allied Health Sciences, Sharda University, Greater Noida, Uttar Pradesh, India

Balwant Singh Department of Botany, K.S. Saket PG College Ayodhya, Dr. Ram Manohar Lohia Avadh University Ayodhya, Faizabad, Uttar Pradesh, India

Gajendra B. Singh University Institute of Biotechnology, Chandigarh University, Mohali, Punjab, India

Tejveer Singh Division of Biochemistry, ICAR- Indian Agricultural Research Institute, New Delhi, Delhi, India

Vanapalli Lohitha Sai Sree Division of Plant Pathology, ICAR-IARI, New Delhi, India

J. Sushmitha Division of Biochemistry, ICAR- Indian Agricultural Research Institute, New Delhi, Delhi, India

Muhammad Salman Tahir Key Laboratory of Animal Disease and Human Health of Sichuan Province, College of Veterinary Medicine, Sichuan Agricultural University, Chengdu, People's Republic of China

Babita Thakur Department of Biotechnology, Chandigarh University, Mohali, Punjab, India

Sucila Thangam Ganesan Department of Microbiology-VRDL, Government Theni Medical College, Theni, Tamil Nadu, India

Manikant Tripathi Biotechnology Program, Dr. Rammanohar Lohia Avadh University, Ayodhya, Uttar Pradesh, India

Ramesh Chandra Tripathi Department of Biological Sciences, Mahatma Gandhi Chitrakoot Gramodaya Vishwavidyalaya Chitrakoot, Chitrakoot, Madhya Pradesh, India

Payal Trivedi Department of Biotechnology, University Institute of Biotechnology, Chandigarh University, Mohali, Punjab, India

T. Vinutha Division of Biochemistry, ICAR- Indian Agricultural Research Institute, New Delhi, Delhi, India

Gatta Vishnu Vyas Department of Plant Biotechnology, University of Agricultural Sciences Bangalore, GKVK, Bengaluru, Karnataka, India

Nitesh Kumar Yadav Department of Aquaculture, College of Fisheries, Central Agricultural University (Imphal), Lembucherra, Tripura, India

Yilbong Yirang ICAR—Central Institute of Fisheries Education, Mumbai, Maharashtra, India

Microbial Metabolomics: A Futuristic Approach in Biotechnology

Babita Thakur, Sukhminderjit Kaur, Hippolyte Tene Mouafo, Sunny Dhiman, Kiran Kharb, and Manikant Tripathi

Abstract

This chapter delves into the transformative role of microbial metabolomics in biotechnology, analyzing how it provides insights into microbial metabolic pathways, interactions, and applications in a variety of sectors. Beyond the analysis of genomic or transcriptomics data, microbial metabolomics provides a deeper knowledge of microbial systems by analyzing bioactive molecules generated by microbes. The chapter deals with the key analytical methods that allow for accurate metabolite identification and quantification, such as nuclear magnetic resonance (NMR) and gas and liquid chromatography combined with mass spectrometry (GC-MS, LC-MS). It also discusses how strain improvement, bioprocess optimization, biofuel generation, and pharmaceutical development are all influenced by metabolomic data. The promise of metabolomics in synthetic biology and systems biotechnology is highlighted, along with its integration with other omics techniques like proteomics and genomics. In environmental biotechnology, microbial metabolomics serves as an essential tool for agricultural applications that enhance crop yield and soil health, as well as for waste treatment and bioremediation.The chapter concludes with future prospects, including the growing use of artificial intelligence (AI) in metabolomics data analysis, which is expected to enhance data interpretation and open new avenues for customized biotechnology applications.

B. Thakur · S. Kaur (✉) · S. Dhiman · K. Kharb
Department of Biotechnology, Chandigarh University, Mohali, Punjab, India
e-mail: sukhminderjit.uibt@cumail.in

H. T. Mouafo
Institute of Medical Research and Medicinal Plant Studies, Yaounde, Cameroon

M. Tripathi
Biotechnology Program, Dr. Rammanohar Lohia Avadh University, Ayodhya, Uttar Pradesh, India

S. Kaur et al. (eds.), *Microbial Metabolomics*, https://doi.org/10.1007/978-981-96-4824-5_1

Keywords

Biotechnology · Microbial metabolomics · Bioactive compounds · Analytical techniques

1 Introduction

Microbial metabolomics as a new and advanced scientific discipline focuses on metabolome analysis including metabolites produced by microorganisms. Metabolites, such as sugars, amino acids, lipids, nucleotides, and other end products of cellular activities, reflect the chemistry and biochemistry events within a microbial system (Muhamadali et al. 2023). Understanding these metabolites and their structures provides important information related to microbes' function and physiology (Sarkar et al. 2023). Unlike genomics or transcriptomics, in which data are obtained regarding possible activities in a certain gene sequence or mRNA expression, metabolomics allows the measuring of active chemical processes and therefore how the specific organism functions in given conditions. This knowledge is critical in biotechnologically based systems involving the use of microbes as biological factories (Muhamadali et al. 2023).

In biotechnology, due to their ability to perform complex biochemical reactions effortlessly, microbes have proven to be important components in a variety of industrial applications. Metabolomics and more specifically its application in microbial metabolomics provide useful insights on what processes and strains to optimize to enhance microbial strain-based production of valuable metabolites, enzymes, antibiotics, alcohols, biofuels, and other biochemical processes (Horak et al. 2019). For instance, microbial fermentation which constitutes a vital process in the manufacture of alcohols, organic acids, and pharmaceuticals can be optimized by characterizing and controlling the metabolic routes responsible for the occurrence. In environmental biotechnology, metabolomics is crucial in the study of microbial activities such as bioremediation and the microbial degradation of wastes such as heavy metals and hydrocarbons (Pinu and Villas-Boas 2017). Thus, when the researchers know or estimate the products participating in these processes, they can design the actual means of environmental remediation. In agriculture, microbial metabolomics is used to understand the mechanisms of action of beneficial microbes including growth-promoting rhizobacteria affecting plant health and stress tolerance, as well as in designing improved and sustainable agriculture (Ibrahim et al. 2021).

Global and targeted studies are also part of metabolomics in microbiology. Global metabolomics is also known as metabonomics and is similar to transcriptomics in that it quantifies the entire complement of metabolites within a microbial system, which gives a snapshot of the organism's metabolic processes under various conditions or stressors (Wu et al. 2022b). In contrast, targeted metabolomics aims at the measurement of certain individual metabolites linked with specific metabolic pathways or functions like microbial antibiotic biosynthesis or biodegradation of environmental pollutants. Having both of these features allows researchers to gain a view of the microbial system on a whole, metabolic level, as well as separately concentrate on biochemical pathways of interest (Akyol et al. 2023). The metabolomics workflow usually involves harvesting metabolite samples from microbial cultures and utilizing complex analytical tools like GC-MS, LC-MS, or nuclear magnetic resonance spectroscopy to discover and profile metabolites (Fiehn 2016).

The data collected are then analyzed bioinformatically to determine significant changes in metabolites and pathways. This opinion can be supported by identifying the development stages of the field of metabolomics which has undergone drastic changes since its conception. Whereas in the initial days of metabolic studies, efforts were directed toward the identification of simple primary metabolites such as sugars and organic acids, metabolomics is a broader concept today that encompasses complex secondary metabolites some of which have tremendous ecological and industrial relevance (Chen et al. 2022).

Recent improvements in the measurement technologies such as high-resolution accurate-mass spectrometry and NMR have greatly impacted the field by allowing for the identification and quantitation of thousands of metabolites. The combination of these technologies with microbial genomics has also advanced the knowledge of microbial metabolism and enabled researchers to correlate certain metabolic signatures to certain genes and pathways (Emwas et al. 2019). The advent of systems in the first decade of the twenty-first century that focus on the analysis of biological components concerning systems has made metabolomics central to the understanding of microbial metabolism (Marshall and Powers 2017). It is now often integrated with genomics, transcriptomics, and proteomics to reflect a diverse aspect of microbial functions at various organizational levels. With the progression of the discipline, microbial metabolomics is increasingly contributing to synthetic biology, in which microorganisms with specific synthetic capabilities are created to manufacture certain substrates for commercial use (Nikoloudaki et al. 2024). This entails biofuel generation, biopolymers, drugs, and other superior added-value products. Metabolomics is used in synthetic biology to find out how the modified metabolic pathways perform in realistic conditions and how the modified pathways can be improved in terms of yield and efficiency (Alarcon-Barrera et al. 2022).

Looking at the future of microbial metabolomics, the combination of microbial metabolomics data with other omics data is also discussed. When integrated with genomic, transcriptomic, and proteomic data, metabolomics gives researchers a global view of the metabolic networks and the regulation of microbial behaviors (Manickam et al. 2023). Furthermore, the integration of artificial intelligence and machine learning into the metabolomics data analysis is revealing new horizons as it allows the prediction of some metabolic effects and thus helps in planning better microbial processes (González-Plaza et al. 2022).

Generally, microbial metabolomics is not only reinventing biotechnology but also has implications for human health and medicine. The investigation of microbial metabolites within the context of the human microbiome and, more specifically, the gut microbiota is providing key insights into how microbial metabolism impacts human health. For instance, as parts of microbiota-derived products, specific metabolites like short-chain fatty acids (SCFAs) and vitamins are valued for their functionality in supporting gut and immunologic homeostasis. Metabolomics also assists in the identification of other microbial metabolites with possible therapeutic uses, such as new antibiotics and anticancer compounds (Rahman et al. 2023). Moreover, to solve other emerging issues such as antibiotic resistance, microbial metabolomics studies have enabled researchers to identify associated metabolites essential in resistance mechanisms. Traditionally, microbial metabolomics has been enhanced by technological enhancements and an understanding of its capacity to transform several areas of biotechnology (Kaur et al. 2024b).

Most studies at the beginning of the twentieth century had been devoted to the identification of isolated metabolites in microbial cultures; however, the discipline only emerged in the early 2000s with bioinformatics and the constant development and improvement of mass spectrometry and NMR technologies (Martinez et al. 2017). Metabolomic data combined with genomic and proteomic data in addition to improvements in computational biology enable more extensive and systems-based analysis of microbial metabolism. Today, metabolomics is recognized as one of the main sources of systems biology approaches, especially in microbial studies. In the future, the field of microbial metabolomics will remain a valuable component of the biotechnological process (Kaur et al. 2024a). There are expectations that high-throughput technologies are likely to expand the scale and rate at which metabolomics is conducted while artificial intelligence in data analysis will improve the way different metabolomic datasets are analyzed (Eicher et al. 2020). The merging of technology with synthetic biology is found to be an area of potentially high growth known as personalized metabolomics where a microbial system can be designed to produce specific metabolites in accord with the needs of humans in matters of health care, nutrition, agriculture, and many industrial applications. In light of recent developments, microbial metabolomics will continue to be at the cutting edge of developments in biotechnology due to the profound insights into consuming microbial metabolic functions and offering exact controls over metabolic pathways (Khanijou et al. 2022).

2 Microbial Metabolomics

Microbial metabolomics is the study of the bioactive compounds that are produced by microorganisms and the role of these bioactive compounds in the metabolic routes (Ye et al. 2022). This section discusses some general principles of microbial metabolism and describes the pathways by which it occurs, the types of products that result, and how microbial metabolism can be affected by factors in the environment.

2.1 Metabolic Pathways in Microorganisms

In the microbial system, metabolic pathways are defined as the systematic array of various biochemical sequences that enable living organisms to survive. These pathways help the microorganism to utilize nutrients available in their environment to produce energy, biomass, and additional metabolites for existence and reproduction (Rowland et al. 2018). Metabolic pathways can be divided into two major types: catabolic and anabolic pathways.

Catabolic Pathways: These are processes by which energy is released from complex molecules such as carbohydrates, proteins, and lipids through the conversion into simpler molecules. For instance, glycolysis is an anaerobic catabolic process where glucose gets converted to pyruvate yielding ATP which is used by microorganisms (Orsi et al. 2022).

Anabolic Pathways: Anabolic processes on the other hand are the breakdown of larger complex molecules into smaller and simpler substances, a process that involves the use of energy. For instance, the synthesis of nucleotides, amino acids, and lipids which are requirements for both cell division and growth is a anabolic activity in microorganisms (Mashego et al. 2007).

Central Carbon Metabolism: These include glycolysis, tricarboxylic acid (TCA) cycle, and pentose phosphate pathway among others. These pathways represent the basis of microbial metabolism and include energy-producing processes, biosynthesis of numerous compounds, and redox processes in cellular organisms (Wu et al. 2023).

It is a real fact that the metabolic potential of microorganisms may differ significantly depending on their habitat. Some microorganisms are very versatile and can use a large number of substrates either aerobically or anaerobically, for example, *Escherichia coli*, while others are limited to certain habitats like extreme thermophiles which are adapted to high-temperature habitats (Poli et al. 2017). Knowledge of these pathways is used in metabolic engineering as sophistication involves the alteration of such pathways to improve the yield of the desired products such as biofuels or pharmaceuticals (Wu et al. 2023).

2.2 Primary vs. Secondary Metabolites

Microbial metabolites are broadly classified into two categories: major products and minor products which are both vital products important in the life cycles of microorganisms. Global metabolites are those that are used in cellular processes of microbial growth, development, and reproduction. These compounds are in many ways synthesized during the logarithmic or exponential phase when microbial cells are replicated (Fadlelmoula et al. 2022). Amino acids, nucleotides, lipids, and organic acids are some examples of organic molecules that are crucial in the human body's functions. Amino acids are important in protein synthesis and nucleotides in the synthesis of DNA and RNA, while lipids are important in the formation of membrane structures. Primary metabolites are synthesized in cells, rigorously controlled in manufacture, and are widely used, for instance, in the production of ethanol and organic acids by fermentation (Sreenivasulu and Fernie 2022).

On the other hand, secondary metabolites are compounds produced by microorganisms that are not essential for their basic life processes but are products that are important in the interaction with the environment and competition with other microorganisms. These compounds are typically synthesized during the stationary phase of microorganisms and when there are low levels of primary metabolites (Calvo et al. 2002). Some are antibiotic agents like penicillin and streptomycin that prevent the development of other microorganisms and pigments, for instance, melanin, which protect microbes against UV radiation; toxins for deterring enemies or competitors; and magnesium chelators such as siderophores which assist microbes to get iron in circumstances where there is little or no iron. It has been seen that most of the secondary metabolites depend upon environmental factors, including nutrient dosage and stress conditions. In biotechnology, secondary metabolites are very significant and are widely used in the manufacturing of antibiotics, immunosuppressants, and anticancer drugs (Rusu et al. 2023).

2.3 Key Metabolites and Their Functions in Microbial Communities

In microbial communities, metabolites can serve as signaling molecules, a source of nutrients, or a nucleophile, each playing a role in governing interactions among different microbial species or microbes with the surrounding environment. Some key metabolites and their functions in microbial communities include the following:

Short-chain fatty acids (SCFAs) are generated through fermentative bacteria, especially within the gut, and butyrate, acetate, and propionate SCFAs are of importance within the gut because they assist in controlling inflammation and act as energy sources for colonic cells. In other microbes, SCFAs also act as substrates to other microbes, doping a mutual metabolizing bureaucracy within the microbial ecosystem. Some of the small molecules are quorum-sensing molecules, for instance, the acyl-homoserine lactones (AHLs) (Zhang et al. 2023a). Quorum sensing makes it possible for bacteria to coordinate population behavior on the gene level and accomplish tasks on behalf of the whole community, such as biofilm formation, virulence, and sporulation. These signaling molecules are essential in pathogenic bacteria and are possible invasive targets (Rutherford and Bassler 2012). In competitive environments, microbes also secrete substances that affect the rate of growth of other microbial species. They provide a selective advantage to the producing organism and can exert a strong influence on the composition of the microbial consortia. For instance, *Streptomyces* species are producers of antibiotics and important members of soil communities where inhibitory effects on pathogenic bacteria and fungi are exerted (Fusco et al. 2023).

Biosurfactants are effective compounds that lower microbial growth surface tension, enable biofilm formation, and promote solubilization of hydrophobic compounds and microbial activeness. In microbial communities, biosurfactants can also act as an agent that affects the ability of microbes to colonize a surface and microbial interactions within soil, water, or even body tissue (Thakur et al. 2024). These metabolites have critical roles in fulfilling the requirements of lone microbes for their existence and growth, as well as preserving the general architecture and performance of microbial populations (Thakur et al. 2023a, b).

2.4 Interactions Between Microbial Metabolism and Environmental Factors

Microbial metabolism is therefore highly sensitive to specific influences like climate including temperature and pH, relative nutrient access, availability of oxygen, and the presence of toxins or pollutants. These environmental conditions define the form of metabolic networks and regulate the formation of metabolites; therefore understanding these interactions is important for biotechnological use and environmental sciences. Microbes are capable of adapting themselves depending on the amounts of nutrients present to them at a particular time (Gupta et al. 2017). For instance, when nitrogen or carbon sources are scarce, the outcome is that most microorganisms change from primary metabolism which is growth and reproduction to secondary metabolism which is survival and competition. It is worth mentioning that a range of secondary metabolites—such as antibiotics or pigments—is generally overproduced under nutrient-limited conditions. Indeed, the availability of oxygen has a large effect on microbial metabolic activity (Krysenko and Wohlleben 2024). Aerobic microorganisms use oxygen as the last in the list of acceptors of electrons in the electron transport system, while the anaerobes on the other hand use other acceptors and/or fermentation. *E. coli* is an example of facultative anaerobes—it can adapt either to aerobic or anaerobic condition depending on the oxygen available to it. This adaptability illustrates how microbes respond to the changes in their environmental conditions, adjusting their metabolic processes. Substrate and enzyme specificities are controlled by the natural environment's pH and temperature (Perez-Garcia et al. 2016). Thus, it is possible to state that any additional increase in metabolic activity is promoted by enzyme activation and effectuation of the conditions met by the enzymes in the cells, whereas any negative impact on the required conditions hampers the cell's metabolic processes. For instance, thermophilic bacteria adapt well to high temperatures mainly because of their ability to synthesize heat-stable enzymes which are helpful in industries such as the biofuel industry (Wortel et al. 2018). The internal environment conditions, including high concentrations of salts or other osmotic pressures, oxidative stress, or the presence of heavy metals, induce protectant metabolites. For instance, under oxidative stress, the microorganisms alter their rate of synthesis of antioxidant molecules, especially glutathione or catalase (Jan et al. 2015).

Such interactions are crucial in the existence of microbes in different ecological provinces. In metabolomics, the ability of microbes to alter their metabolites and the metabolic routes as a result of the changes experienced in the environment is well documented—this adaptability is particularly crucial for biotechnological processes such as bioremediation where microbes are tailored to live in habitats polluted with toxins and are expected to break them down (Wang et al. 2021).

3 Methodologies and Technologies in Microbial Metabolomics

Microbial metabolomics uses various novel techniques and technologies that allow the identification of microbial metabolites. These tools enable the researchers to dissect the intricate nature of microbial metabolism and understand the behavior of the microbes, their interactions, and mechanisms of metabolism under various stress conditions arising due to changes in niches (Mayers et al. 2024). Metabolomics, from sample collection to data analysis, combines advanced biochemical analytic technology with sophisticated data processing to generate meaningful insights. This approach provides a powerful strategy for understanding complex biochemical processes (Bishop et al. 2022). Here, we further discuss the strategies employed in microbial metabolomics such as the analytical platforms of measurement, sampling strategies, data collection, and data analysis as well as the use of metabolomics databases and tools.

3.1 Analytical Techniques

In this context, metabolomics utilizes several analytical methods that can be used in the detection, identification, and quantification of metabolites in complex organisms generated by microorganisms with high sensitivity and selectivity (Salem et al. 2020). The most popular approaches which are employed are gas chromatography-mass spectrometry (GC-MS), liquid chromatography-mass spectrometry (LC-MS), nuclear magnetic resonance (NMR), and Fourier transform infrared spectroscopy (FTIR) as shown in Fig. 1.

All of these technologies are rather valuable, but each of them has strengths and weaknesses according to the type of metabolites under investigation and the objective of the study (Liu et al. 2021). Gas chromatography-mass spectrometry (GC-MS) is particularly suitable for the determination of volatile and semi-volatile

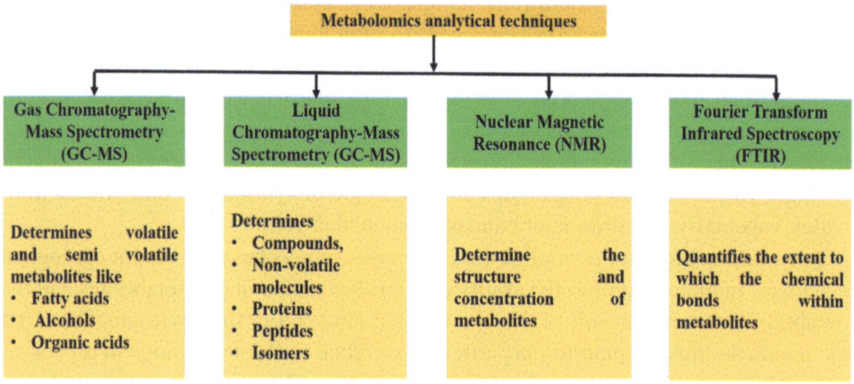

Fig. 1 Distinct types of metabolomics analytical techniques

metabolites which are small, nonpolar molecules, including fatty acids, alcohols, and organic acids. In this technique, the metabolite constituent of the sample is first introduced to a vapor phase and then proceeds through a gas chromatography column, in which metabolites with optimal volatility and specificity toward the sample's stationary phase are selected (Fiehn 2016). After separation, the metabolites are ionized by mass spectrometry, information derived from which will give details on the molecular weight of the metabolites and their structure. GC-MS analysis is very sensitive and provides good resolution, hence recommended for general metabolomics where an array of metabolites is to be quantified (Kumar and Baggi 2022). Nevertheless, GC-MS is not without its drawbacks; for example, nonvolatile metabolites such as sugars and amino acids must in turn be derivatized before analysis—an added step in sample preparation. This technology that can quantitatively measure a large number of metabolites irrespective of their polarity, volatility, and stability makes it one of the most popular techniques in metabolomics (Gould et al. 2023). In LC-MS, the sample is first introduced to a liquid chromatography system with substrates being separated depending on polarity and interaction with the "stationary" phase. The metabolites produced are ionized and quantitatively and qualitatively examined through mass spectrometry. Another benefit of LC-MS is that it has high sensitivity due to its ability to identify low-concentrated metabolites (Thomas et al. 2022). For the concentration determination of primary metabolites like sugars, amino acids, nucleotides, complex lipids, and secondary metabolites, it is particularly valuable. LC-MS also provides both hypothesis-based and non-hypothesis-driven metabolomics techniques where the researcher can choose the metabolic pathway of interest or not at all. Innovations in the LC-MS techniques used in recent years have included Orbitrap and Q-TOF systems in high-resolution mass spectrometry (Farag et al. 2023).

Nuclear magnetic resonance (NMR) spectroscopy is a nondestructive method used to determine the structure and concentration of metabolites according to their magnetism. NMR defines the resonance frequencies directly associated with atomic nuclei in a material through an external magnetic field, which is generally a proton (^{1}H) or carbon (^{13}C). Chosen chemical shifts and coupling patterns, which can be identified in the NMR spectrum, contain much information about the chemical structure of the metabolites (Wishart et al. 2022). Another advantage of NMR is that it offers structural data even if the sample metabolites are not pre-separated, which is essential for complying with the growing demands of metabolomic studies. NMR is also very reproducible and semiquantitative which enables the determination of the relative concentration of metabolites (Ocampos et al. 2024). Nevertheless, NMR has lower sensitivity compared with other mass spectrometry-based methods, which implies that it is not suited for the detection of rare metabolites. However, NMR is applied in over-targeted and non-targeted metabolomics and is nondestructive that retains samples for other studies (Tampieri et al. 2021). Fourier transform infrared spectroscopy (FTIR) is a technique that quantifies the extent to which the chemical bonds within metabolites absorb infrared light. Each type of bond absorbs radiation at a particular wavelength, and thus each compound or metabolite has its own spectral signature which can be used to determine the nature of a metabolite. FTIR is especially valuable for determining the contents of functional groups including hydroxyl, carbonyl, phosphate, etc., which are common in primary and secondary metabolites (Al-Kelani and Buthelezi 2024). FTIR is a fast and noninvasive methodological approach that does not necessitate extensive sample pretreatment. They are also commonly used in the identification of metabolic behavior of microbial cultures and biofilm formation, as well as in overall biochemical profiling. Nevertheless, FTIR offers lower sensitivity as compared with MS-based techniques and is employed in combination with other methods for further metabolomics studies. These analytical techniques are basic to any metabolomics approach, offering the instruments for creating extensive overviews of microbial metabolites. The use of several approaches would limit the shortcomings of individual procedures and provide a better perspective of the microbial metabolism (Fadlelmoula et al. 2022).

3.2 Sampling and Sample Preparation Techniques

Sample collection and sample preparation protocols have large direct effects on the quality and accuracy of metabolomics data. As microbial metabolites may be very labile and may change quickly depending on the samples' collected and processed conditions, special care should be taken to ensure that the metabolites are not degraded. Appropriate sample preparation secures the metabolites that are in focus while at the same time providing merit data from the analytical approaches (Rakusanova and Cajka 2024). Metabolomics starts by sampling: the sample may be colonies of microorganisms, biofilm, or a microbial sample taken from any environment, for example, soil, water supply, or clinical materials. This is particularly crucial for microbial metabolomics because the correct timing of sampling is critical when studying the dynamics of such processes as growth phases or the effect of certain stress factors. Sampling may include taking microbial cells from culture media, filtering environmental samples to separate microbial populations, or removing biofilms from the substratum (Vuckovic 2012).

With regard to microbial metabolomics, one of the problems is to avoid contamination of the samples and the subsequent interference with the actual metabolism of the microorganisms. For instance, variation in temperature, pH, or oxygen during sample collection can lead to a stress response that affects the metabolite profile. To overcome these problems, samples are immediately quenched through cooling or freezing techniques to stop metabolic activity and thereby maintain the inclusion of native metabolites (Scalbert et al. 2009). After sampling, the second process that is exercised is stopping microbial activity so that there is no alteration of metabolite concentrations. This is normally done by pouring the samples into liquid nitrogen or some ice-cold solvents well or by quick cold shock. Furthermore, the extraction of intracellular and extracellular metabolites is needed for further analysis. Following the type of metabolites bound, for extraction the different solvents are used. For instance, when working with metabolites like sugars, amino acids, and nucleotides, aqueous solvents like methanol, water, and ethanol are used (Pinu et al. 2019). These solvents effectively isolate the polar metabolites from the microbial cells and the simultaneous perturbation of the lipids and other hydrophobic substances. Lipids and nonpolar secondary metabolites are obtained using solvents like chloroform or hexane. The type and concentration of the solvent to be used can be dependent on the particular metabolite of interest and the compatibility of the solvent to the ensuing separation/negative analysis. Some samples may require further preparation after the extraction process is complete, for instance, filtration, centrifugation, or derivatization (Arteaga-Clemente et al. 2024). Particulate material or cell debris are commonly separated from the solution through filtration or centrifugation so that only the metabolites are measured. In GC-MS analysis, the metabolites associated may have to be derivatized chemically to enhance their volatility and improve their phasing within the GC column. The primary objective of sample preparation is to be able to prepare a good and consistent sample that will represent the metabolism of the microbial system of interest. Maintenance of quality samples is very crucial in improving the reliability of metabolomics data (Li et al. 2022).

3.3 Data Acquisition and Processing in Metabolomics

After the required sample preparation and analysis with suitable analytical tools and methods, the data acquisition and processing starts. Metabolomics results in the production of very large datasets, ranging from several hundreds of metabolites to thousands, for the analysis and/or interpretation of which special programs and methods have to be used. In the course of the analysis, analytical instruments produce raw data, for example, mass spectra or NMR spectra that correspond to detected metabolites (Saoi and Britz-McKibbin 2021). Instrumentation data acquisition in case of MS-based techniques such as GC-MS and LC-MS hence entails the measurement of mass to charge ratio (m/z) of the ionized metabolites providing a benchmark for molecular size and forms. In NMR, the acquired data reflect the chemical shift of an atom within the metabolite meaning structural information can be obtained from the data acquired. High-throughput instruments especially mass spectrometers produce the large number of dimensions with high accuracy (Alves et al. 2024). High-throughput analysis produces vast amount of high-dimensional data, particularly with today's instruments especially high-resolution mass spectrometers, which produce extensive and well-quantified datasets. However, raw data must undergo some processing to make the data usable for further analysis. Sample preparation and pretreatment of metabolomics data are part of preprocessing, a major step which includes the process of signal noise reduction, baseline correction, as well as peak matching across the different samples (Deschamps et al. 2023). In MS-based analyses, the peaks obtained based on m/z ratio and, in the case of LC-MS, the retention time are obtained due to metabolites. Peak alignment activity involves confirmation of the right match in metabolites from one sample and that of another (Santamaria and Pinto 2024).

Thus, metabolomics data are usually normalized to correct variability arising from sample preparation, extraction efficiency, and instrument sensitivity. These include the use of the total ion current (TIC) or internal standard as a means of normalizing the data for technical variation to attain a good comparison of the metabolic profiles between the samples. In this case, the data are further passed through a process to determine the metabolite profiles in the sample. Regarding MS-based technologies, metabolites are distinguished from one another according to their m/z ratios and compared with mass spectra from metabolomic databases (Zhang et al. 2023b). In NMR-based approach, the chemical shifts are compared with reference spectra to come up with corresponding metabolites. One of the common ways to quantify is internal standards which are added to the sample at certain concentrations and they help in calculating concentrations of pertinent metabolites. The other is the relative quantification in which the intensity of metabolite peaks is compared between sample groups. After its metabolites are detected and measured, authorization turns to analytical techniques, for interpretation of the data. Some of the multivariate statistical techniques applied widely in the field include PCA (principal component analysis) and PLS-DA (partial least squares discriminant analysis) for comparison of the sample's distinctiveness (Nagana Gowda and Raftery 2021). These methods enable the researchers to run 2-D plots to determine how metabolites are related to one another and, in turn, pinpoint which metabolites are significantly different between the experimental conditions. Heat maps, volcano plots, and pathway enrichment analysis are applied to endeavor the data and to portray the variations in metabolite concentration. These analyses enable scientists to assess the biological meaning of the revealed metabolite patterns and to gain a vision of metabolic pathways altered owing to conditions or gene mutations (Álvarez-Sánchez et al. 2010).

3.4 Metabolomics Databases and Resources

Metabolic profiling and analysis is therefore dependent on extensive databases and resources of metabolites including identification and quantification datasets. These databases offer reference spectra, structures, and biochemical data on metabolites, enabling the identification and analysis of metabolomics data.

KEGG Kyoto Encyclopedia of Genes and Genomes is one of the most popular and frequently used databases in metabolomics. It enables one to understand the metabolic routes, enzyme usage, and other biochemical transformations; hence researchers can overlay the metabolites onto characterized metabolic structures. The KEGG database is valuable for pathway mapping using metabolomics data, known as ontology data, which lacks functional annotations unlike genomics and proteomics data. It helps in studying the functional roles of metabolites in cellular processes (Huckvale and Moseley 2024). While primarily restricted to human metabolites, the Human Metabolome Database (HMDB) can also be used to search for microbial metabolites with emphasis on those affecting host-microbe interactions. HMDB covers databases regarding the structural and functional characteristics of metabolites and links to disease associations, which are of relevance to microbial communities in the clinic (Jin et al. 2023).

METLIN METLIN is a database of metabolite mass spectra, primarily covering MS metabolomics. It has over 450, metabolite entries mainly in humans, 50% of which are experimental spectra and the rest predicted ones, providing high confidence in the identification of individual metabolites from MS data. METLIN also provides support for untargeted metabolomics experiments where the unknown metabolites are used to search against the database for discovery purposes (Giera et al. 2024).

ChEBI ChEBI (Chemical Entities of Biological Interest) is a chemical database containing information on metabolites as to their structures, characteristics, and functions. It is extensively used in metabolomics, for the identification and annotation of compounds most especially when working with matrices containing multiple primary and secondary metabolites. MetaboAnalyst is a web-based tool for the analytical and graphical analysis of metabolomics data (Yones et al. 2021). It supports data preprocessing (resampling, normalization, baseline correction, and transformation), multivariate analysis, clustering evaluation based on associations and correlations, and grouped data analysis for pathway enrichment and biomarker identification. To perform pathway analysis and interpret metabolomics results, MetaboAnalyst provides ready-for-use integration with other databases—e.g., KEGG and HMDB. Utilizing these databases and resources, researchers will improve the accuracy of metabolite identification and gain a better understanding of the biological implications of their metabolomics data (Cai et al. 2023).

Microbial metabolomics provides valuable methodologies and technologies to probe the metabolic complexity of various microorganisms. Together, the innovative analytical strategies, coupled with powerful data processing workflows and comprehensive databases, provide scientists with new avenues to explore microbial metabolism for biotechnological applications (Meng et al. 2023). The advancement of these technologies should not only allow for improved understanding of the mechanisms that govern the structure and function of these essential microbial communities in industrial, agricultural, and environmental processes but also capitalize upon their functional and structural properties to further broaden their application (Khanijou et al. 2022).

4 Applications of Microbial Metabolomics in Biotechnology

Microbial metabolomics plays a critical role in various biotechnological applications, offering valuable insights into the metabolic states of microorganisms, optimizing industrial processes, and aiding in the development of new products. By profiling and analyzing metabolites produced by microorganisms, researchers can enhance microbial productivity, improve metabolic pathways, and better understand microbial interactions with their environments (Tanaka et al. 2024). This section explores the wide-ranging applications of microbial metabolomics in

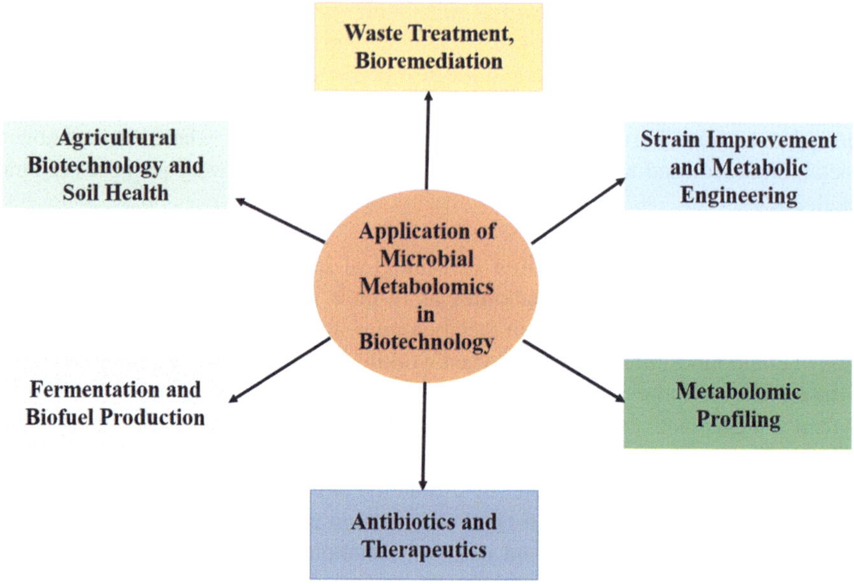

Fig. 2 Application of microbial metabolomics in biotechnology

biotechnology, spanning strain improvement, biofuel production, pharmaceutical development, agriculture, and environmental sustainability as shown in Fig. 2.

4.1 Strain Improvement and Metabolic Engineering

Microbial metabolomics in relation to biotechnology has several main focuses, such as strain improvement and metabolic engineering. Metabolomics are then mapped against the metabolic pathways of the microorganisms and metabolic inefficiencies are then explored and targeted for genetic modification. Metabolomics facilitates a snapshot of organismal metabolites and enables analysis of which pathways are active, not being utilized, or found to be blocked. Metabolomics can provide context for the introduction or modification of genes in metabolic engineering such that desired metabolites, including amino acids, organic acids, or biofuels, are produced at optimum levels (Dromms and Styczynski 2012). Metabolomics also facilitates the design for the optimal microbes for industrial purposes through the rational engineering of the metabolic pathways to redirect metabolic fluxes toward the synthesis of desired products. Metabolomics can be used to modify strains of *Escherichia coli* or *Saccharomyces cerevisiae* to overexpress or silence certain genes and, through this, to increase the yield of bio-based products in industrial fermentation (Joshi et al. 2022). Metabolic profiling can be used to identify and make modifications in pathway bottlenecks to produce higher productivity while reducing by-product formation. Similarly, metabolomics is used in strain improvement to create microbes that can tolerate stressors, like high temperatures, pH changes, or toxic by-products that can inhibit industrial production processes (Iman et al. 2022).

4.2 Bioprocess Optimization Using Metabolomic Profiling

One major use of microbial metabolomics is bioprocess optimization, especially in the case of large-scale industrial fermentations. The microbial metabolomics tool allows real-time monitoring of metabolite concentrations during fermentation or similar bioprocesses, providing valuable insights into the physiological state of the microorganisms. This knowledge is indispensable to optimize the temperature, pH, oxygen, and nutrient levels to maximize not only microbial productivity but also the product yield (Reiter et al. 2022). Metabolomics can serve as a signaling tool to assess the impact of process parameters on the metabolic behavior of microorganisms, such as in bioprocess development for enzyme or bio-based chemical production. Metabolic profiling enables researchers to find fermentation conditions that enhance desired product yield and help suppress by-products of unwanted products. The ability to monitor in real time can also be used to identify early signs of metabolic stress, which can then drive adjustments to fermentation conditions to prevent culture failure or suboptimal performance (Singh et al. 2024).

Furthermore, metabolomics-based process control can be coupled with other omics data, such as genomics and transcriptomics data, leading to integrated process models that predict the response of microbial metabolism to changes in operating conditions. This high-throughput approach, systems biotechnology, increases specificity during bioprocess optimization and consequently results in more efficient and cost-effective industrial processes (Cortada-Garcia et al. 2024).

4.3 Fermentation and Biofuel Production

Microbial metabolomics is essential for the production of biofuels and fermentation products by elucidating the metabolic pathways that involve the conversion of biofuel substrates into biofuels. Microbial fermentation is the main approach used by the biofuel industry to convert renewable sources of biomass using fermentation microbes (e.g., plant waste) into bioethanol, biodiesel, and biogas. Researchers can follow the metabolic fluxes taking place in microorganisms used for biofuel production to find out which intermediate products are of key interest and which pathways can be adjusted to increase the fuel yield (Hollinshead et al. 2014). For example, if we consider the production of bioethanol, microbial metabolomics can be applied to investigate the catabolic responses of yeast strains to a variety of substrates and fermentation environments. The metabolic bottlenecks can then be found and used to genetically engineer yeast to better convert sugars to ethanol. Metabolomics develops powerful tools for finding specific and novel microbial strains that may be able to utilize new alternative feedstocks like lignocellulosic biomass, potentially increasing biofuel production efficiency. In biodiesel production, for example, metabolomics can allow the metabolic pathways for the production of lipids (the raw material, ultimately, of biodiesel) in algal or bacterial cultures to be optimized (Azhar et al. 2017). Surveying the lipid metabolism of these microorganisms allows researchers to find strains that have an inherent high capacity to produce lipids, or engineering microbial cells to modulate metabolic pathways to promote fat accumulation. In summary, microbial metabolomics can be a useful approach for a more efficient and sustainable biofuel-producing process toward renewable energy transition (Adegboye et al. 2021).

4.4 Pharmaceutical Applications: Antibiotics and Therapeutics

The pharmaceutical industry could also be greatly affected, especially in antibiotic and therapeutic agent discovery and development, from microbial metabolomics. Bacteria and fungi, in particular, also release several secondary metabolites that have different properties such as antimicrobial, antifungal, or anticancer activity. Many of these secondary metabolites, whose production is often upregulated in response to environmental stressors, are the precursors for various pharmaceutical products. In the search for new antibiotics, it is metabolomics that enables strong comparison between the metabolomes of diverse microbial species in different environmental conditions at the forefront of uncovering such new antibiotics (Damavandi et al. 2023). Changes in metabolite production can reveal unexplored compounds to be tested for a biological effect. Detection and characterization of secondary metabolites from *Streptomyces* and *Penicillium* species, for instance, which form the basis of important drugs including streptomycin and penicillin, respectively, have a great impact on their development. Besides the identification of antibiotics, metabolomics is one of the key components in the production of immunosuppressant, antitumor agents, and antifungal drugs (Tangerina et al. 2020). Metabolomics enables the optimization of microbial strains and fermentation conditions, enhancing the bioactivity yields and potency of such bioactive compounds. Finally, metabolomics can assist in the running of those kinds of studies for the drug resistance mechanism in pathogenic microorganisms and lead to the discovery of novel approaches to counteract drug resistance as well as enhance the effectiveness of currently used drugs (Ramírez-Rendon et al. 2022).

4.5 Agricultural Biotechnology and Soil Health

Microbial metabolomics is a valuable tool for enhanced crop productivity, soil health, and pest management in agriculture. Analysis of metabolites produced by plant-associated microbes, such as plant growth-promoting rhizobacteria (PGPR), to identify compounds involved in plant development, disease resistance, and nutrient uptake (Rani et al. 2023). The following are examples: Soil microbiome metabolomics can provide insights into how microbial community composition and functioning affect nutrient cycling, soil fertility, and plant health. Profiling the metabolic cross-talk between plants and associated microbes can help scientists develop microbial inoculants or biofertilizers that aid in sustainable agriculture (Dutta et al. 2022). These biofertilizers usually comprise plant-beneficial microbes that secrete a wide range of secondary metabolites, e.g., siderophores and phytohormones that enhance plant growth and stress tolerance. It is also useful for developing biopesticides, which are pest- and disease-control products based on naturally occurring microbial metabolites (Yadav et al. 2024). This can be beneficial when developing biological control agents that can serve as sustainable replacements for chemical pesticides due to their pest-inhibiting effects; by screening for insecticidal or fungicidal metabolites, scientists can potentially develop environmentally sustainable biopesticides (Thakur and Kaur 2024). Metabolomics can also monitor the metabolic activities of soil microbes and how agricultural practices, such as crop rotation and organic farming, impact the microbial communities and their role in soil fertility and soil health (Olanrewaju et al. 2024).

4.6 Microbial Metabolomics in Environmental Biotechnology (Waste Treatment, Bioremediation)

Microbial metabolomics has significant implications in environmental biotechnology, notably waste treatment and bioremediation. Since microorganisms play a significant role in pollutant degradation and nutrient cycling in the environment, metabolomics sheds light on such aspects but from the metabolic viewpoint. In the treatment of wastewater, for example, organic pollutants are metabolized by microbial communities, degrading them into less harmful compounds (Wu et al. 2022a). Metabolomics also enables monitoring of the metabolic pathways associated with the degradation of certain pollutants such as heavy metals, pharmaceuticals, and industrial chemicals as shown in Fig. 3. These insights can be used to optimize the composition of microbial consortia, as well as bioreactor operational conditions to enhance the effectiveness of waste treatment processes. Similarly, bioremediation is the use of microbes to decontaminate sites with oil spills and heavy metals and soils or waters polluted with pesticides. Microbial metabolomics is an approach that can ardently address this problem by identifying metabolic activation of microbial communities by pollutants and specific microorganisms performing contaminant degradation and revealing the underlying metabolic pathways. This knowledge would allow the isolation of strains that more efficiently degrade pollutants or engineer strains with expanded degradation capabilities for bioremediation purposes (Ayilara and Babalola 2023).

Additionally, in the area of developing microbial fuel cells (MFCs) that utilize microbial metabolism to produce electricity from organic waste, metabolomics is utilized. Metabolomics accelerates the development of sustainable energy technologies that can help clean the environment by optimizing metabolic pathways that convert waste to energy (Alidoosti et al. 2024).

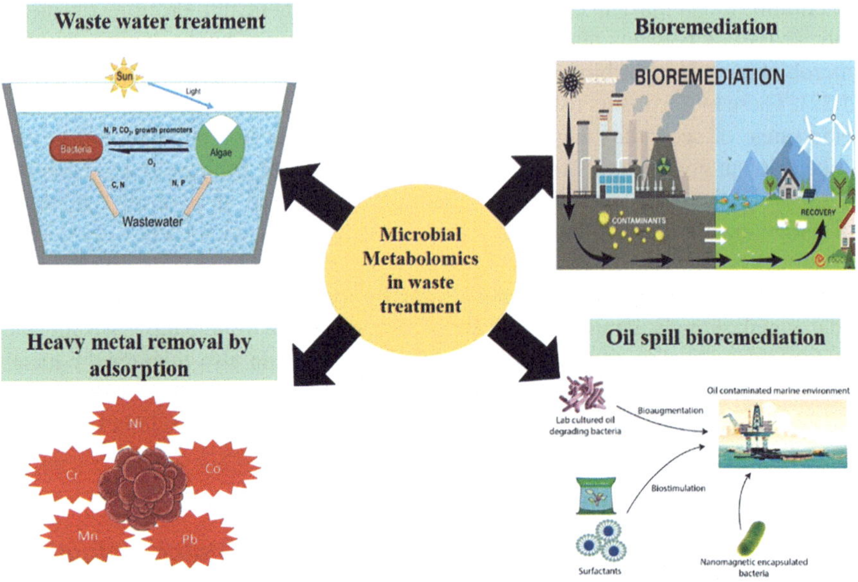

Fig. 3 Microbial metabolomics in waste treatment

5 Role of Microbial Metabolomics in Understanding Microbial Communities

Microbial metabolomics has become an essential tool for exploring the complex interactions within microbial communities, providing insights into how microorganisms influence each other and their environments. This field is particularly crucial in understanding the microbiome, microbial ecology, and how microbes contribute to global biochemical cycles (Bhosle et al. 2022). Metabolomics enables researchers to link specific microbial metabolic activities with their ecological roles, revealing how microbial communities adapt to various environmental conditions and how their collective metabolic outputs influence human health, agriculture, and industry.

5.1 Microbiome Studies and Host-Microbe Interactions

Microbial metabolomics is an integral part of microbiome investigation, especially in revealing the metabolic interplay between microbes and their hosts. For example, in the case of the human microbiome, metabolomics is applied to better profile microbiota-derived metabolites and their influence on human health. Gut bacteria can produce short-chain fatty acids (SCFAs) during the fermentation of dietary fibers, which in turn have been shown to exert effects on immune modulation, energy metabolism, as well as neurological functions (Shaffer et al. 2017). Metabolomics can play a vital role in deciphering the influence of host-microbe interactions on the progression of diseases or the maintenance of health. It assists in determining microbial metabolites that may modulate host signaling pathways and may be involved in inflammation or nutrient absorption. Additionally, metabolomics plays an essential role in the correlation between dysbiosis (i.e., an imbalance in the microbiome) and metabolic diseases like obesity, diabetes, and inflammatory bowel disease (IBD). Insights into these interactions might inform novel therapies such as probiotics, prebiotics, and dietary interventions to restore a healthy microbiome (Puig-Castellví et al. 2023).

5.2 Metabolomics in Studying Microbial Ecology

Metabolomics gives high-resolution systemic views of the activity in microbial communities and is one of the best tools available to explore the metabolic activity of microbes in the context of certain environmental conditions. Profiling metabolites from diverse environments—from soils and oceans to extreme habitats such as hot springs and polar ice caps—allows scientists to infer the contributions of microbial communities to nutrient cycling, carbon sequestration, and pollutant degradation (Robert William Brown 2021). This type of metabolomics allows for the investigation of the community-level metabolism of a microbiome, indicating how various microbial species may work together or compete against one another for resources. Metabolomics can reveal metabolites involved in the exchange of nutrients, for example, between plants and nitrogen-fixing and mycorrhizal fungi in soil ecosystems. Such information can be crucial in devising sustainable agricultural strategies that utilize beneficial microbial interactions to enhance host plant growth and soil quality (Dhiman et al. 2024). Additionally, metabolomics can be used to determine the effects of environmental stressors like climate change, pollution, or nutrient depletion on microbial communities and their metabolic functions. This is especially crucial when investigating biogeochemical cycles and the contributions of microbial metabolism to climate change through the production or consumption of greenhouse gases such as methane and carbon dioxide (Gupta et al. 2022).

5.3 Applications in Synthetic Biology

A potential approach for synthetic biology is rooted in microbiology metabolomics, which aims to understand the cellular mechanisms driving the production of high-value biochemicals, biofuels, and pharmaceuticals by designing and modifying microbes with top-performing, high-titer pathways. Synthetic biologists, using metabolomics, can gain insights into the metabolism of the microbial networks allowing them to either subsequently implement a novel metabolic network into a microorganism or augment the existing pathways to produce a microorganism with a specific function. Such tools help, for example, in metabolic engineering of microbes for the production of bio-based materials or chemicals that are particularly hard to produce otherwise (i.e., by chemical synthesis) (Kato et al. 2022). In synthetic biology, metabolic pathway engineering frequently uses metabolomics to find key regulatory metabolites to enable flux through desired biosynthetic routes. Metabolomics, applied in this context, is important for developing good microbial cell factories for biofuels, bioplastics, and therapeutic molecules. Furthermore, biosensing element can lead to the creation of engineered microbes capable of detection of and responding to specific environmental or metabolic signals (such as toxic/harmful compounds with specific detection thresholds) by providing measurable outputs (e.g., fluorescence or color change). They have huge applications in environmental monitoring, diagnostics in health care, and control of industrial processes (Amer and Baidoo 2021).

5.4 Predicting Functional Traits in Microbial Communities

Predicting functional traits of microbial communities from metabolic profiles is one of the most exciting applications of metabolomics. While metagenomics provides sets of genes associated with known metabolic pathways from which inferences can be gleaned, metabolomics allows the observed expression of these pathways within innovating species within a community to be dissected, linking genotypes to phenotypes and mapping biochemistries to ecosystem function (Navgire et al. 2022). For example, mass spectrometry-based metabolomics can directly identify the microbes involved in the degradation of complex organic compounds, synthesis of microbial biopolymers, or biotransformation and detoxification of xenobiotics in environmental samples. Metabolomics can be useful in health care for predicting the functional roles of gut microbiota, which may provide beneficial information on the potential to influence drug metabolism, immune responses, or even disease susceptibility. However, the most powerful application for this predictive capability is the implementation of personalized precision medicine guided by the metabolic potential of an individual's microbiome (Jariyasopit and Khoomrung 2023). Metabolomics can also be used to predict how a microbial community will respond to environmental perturbations, such as the addition of novel nutrients, global warming, or exposure to toxins. Such predictive power is essential in areas such as bioremediation, waste treatment, and agriculture in which microbial activities are essential for ecosystem health (Aguiar-Pulido et al. 2016).

6 Metabolomics and Microbial Biotechnology for Health and Medicine

Microbial metabolomics has vast potential in advancing health and medicine, especially in understanding the interactions between microbes and the human body. By exploring the metabolomes of microorganisms, scientists are uncovering new therapeutic targets, developing novel probiotics, and gaining insights into how microbial metabolites influence human diseases (Yu et al. 2024). This section explores how microbial metabolomics contributes to drug discovery, disease management, and the growing field of personalized medicine.

6.1 Probiotics, Prebiotics, and Gut Microbiota

Metabolomics, the study of microbial metabolites that mediate gut homeostasis, has emerged as an important approach to understanding the interaction of probiotics, prebiotics, and gut ecosystems. Probiotics are live beneficial microorganisms that, when consumed in an adequate amount, confer a health benefit, typically by producing metabolites that promote intestinal health. Prebiotics, however, are nondigestible food ingredients that promote the growth of probiotics (Thakur et al. 2024). This omics approach makes it possible to delve into the influence of these probiotics and prebiotics upon gut metabolite production, including SCFAs, which play crucial roles in preserving gut barrier integrity, downregulating several pro-inflammatory pathways, and modulating immune responses. Research that profiles how the gut metabolome changes in response to probiotic or prebiotic interventions is essential to designing better functional food or supplements for gut health (Thakur et al. 2023a, b).

6.2 Microbial Metabolites in Drug Discovery and Development

The use of microbial metabolomics has been critical in the discovery of new drugs, most notably antibiotics, antivirals, and anticancer agents. Bioactive secondary metabolites have been farmed from microorganisms for decades, but especially unique or harsh environments have been previously associated with an array of bioactive secondary metabolites. The metabolites can be screened and identified quickly with metabolomics enabling the finding of new therapeutic compounds (Kaur et al. 2024b). Microbial metabolomics, for example, can discover new mechanism-of-action antibiotics, which is particularly timely in light of the concerns surrounding antibiotic resistance. Identifications of metabolic profiling of natural microbial sources are also simultaneously uncovering microbial metabolites, including those from immunosuppressants to anticancer agents, which represent promising targets of drug development (Pham et al. 2019).

6.3 Microbial Metabolomics in Cancer and Chronic Diseases

Recent advances in metabolomics have shown how microbial metabolites contribute to cancer progression, metabolic disorders, and other chronic diseases. Microbial communities, particularly in the gut, influence host metabolism by producing bioactive compounds that can either promote or inhibit disease processes. For example, gut microbiota-derived metabolites such as secondary bile acids have been linked to colorectal cancer, while others like indole and butyrate are considered protective against cancer progression (Ganesan et al. 2022). By profiling the metabolites present in cancer patients' microbiomes, researchers can identify potential biomarkers for early detection and treatment. Additionally, microbial metabolomics offers new insights into how microbiota contribute to chronic inflammation, autoimmune diseases, and metabolic disorders like type 2 diabetes (Schmidt et al. 2021).

6.4 Metabolomic Insights into Antibiotic Resistance Mechanisms

The rise of antibiotic resistance is one of the most pressing global health challenges, and microbial metabolomics provides crucial insights into how microbes evolve resistance mechanisms. By profiling the metabolites produced by resistant bacterial strains, metabolomics can reveal the biochemical pathways responsible for inactivating antibiotics, altering drug targets, or increasing efflux pump activity. Metabolomics also helps in identifying resistance-associated metabolites, which can be targeted for developing novel antibiotic adjuvant compounds that restore the efficacy of existing antibiotics by inhibiting resistance pathways (Maeda and Furusawa 2024). Through these insights, microbial metabolomics aids in the fight against drug-resistant pathogens by enabling the development of new therapeutic strategies and diagnostics to manage antibiotic resistance more effectively (Muteeb et al. 2023).

7 Challenges and Limitations in Microbial Metabolomics

Despite the immense potential of microbial metabolomics, several challenges and limitations hinder its widespread application, particularly in studying complex microbial systems. One major hurdle is the sheer diversity and complexity of microbial metabolites, which range from small molecules like amino acids to intricate secondary metabolites. Current analytical techniques, though powerful, struggle to capture all metabolites with equal sensitivity and accuracy. The complexity of the microbial environment—whether in soil, the human gut, or the aquatic ecosystem—adds further difficulty (Raajaraam and Raman 2024). Environmental samples often contain mixed microbial populations, making it challenging to associate specific metabolites with particular species. Additionally, microbial communities are dynamic, with different organisms dominating under varying conditions, complicating the creation of reproducible metabolomic profiles. Metabolite identification and quantification are significant bottlenecks. While high-resolution techniques like mass spectrometry (MS) and nuclear magnetic resonance (NMR) provide detailed data, the identification of unknown metabolites remains daunting, as many metabolites are not yet cataloged in public databases (Robert W Brown et al. 2024). Even for known metabolites, accurate quantification can be problematic due to matrix effects, where the complex mixture in samples interferes with signal detection. This is particularly challenging in environmental microbiology, where sample composition can vary widely. Another major challenge is the integration and interpretation of the large, multidimensional datasets generated by metabolomics. These datasets are often combined with other omics data such as genomics, transcriptomics, and proteomics to form a comprehensive view of microbial functions. However, each type of data requires different analytical approaches, and combining them involves dealing with disparate timescales and biological processes (Bauermeister et al. 2022). Many metabolites also have pleiotropic roles, participating in multiple biological pathways, which makes it difficult to decipher their exact functions in a given context. Cost, time, and technological barriers further complicate the process. Metabolomics research, especially on a large scale, can be expensive. Instruments like mass spectrometers and NMR machines are costly to purchase and maintain, and the reagents and standards for metabolomic analysis are also expensive. Moreover, metabolomic studies, especially those that are large-scale or longitudinal, can take a long time to complete due to the extensive sample collection, processing, and data analysis involved. The processing of large datasets requires significant computational resources and expert interpretation, often extending project timelines considerably (Qiu et al. 2023).

8 Future Trends in Microbial Metabolomics

Future trends in microbial metabolomics are poised to significantly enhance our understanding of microbial systems, largely through the integration of multiple omics approaches. By combining genomics, transcriptomics, proteomics, and metabolomics, researchers can gain a holistic view of microbial functions, enabling deeper insights into metabolic pathways, regulatory networks, and microbial interactions with their environments. This integration will bridge the gap between genetic information and metabolite production, providing a more complete picture of microbial behavior. Another promising trend is the use of artificial intelligence (AI) and machine learning (ML) in metabolomic data analysis. These advanced computational techniques can process the vast, complex datasets generated by metabolomics more efficiently, identifying patterns and correlations that may not be immediately apparent to human analysts. AI and ML have the potential to accelerate discoveries, predict metabolic responses, and improve the precision of microbial metabolomics applications. High-throughput technologies are also evolving rapidly, enabling faster and more comprehensive metabolomic profiling. These advancements allow for the analysis of thousands of samples simultaneously, significantly reducing the time and cost associated with large-scale studies. This is particularly beneficial for applications in environmental and industrial biotechnology, where monitoring microbial communities over time is essential. Lastly, personalized metabolomics is emerging as a key trend in precision biotechnology, where individual microbial profiles can be tailored for specific applications, such as optimizing probiotics for personalized health or designing microbial consortia for customized biofuel production. Together, these innovations are set to propel microbial metabolomics into a new era of precision, efficiency, and broad applicability across biotechnology, health care, and environmental sciences.

References

Adegboye MF, Ojuederie OB, Talia PM, Babalola OO (2021) Bioprospecting of microbial strains for biofuel production: metabolic engineering, applications, and challenges. Biotechnol Biofuels 14(1):5

Aguiar-Pulido V, Huang W, Suarez-Ulloa V, Cickovski T, Mathee K, Narasimhan G (2016) Metagenomics, metatranscriptomics, and metabolomics approaches for microbiome analysis: supplementary issue: bioinformatics methods and applications for big metagenomics data. Evol Bioinforma 12:EBO–S36436

Akyol S, Ashrafi N, Yilmaz A, Turkoglu O, Graham SF (2023) Metabolomics: an emerging "omics" platform for systems biology and its implications for Huntington disease research. Meta 13(12):1203

Alarcon-Barrera JC, Kostidis S, Ondo-Mendez A, Giera M (2022) Recent advances in metabolomics analysis for early drug development. Drug Discov Today 27(6):1763–1773

Alidoosti F, Giyahchi M, Moien S, Moghimi H (2024) Unlocking the potential of soil microbial communities for bioremediation of emerging organic contaminants: omics-based approaches. Microb Cell Factories 23(1):210

Al-Kelani M, Buthelezi N (2024) Advancements in medical research: exploring fourier transform infrared (FTIR) spectroscopy for tissue, cell, and hair sample analysis. Skin Res Technol 30(6):e13733

Álvarez-Sánchez B, Priego-Capote F, de Castro MDL (2010) Metabolomics analysis II. Preparation of biological samples prior to detection. TrAC Trends Anal Chem 29(2):120–127

Alves MF, Katchborian-Neto A, Bueno PCP, Carnevale-Neto F, Casoti R, Ferreira MS, Murgu M, de Paula ACC, Dias DF, Soares MG, Chagas-Paula DA (2024) LC-MS/DIA-based strategy for comprehensive flavonoid profiling: an Ocotea spp. applicability case. RSC Adv 14(15):10481–10498

Amer B, Baidoo EEK (2021) Omics-driven biotechnology for industrial applications. Front Bioeng Biotechnol 9:613307

Arteaga-Clemente G, García-González MA, González-González M (2024) Soil lipid analysis by chromatography: a critical review of the current state in sample preparation. J Chromatogr Open 6:100173

Ayilara MS, Babalola OO (2023) Bioremediation of environmental wastes: the role of microorganisms. Front Agron 5:1183691

Azhar SHM, Abdulla R, Jambo SA, Marbawi H, Gansau JA, Faik AAM, Rodrigues KF (2017) Yeasts in sustainable bioethanol production: a review. Biochem Biophys Rep 10:52–61

Bauermeister A, Mannochio-Russo H, Costa-Lotufo LV, Jarmusch AK, Dorrestein PC (2022) Mass spectrometry-based metabolomics in microbiome investigations. Nat Rev Microbiol 20(3):143–160

Bhosle A, Wang Y, Franzosa EA, Huttenhower C (2022) Progress and opportunities in microbial community metabolomics. Curr Opin Microbiol 70:102195

Bishop SL, Drikic M, Wacker S, Chen YY, Kozyrskyj AL, Lewis IA (2022) Moving beyond descriptive studies: harnessing metabolomics to elucidate the molecular mechanisms underpinning host-microbiome phenotypes. Mucosal Immunol 15(6):1071–1084

Brown RW (2021) Metabolomic and volatilomic profiling for the assessment of soil carbon cycling and biological quality. Bangor University, Bangor

Brown RW, Reay MK, Centler F, Chadwick DR, Bull ID, McDonald JE, Evershed RP, Jones DL (2024) Soil metabolomics-current challenges and future perspectives. Soil Biol Biochem 193:109382

Cai Y, Zhou Z, Zhu Z-J (2023) Advanced analytical and informatic strategies for metabolite annotation in untargeted metabolomics. TrAC Trends Anal Chem 158:116903

Calvo AM, Wilson RA, Bok JW, Keller NP (2002) Relationship between secondary metabolism and fungal development. Microbiol Mol Biol Rev 66(3):447–459

Chen Y, Li E-M, Xu L-Y (2022) Guide to metabolomics analysis: a bioinformatics workflow. Meta 12(4):357

Cortada-Garcia J, Haggarty J, Weidt S, Daly R, Arnold SA, Burgess K (2024) On-line targeted metabolomics for real-time monitoring of relevant compounds in fermentation processes. Biotechnol Bioeng 121(2):683–695

Damavandi MS, Shojaei H, Esfahani BN (2023) The anticancer and antibacterial potential of bioactive secondary metabolites derived from bacterial endophytes in association with Artemisia absinthium. Sci Rep 13(1):18473

Deschamps E, Calabrese V, Schmitz I, Hubert-Roux M, Castagnos D, Afonso C (2023) Advances in ultra-high-resolution mass spectrometry for pharmaceutical analysis. Molecules 28(5):2061

Dhiman S, Kaur P, Narang J, Mukherjee G, Thakur B, Kaur S, Tripathi M (2024) Fungal bioprocessing for circular bioeconomy: exploring lignocellulosic waste valorization. Mycology 15:1–26

Dromms RA, Styczynski MP (2012) Systematic applications of metabolomics in metabolic engineering. Meta 2(4):1090–1122

Dutta P, Muthukrishnan G, Gopalasubramaiam SK, Dharmaraj R, Karuppaiah A, Loganathan K, Periyasamy K, Pillai MA, Upamanya GK, Boruah S, Deb L, Kumari A, Mahanta M, Heisnam P, Mishra AK (2022) Plant growth-promoting rhizobacteria (PGPR) and its mechanisms against plant diseases for sustainable agriculture and better productivity. Biocell 46(8):1843–1859

Eicher T, Kinnebrew G, Patt A, Spencer K, Ying K, Ma Q, Machiraju R, Mathé EA (2020) Metabolomics and multi-omics integration: a survey of computational methods and resources. Meta 10(5):202

Emwas A-H, Roy R, McKay RT, Tenori L, Saccenti E, Gowda GAN, Raftery D, Alahmari F, Jaremko L, Jaremko M, Wishart DS (2019) NMR spectroscopy for metabolomics research. Meta 9(7):123

Fadlelmoula A, Pinho D, Carvalho VH, Catarino SO, Minas G (2022) Fourier transform infrared (FTIR) spectroscopy to analyse human blood over the last 20 years: a review towards lab-on-a-chip devices. Micromachines 13(2):187

Farag MA, Baky MH, von Bergen M, Hegazi NM (2023) The use of omics in monitoring food gut microbiota interaction outcomes: a review of novel trends and technologies. Curr Opin Food Sci 52:101064

Fiehn O (2016) Metabolomics by gas chromatography–mass spectrometry: Combined targeted and untargeted profiling. Curr Protoc Mol Biol 114(1):30–34

Fusco W, Lorenzo MB, Cintoni M, Porcari S, Rinninella E, Kaitsas F, Lener E, Mele MC, Gasbarrini A, Collado MC, Cammarota G, Ianiro G (2023) Short-chain fatty-acid-producing bacteria: key components of the human gut microbiota. Nutrients 15(9):2211

Ganesan R, Jeong J-J, Kim DJ, Suk KT (2022) Recent trends of microbiota-based microbial metabolites metabolism in liver disease. Front Med 9:841281

Giera M, Aisporna A, Uritboonthai W, Hoang L, Derks RJE, Joseph KM, Baker ES, Siuzdak G (2024) XCMS-METLIN: data-driven metabolite, lipid, and chemical analysis. Mol Syst Biol 20:1–3

González-Plaza JJ, Furlan C, Rijavec T, Lapanje A, Barros R, Tamayo-Ramos JA, Suarez-Diez M (2022) Advances in experimental and computational methodologies for the study of microbial-surface interactions at different omics levels. Front Microbiol 13:1006946

Gould O, Nguyen N, Honeychurch KC (2023) New applications of gas chromatography and gas chromatography-mass spectrometry for novel sample matrices in the forensic sciences: a literature review. Chemosensors 11(10):527

Gupta A, Gupta R, Singh RL (2017) Microbes and environment. In: Principles and applications of environmental biotechnology for a sustainable future. Springer, Singapore, pp 43–84

Gupta S, Schillaci M, Roessner U (2022) Metabolomics as an emerging tool to study plant–microbe interactions. Emerg Top Life Sci 6(2):175–183

Hollinshead W, He L, Tang YJ (2014) Biofuel production: an odyssey from metabolic engineering to fermentation scale-up. Front Microbiol 5:344

Horak I, Engelbrecht G, van Rensburg PJJ, Claassens S (2019) Microbial metabolomics: essential definitions and the importance of cultivation conditions for utilizing Bacillus species as bionematicides. J Appl Microbiol 127(2):326–343

Huckvale ED, Moseley HNB (2024) Predicting the pathway involvement of all pathway and associated compound entries defined in the Kyoto Encyclopedia of Genes and Genomes. Meta 14(11):582

Ibrahim M, Raajaraam L, Raman K (2021) Modelling microbial communities: harnessing consortia for biotechnological applications. Comput Struct Biotechnol J 19:3892–3907

Iman MN, Herawati E, Fukusaki E, Putri SP (2022) Metabolomics-driven strain improvement: a mini review. Front Mol Biosci 9:1057709

Jan AT, Azam M, Siddiqui K, Ali A, Choi I, Haq QMR (2015) Heavy metals and human health: mechanistic insight into toxicity and counter defense system of antioxidants. Int J Mol Sci 16(12):29592–29630

Jariyasopit N, Khoomrung S (2023) Mass spectrometry-based analysis of gut microbial metabolites of aromatic amino acids. Comput Struct Biotechnol J 21:4777

Jin Z, Sato Y, Kawashima M, Kanehisa M (2023) KEGG tools for classification and analysis of viral proteins. Protein Sci 32(12):e4820

Joshi A, Verma KK, Rajput DV, Minkina T, Arora J (2022) Recent advances in metabolic engineering of microorganisms for advancing lignocellulose-derived biofuels. Bioengineered 13(4):8135–8163

Kato Y, Inabe K, Hidese R, Kondo A, Hasunuma T (2022) Metabolomics-based engineering for biofuel and bio-based chemical production in microalgae and cyanobacteria: a review. Bioresour Technol 344:126196

Kaur G, Thakur B, Gill R, Kaur R, Kaur S (2024a) Biopolymeric conjugation with food additives. Phys Sci Rev 9(9):2899–2923

Kaur S, Thakur B, Kaur G, Kaur R (2024b) Phytochemicals and phyto-essential oils as potent antiviral agents. In: Promising antiviral herbal and medicinal plants. CRC Press, pp 62–78

Khanijou JK, Kulyk H, Bergès C, Khoo LW, Ng P, Yeo HC, Helmy M, Bellvert F, Chew W, Selvarajoo K (2022) Metabolomics and modelling approaches for systems metabolic engineering. Metab Eng Commun 15:e00209

Krysenko S, Wohlleben W (2024) Role of carbon, nitrogen, phosphate and sulfur metabolism in secondary metabolism precursor supply in Streptomyces spp. Microorganisms 12(8):1571

Kumar S, Baggi TR (2022) Analytical methods for herbal products containing synthetic cannabinoids: a review. Forensic Chem 27:100396

Li B, Liu K, Kwok L-Y, Guo S, Bai L, Yang X, Chen Y (2022) Development of a non-target metabolomics-based screening method for elucidating metabolic and probiotic potential of bifidobacteria. Innov Food Sci Emerg Technol 77:102971

Liu R, Bao Z-X, Zhao P-J, Li G-H (2021) Advances in the study of metabolomics and metabolites in some species interactions. Molecules 26(11):3311

Maeda T, Furusawa C (2024) Laboratory evolution of antimicrobial resistance in bacteria to develop rational treatment strategies. Antibiotics 13(1):94

Manickam S, Rajagopalan VR, Kambale R, Rajasekaran R, Kanagarajan S, Muthurajan R (2023) Plant metabolomics: current initiatives and future prospects. Curr Issues Mol Biol 45(11):8894–8906

Marshall DD, Powers R (2017) Beyond the paradigm: combining mass spectrometry and nuclear magnetic resonance for metabolomics. Prog Nucl Magn Reson Spectrosc 100:1–16

Martinez KB, Leone V, Chang EB (2017) Microbial metabolites in health and disease: navigating the unknown in search of function. J Biol Chem 292(21):8553–8559

Mashego MR, Rumbold K, De Mey M, Vandamme E, Soetaert W, Heijnen JJ (2007) Microbial metabolomics: past, present and future methodologies. Biotechnol Lett 29:1–16

Mayers JR, Varon J, Zhou RR, Daniel-Ivad M, Beaulieu C, Bhosle A, Glasser NR, Lichtenauer FM, Ng J, Vera MP, Huttenhower C, Perrella MA, Clish CB, Zhao SD, Baron RM, Balskus EP (2024) A metabolomics pipeline highlights microbial metabolism in bloodstream infections. Cell 187(15):4095–4112

Meng X, Li X, Yang L, Yin R, Qi L, Guo QI (2023) Review on microbial metabolomics of probiotics and pathogens: methodologies and applications. Biocell 47(1):91–107

Muhamadali H, Winder CL, Dunn WB, Goodacre R (2023) Unlocking the secrets of the microbiome: exploring the dynamic microbial interplay with humans through metabolomics and their manipulation for synthetic biology applications. Biochem J 480(12):891–908

Muteeb G, Rehman MT, Shahwan M, Aatif M (2023) Origin of antibiotics and antibiotic resistance, and their impacts on drug development: a narrative review. Pharmaceuticals 16(11):1615

Nagana Gowda GA, Raftery D (2021) NMR-based metabolomics. Cancer metabolomics: methods and applications. Springer, New York, pp 19–37

Navgire GS, Goel N, Sawhney G, Sharma M, Kaushik P, Mohanta YK, Mohanta TK, Al-Harrasi A (2022) Analysis and Interpretation of metagenomics data: an approach. Biol Proced Online 24(1):18

Nikoloudaki O, Aheto F, Di Cagno R, Gobbetti M (2024) Synthetic microbial communities: A gateway to understanding resistance, resilience, and functionality in spontaneously fermented food microbiomes. Food Res Int 114780:114780

Ocampos FMM, de Souza AJB, Ribeiro GH, Almeida LS, Cônsolo NRB, Colnago LA (2024) NMR-based plant metabolomics protocols: a step-by-step guide. Front Nat Prod 3:1414506

Olanrewaju OS, Glick BR, Babalola OO (2024) Metabolomics-guided utilization of beneficial microbes for climate-resilient crops. Curr Opin Chem Biol 79:102427

Orsi E, Claassens NJ, Nikel PI, Lindner SN (2022) Optimizing microbial networks through metabolic bypasses. Biotechnol Adv 60:108035

Perez-Garcia O, Lear G, Singhal N (2016) Metabolic network modeling of microbial interactions in natural and engineered environmental systems. Front Microbiol 7:673

Pham JV, Yilma MA, Feliz A, Majid MT, Maffetone N, Walker JR, Kim E, Cho HJ, Reynolds JM, Song MC, Park SR, Yoon YJ (2019) A review of the microbial production of bioactive natural products and biologics. Front Microbiol 10:1404

Pinu FR, Villas-Boas SG (2017) Extracellular microbial metabolomics: the state of the art. Meta 7(3):43

Pinu FR, Beale DJ, Paten AM, Kouremenos K, Swarup S, Schirra HJ, Wishart D (2019) Systems biology and multi-omics integration: viewpoints from the metabolomics research community. Meta 9(4):76

Poli A, Finore I, Romano I, Gioiello A, Lama L, Nicolaus B (2017) Microbial diversity in extreme marine habitats and their biomolecules. Microorganisms 5(2):25

Puig-Castellví F, Pacheco-Tapia R, Deslande M, Jia M, Andrikopoulos P, Chechi K, Bonnefond A, Froguel P, Dumas M-E (2023) Advances in the integration of metabolomics and metagenomics for human gut microbiome and their clinical applications. TrAC Trends Anal Chem 167:117248

Qiu S, Cai Y, Yao H, Lin C, Xie Y, Tang S, Zhang A (2023) Small molecule metabolites: discovery of biomarkers and therapeutic targets. Signal Transduct Target Ther 8(1):132

Raajaraam L, Raman K (2024) Modeling microbial communities: perspective and challenges. ACS Synth Biol 13(8):2260–2270

Rahman S, O'connor AL, Becker SL, Patel RK, Martindale RG, Tsikitis VL (2023) Gut microbial metabolites and its impact on human health. Ann Gastroenterol 36(4):360

Rakusanova S, Cajka T (2024) Tips and tricks for LC–MS-based metabolomics and lipidomics analysis. TrAC Trends Anal Chem 117940:117940

Ramírez-Rendon D, Passari AK, Ruiz-Villafán B, Rodríguez-Sanoja R, Sánchez S, Demain AL (2022) Impact of novel microbial secondary metabolites on the pharma industry. Appl Microbiol Biotechnol 106(5):1855–1878

Rani P, Rajput S, Thakur B, Kaur S (2023) Immobilization and co-mobilization: an unexploited biotechnological tool for enhancing efficiency of biofertilizers. In: Metabolomics, proteomes and gene editing approaches in biofertilizer industry. Springer, pp 219–236

Reiter A, Asgari J, Wiechert W, Oldiges M (2022) Metabolic footprinting of microbial systems based on comprehensive in silico predictions of MS/MS relevant data. Meta 12(3):257

Rowland I, Gibson G, Heinken A, Scott K, Swann J, Thiele I, Tuohy K (2018) Gut microbiota functions: metabolism of nutrients and other food components. Eur J Nutr 57:1–24
Rusu AV, Trif M, Rocha JM (2023) Microbial secondary metabolites via fermentation approaches for dietary supplementation formulations. Molecules 28(16):6020
Rutherford ST, Bassler BL (2012) Bacterial quorum sensing: its role in virulence and possibilities for its control. Cold Spring Harb Perspect Med 2(11):a012427
Salem MA, de Souza L, Serag A, Fernie AR, Farag MA, Ezzat SM, Alseekh S (2020) Metabolomics in the context of plant natural products research: from sample preparation to metabolite analysis. Meta 10(1):37
Santamaria G, Pinto FR (2024) Bioinformatic analysis of metabolomic data: from raw spectra to biological insight. Biochemist 4(2):90–114
Saoi M, Britz-McKibbin P (2021) New advances in tissue metabolomics: a review. Meta 11(10):672
Sarkar S, Sha SP, Ghatani K (2023) Metabolomics of ethnic fermented foods and beverages: understanding new aspects through Omic techniques. Front Sustain Food Syst 7:1040567
Scalbert A, Brennan L, Fiehn O, Hankemeier T, Kristal BS, van Ommen B, Pujos-Guillot E, Verheij E, Wishart D, Wopereis S (2009) Mass-spectrometry-based metabolomics: limitations and recommendations for future progress with particular focus on nutrition research. Metabolomics 5:435–458
Schmidt DR, Patel R, Kirsch DG, Lewis CA, Vander Heiden MG, Locasale JW (2021) Metabolomics in cancer research and emerging applications in clinical oncology. CA Cancer J Clin 71(4):333–358
Shaffer M, Armstrong AJS, Phelan VV, Reisdorph N, Lozupone CA (2017) Microbiome and metabolome data integration provides insight into health and disease. Transl Res 189:51–64
Singh R, Fatima E, Thakur L, Singh S, Ratan C, Kumar N (2024) Advancements in CHO metabolomics: techniques, current state and evolving methodologies. Front Bioeng Biotechnol 12:1347138
Sreenivasulu N, Fernie AR (2022) Diversity: current and prospective secondary metabolites for nutrition and medicine. Curr Opin Biotechnol 74:164–170
Tampieri A, Szabó M, Medina F, Gulyás H (2021) A brief introduction to the basics of NMR spectroscopy and selected examples of its applications to materials characterization. Phys Sci Rev 6(1):20190086
Tanaka K, Bamba T, Kondo A, Hasunuma T (2024) Metabolomics-based development of bioproduction processes toward industrial-scale production. Curr Opin Biotechnol 85:103057
Tangerina MMP, Furtado LC, Leite VMB, Bauermeister A, Velasco-Alzate K, Jimenez PC, Garrido LM, Padilla G, Lopes NP, Costa-Lotufo LV, Pena Ferreira MJ (2020) Metabolomic study of marine Streptomyces sp.: secondary metabolites and the production of potential anticancer compounds. PLoS One 15(12):e0244385
Thakur B, Kaur S (2024) The role of artificial intelligence in biofertilizer development. In: Metabolomics, proteomics and gene editing approaches in biofertilizer industry: volume II. Springer, pp 157–176
Thakur B, Kaur S, Rani N, Kaur R, Upadhyay SK, Tripathi M (2023a) Exploring microbial contributions to nutraceutical production: from natural to designed foods. Mol Biotechnol:1–16
Thakur B, Kaur S, Tripathi M, Upadhyay SK (2023b) Exploring the potential of lactic acid bacteria and its molecular mechanism of action in the development of biosurfactants: current finding and future outlook. Biotechnol Genet Eng Rev 00(00):1–32. https://doi.org/10.1080/02648725.2023.2216421
Thakur B, Kaur S, Dwibedi V, Albadrani GM, Al-Ghadi MQ, Abdel-Daim MM (2024) Unveiling the antimicrobial and antibiofilm potential of biosurfactant produced by newly isolated Lactiplantibacillus plantarum strain 1625. Front Microbiol 15:1459388
Thomas SN, French D, Jannetto PJ, Rappold BA, Clarke WA (2022) Liquid chromatography–tandem mass spectrometry for clinical diagnostics. Nat Rev Method Primers 2(1):96
Vuckovic D (2012) Current trends and challenges in sample preparation for global metabolomics using liquid chromatography–mass spectrometry. Anal Bioanal Chem 403(6):1523–1548

Wang Y, Wu J, Lv M, Shao Z, Hungwe M, Wang J, Bai X, Xie J, Wang Y, Geng W (2021) Metabolism characteristics of lactic acid bacteria and the expanding applications in food industry. Front Bioeng Biotechnol 9:612285

Wishart DS, Cheng LL, Copié V, Edison AS, Eghbalnia HR, Hoch JC, Gouveia GJ, Pathmasiri W, Powers R, Schock TB, Sumner LW, Uchimiya M (2022) NMR and metabolomics—a roadmap for the future. Meta 12(8):678

Wortel MT, Noor E, Ferris M, Bruggeman FJ, Liebermeister W (2018) Metabolic enzyme cost explains variable trade-offs between microbial growth rate and yield. PLoS Comput Biol 14(2):e1006010

Wu C, Ma Y, Wang D, Shan Y, Song X, Hu H, Ren X, Ma X, Luo J, Cui J, Ma Y (2022a) Microbiology combined with metabonomics revealing the response of soil microorganisms and their metabolic functions exposed to phthalic acid esters. Ecotoxicol Environ Saf 233:113338

Wu M, Zuo S, Maiorano G, Kosobucki P, Stadnicka K (2022b) How to employ metabolomic analysis to research on functions of prebiotics and probiotics in poultry gut health? Front Microbiol 13:1040434

Wu Z, Liang X, Li M, Ma M, Zheng Q, Li D, An T, Wang G (2023) Advances in the optimization of central carbon metabolism in metabolic engineering. Microb Cell Factories 22(1):76

Yadav Y, Thakur B, Kaur S (2024) Nanobiofertilizers: a new paradigm for sustainable agriculture. In: Metabolomics, proteomics and gene editing approaches in biofertilizer industry: Volume II. Springer, pp 201–221

Ye D, Li X, Shen J, Xia X (2022) Microbial metabolomics: from novel technologies to diversified applications. TrAC Trends Anal Chem 148:116540

Yones SA, Csombordi R, Komorowski J, Diamanti K (2021) MetaFetcheR: an R package for complete mapping of small-compound data. Meta 11(11):743

Yu J, Li L, Tao X, Chen Y, Dong D (2024) Metabolic interactions of host-gut microbiota: new possibilities for the precise diagnosis and therapeutic discovery of gastrointestinal cancer in the future—a review. Crit Rev Oncol Hematol 203:104480

Zhang D, Jian Y-P, Zhang Y-N, Li Y, Gu L-T, Sun H-H, Liu M-D, Zhou H-L, Wang Y-S, Xu Z-X (2023a) Short-chain fatty acids in diseases. Cell Commun Signal 21(1):212

Zhang Y, Fan S, Wohlgemuth G, Fiehn O (2023b) Denoising autoencoder normalization for large-scale untargeted metabolomics by gas chromatography–mass spectrometry. Meta 13(8):944

Microbial Metabolomics: Developments, Applications and Future Prospects

Kiran Kharb, Sukhminderjit Kaur, Babita Thakur, Rajinder Kaur, Sunny Dhiman, and Manikant Tripathi

Abstract

Microbial metabolomics is a study of a microbe's metabolism and it reflects the actual physiological state of the microorganism. Its analysis focuses on intracellular and extracellular metabolites, signalling molecules and metabolic pathways. Moreover, metabolomics permits the identification of specific biomarker molecules associated with structural and metabolic states. Microbial metabolomics is a crucial component of the "omics" approaches applied in systems biology. Its approaches are targeted and untargeted depending on the research objectives and questions. Accessing the hidden diversity of natural products has become easier because of advancements in technologies and post-genomic developments. Currently, microbial metabolomics is commonly used in diagnostic applications, drug development, agriculture, bioremediation, environmental ecosystem and microbial metabolomics reconstructions. The chapter aims to provide an overview of microbe metabolomics and its challenges, as well as the development of new technologies, applications and future prospects for metabolomics.

Keywords

Approaches · Genomics · Microbial metabolomics · Omics · Technology

K. Kharb · S. Kaur (✉) · B. Thakur · S. Dhiman
Department of Biotechnology, Chandigarh University, Mohali, Punjab, India
e-mail: sukhminderjit.uibt@cumail.in

R. Kaur
Department of Plant Sciences, University of Idaho Moscow, Moscow, ID, USA

M. Tripathi
Biotechnology Program, Dr. Rammanohar Lohia Avadh University, Ayodhya, Uttar Pradesh, India

S. Kaur et al. (eds.), *Microbial Metabolomics*,
https://doi.org/10.1007/978-981-96-4824-5_2

1 Introduction

Systems biology provides invaluable insights into fundamental biological questions of the biosphere by examining dynamic interactions within living organisms. Data is collected and integrated to create a comprehensive understanding of biological systems. With advances in biotechnology, researchers have developed new technologies to investigate the occurrence of disease and diagnose disease in humans. These advances led to the emergence of the concept of "omics", which gave rise to genomics, proteomics, transcriptomics and metabolomics. Genomics focuses on the physical structure of the genome. It aims to identify, map and organise genome elements including chromosomes and cellular assessment at the transcriptomic and translational levels. This paved the way for proteomics and transcriptomics. As shown in Fig. 1, these tools advance the science of metabolism. This serves as an important addition to our understanding of genetic functioning. Metadata is processed to simplify and synthesise interconnected data to create functionally integrated web systems. These are instrumental in developing metabolomics, an important complement for evaluating genetic function. To reduce and integrate information and create a functionally integrated network system, metadata is processed (Mohd Kamal et al. 2022).

Metabolomics and metabonomics are two terminologies mentioned in the literature and are interchangeable. Distinct studies proved that these two terms are physiologically distinct, but technically they are the same. However, we preferred to use the term metabolomics. Microbial metabolomics is an analytical approach that constitutes an integrated component of systems biology and a unique contribution as one of their omics concepts. It states the accurate and actual physiological state of the cell by examining the complete sets of metabolites within a microorganism. Metabolites are the group of small molecules created during a biological process. They can be created by catabolic or anabolic mechanisms during various cellular

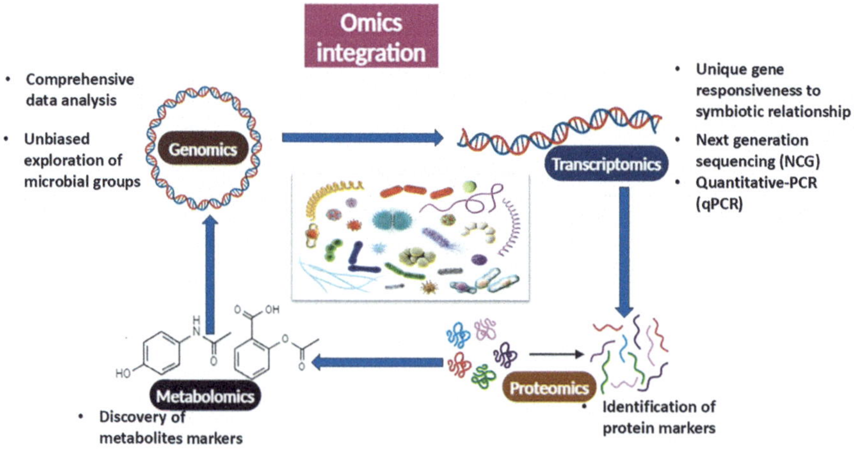

Fig. 1 Omics integration, i.e. genomics, transcriptomics, proteomics and metabolomics

reactions, i.e. Krebs cycle and protein synthesis. Metabolomics is also considered as the link between phenotype and genotype. For standard metabolomics analyses, mainly blood, urine and faeces are used as the sample for studying the metabolites of various metabolic pathways (Ju et al. 2019). After the extraction and isolation of metabolites from the biological sample, metabolite profiling is done. Metabolite profiling mainly uses two analytical techniques, i.e. nuclear magnetic resonance (NMR) spectroscopy and mass spectrometry (MS) for data interpretation. Microbial metabolomics mainly uses two approaches targeted and untargeted depending on the research objectives (Oyedeji et al. 2021). The targeted metabolomics method provides a thorough description of the metabolic enzymes and their kinetics by qualitatively or quantitatively measuring specific or targeted known substances. Meanwhile, the untargeted metabolomics approach aims to identify the cellular metabolites. Both approaches can be used for biomarker identification, but targeted ones have significantly better selectivity and sensitivity than untargeted metabolomics approaches. Targeted metabolomics approaches can analyse 20 metabolites at a time, and these metabolites are chemically characterised before the start of data acquisition as shown in Fig. 2. Sample preparation is one of the main obstacles in metabolic research, for reliable data production knowing real physiological changes or the state of a good metabolomic experimental design is incredibly important (Rampler et al. 2020). Moreover, information on metabolite preparation methods has not been clearly described for targeted or untargeted approaches.

In the post-genomic era, microbial metabolomics has been used increasingly in scientific fields, such as microbiology, food technology, plants, drug discovery and development, metabolic engineering and many more. At the cellular level, microbial metabolomics provides an overview of detailed characterisation of significant metabolic changes of the phenotype. This approach helps in refining the quality and safety of food products. Microbial metabolomics has been used to determine regulations and relationships between components of plant systems (Kumar et al. 2017). Furthermore, metabolomics provides crucial information to elucidate the pathophysiology of diseases and identify biomarkers that allow precision medication. Studies of bacterial systems, including bacterial strain identification and differentiation, changes in metabolic rate during antimicrobial therapy and the drug's mechanism of action, are made possible by the readily available databases on bacterial gene regulators and metabolic pathways (Han et al. 2018). Many metabolites have now been identified as biomarkers for a wide range of illnesses, including cancer, toxin-related diseases, diabetes, coronary heart disease and mental disorders, and used in oncology, thanks to advancements in technology (Tang 2011). Significant technical obstacles still exist in microbial metabolomics, nevertheless, including problems with sample preparation, biomarker identification and mechanism interpretation. Additionally, there aren't many thorough studies that summarise the methods and new approaches being used in the field of microbial metabolomics.

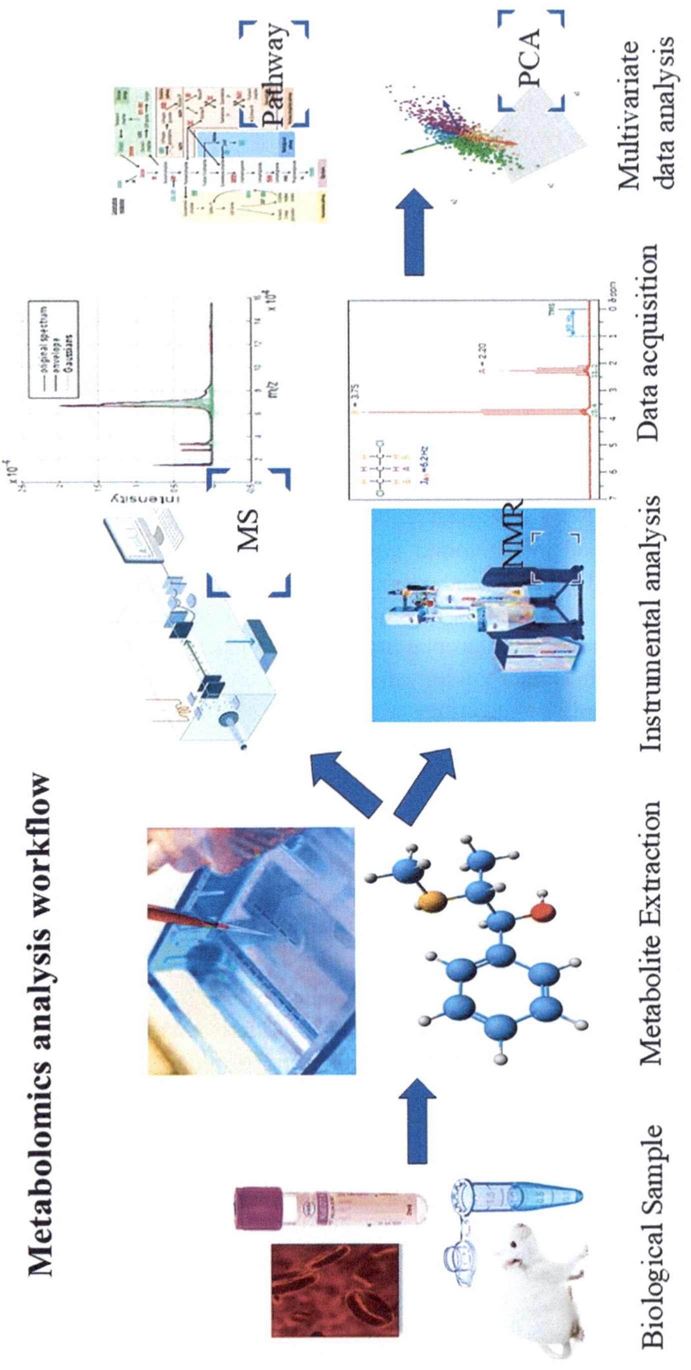

Fig. 2 A standard metabolomics analysis workflow

2 The Challenges of Microbial Metabolomics

Metabolomics is the comprehensive study of metabolite profiles of a living system under a given set of conditions. It deals with the identification and quantification of the group of small molecules of metabolites which are known as metabolome. Metabolome complexity poses many challenges to metabolomics. The main challenge of metabolomics is the ability to identify and quantify the whole collection of intracellular and extracellular metabolites with a molecular mass lower than 1 kilodalton (Arun et al. 2023). Many metabolites vary among different organisms and in many cases they are unidentified. During the transcriptomic- and proteomic-level data analyses, products generated from metabolic reactions are highly complex in their chemical structure and properties in contrast to the genome. These generated metabolites' nature is dynamic and simultaneously it's challenging to determine the entire set of metabolites at a given physiological state. Because metabolites are dynamic, they change in composition and concentration in response to external stimuli. For example, the link between metabolites and genes becomes indirect when bacteria have fewer metabolites than genes, resulting in complicated and nonlinear relationships between the two. Additionally, the activities of some genes can be revealed by their relationship with one or more changed metabolites (Vitale et al. 2024). The intrinsic complexity and difficulties of metabolomics research are a result of these elements taken together.

3 Recent Developments in Microbial Metabolomics

Although microbial metabolomics has made significant progress in different areas of life sciences, still it is considered in its infancy or at an emerging stage by some researchers (Pinu et al. 2019). Metabolomics has two approaches: either targeted or untargeted approaches. Targeted approaches require prior knowledge of the molecule(s) of interest to design a specific analytical method. The primary advantage of this approach lies in its ability to accurately identify and quantify specific metabolites, providing precise insights into their role and significance. If several metabolites need to be identified with subsequent annotation in a given sample, it comes under untargeted approaches. The approach includes MS databases (e.g. METLIN and MS-DIAL) and integrated software solutions (Han et al. 2021). Significant developments in three domains have made the metabolome research achievable.

1. *Analytical Techniques Improvement*

Microbe metabolomics is a technology-driven discipline. Several technologies are used to run sample metabolite profiling. Some primary analytical technologies used in metabolomics are NMR and MS, single-stage mass spectrometry coupled with liquid chromatography-mass spectrometry (LC-MS) and UV detection coupled with high-performance liquid chromatography/ultra-performance liquid chromatography (HPLC/UPLC). According to Emwas et al. (2019), every methodology has its pros and cons, and no single method can completely detect and quantify every metabolite present in a typical biological sample. To properly handle these issues, it is imperative to improve analytical approaches, procedures and instruments.

2. *Genomic Sequence Availability*
   The only way to assess an organism's whole metabolic capacity is to use its genomic sequence. The Human Genome Project (HGP), which was completed in 2003, is often regarded as the "crowning achievement in biology" history. *Haemophilus influenzae* was the first free-living bacterium to have its whole genome sequenced in 1995 (Bennett and Baker 2019). Genomic sequence analyses revealed microbial diversity which was previously unknown. A billion base pairs were added to the sequence database by several metagenomic projects and consortia. By using metagenomic sequencing, the microbial communities associated with humans have also been explored. In the study of microbial communities, the biodiversity of habitats and their functional capabilities are analysed through whole-genome sequencing and whole-genome shotgun (WGS) metagenomics. In contrast to WGS, marker gene analyses involve sequencing a specific gene region to reveal the biodiversity and composition of specific taxonomic groups. The marker genes are the 16 s rRNA gene used to analyse the presence of archaea and bacteria. Biases may be introduced into the metagenomic data produced by WGS metagenomics when whole-genome amplification techniques are used for low-concentration DNA samples (Pérez-Cobas et al. 2020). Nevertheless, by integrating metagenomic data with other "meta-omics" information, it makes it possible to rebuild system-wide networks in microbial communities, which advance ecosystem biology.

3. *Data Integration, Standards and Models*
   Multi-omics approaches enable the integration of metabolomics data, which becomes the foundation of whole systems biology. Numerous annotation and statistical tools, databases and network analysis software have been developed. Clusters of Orthologous Groups (COG), Entrez Gene, Kyoto Encyclopedia of Genes and Genomes (KEGG), TransportDB and UniProtKB are some bioinformatics resources with metabolomics applicability. The goal is to establish common interchangeable standard datasets that will increase data sharing and integration between the genomics and proteomics communities (Deutsch et al. 2017). Information from these datasets aids in reconstructing metabolic pathways and engineering in various microbes. Furthermore, novel models are being developed to address specific challenges posed by metabolomics.

4 Applications of Microbial Metabolomics

Since microbes may readily alter the ecosystem and human health, they are essential to systems biology research. For integrating information about biology into systems microbiology to facilitate the understanding of their interactions with each other and their cellular function, microbial metabolomics acts as a key platform within systems microbiology. Metabolomics has become an invaluable tool for the biosphere and influencing health determinants due to its powerful phenotyping ability (Shen et al. 2023). With the continuous development of metabolomics research technology, numerous clinical or drug researchers have contributed innovative concepts and methods. Currently, microbial metabolomics is commonly used in diagnostic applications, drug development, agriculture, bioremediation, environmental ecosystem and microbial metabolomics reconstructions.

1. *Diagnostic Applications (Disease Diagnosis)*

Microbial metabolomics has become an indispensable tool for disease diagnosis in recent years. This tool helps in early disease detection, classification of disease, monitoring treatment responses and discovery of biomarkers (Chasapi et al. 2022). Early disease detection is a prominent application area. With the identification of early signs of various diseases such as cancer, diabetes and cardiovascular disorders, microbial metabolomics can detect changes in metabolites within the body (Al-Sulaiti et al. 2023). Early diagnosis offers better treatment opportunities and increases the probability of favourable results. Furthermore, microbial metabolomics contributes to disease classification and differentiation subtypes and a better understanding of individual disease variation and diversity, and it helps in precise medication. During treatment, microbial metabolomics is also used to monitor disease changes and treatment effectiveness (Zhou et al. 2023). This helps doctors to adjust the treatment plans of their patients for the best possible outcomes. Moreover, microbial metabolomics plays a crucial role in the identification of biomarkers as well. With the ability for disease diagnosis, prognosis and monitoring, these biomarkers provide crucial information for medical decision-making.

In conclusion, microbial metabolomics application in disease detection signifies a breakthrough in the medical domain. Microbial metabolomics holds promise for enhancing the quality of patient life, improving disease management and offering new opportunities for medical research.

2. *Drug Development*

The pharmaceutical sector has embraced proteomics, genomics and other omics technologies for medication development. Metabolomics was not as advanced in drug discovery as it is now, but it has become a more sophisticated and quantitative biochemical technique in recent decades. The process of drug development can be divided into five stages: basic research, preclinical and clinical stages, review process and post-marketing evaluation (Mohs and Greig 2017). According to the first step, the process involves the selection of novel targets for disease modification, linking specific alternations with disease phenotypes. Novel targets are easily identified based onomics data, such as transcriptomics and genomics (El-Husseini et al. 2020). For instance, a recent study showed that carriers of the hydroxysteroid 17-β dehydrogenase 13 (HSD17B13) gene variant have a lower risk of cirrhosis and

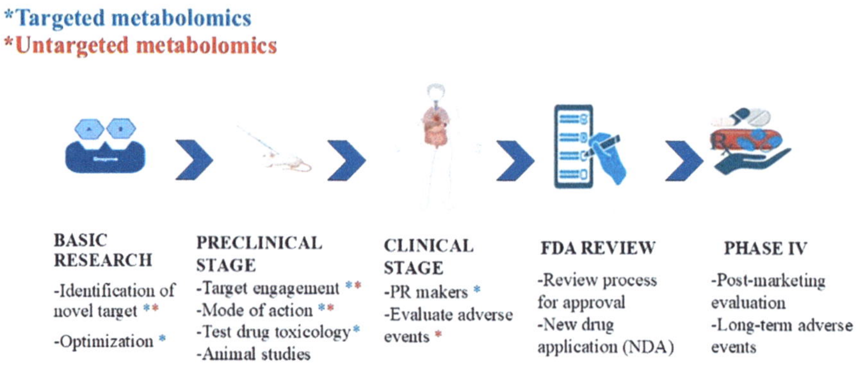

Fig. 3 Process of a new drug development

non-alcoholic steatohepatitis (Luukkonen et al. 2020). These targets, identified by population-wide screens and associations, without prior knowledge of their biochemical function, enter the drug development pipeline. The next step is the preclinical stage, where the drug is injected into an animal for drug testing. With the help of the preclinical stage, information regarding the biochemical processes, the mode of action (MoA) and target engagement (TE) marker identification can be acquired. Identifying these markers is the main key to the ultimate success of metabolomics approaches. The next step is the clinical stage, which describes the pharmacodynamics (PD) and pharmacokinetics (PK) with the physiological responses of the investigational new drug. Within the next step, a new drug application (NDA) gets reviewed for approval. In the final step, post-marketing evaluation is done. Thus, a new drug is developed, launched in the market and provided to the people as shown in Fig. 3.

In contrast to target-based drug development (TDD), in which largely screening is based upon a known target, phenotypic drug development (PDD) is a target-agnostic process, widely accepted in the pharmaceutical industry. The overall goal of these processes is to identify and optimise new targets and chemical entities (NCEs), particularly those with pharmacological properties (distribution, absorption and druggable compound metabolism), before moving them to clinical trials as investigational new drugs (INDs) (Nemmani 2021). The emergence of PDD and genetically generated targets has caused metabolomics approaches to become increasingly important in the first stages of drug development. However, processing high data quantities, missing pathways during biological interpretation and the difficulty of distinguishing between target-specific and general metabolic modifications are some of the major obstacles this technique confronts. Biologically connecting drug-induced pathway regulation (PR) to the target can also be difficult. For example, studying an alkaloid with an unknown target in a rat model of non-alcoholic fatty liver caused by a high-fat diet presents difficulties (Cui et al. 2020). Almost 20 significantly altered serum metabolites from different pathways are analysed based on untargeted metabolomics. Therefore, it is impossible to distinguish whether the observed metabolic changes are a primary and secondary effect of treatment and which specific factors are underlying the reported phenotypic alternations.

3. *Bioremediation*

It is the process of clearing hazardous contaminants with the help of biological agents such as microbes or plants. Microorganisms are one of the best natural tools that can be used to eradicate toxic pollutants. Bioremediation is a safe, economical and sustainable remediation technique. Moreover, its viability depends on the nature, locality and intensity of the pollution (Patel et al. 2022). Microbes are ubiquitous in nature, and they can remediate pollutants and utilise them as carbon sources as shown in Table 1.

4. *Agriculture*

Microbial metabolomics is a powerful tool in agriculture research, particularly used for crop improvement and development, differentiation processes, fruit maturation processes, pest and disease management, genome editing and metabolic profiling and in understanding plant-pathogen interactions and plant adverse environmental stress factors. By analyzing the metabolic changes during development, growth, and differentiation; examined population differentiation; elucidated biosynthetic pathways; characterized cultivars; and conducted non-targeted research to gain insights into genetic variability, physiological adaptations, and biochemical mechanisms influences the overall metabolic functions. Normally, plant-pathogen interaction is adversely affected, but if both obtain benefits, it's called symbiosis. Adversely, pathogen attack leads to yield losses although occasionally microbes can help the plant to improve nutrient uptake, e.g. legumes' interaction with nitrogen-fixing bacteria. Production of a plant's primary and secondary metabolites is the key response to pathogen attacks.

The metabolic profiling can be done to study the impact of heredity on the crops. Currently, research on genetic improvement of crops is being conducted to increase the quality and quantity of crop yield, protect from pest-created diseases and prevent resistance to several environmental factors. The chemical compounds in agriculture metabolomics can be detected with the help of several analytical techniques (Ibarra-Estrada et al. 2016). Currently, there are many examples of metabolomic profiling studies that have been applied to agriculture with various purposes as mentioned in Table 2.

5. *Environmental Ecosystems*

The study of microbial communities, which are an essential part of ecosystems, was traditionally restricted to those that could be grown. New developments in next-generation sequencing have greatly improved the comprehension of forest microbial diversity. By enabling the examination of diverse environmental samples straight from their native habitats, this method provides the benefit of massively parallel sequencing (Dubey et al. 2022). Further understanding of ecosystem functioning is made possible by metabolomics, which also makes it possible to profile metabolites linked to biotic stress and their biological activities.

Table 1 Use of microbes in bioremediation

Sr. no.	Microbes	Pollutant remediated	References
1.	*Pseudomonas aeruginosa*	Lead (Pb)	Nath et al. (2018)
2.	*Bacillus cereus*	Cadmium (Cd)	Nath et al. (2018)
3.	*Staphylococcus species*	Endosulfans	Liu et al. (2018)
4.	*Pseudomonas aeruginosa*	Cadmium (Cd)	Chellaiah (2018)
5.	*Corynebacterium propinquum*, *Bacillus subtilis* and *Alcaligenes odorans*	Phenol	Gaur et al. (2018)
6.	*Microbacterium* species, *Ralstonia* species, *Acinetobacter* species and *pseudomonas* species	Aromatic hydrocarbons	Basu et al. (2018)
7.	*Bacillus* species	Nickel (Ni)	Taran et al. (2019)
8.	*Oudemansiella canarii*	Congo red dye	Iark et al. (2019)
9.	*Rhodococcus* species	Aluminium (Al), lead (Pb), cadmium (Cd), and copper (Cu)	Nanda et al. (2019)
10.	*Cupriavidus* species	Cadmium (Cd)	Li et al. (2019)
11.	*Oscillatoria laete-virens*, *Spirogyra insignis*, *Pseudochlorococcum typicum* and *Arthrospira platensis*	Lead (Pb)	Song et al. (2019)
12.	*Pseudomonas* species	Copper (Cu) and lead (Pb)	Nanda et al. (2019)
13.	*Bacillus subtilis*	Lead (Pb)	Qiao et al. (2019)
14.	*Rhodopirellula* species, *Rhodovibrio* species, *Formosa* species	Hydrocarbon	Machado et al. (2019)
15.	*Shigella* species, *microbacterium* species and *micrococcus* species	Arsenic (Ar) and uranium (U)	Bhakat et al. (2019)
16.	*Escherichia coli*	Hexavalent chromium	Mohamed et al. (2020)
17.	*Alcaligenes odorans*, *Fusarium* species, *Pseudomonas aeruginosa* and *Corynebacterium propinquum*	Oils	Pande et al. (2020)
18.	*Ulothrix tenuissima*, *spirulina* species and *Chlamydomonas reinhardtii*	Chromium (Cr)	Aregbesola et al. (2020)
19.	*Lysinibacillus sphaericus*	Azo dye	Lu et al. (2020)
20.	*Pseudochlorococcum typicum*, *Chlamydomonas reinhardtii*, *Cyclotella cryptica* and *spirogyra hyaline*	Mercury (Hg)	Shah and Jain (2020)
21.	*Staphylococcus aureus*, *Klebsiella oxytoca*, *Bacillus firmus* and *Bacillus macerans*	Vat dyes	Sangkharak et al. (2020)
22.	*Desulfovibrio desulfuricans*	Cadmium (Cd), copper (Cu) and zinc (Zn)	Thakare et al. (2021)

(continued)

Table 1 (continued)

Sr. no.	Microbes	Pollutant remediated	References
23.	*Shewanella* species, *Thermus* species and *Enterobacter* species	Chromium (Cr)	Mousavi et al. (2021)
24.	*Bacillus licheniformis*	Dyes	Mousavi et al. (2021)
25.	*Spirulina* species and *chlorella* species	Nickel (Ni), lead (Pb) and dichromate	Geetha et al. (2021)
26.	*Cladosporium* species, *Trichoderma* species, *aspergillus versicolor*, *Paecilomyces* species and *aspergillus fumigatus*	Cadmium (Cd)	Unuofin et al. (2021)
27.	*Cunninghamella elegans* and *Saccharomyces cerevisiae*	Mercury (Hg) and heavy metals	Duc et al. (2021)
28.	*Sulfolobus solfataricus*	Copper (Cu)	Thakare et al. (2021)
29.	*Pseudoalteromonas* species and *Agarivorans* species	Hydrocarbons	Dell'Anno et al. (2021)
30.	*Lysinibacillus sphaericus*	Chromium (Cr), copper, crude oil, lead (Pb) and cobalt (Co)	Kharangate-Lad and D'Souza (2021)
31.	*Myceliophthora thermophila* and *Burkholderia* species	Polycyclic aromatic hydrocarbons and N,N-dimethyl-p-phenylenediamine	Mohapatra and Phale (2021)
32.	*Bacillus licheniformis* and *bacillus sonorensis*	Naphthalene	Rabani et al. (2022)
33.	*Burkholderia cepacia*, *Bacillus coagulans*, *Bacillus cereus* and *Serratia ficaria*	Diesel oil	Miri et al. (2022)
34.	*Pseudomonas aeruginosa* and *Aeromonas* species	Chromium (Cr), uranium (U), nickel (Ni) and copper (Cu)	Gaur et al. (2022)
35.	*Pseudomonas aeruginosa*	Crude oil	Mukjang et al. (2022)
36.	*Dehalococcoides* species	Chloroethenes	Dutta et al. (2022)
37.	*Pseudomonas fluorescence* and *Bacillus subtilis*	Zinc (Zn) and iron (Fe)	Širić et al. (2022)
38.	*Rhodopseudomonas palustris* and *Aerococcus* species	Lead (Pb), cadmium (Cd) and chromium (Cr)	Sravya and Sangeetha (2022)

6. *Microbial Reconstruction*

Systems biology has an intriguing chance to recreate and investigate microbial metabolism. In the post-genomic age, a paradigm change has occurred in the evaluation of biological information, ranging from genes to proteins. In cellular processing, the fundamental principles of molecular biology are replaced by dynamic

Table 2 Applications of advanced analytical techniques in different plant species and crops

Sr. no.	Plant species/crop	Analytical technique and its purpose	References
1.	*Aporosa cardiosperma* leaves	GC-MS/FT-IR Determined metabolite constituents and quantified flavonoid, phenol and tannin content which neutralise the free radicals	Abdul et al. (2024)
2.	*Zea mays*	MALDI-TOF MS Analysed effect of phosphorous (P) deficiency	Ibarra-Estrada et al. (2016)
3.	*Zea mays* L. leaves	GC-MS To carry out the genetic mapping from metabolic profiling of maize leaves	Ibarra-Estrada et al. (2016)
4.	*Oryza sativa* L.	CE-MS Identification of genomic regions responsible for high iron and zinc content	Dixit et al. (2019)
5.	*Solanum tuberosum* L.	LC-MS Identification of metabolites 36–40, i.e. amino acids, aromatic amines and organic acids	Munyai et al. (2022)
6.	*Persea americana*	GC-APCI-TOF-MS Investigating fruit maturation through its metabolomic profile	Ibarra-Estrada et al. (2016)
7.	*Solanum lycoperscium*	LC-MS Metabolite profile and distinct changes in taste quality	Cai et al. (2024)
8.	*Hordeum vulgare*	GC-MS Metabolic responses to salt stress	Shelden et al. (2016)
9.	*Vigna radiata*	GC-MS Metabolite profiling to investigate metabolic changes that occur during sprouting	Na Jom et al. (2011)
10.	*Capsicum annum* species	GC-MS Metabolite profiling and biological activities	Ahmad et al. (2022)

interactions and many linkages between metabolites. According to Dos Santos et al. (2021), this demonstrates the complexity of metabolic networks and the ability of metabolites to interact with macromolecules through competitive and allosteric binding to the active sites of enzymes, resulting in changes in enzyme activity. As previously stated, metabolomics complements transcriptomics and proteomics. It demonstrates that metabolomes, which properly reflect microbial phenotypes, are located further downstream in the hierarchical functional analyses. The computation of fluxes via metabolic pathways is known as metabolic fluxes. Furthermore, as transcripts and proteins are unable to compute metabolic fluxes on their own, metabolomics inputs are necessary for the elucidation of cellular networks (Emwas et al. 2022).

Large quantities of "omics"-related data, together with a variety of computational methods and tools, may be used to analyse biological activities. Even for bacteria about which there is little information in the literature, metabolic networks have been reconstructed at the genomic level. An annotated genome usually serves as the starting point for the methodical process of rebuilding a metabolic network, which ends with a predictive model of microbial physiology. The biological components of all systems have an impact on the reconstruction of metabolic pathways, and different networks aid in the diversification and refinement of the metabolic content. By combining automated analysis with manual curation, this technique fills in the gaps and reconciles genomic and biochemical data with known metabolic activities (Noriega-Atala et al. 2021).

Metabolic engineering is a real-world example of this. Microorganisms are prime candidates for the industrial production of essential products, ranging from pollution degradation to renewable energy generation. Rather than rationally modifying an organism through genetics to obtain improved performance, a thorough understanding of the many stages in the hierarchical structure (from genome to metabolome) can lead to an accurate design and management of biological systems.

5 Conclusion

In this chapter, we have discussed microbial metabolomics, a study of a microbe's metabolism that reflects the actual physiological state of the microorganism. With the help of novel analytical technologies, our analysis focuses on intracellular and extracellular metabolites, signalling molecules and metabolic pathways of the microorganism in diversified microbe metabolomics applications. Main methods like NMR and MS are used for the identification and characterisation of many metabolites in microbes. The study of spontaneous disease has a lot of potential, and metabolomics could help identify biomarkers and expand our knowledge about disease causes. Developing new drugs, monitoring treatment response and carrying out toxicological studies are further opportunities. Even though microbial metabolomics is a relatively new field, in the last few years, it has achieved breakthroughs in various fields of systems biology. Microbes are no longer regarded as isolated organisms existing in a system, but rather an integrated component for understanding functional biology. Different analytical techniques have been developed and improved for easier interpretation and result analysis. The development of analytical techniques implemented in microbe metabolomics allows us to be a step ahead towards new applications in systems biotechnology.

References

Abdul U, Manikandan DB, Arumugam M, Alomar SY, Manoharadas S, Ramasamy T (2024) GC–MS based metabolomic profiling of Aporosa cardiosperma (Gaertn.) Merr. leaf extracts and evaluating its therapeutic potential. Sci Rep 14:16010

Ahmad R, Alqathama A, Aldholmi M, Riaz M, Abdalla AN, Mostafa A, Al-Said HM, Alqarni AM, Ullah R, Asgher SS, Amir M, Shaaban H, Ahmad W (2022) Gas chromatography-mass spectrometry (GC-MS) metabolites profiling and biological activities of various Capsicum annum cultivars. Plan Theory 11:1022

Al-Sulaiti H, Almaliti J, Naman CB, Al Thani AA, Yassine HM (2023) Metabolomics approaches for the diagnosis, treatment, and better disease management of viral infections. Meta 13:948

Aregbesola OA, Kumar A, Mokoena MP, Olaniran AO (2020) Cloning, overexpression, purification, characterization and structural modelling of a metabolically active Fe2+ dependent 2,6-dichloro-p-hydroquinone 1,2-dioxygenase (CpsA) from Bacillus cereus strain AOA-CPS_1. Int J Biol Macromol 161:247–257

Arun A, Dutta S, Sarkar R, Sinha E, Panwar K, Venu G, Prajapati SK (2023) Metabolomics and its application in advance diagnostic microbiology field. Molecules 3:4

Basu S, Rabara RC, Negi S, Shukla P (2018) Engineering PGPMOs through gene editing and systems biology: a solution for phytoremediation? Trends Biotechnol 36:499–510

Bennett RJ, Baker KS (2019) Looking backward to move forward: the utility of sequencing historical bacterial genomes. J Clin Microbiol 57:10–1128

Bhakat K, Chakraborty A, Islam E (2019) Characterization of arsenic oxidation and uranium bioremediation potential of arsenic resistant bacteria isolated from uranium ore. Environ Sci Pollut Res 26:12907–12919

Cai W, Jiang P, Liu Y, Miao X, Liu A (2024) Distinct changes of taste quality and metabolite profile in different tomato varieties revealed by LC-MS metabolomics. Food Chem 442:138456

Chasapi SA, Karagkouni E, Kalavrizioti D, Vamvakas S, Zompra A, Takis PG, Goumenos DS, Spyroulias GA (2022) NMR-based metabolomics in differential diagnosis of chronic kidney disease (CKD) subtypes. Meta 12:490

Chellaiah ER (2018) Cadmium (heavy metals) bioremediation by Pseudomonas aeruginosa: a minireview. Appl Water Sci 8:154

Cui H, Li Y, Cao M, Liao J, Liu X, Miao J, Fu H, Song R, Wen W, Zhang Z, Wang H (2020) Untargeted metabolomic analysis of the effects and mechanism of nuciferine treatment on rats with nonalcoholic fatty liver disease. Front Pharmacol 11:858

Dell'Anno F, Rastelli E, Tangherlini M, Corinaldesi C, Sansone C, Brunet C, Balzano S, Ianora A, Musco L, Montereali MR, Dell'Anno A (2021) Highly contaminated marine sediments can host rare bacterial taxa potentially useful for bioremediation. Front Microbiol 12:584850

Deutsch EW, Orchard S, Binz P-A, Bittremieux W, Eisenacher M, Hermjakob H, Kawano S, Lam H, Mayer G, Menschaert G, Jones AR (2017) Proteomics standards initiative: fifteen years of progress and future work. J Proteome Res 16:4288–4298

Dixit S, Singh UM, Abbai R, Ram T, Singh VK, Paul A, Virk PS, Kumar A (2019) Identification of genomic region (s) responsible for high iron and zinc content in rice. Sci Rep 9:8136

dos Santos G, Renovato-Martins M, de Brito NM (2021) The remodel of the "central dogma": a metabolomics interaction perspective. Metabolomics 17:1–15

Dubey A, Malla MA, Kumar A (2022) Role of next-generation sequencing (NGS) in understanding the microbial diversity. In: Molecular genetics and genomics tools in biodiversity conservation. Springer, pp 307–328

Duc HD, Hung NV, Oanh NT (2021) Anaerobic degradation of Endosulfans by a mixed culture of Pseudomonas sp. and Staphylococcus sp. Appl Biochem Microbiol 57:327–334

Dutta N, Usman M, Ashraf MA, Luo G, Zhang S (2022) Efficacy of emerging technologies in addressing reductive dechlorination for environmental bioremediation: a review. J Hazard Mater Lett 3:100065

El-Husseini ZW, Gosens R, Dekker F, Koppelman GH (2020) The genetics of asthma and the promise of genomics-guided drug target discovery. Lancet Respir Med 8:1045–1056

Emwas A-H, Roy R, McKay RT, Tenori L, Saccenti E, Gowda GAN, Raftery D, Alahmari F, Jaremko L, Jaremko M, Wishart DS (2019) NMR spectroscopy for metabolomics research. Metabolites 9:123

Emwas A-H, Szczepski K, Al-Younis I, Lachowicz JI, Jaremko M (2022) Fluxomics-new metabolomics approaches to monitor metabolic pathways. Front Pharmacol 13:805782

Gaur N, Narasimhulu K, PydiSetty Y (2018) Recent advances in the bio-remediation of persistent organic pollutants and its effect on environment. J Clean Prod 198:1602–1631

Gaur VK, Tripathi V, Manickam N (2022) Bacterial-and fungal-mediated biodegradation of petroleum hydrocarbons in soil. In: Development in wastewater treatment research and processes. Elsevier, pp 407–427

Geetha N, Bhavya G, Abhijith P, Shekhar R, Dayananda K, Jogaiah S (2021) Insights into nano-mycoremediation: secretomics and mycogenic biopolymer nanocomposites for heavy metal detoxification. J Hazard Mater 409:124541
Han M-L, Zhu Y, Creek DJ, Lin Y-W, Anderson D, Shen H-H, Tsuji B, Gutu AD, Moskowitz SM, Velkov T, Li J (2018) Alterations of metabolic and lipid profiles in polymyxin-resistant Pseudomonas aeruginosa. Antimicrob Agents Chemother 62:10–1128
Han S, Van Treuren W, Fischer CR, Merrill BD, DeFelice BC, Sanchez JM, Higginbottom SK, Guthrie L, Fall LA, Dodd D, Fischbach MA, Sonnenburg JL (2021) A metabolomics pipeline for the mechanistic interrogation of the gut microbiome. Nature 595:415–420
Iark D, dos Reis Buzzo AJ, Garcia JAA, Côrrea VG, Helm CV, Corrêa RCG, Peralta RA, Moreira R d FPM, Bracht A, Peralta RM (2019) Enzymatic degradation and detoxification of azo dye Congo red by a new laccase from Oudemansiella canarii. Bioresour Technol 289:121655
Ibarra-Estrada E, Soto-Hernández RM, Palma-Tenango M (2016) Metabolomics as a tool in agriculture. IntechOpen, London
Ju R, Liu X, Zheng F, Zhao X, Lu X, Zeng Z, Lin X, Xu G (2019) Removal of false positive features to generate authentic peak table for high-resolution mass spectrometry-based metabolomics study. Anal Chim Acta 1067:79–87
Kharangate-Lad A, D'Souza NC (2021) Current approaches in bioremediation of toxic contaminants by application of microbial cells; biosurfactants and bioemulsifiers of microbial origin. Rhizobiont Bioremediation Hazard Waste:217–263
Kumar R, Bohra A, Pandey AK, Pandey MK, Kumar A (2017) Metabolomics for plant improvement: status and prospects. Front Plant Sci 8:1302
Li F, Zheng Y, Tian J, Ge F, Liu X, Tang Y, Feng C (2019) Cupriavidus sp. strain Cd02-mediated pH increase favoring bioprecipitation of Cd2+ in medium and reduction of cadmium bioavailability in paddy soil. Ecotoxicol Environ Saf 184:109655
Liu Z, Shao B, Zeng G, Chen M, Li Z, Liu Y, Jiang Y, Zhong H, Liu Y, Yan M (2018) Effects of rhamnolipids on the removal of 2,4,2,4-tetrabrominated biphenyl ether (BDE-47) by Phanerochaete chrysosporium analyzed with a combined approach of experiments and molecular docking. Chemosphere 210:922–930
Lu J, Zhang C, Leong HY, Show PL, Lu F, Lu Z (2020) Overproduction of lipoxygenase from Pseudomonas aeruginosa in Escherichia coli by auto-induction expression and its application in triphenylmethane dyes degradation. J Biosci Bioeng 129:327–332
Luukkonen PK, Tukiainen T, Juuti A, Sammalkorpi H, Haridas PAN, Niemelä O, Arola J, Orho-Melander M, Hakkarainen A, Kovanen PT, Dwivedi O, Groop L, Hodson L, Gastaldelli A, Hyötyläinen T, Orešič M, Yki-Järvinen H (2020) Hydroxysteroid 17-β dehydrogenase 13 variant increases phospholipids and protects against fibrosis in nonalcoholic fatty liver disease. JCI Insight 5:20200312
Machado LF, de Assis Leite DC, da Costa Rachid CTC, Paes JE, Martins EF, Peixoto RS, Rosado AS (2019) Tracking mangrove oil bioremediation approaches and bacterial diversity at different depths in an in situ mesocosms system. Front Microbiol 10:2107
Miri S, Rasooli A, Brar SK, Rouissi T, Martel R (2022) Biodegradation of p-xylene—a comparison of three psychrophilic Pseudomonas strains through the lens of gene expression. Environ Sci Pollut Res:1–15
Mohamed MSM, El-Arabi NI, El-Hussein A, El-Maaty SA, Abdelhadi AA (2020) Reduction of chromium-VI by chromium-resistant Escherichia coli FACU: a prospective bacterium for bioremediation. Folia Microbiol (Praha) 65:687–696
Mohapatra B, Phale PS (2021) Microbial degradation of naphthalene and substituted naphthalenes: metabolic diversity and genomic insight for bioremediation. Front Bioeng Biotechnol 9:602445
Mohd Kamal K, Mahamad Maifiah MH, Abdul Rahim N, Hashim YZH-Y, Abdullah Sani MS, Azizan KA (2022) Bacterial metabolomics: sample preparation methods. Biochem Res Int 2022:9186536
Mohs RC, Greig NH (2017) Drug discovery and development: Role of basic biological research. Alzheimer's Dement Transl Res Clin Interv 3:651–657

Mousavi SM, Hashemi SA, Iman Moezzi SM, Ravan N, Gholami A, Lai CW, Chiang W-H, Omidifar N, Yousefi K, Behbudi G (2021) Recent advances in enzymes for the bioremediation of pollutants. Biochem Res Int 2021:5599204

Mukjang N, Chitov T, Mhuantong W, Champreda V, Pathom-Aree W, Sattayawat P, Bovonsombut S (2022) Bacterial communities associated with crude oil bioremediation through composting approaches with indigenous bacterial isolate. Life 12:1712

Munyai R, Raletsena MV, Modise DM (2022) LC-MS based metabolomics analysis of potato (Solanum tuberosum L.) cultivars irrigated with quicklime treated acid mine drainage water. Meta 12:221

Na Jom K, Frank T, Engel K-H (2011) A metabolite profiling approach to follow the sprouting process of mung beans (Vigna radiata). Metabolomics 7:102–117

Nanda M, Kumar V, Sharma DK (2019) Multimetal tolerance mechanisms in bacteria: the resistance strategies acquired by bacteria that can be exploited to 'clean-up' heavy metal contaminants from water. Aquat Toxicol 212:1–10

Nath S, Deb B, Sharma I (2018) Isolation of toxic metal-tolerant bacteria from soil and examination of their bioaugmentation potentiality by pot studies in cadmium-and lead-contaminated soil. Int Microbiol 21:35–45

Nemmani KVS (2021) Pharmacological screening: drug discovery. Drug Discov Dev From Targets Mol to Med:211–233

Noriega-Atala E, Lovett PM, Morrison CT, Surdeanu M (2021) Neural architectures for biological inter-sentence relation extraction. arXiv Prepr. arXiv2112.09288

Oyedeji AB, Green E, Adebiyi JA, Ogundele OM, Gbashi S, Adefisoye MA, Oyeyinka SA, Adebo OA (2021) Metabolomic approaches for the determination of metabolites from pathogenic microorganisms: a review. Food Res Int 140:110042

Pande V, Pandey SC, Sati D, Pande V, Samant M (2020) Bioremediation: an emerging effective approach towards environment restoration. Environ Sustain 3:91–103

Patel AK, Singhania RR, Albarico FPJB, Pandey A, Chen C-W, Dong C-D (2022) Organic wastes bioremediation and its changing prospects. Sci Total Environ 824:153889

Pérez-Cobas AE, Gomez-Valero L, Buchrieser C (2020) Metagenomic approaches in microbial ecology: an update on whole-genome and marker gene sequencing analyses. Microb Genomics 6:e000409

Pinu FR, Goldansaz SA, Jaine J (2019) Translational metabolomics: current challenges and future opportunities. Meta 9:108

Qiao W, Zhang Y, Xia H, Luo Y, Liu S, Wang S, Wang W (2019) Bioimmobilization of lead by Bacillus subtilis X3 biomass isolated from lead mine soil under promotion of multiple adsorption mechanisms. R Soc Open Sci 6:181701

Rabani MS, Sharma R, Singh R, Gupta MK (2022) Characterization and identification of naphthalene degrading bacteria isolated from petroleum contaminated sites and their possible use in bioremediation. Polycycl Aromat Compd 42:978–989

Rampler E, El Abiead Y, Schoeny H, Rusz M, Hildebrand F, Fitz V, Koellensperger G (2020) Recurrent topics in mass spectrometry-based metabolomics and lipidomics—standardization, coverage, and throughput. Anal Chem 93:519–545

Sangkharak K, Choonut A, Rakkan T, Prasertsan P (2020) The degradation of phenanthrene, pyrene, and fluoranthene and its conversion into medium-chain-length polyhydroxyalkanoate by novel polycyclic aromatic hydrocarbon-degrading bacteria. Curr Microbiol 77:897–909

Shah H, Jain S (2020) Bioremediation: an approach for environmental pollutants detoxification. In: Waste to energy: prospects and applications. Springer, Singapore, pp 121–142

Shelden MC, Dias DA, Jayasinghe NS, Bacic A, Roessner U (2016) Root spatial metabolite profiling of two genotypes of barley (Hordeum vulgare L.) reveals differences in response to short-term salt stress. J Exp Bot 67:3731–3745

Shen Y, Wu S-D, Chen Y, Li X-Y, Zhu Q, Nakayama K, Zhang W-Q, Weng C-Z, Zhang J, Wang H-K, Wu J, Jiang W (2023) Alterations in gut microbiome and metabolomics in chronic hepatitis B infection-associated liver disease and their impact on peripheral immune response. Gut Microbes 15:2155018

Širić I, Eid EM, Taher MA, El-Morsy MHE, Osman HEM, Kumar P, Adelodun B, Abou Fayssal S, Mioč B, Andabaka Ž, Choi KS, Kumar V (2022) Combined use of spent mushroom substrate biochar and PGPR improves growth, yield, and biochemical response of cauliflower (Brassica oleracea var. botrytis): a preliminary study on greenhouse cultivation. Horticulturae 8:830
Song J, Zhang S, Xie Y, Li Q (2019) Purification and characteristics of an aflatoxin B1 degradation enzyme isolated from Pseudomonas aeruginosa. FEMS Microbiol Lett 366:fnz034
Sravya K, Sangeetha S (2022) Feasibility study on bioremediation techniques to contaminated soils. Mater Today Proc 51:2556–2560
Tang J (2011) Microbial metabolomics. Curr Genomics 12:391–403
Taran M, Fateh R, Rezaei S, Gholi MK (2019) Isolation of arsenic accumulating bacteria from garbage leachates for possible application in bioremediation. Iran J Microbiol 11:60
Thakare M, Sarma H, Datar S, Roy A, Pawar P, Gupta K, Pandit S, Prasad R (2021) Understanding the holistic approach to plant-microbe remediation technologies for removing heavy metals and radionuclides from soil. Curr Res Biotechnol 3:84–98
Unuofin JO, Falade AO, Aladekoyi OJ (2021) Applications of microbial laccases in bioremediation of environmental pollutants: potential issues, challenges, and prospects. Bioremediation Environ Sustain:519–540
Vitale GA, Geibel C, Minda V, Wang M, Aron AT, Petras D (2024) Connecting metabolome and phenotype: recent advances in functional metabolomics tools for the identification of bioactive natural products. Nat Prod Rep 41:885
Zhou S, Xu H, Zhu J, Fan X, Zhang J (2023) Clinical efficacy and metabolomics study of Wendan decoction in the treatment of phlegm-dampness obstructive sleep apnea-hypopnea syndrome with type 2 diabetes mellitus. J Ethnopharmacol 317:116775

Microbial Metabolite Databases

K. K. Chetan, J. H. Ashwini, T. Kavya,
Vanapalli Lohitha Sai Sree, Shashank Goud,
and Rashmitha

Abstract

Microbial metabolites, the diverse array of small molecules produced by microorganisms that play a pivotal role in various biological processes, hold immense potential for advancements in medicine, biotechnology and environmental science. To harness this potential, comprehensive databases are essential for cataloguing, characterizing and analysing these compounds. This chapter provides a comprehensive overview of existing microbial metabolite databases, exploring their scope, structure and utility. Key databases such as Kyoto Encyclopedia of Genes and Genomes (KEGG), BioCyc, MetaCyc, MiMeDB, Human Metabolome Database (HMDB) and MPMdb are discussed, highlighting their critical role in connecting microbial diversity with metabolic function. By understanding the landscape of available databases, researchers can effectively explore the complex world of microbial metabolites and unlock their full potential. Additionally, we highlight the importance of data integration, quality and continuous updates in maximizing the value of these databases for advancing our knowledge of microbial metabolism and its implications for the environment.

Keywords

Microbial metabolites · Databases · KEGG · Metabolic pathways

K. K. Chetan (✉) · J. H. Ashwini · V. L. S. Sree
Division of Plant Pathology, ICAR-IARI, New Delhi, India

T. Kavya · S. Goud
Division of Microbiology, ICAR-IARI, New Delhi, India

Rashmitha
Division of Seed Science and Technology, ICAR-IARI, New Delhi, India

S. Kaur et al. (eds.), *Microbial Metabolomics*,
https://doi.org/10.1007/978-981-96-4824-5_3

1 Introduction

Microbial metabolites, encompassing both primary and secondary metabolites, are pivotal in various biological processes (Luckner 2013). Primary metabolites, such as amino acids, nucleotides and carbohydrates, are essential for growth and reproduction, while secondary metabolites, including antibiotics, pigments and toxins, contribute to microbial survival and interactions with their environment (Bills and Gloer 2016). These metabolites have profound implications for medicine, biotechnology and environmental science. For instance, microbial-derived antibiotics have revolutionized the treatment of bacterial infections (Hutchings et al. 2019), and biotechnological applications harness these metabolites for the production of biofuels and biodegradable plastics (Kumar et al. 2005). In environmental science, microbial metabolites are instrumental in bioremediation efforts, aiding in the detoxification of pollutants (Kumar et al. 2005). Databases play a crucial role in the study of these metabolites by cataloguing, characterizing and analysing them, thus providing essential tools for understanding metabolic processes and their applications.

A database serves as a structured collection of data. Twenty-five years ago, few biologists saw the need for databases because biological data was relatively scarce, aside from biological literature. However, the landscape changed dramatically with the advent of genome sequencing technologies. The amount of biological data began to expand rapidly as genome sequencing became more common, and then it surged with the widespread adoption of next-generation sequencing (NGS) (Metzker 2010), fuelling the "omics" revolution. As a result, the significance of databases in microbiology has grown substantially and will continue to increase in the foreseeable future (Howe et al. 2008). The databases, which provide organized repositories of information on metabolite structures, functions and pathways, are critical for generating hypotheses, designing experiments and conducting comparative studies, thereby advancing research in microbial metabolism. This chapter focuses specifically on databases relevant to microbiology, rather than attempting to cover the vast array of biological databases available today. However, it is important to note that the boundaries between disciplines are often unclear, so some comprehensive databases that serve broader biological fields will be mentioned in the context of microbiology. Hence, this chapter aims to offer an in-depth review of major microbial databases, discussing their scope, structure and usefulness, and to emphasize the importance of data integration, quality and regular updates in maximizing the research value of these resources. It is important to remember that although many biological databases may contain similar data (e.g. identical genomes), they are developed for different purposes, include diverse types of data, use different tools for access and are curated by different individuals. Thus, no two databases are exactly the same.

2 Overview of Key Microbial Metabolite Databases

Databases can be classified in various ways, but no single, standardized method exists. For the purposes of this book chapter, we have grouped them into three categories: general-purpose databases, special-purpose protein databases and metabolite-specific databases.

2.1 General-Purpose Databases

General resources are composed of interconnected databases and tools, aiming to provide a comprehensive platform for a wide range of users. The National Center for Biotechnology Information (NCBI) at the US National Institutes of Health and the European Molecular Biology Laboratory's European Bioinformatics Institute (EMBL-EBI) are recognized leaders, offering some of the world's most extensive collections of genomic and molecular biology data making them accessible to users through various text-based and sequence-based search tools. The primary features of these resources are outlined by their developers and curators in corresponding publications (NCBI Resource Coordinators 2015 and Brooksbank et al. 2014 respectively).

NCBI is an essential resource for microbiologists, just as it is for other biologists. Microbiologists rely on NCBI literature resources, including PubMed and PubMed Central, which provide access to peer-reviewed journal articles, and the NCBI Bookshelf, which offers free access to the full text of books and reports. A central component of the NCBI's offerings is its non-redundant databases, which include nucleotide and protein sequences, along with their curated subset known as Reference Sequences, or RefSeq (Pruitt et al. 2012). The NCBI protein database compiles protein sequencing projects and provides links to related records in databases. The BLAST tool, most widely used sequence database search tool, now allows users to search for protein sequence similarities within specific taxonomic groups directly from its NCBI web page (Altschul et al. 1997).

Similarly, *EMBL-EBI* has adopted a user-centric approach for its interconnected databases and tools (Brooksbank et al. 2014). The databases at EMBL-EBI are organized into several categories: DNA and RNA (covering genes, genomes and expression data), proteins (including sequences, families and structures), metabolites (with a focus on chemogenomics and metabolomics) and systems (encompassing reactions, interactions and pathways). The Universal Protein Resource (UniProt) and its curated knowledge base, UniProtKB, are among the most esteemed databases provided by EMBL-EBI (UniProt Consortium 2014). For microbiologists, EMBL-EBI offers a range of useful resources scattered throughout its extensive database collection. Notably, the Metagenomics portal stands out as a valuable tool; it enables researchers to submit, store and analyse genomic data from diverse environments, with an analysis pipeline that facilitates feature (e.g. genes and rRNAs), functional (e.g. families, structures and ontologies) and taxonomic (e.g. operational taxonomic units) predictions (Hunter et al. 2014).

2.2 Special-Purpose Protein Databases

Microorganisms may have hundreds to over ten thousand proteins encoded in their genomes. Classifying these proteins into distinct families based on conserved domains and evolutionary relationships helps to elucidate their biological functions. Several resources are dedicated to this classification, focusing on either sequence or structural data. For example, Pfam (Finn et al. 2014), SMART (Letunic et al. 2015) and TIGRFAM (Haft et al. 2003) are well-curated collections that identify conserved proteins and domains. The Clusters of Orthologous Groups of proteins (COGs) database (Tatusov et al. 2003) offers a phylogenetic classification of proteins, and its latest version has significantly improved with increased coverage of microbial genomes (Galperin et al. 2015). The Conserved Domain Database (CDD) at NCBI consolidates data from these databases, as well as its own models, into a unified collection (Marchler-Bauer et al. 2015). Similarly, the InterPro database at EMBL-EBI integrates data from multiple sources to provide extensive information on protein domains, regions, motifs and functional sites (Mitchell et al. 2015).

Understanding the structure of biological macromolecules is crucial for comprehending their functions. The Research Collaboratory for Structural Bioinformatics (RCSB) Protein Data Bank (PDB) serves as a major repository for atomic coordinates and information on the three-dimensional structures of proteins, nucleic acids and their complexes (Rose et al. 2015). Additionally, the SCOP (Structural Classification of Proteins) and CATH (Class, Architecture, Topology and Homology) databases (Andreeva et al. 2014 and Sillitoe et al. 2015 respectively) are comprehensive resources for the classification of protein structures. Identifying orthologs (genes and their products in different species that originated from a common ancestral gene) and paralogs (genes and their products resulting from gene duplication) is also essential for predicting protein functions. Alongside COGs, the OrthoDB database provides detailed information on gene and protein orthology (Kriventseva et al. 2015).

2.3 Metabolite-Specific Databases

Reconstructing metabolic pathways from genome sequences is a fundamental task in modern biology, combining experimental biochemistry with comparative genome analysis. The Kyoto Encyclopedia of Genes and Genomes (KEGG) is a leading resource in this area, including KEGG PATHWAY, KEGG GENES and KEGG COMPOUND (Kanehisa et al. 2014). KEGG provides comprehensive information on metabolic pathways, enzyme functions and chemical compounds, linking metabolic processes with gene and protein data. Its pathway maps are highly valued for their clarity and integration with other biological data, though some pathways and metabolites may be underrepresented. KEGG's applications are diverse, ranging from pathway analysis to gene function prediction. In contrast, BioCyc (Caspi et al. 2014) offers a range of pathway/genome databases for various organisms, including bacteria and eukaryotes. It is particularly strong in providing organism-specific pathway information and interactive tools for visualizing metabolic networks. However, the depth of data can vary, and the interface may be less intuitive for some users. MetaCyc (Caspi et al. 2014), a metabolic branch of the larger BioCyc collection, includes over 2000 experimentally determined metabolic pathways from more than 2000 organisms across all 3 domains of life. Protein-protein interactions are vital for numerous cellular functions, including metabolism, transport and signal transduction. Despite its comprehensive coverage, some pathways may suffer from inconsistencies or incomplete annotations. The STRING database is a key resource for functional protein networks (Franceschini et al. 2013), while the BioGRID database provides information on protein and genetic interactions across many model organisms, including *E. coli*, *B. subtilis*, eukaryotic microorganisms and viruses (Chatr-Aryamontri et al. 2015). MiMeDB, specializing in microbial metabolites, offers targeted information on metabolites produced by microorganisms. Its focused approach provides detailed annotations, although it may have limited coverage compared to larger databases.

Two specialized databases, the Transporter Classification Database (TCDB) and TransportDB, provide extensive overviews, curated annotations and detailed genomic comparisons of membrane transport proteins across various bacterial, archaeal and eukaryotic genomes (Ren et al. 2007; Saier et al. 2014). More focused community databases target specific transport systems, including archaeal and bacterial ABC transporters (Fichant et al. 2006), β-barrel outer membrane proteins (Tsirigos et al. 2011) and type III (Wang et al. 2012), type IV (Souza et al. 2012) and type VI secretion systems (Li et al. 2015). Microme is a European platform dedicated to microbial metabolism, designed to facilitate large-scale inference of metabolic fluxes directly from genome sequences. It includes the well-regarded MicroScope platform (Vallenet et al. 2013) for microbial genome annotation and analysis and acts as a gateway to various analytical and data mining tools.

Microorganisms possess a diverse array of enzymes involved in the assembly, modification and breakdown of oligo- and polysaccharides. The Carbohydrate-Active Enzymes (CAZy) database provides a classification system that connects enzyme sequences with their specificities and three-dimensional structures (Lombard et al. 2014). Additionally, numerous community databases are dedicated to specific metabolic processes in microorganisms. For example, AromaDeg focuses on the aerobic degradation of aromatic compounds by bacteria (Duarte et al. 2014). The DEOP database offers curated information on osmo-protectants and related pathways for various bacterial species (Bougouffa et al. 2014), while BacMet serves as a resource for antibacterial biocide and metal resistance genes (Pal et al. 2014). The ClusterMine360 database provides information on microbial polyketide and non-ribosomal peptide gene clusters (Conway and Boddy 2013). The BCSDB database details bacterial carbohydrate structures (Toukach 2011), and the ProGlycProt repository (Bhat et al. 2012) lists experimentally characterized glycoproteins from bacteria and archaea. CyanoLyase (Bretaudeau et al. 2013) and mVOC (Lemfack et al. 2014) are curated databases for phycobiliproteins and microbial volatile compounds, respectively. The Human Metabolome Database (HMDB) contains extensive data on human metabolites, including those influenced by the microbiome, and integrates clinical data on metabolite concentrations. While it offers valuable insights into microbiome interactions, its primary focus remains on human metabolites. Lastly, MPMdb is dedicated to metabolites produced by microbial pathogens, providing data on compounds involved in pathogenicity and disease mechanisms. Although it offers specialized information, its focus on pathogens limits coverage of non-pathogenic microorganisms.

The Microbial Signal Transduction (MiST) database provides extensive information on receptors, kinases, response regulators and transcription factors across thousands of bacterial and archaeal genomes (Ulrich and Zhulin 2010). A subset of these proteins is also catalogued in the P2CS (prokaryotic two-component systems) database (Ortet et al. 2015). The Quorumpeps database focuses on a collection of quorum-sensing peptides, which are specific signaling molecules (Wynendaele et al. 2013). Transcriptional regulation is a primary mechanism through which bacteria respond to signals, and several databases are dedicated to cataloguing known and predicted transcriptional regulators and their targets. The SwissRegulon database offers genome-wide annotations of regulatory sites for both eukaryotic and prokaryotic model organisms, including *E. coli*, *B. subtilis*, *S. aureus*, *Vibrio cholerae* and *M. tuberculosis* (Pachkov et al. 2012). RegTransBase contains a curated collection of regulatory sequences and interactions, based on thousands of scientific publications (Cipriano et al. 2013), while RegPrecise provides inferred regulatory interactions across hundreds of bacterial genomes, with a focus on phylogenetic, structural and functional aspects (Novichkov et al. 2013).

The Network Portal offers analysis and visualization tools for gene regulatory networks in several bacterial species, such as *E. coli*, *Bacillus subtilis*, *Mycobacterium tuberculosis*, *Pseudomonas aeruginosa* and *Campylobacter jejuni* (Turkarslan et al. 2014). It functions as a modular database for both user-uploaded and public data analysis. The ProOpDB database includes predictions of operons for over 1200 bacterial and archaeal genomes (Taboada et al. 2012). The CollecTF database provides a collection of experimentally verified transcription factor-binding sites (Kilic et al. 2014). Additionally, WebGeSTer DB is the largest resource for intrinsic transcription terminators identified in more than a thousand bacterial genomes (Mitra et al. 2011).

3 Database Structure and Utility

Microbial metabolite databases utilize various data formats and organizational structures. Data is meticulously organized into records representing metabolites, pathways, enzymes and related information (Zhou et al. 2022). Key components include detailed metabolite entries, visual pathway maps and links to associated enzymes and genes (Huerta-Cepas et al. 2016). Search and retrieval functionalities are critical for effective database use, often featuring keyword searches, advanced filters and interactive tools for pathway visualization. These features enable researchers to efficiently access and analyse data, facilitating comprehensive studies of metabolic networks (Zhou et al. 2022). Data integration from multiple sources enriches the understanding of complex metabolic processes, while visualization tools aid in interpreting intricate pathways and interactions (Karp et al. 2019). Comparative analysis across different organisms or conditions is made possible, providing insights into metabolic diversity and functional differences. The integration of these databases with other omics data, such as genomics and proteomics, further enhances their utility, supporting in-depth research and modeling of metabolic pathways, as in the case of BlastKOALA and GhostKOALA (Kanehisa et al. 2016).

4 Challenges in Usage of Databases

- *Data integration and standardization*: One of the primary challenges in microbial metabolite databases is integrating diverse data types from various sources (Fondi and Lio 2015). Metabolite data often come from different platforms with varying formats, making it difficult to unify and standardize information across databases. Ensuring consistent data formats and ontologies is crucial for effective integration.
- *Data quality and completeness*: Many microbial metabolite databases suffer from incomplete or low-quality data (Jurado-Rueda et al. 2023). The accuracy of metabolite annotations and functional assignments can be inconsistent due to variations in experimental methods and data curation practices. Improving the accuracy and completeness of data is essential for reliable research outcomes.
- *Scalability and data management*: The rapid growth in microbial genome sequencing and metabolomic studies leads to an ever-increasing volume of data (Franzosa et al. 2015). Managing and updating large-scale databases to accommodate new information while maintaining performance and accessibility poses significant technical challenges.
- *Coverage of microbial diversity*: Existing databases may not comprehensively cover the full diversity of microbial species and their metabolites (Heyer et al. 2017). Many databases focus on well-studied organisms, potentially overlooking less-characterized microbes with unique or novel metabolites. Expanding coverage to include a broader range of microbial species is necessary for a more complete understanding of microbial metabolism.
- *Integration with functional data*: Linking metabolite data with functional genomics and proteomics data remains a challenge (Heyer et al. 2017). Establishing connections between metabolites, their biosynthetic pathways and associated genetic and protein information can enhance our understanding of metabolic functions and their biological implications.

5 Future Directions

- *Enhanced data integration*: Future efforts should focus on developing robust frameworks and tools for integrating data across multiple databases and research platforms. Implementing standardized data formats and ontologies, along with advanced data harmonization techniques, will improve the usability and interoperability of metabolite databases.
- *Improved data quality and annotation*: Investments in automated and manual curation processes, along with advanced computational tools for data validation and annotation, will enhance the quality and reliability of metabolite information. Incorporating community-driven efforts for data validation can also contribute to more accurate and comprehensive databases.
- *Expansion of database coverage*: To address gaps in microbial diversity, future databases should aim to include a wider range of microorganisms, including those from underrepresented environments and less-studied taxa. This expansion will provide a more holistic view of microbial metabolism and its applications.
- *Integration with omics data*: Developing strategies for integrating metabolite data with other omics data, such as genomics, proteomics and transcriptomics, will offer a more comprehensive understanding of microbial metabolic networks and their regulatory mechanisms. Multi-omics approaches can reveal novel insights into microbial functions and interactions.
- *Utilization of advanced technologies*: Leveraging emerging technologies, such as artificial intelligence and machine learning, can enhance data analysis, prediction and interpretation in microbial metabolite research. These technologies can aid in identifying novel metabolites, predicting their functions and discovering new applications.
- *Enhanced user accessibility and tools*: Future developments should focus on creating user-friendly interfaces and advanced analytical tools that enable researchers to efficiently explore and analyse metabolite data. Improved visualization and querying capabilities will facilitate more effective data interpretation and application.

6 Conclusion

The extensive knowledge about microorganisms contained in electronic journal articles will increasingly be combined with genomic and other omics data to enable more thorough and integrated analyses of microbes. As we embark on this journey, it is crucial for microbiologists to effectively utilize genomic resources, such as databases and computational tools to advance their research and to archive and share their findings in a robust and accessible manner. One straightforward approach is to deposit experimental research results, particularly high-throughput data, into public repositories. Authors should also use database accession numbers in their research papers to link genes, proteins and other data sets described in their work to databases. Many electronic journal editions now offer hyperlinks to these metabolic databases, allowing readers to access relevant data with a single click. While some biological databases may evolve or disappear over time, and others may change in form and function, it is evident that biological databases ranging from large-scale repositories and comprehensive information portals to niche community databases will continue to play an increasingly vital role in biological research and discovery.

References

Altschul SF, Madden TL, Schaffer AA, Zhang J, Zhang Z, Miller W, Lipman DJ (1997) Gapped BLAST and PSI-BLAST: a new generation of protein database search programs. Nucleic Acids Res 25:3389–3402. https://doi.org/10.1093/nar/25.17.3389

Andreeva A, Howorth D, Chothia C, Kulesha E, Murzin AG (2014) SCOP2 prototype: a new approach to protein structure mining. Nucleic Acids Res 42:D310–D314. https://doi.org/10.1093/nar/gkt1242

Bhat AH, Mondal H, Chauhan JS, Raghava GP, Methi A, Rao A (2012) ProGlycProt: a repository of experimentally characterized prokaryotic glycoproteins. Nucleic Acids Res 40:D388–D393. https://doi.org/10.1093/nar/gkr911

Bills GF, Gloer JB (2016) Biologically active secondary metabolites from the fungi. Microbiol Spectr 4(6):10–1128

Bougouffa S, Radovanovic A, Essack M, Bajic VB (2014) DEOP: a database on osmoprotectants and associated pathways. Database (Oxford) 2014:bau100. https://doi.org/10.1093/database/bau100

Bretaudeau A, Coste F, Humily F, Garczarek L, Le Corguille G, Six C, Ratin M, Collin O, Schluchter WM, Partensky F (2013) CyanoLyase: a database of phycobilin lyase sequences, motifs and functions. Nucleic Acids Res 41:D396–D401. https://doi.org/10.1093/nar/gks1091

Brooksbank C, Bergman MT, Apweiler R, Birney E, Thornton J (2014) The European Bioinformatics Institute's data resources 2014. Nucleic Acids Res 42:D18–D25. https://doi.org/10.1093/nar/gkt1206

Caspi R, Altman T, Billington R, Dreher K, Foerster H, Fulcher CA, Holland TA, Keseler IM, Kothari A, Kubo A, Krummenacker M, Latendresse M, Mueller LA, Ong Q, Paley S, Subhraveti P, Weaver DS, Weerasinghe D, Zhang P, Karp PD (2014) The MetaCyc database of metabolic pathways and enzymes and the BioCyc collection of pathway/genome databases. Nucleic Acids Res 42:D459–D471. https://doi.org/10.1093/nar/gkt1103

Cipriano MJ, Novichkov PN, Kazakov AE, Rodionov DA, Arkin AP, Gelfand MS, Dubchak I (2013) RegTransBase–a database of regulatory sequences and interactions based on literature: a resource for investigating transcriptional regulation in prokaryotes. BMC genomics 14:1-8. https://doi.org/10.1186/1471-2164-14-213

Chatr-Aryamontri A, Breitkreutz BJ, Oughtred R, Boucher L, Heinicke S, Chen D, Stark C, Breitkreutz A, Kolas N, O'Donnell L, Reguly T, Nixon J, Ramage L, Winter A, Sellam A, Chang C, Hirschman J, Theesfeld C, Rust J, Livstone MS, Dolinski K, Tyers M (2015) The BioGRID interaction database: 2015 update. Nucleic Acids Res 43:D470–D478. https://doi.org/10.1093/nar/gku1204

Conway KR, Boddy CN (2013) ClusterMine360: a database of microbial PKS/NRPS biosynthesis. Nucleic Acids Res 41:D402–D407. https://doi.org/10.1093/nar/gks993

Duarte M, Jauregui R, Vilchez-Vargas R, Junca H, Pieper DH (2014) AromaDeg, a novel database for phylogenomics of aerobic bacterial degradation of aromatics. Database (Oxford) 2014:bau118. https://doi.org/10.1093/database/bau118

Fichant G, Basse MJ, Quentin Y (2006) ABCdb: an online resource for ABC transporter repertories from sequenced archaeal and bacterial genomes. FEMS Microbiol Lett 256:333–339. https://doi.org/10.1111/j.1574-6968.2006.00139.x

Finn RD, Bateman A, Clements J, Coggill P, Eberhardt RY, Eddy SR, Heger A, Hetherington K, Holm L, Mistry J, Sonnhammer EL, Tate J, Punta M (2014) Pfam: the protein families database. Nucleic Acids Res 42:D222–D230. https://doi.org/10.1093/nar/gkt1223

Fondi M, Lio P (2015) Multi-omics and metabolic modelling pipelines: challenges and tools for systems microbiology. Microbiol Res 171:52–64

Franceschini A, Szklarczyk D, Frankild S, Kuhn M, Simonovic M, Roth A, Lin J, Minguez P, Bork P, von Mering C, Jensen LJ (2013) 2464 jb.asm.org STRINGv9.1: protein-protein interaction networks, with increased coverage and integration. Nucleic Acids Res 41:D808–D815. https://doi.org/10.1093/nar/gks1094

Franzosa EA, Hsu T, Sirota-Madi A, Shafquat A, Abu-Ali G, Morgan XC, Huttenhower C (2015) Sequencing and beyond: integrating molecular 'omics' for microbial community profiling. Nat Rev Microbiol 13(6):360–372

Galperin MY, Makarova KS, Wolf YI, Koonin EV (2015) Expanded microbial genome coverage and improved protein family annotation in the COG database. Nucleic Acids Res 43:D261–D269. https://doi.org/10.1093/nar/gku1223

Haft DH, Selengut JD, White O (2003) The TIGRFAMs database of protein families. Nucleic Acids Res 31:371–373. https://doi.org/10.1093/nar/gkg128

Heyer R, Schallert K, Zoun R, Becher B, Saake G, Benndorf D (2017) Challenges and perspectives of metaproteomic data analysis. J Biotechnol 261:24–36

Howe D, Costanzo M, Fey P, Gojobori T, Hannick L, Hide W, Hill DP, Kania R, Schaeffer M, St Pierre S, Twigger S, White O, Rhee SY (2008) Big data: the future of biocuration. Nature 455:47–50. https://doi.org/10.1038/455047a

Huerta-Cepas J, Szklarczyk D, Forslund K, Cook H, Heller D, Walter MC, Rattei T, Mende DR, Sunagawa S, Kuhn M, Jensen LJ (2016) eggNOG 4.5: a hierarchical orthology framework with improved functional annotations for eukaryotic, prokaryotic and viral sequences. Nucleic Acids Res 44(D1):D286–D293

Hunter S, Corbett M, Denise H, Fraser M, Gonzalez-Beltran A, Hunter C, Jones P, Leinonen R, McAnulla C, Maguire E, Maslen J, Mitchell A, Nuka G, Oisel A, Pesseat S, Radhakrishnan R, Rocca-Serra P, Scheremetjew M, Sterk P, Vaughan D, Cochrane G, Field D, Sansone SA (2014) EBI metagenomics–a new resource for the analysis and archiving of metagenomic data. Nucleic Acids Res 42:D600–D606. https://doi.org/10.1093/nar/gkt961

Hutchings MI, Truman AW, Wilkinson B (2019) Antibiotics: past, present and future. Curr Opin Microbiol 51:72–80

Jurado-Rueda F, Alonso-Guirado L, Perea-Chamblee TE, Elliott OT, Filip I, Rabadán R, Malats N (2023) Benchmarking of microbiome detection tools on RNA-seq synthetic databases according to diverse conditions. Bioinf Adv 3(1):14

Kanehisa M, Goto S, Sato Y, Kawashima M, Furumichi M, Tanabe M (2014) Data, information, knowledge and principle: back to metabolism in KEGG. Nucleic Acids Res 42:D199–D205. https://doi.org/10.1093/nar/gkt1076

Kanehisa M, Sato Y, Morishima K (2016) BlastKOALA and GhostKOALA: KEGG tools for functional characterization of genome and metagenome sequences. J Mol Biol 428:726–731

Karp PD, Billington R, Caspi R, Fulcher CA, Latendresse M, Kothari A, Keseler IM, Krummenacker M, Midford PE, Ong Q, Ong WK (2019) The BioCyc collection of microbial genomes and metabolic pathways. Brief Bioinform 20(4):1085–1093

Kilic S, White ER, Sagitova DM, Cornish JP, Erill I (2014) CollecTF: a database of experimentally validated transcription factor-binding sites in Bacteria. Nucleic Acids Res 42(D1):D156-D160. https://doi.org/10.1093/nar/gkt1123

Kriventseva EV, Tegenfeldt F, Petty TJ, Waterhouse RM, Simao FA, Pozdnyakov IA, Ioannidis P, Zdobnov EM (2015) OrthoDB v8: update of the hierarchical catalog of orthologs and the underlying free software. Nucleic Acids Res 43:D250–D256. https://doi.org/10.1093/nar/gku1220

Kumar K, Gupta SC, Chander Y, Singh AK (2005) Antibiotic use in agriculture and its impact on the terrestrial environment. Adv Agron 87:1–54

Lemfack MC, Nickel J, Dunkel M, Preissner R, Piechulla B (2014) mVOC: a database of microbial volatiles. Nucleic Acids Res 42(D744):D748. https://doi.org/10.1093/nar/gkt1250

Letunic I, Doerks T, Bork P (2015) SMART: recent updates, new developments and status in 2015. Nucleic Acids Res 43:D257–D260. https://doi.org/10.1093/nar/gku949
Li J, Yao Y, Xu HH, Hao L, Deng Z, Rajakumar K, Ou HY (2015) SecReT6: a web-based resource for type VI secretion systems found in bacteria. Environ Microbiol 17:2196. https://doi.org/10.1111/1462-2920.12794
Lombard V, Ramulu H, Drula E, Coutinho PM, Henrissat B (2014) The carbohydrate-active enzymes database (CAZy) in 2013. Nucleic Acids Res 42:D490–D495. https://doi.org/10.1093/nar/gkt1178
Luckner M (2013) Secondary metabolism in microorganisms, plants and animals. Springer Science & Business Media
Marchler-Bauer A, Derbyshire MK, Gonzales NR, Lu S, Chitsaz F, Geer LY, Geer RC, He J, Gwadz M, Hurwitz DI, Lanczycki CJ, Lu F, Marchler GH, Song JS, Thanki N, Wang Z, Yamashita RA, Zhang D, Zheng C, Bryant SH (2015) CDD: NCBI's conserved domain database. Nucleic Acids Res 43:D222–D226. https://doi.org/10.1093/nar/gku1221
Metzker ML (2010) Sequencing technologies- the next generation. Nat Rev Genet 11:31–46. https://doi.org/10.1038/nrg2626
Mitchell A, Chang HY, Daugherty L, Fraser M, Hunter S, Lopez R, McAnulla C, McMenamin C, Nuka G, Pesseat S, Sangrador-Vegas A, Scheremetjew M, Rato C, Yong SY, Bateman A, Punta M, Attwood TK, Sigrist CJ, Redaschi N, Rivoire C, Xenarios I, Kahn D, Guyot D, Bork P, Letunic I, Gough J, Oates M, Haft D, Huang H, Natale DA, Wu CH, Orengo C, Sillitoe I, Mi H, Thomas PD, Finn RD (2015) The InterPro protein families database: the classification resource after 15 years. Nucleic Acids Res 43:D213–D221. https://doi.org/10.1093/nar/gku1243
Mitra A, Kesarwani AK, Pal D, Nagaraja V (2011) WebGeSTer DB—a transcription terminator database. Nucleic Acids Res 39(1):D129-D135. https://doi.org/10.1093/nar/gkq971
NCBI Resource Coordinators (2015) Database resources of the National Center for biotechnology information. Nucleic Acids Res 43:D6–D17. https://doi.org/10.1093/nar/gku1130
Novichkov PS, Kazakov AE, Ravcheev DA, Leyn SA, Kovaleva GY, Sutormin RA, Kazanov MD, Riehl W, Arkin AP, Dubchak I, Rodionov DA (2013) RegPrecise 3.0–a resource for genome-scale exploration of transcriptional regulation in bacteria. BMC genomics 14:1-12. https://doi.org/10.1186/1471-2164-14-745
Ortet P, Whitworth DE, Santaella C, Achouak W, Barakat M (2015) P2CS: updates of the prokaryotic two-component systems database. Nucleic Acids Res 43(D1):D536-D541. https://doi.org/10.1093/nar/gku968
Pachkov M, Balwierz PJ, Arnold P, Ozonov E, Van Nimwegen E (2012) SwissRegulon, a database of genome-wide annotations of regulatory sites: recent updates. Nucleic Acids Res 41(D1):D214-D220. https://doi.org/10.1093/nar/gks1145
Pal C, Bengtsson-Palme J, Rensing C, Kristiansson E, Larsson DG (2014) BacMet: antibacterial biocide and metal resistance genes database. Nucleic Acids Res 42:D737–D743. https://doi.org/10.1093/nar/gkt1252
Pruitt KD, Tatusova T, Brown GR, Maglott DR (2012) NCBI reference sequences (RefSeq): current status, new features and genome annotation policy. Nucleic Acids Res 40:D130–D135. https://doi.org/10.1093/nar/gkr1079
Ren Q, Chen K, Paulsen IT (2007) TransportDB: a comprehensive database resource for cytoplasmic membrane transport systems and outer membrane channels. Nucleic Acids Res 35:D274–D279. https://doi.org/10.1093/nar/gkl925
Rose PW, Prlic A, Bi C, Bluhm WF, Christie CH, Dutta S, Green RK, Goodsell DS, Westbrook JD, Woo J, Young J, Zardecki C, Berman HM, Bourne PE, Burley SK (2015) The RCSB protein data Bank: views of structural biology for basic and applied research and education. Nucleic Acids Res 43:D345–D356. https://doi.org/10.1093/nar/gku1214
Saier MH Jr, Reddy VS, Tamang DG, Vastermark A (2014) The transporter classification database. Nucleic Acids Res 42:D251–D258. https://doi.org/10.1093/nar/gkt1097
Sillitoe I, Lewis TE, Cuff A, Das S, Ashford P, Dawson NL, Furnham N, Laskowski RA, Lee D, Lees JG, Lehtinen S, Studer RA, Thornton J, Orengo CA (2015) CATH: comprehensive struc-

tural and functional an notation for genome sequences. Nucleic Acids Res 43:D376–D381. https://doi.org/10.1093/nar/gku947

Souza RC, del Rosario Quispe Saji G, Costa MO, Netto DS, Lima NC, Klein CC, Vasconcelos AT, Nicolas MF (2012) AtlasT4SS: a curated database for type IV secretion systems. BMC Microbiol 12:172. https://doi.org/10.1186/1471-2180-12-172

Taboada B, Ciria R, Martinez-Guerrero CE, Merino E (2012) ProOpDB: Prokaryotic Operon Data Base. Nucleic Acids Res 40(D1):D627-D631. https://doi.org/10.1093/nar/gkr1020

Tatusov RL, Fedorova ND, Jackson JD, Jacobs AR, Kiryutin B, Koonin EV, Krylov DM, Mazumder R, Mekhedov SL, Nikolskaya AN, Rao BS, Smirnov S, Sverdlov AV, Vasudevan S, Wolf YI, Yin JJ, Natale DA (2003) The COG database: an updated version includes eukaryotes. BMC Bioinf 4:41. https://doi.org/10.1186/1471-2105-4-41

Toukach PV (2011) Bacterial carbohydrate structure database 3: principles and realization. J Chem Inf Model 51:159–170. https://doi.org/10.1021/ci100150d

Tsirigos KD, Bagos PG, Hamodrakas SJ (2011) OMPdb: a database of beta-barrel outer membrane proteins from gram-negative bacteria. Nucleic Acids Res 39:D324–D331. https://doi.org/10.1093/nar/gkq863

Turkarslan S, Wurtmann EJ, Wu WJ, Jiang N, Bare JC, Foley K, Reiss DJ, Novichkov P, Baliga NS (2014) Network portal: a database for storage, analysis and visualization of biological networks. Nucleic Acids Res 42(D1):D184-D190. https://doi.org/10.1093/nar/gkt1190

Ulrich LE, Zhulin IB (2010) The MiST2 database: a comprehensive genomics resource on microbial signal transduction. Nucleic Acids Res, 38(1):D401-D407. https://doi.org/10.1093/nar/gkp940

UniProt Consortium (2014) Activities at the universal protein resource (UniProt). Nucleic Acids Res 42:D191–D198. https://doi.org/10.1093/nar/gkt1140

Vallenet D, Belda E, Calteau A, Cruveiller S, Engelen S, Lajus A, Le Fevre F, Longin C, Mornico D, Roche D, Rouy Z, Salvignol G, Scarpelli C, Thil Smith AA, Weiman M, Medigue C (2013) MicroScope–an integrated microbial resource for the curation and comparative analysis of genomic and metabolic data. Nucleic Acids Res 41:D636–D647. https://doi.org/10.1093/nar/gks1194

Wang Y, Huang H, Sun M, Zhang Q, Guo D (2012) T3DB: an integrated database for bacterial type III secretion system. BMC Bioinf 13:66. https://doi.org/10.1186/1471-2105-13-66

Wynendaele E, Bronselaer A, Nielandt J, D'Hondt M, Stalmans S, Bracke N, Verbeke F, Van De Wiele C, De Tre G, De Spiegeleer B (2013) Quorumpeps database: chemical space, microbial origin and functionality of quorum sensing peptides. Nucleic Acids Res 41(D1):D655-D659. https://doi.org/10.1093/nar/gks1137

Zhou Z, Tran PQ, Breister AM, Liu Y, Kieft K, Cowley ES, Karaoz U, Anantharaman K (2022) METABOLIC: high-throughput profiling of microbial genomes for functional traits, metabolism, biogeochemistry, and community-scale functional networks. Microbiome 10(1):33

Integrating Metabolomics with Omics Techniques: Exploring Cutting-Edge Technologies in Comprehensive Biological Analysis

Muhammad Asmat Ullah Saleem,
Muhammad Ahsan Ullah Danish, Farhat Bano,
Muhammad Farhan Qadir, and Muhammad Salman Tahir

Abstract

Metabolomics and multi-omics (genomics, transcriptomics, and proteomics) offer a profound comprehension of biological systems. This integration provides a clear view of cellular functions and their interactions. The study of small molecules and investigating their biochemical activities within cells in a system is termed metabolomics. Combined with related omics data, this information enhanced understanding resolution through detailed analysis of biological processes, such as disease mechanisms and potential therapeutic targets. Herein, through this book chapter, we aim to explain advanced technologies in metabolomics, highlight their principles and methodologies involved to address challenges and opportunities in multi-omics integration, and present a case study on their application in health and disease research. Reviewing current literature on the application of mass spectrometry (MS), which identifies and quantifies metabolites, and nuclear magnetic resonance (NMR) spectroscopy, we further

M. A. U. Saleem (✉) · F. Bano (✉)
College of Veterinary Medicine, Northeast Agricultural University, Harbin, People's Republic of China

M. A. U. Danish
Department of Allied Health Sciences, The University of Agriculture, Dera Ismail Khan, Khyber Pakhtunkhwa, Pakistan

M. F. Qadir
School of Life Science, Shanxi University, Taiyuan, People's Republic of China
e-mail: g207@email.sxu.edu.cn

M. S. Tahir
Key Laboratory of Animal Disease and Human Health of Sichuan Province, College of Veterinary Medicine, Sichuan Agricultural University, Chengdu, People's Republic of China

S. Kaur et al. (eds.), *Microbial Metabolomics*,
https://doi.org/10.1007/978-981-96-4824-5_4

enhance our understanding of structural information about metabolites. The application of advanced bioinformatics tools for integrating and interpreting large-scale omics data and their linkage with metabolomics will also be discussed. These tools are crucial for handling the vast amount of data generated and extracting meaningful insights. To illustrate practical applications, we will also include various studies where integrated omics approaches have been successfully applied. We also comprehend how these tools can help improve diagnostic accuracy and develop targeted therapies and personalized medicine. The emerging trends in integrative omics research, the ongoing challenges, and potential solutions will also be discussed. To enhance the effectiveness of this approach, we will include detailed descriptions of figures and tables, such as workflow integration illustrations, metabolic pathways, and visual representations of multi-omics data integration. Taken together, this approach will provide researchers and practitioners with a thorough understanding of the current state and future potential of integrative omics technologies in biological analysis. The insights gained from this information will have significant implications for improving the existing information of the application of the tools to understand biological processes comprehensively.

Keywords

Biological processes · Bioinformatics tools · Metabolomics · Multi-omics · Mass spectrometry

1 Introduction

The consistent inclusion of metabolomics with genomic, transcriptomic, and proteomic data represents a new paradigm shift in systems biology. It will likely have an enormous impact on understanding diseases, their treatment, and personalized medicine (Walsh et al. 2018). Metabolomics is the branch of science that deals with small molecules and metabolites in biological organizations, which form the core units of cellular functions and metabolic networks. It entails the incorporation of data from several omics disciplens, within which biological phenomena are studied as opposed to a single global omics dataset, whereby the relationship between genes, proteins, and metabolites is better assessed at the molecular dimension (Pinu et al. 2019). Although much effort has been made toward research on multi-omics integration, several aspects still need to be better handled, especially dealing with data variety, data analytics, and workflow processes. This chapter provides an overview of available evidence for metabolomics integration with other omics and discusses recent trends in mass spectrometry (MS), nuclear magnetic resonance (NMR), and bioinformatics development (Gowda et al. 2008).

The recent literature has emphasized the significance of a multi-omics approach in primary and secondary diseases. For instance, it has been demonstrated that an integrative omics approach is possible to improve the outcomes of diagnosis and develop new treatment options for diseases such as cancer and metabolic disorders (Chen et al. 2023). Nevertheless, while these studies have added to the understanding of the possible advantages of metabolomics integration, they also reveal a crucial impediment concerning the current knowledge of optimal management and integration of extensive and diverse datasets that are acquired from various sources through these studies (Cavill et al. 2016). The absence of standard practices and technological platforms continues to be a drawback to the reproducibility and extendibility of multi-omics studies. Addressing these gaps is necessary for exploring the possibilities that the branch of metabolomics has in systems biology and personalized medicine to make sure that researchers utilize the full breadth of omics data in improving health (Pinu et al. 2019).

The chapter aims to fill the existing voids in the multi-omics integration by conducting a thorough literature analysis, focusing on currently available technologies and methods. Despite limitations in certain studies previously conducted, literature integration approaches proposed in this review depict how the barriers experienced in metabolomics and multi-omics integration can be overcome (Eicher et al. 2020). In this manner, new ways for improving data architecture, workflow dependencies, and reproducibility in the area are provided. Inconclusion, the chapter contributes to advancing the field of metabolomics and the prospects of developing systems biology and contemporary medicine in metastasis (Han et al. 2021).

2 Principles and Techniques of Metabolomics

2.1 Fundamentals of Metabolomics

Metabolomics is an analytical approach that focuses on identifying and quantifying all relatively small molecules and active metabolites of the biological system (Muthubharathi et al. 2021). This robust area of inquiry allows for an understanding the basic concepts surrounding cells and their biochemical activities. The central component of metabolomics is the imaging of the metabolome, which refers to the area that includes all the metabolites found in a biological specimen, which can be a cell, tissue, or even an organism (Alonso et al. 2015). These metabolites contain several chemicals, metabolites of metabolic intermediates and end products, signaling cascades, and small biomolecules. Through these rapid developments in the metabolome, scientists have come to appreciate the complexity of biochemical reactions that occur in the living states as shown in Fig. 1 (Yang et al. 2019). One of the ways knowledge of the various cellular functions and the biochemical pathways or processes of metabolomics is developed by investigating metebolite patterns (Gowda et al. 2008). Metabolites are end products of many complicated controls and interactions that occur within cells, and changes in their distribution and concentration often reflect either the physiological or pathalogical conditions of that cell or tissue (Johnson and Gonzalez 2012). Through metabolomic approaches, it is possible to determine the distinct metabolic concentrations that correlate with specific cellular activities, such as biosynthesis, signal transduction, and stress response

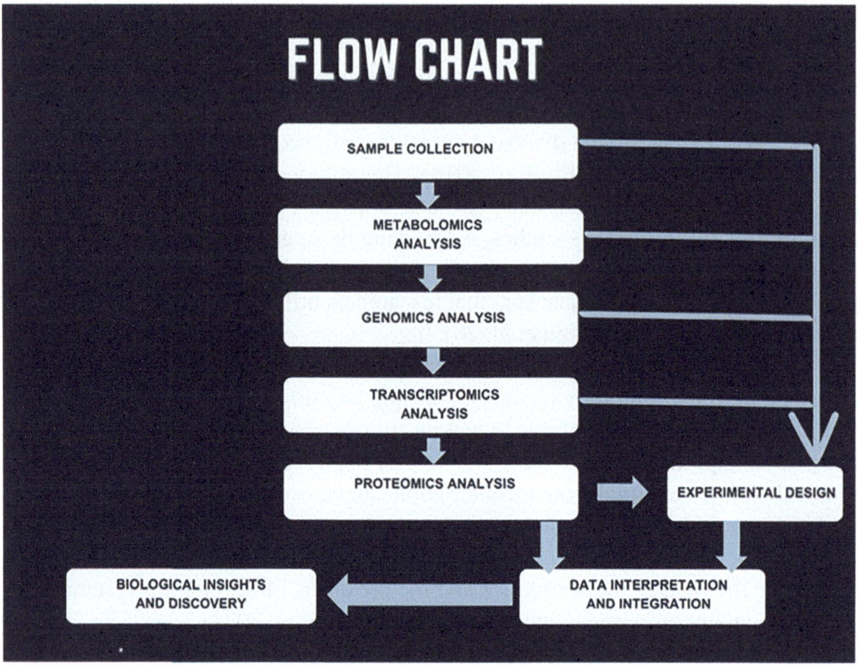

Fig. 1 This flowchart shows how metabolomics, genomes, transcriptomics, and proteomics are integrated into a biological investigation

(Walsh et al. 2018). This can help answer the questions concerning the mechanisms that govern these activities, consequently providing new ways of diagnosing and treating diseases.

Metabolomics primarily depends on high-tech analytical methods, such as mass spectrometry (MS) and nuclear magnetic resonance (NMR) spectroscopy as shown in Table 1 (Gowda et al. 2008). Because of its sensitivity and specificity, MS has proved to be quite effective in targeting and quantifying even a very complex mixtures of metabolites in preclinical samples (Nagana Gowda et al. 2015). Moreover, techniques such as liquid chromatography-mass spectrometry (LC-MS) and gas chromatography-mass spectrometry (GC-MS) enable the separation and identification of metabolites, providing valuable insights into their chemical properties and quantitative aspects as shown in Table 1 (Nandania et al. 2018). On the other hand, NMR represents a noninvasive approach to metabolomics that should be used in combination with, rather than substituted for mass spectrometry as shown in Fig. 1 (Emwas et al. 2019). In the applied techniques, quantitative analysis of the metabolites based on the NMR spectra of specific nuclei is possible, thanks to the unique mobility and molecular conformations of the compounds captured in complex liquids (De Castro et al. 2019). Within this framework, researchers are familiar with the concept of "picturing" the entire metabolic content, something referred to as the metabolite. Mass spectrometry and NMR spectrometry are the primary analytical

Table 1 Summary of analytical technologies in metabolomics

Technology	Advantages	Disadvantages	Reference
Mass spectrometry (MS)	High sensitivity, broad metabolite coverage, can identify unknown compounds	Requires complex sample preparation, expensive equipment, matrix effects	Guo et al. (2020)
Nuclear magnetic resonance (NMR)	Nondestructive, provides structural information and quantification without standards	Lower sensitivity compared to MS, requires larger sample amounts	Yu et al. (2021)
Gas chromatography (GC)	High-resolution, ideal for volatile compounds, robust and reproducible	Limited to volatile and thermally stable compounds, long analysis time	Gruber et al. (2020)
Liquid chromatography (LC)	Versatile, works for a wide range of metabolites, highly sensitive	Can require long run times; sensitivity depends on the detector used	Hemida (2021)
Capillary electrophoresis (CE)	High separation efficiency, suitable for charged compounds, small sample volume	Limited to ionic species, complex operation, lower sensitivity than MS	Fang et al. (2024)

platforms of modern metabolomics and tools for conducting molecular profiles (Cesare Marincola and Mannina 2020). This is done with the aim of revealing basic laws of the molecular biology of cells, cellular biology, and biochemistry (Zhang et al. 2020).

2.2 Mass Spectrometry in Metabolomics

The discipline of metabolomics largely depends on mass spectrometry, which not only serves as a profiling tool but also as an analyzer of metbolic flux within biological samples (Yang et al. 2019). Among the various bioanalytical strategies utilized by metabolomics researchers, liquid chromatography-mass spectrometry and gas chromatography-mass spectrometry are likely the most widely used techniques, each with its own advantages and limitations (Alonso et al. 2015). The LC-MS technique can measure almost any class of metabolites, even highly polar, gaseous, and thermally unstable metabolites (Emwas et al. 2019). Liquid chromatography separates the sample, and individual components are ionized, after which the mass spectrometer detects them (Huang 2021). Such techniques are essential when analyzing complex mixtures of bio-samples since they allow for the determination of both specific structure and concentration (Saurina and Sentellas 2017). Furthermore, the possibility of measuring low concentrations of metabolites with the current LC-MS instruments has turned this instrument into an essential ally in mapping biological pathways and cellular functions (Bueschl et al. 2013).

On the other hand, GC-MS analysis is more appropriate for low-volume, vaporized, and thermally stable materials; hence, each essential element that makes up the sample is fully separated (Bertholf 2006). The procedure consists of gas chromatography, where sample components are separated based on their volatility and boiling points; after this, the components are ionized and detected by a mass spectrometer (Koch 1989). Amino, fatty, and organic acids are the kinds of metabolites that can be accurately and reliably identified and quantified using GC-MS (Jousse and Pujos-Guillot 2013). Moreover, even though sample preparation of biofluids for GC-MS may be more tedious and involve derivatization of some metabolic products, the method provides good separation efficiency and reproducibility, making it an essential comlement to LC-MS in metabolomics (Garcia and Barbas 2011). Multi-omics approaches have also directed our understanding of cellular metabolism and biochemical networks using metabolomics and other omics techniques like genomics, transcriptomics, and proteomics (Pinu et al. 2019). For instance, incorporating various data analyses, such as genomics with metabolomics, could assist in bringing to light the genotype and phenotype, where the genetic variation thus depicts the production and regulation of the metabolites (Kim et al. 2016). Furthermore, the integration of genomics and metabolomics data can help elucidate the relationship between genotypes and phenotypes, where genetic variations may influence the production and regulation of metabolites (Draisma et al. 2015).

2.3 Nuclear Magnetic Resonance (NMR) Spectroscopy in Metabolomics

Nuclear magnetic resonance (NMR) spectroscopy has established itself as an effective and vital tool for metabolomics (Capati et al. 2017). NMR is a noninvasive technique that allows the structural identification of metabolites of any size and is essential for understanding the structure and behavior of small molecules in cells (Lacy et al. 2014). In the field of magnetic resonance spectroscopy (MRS)-based metabolism, one of the advantages is that MRS can provide a sufficient degree of metabolic and morphological characteristics of studied biological systems. NMR spectroscopy can decipher organic molecules and determine the shape and arrangements of atoms inside them (Moco 2022). This step is crucial in helping determine the metabolites that are contained in a sample, which is necessary for metabolomic studies where biochemical networks of cells and cellular functions are being investigated (Dettmer et al. 2007).

In recent years, more possibilities for the practical application of NMR have been opened for metabolomics studies. As a result, it became possible to identify, quantify, and detect ever-smaller amounts of metabolites present even in complex biological systems (Wishart 2019). Furthermore, the use of specific NMR pulse sequences and robust data processing approaches has opened new pathways for metabolomic research, enabling the field to meet growing demands on MRS and expand the range of its applications (Emwas et al. 2019). NMR spectroscopy has distinct benefits when D/L (Dextro/Levo) mass spectrometry, a method that is quite common with metabolomics, is considered (Emwas et al. 2019). While mass spectrometry (MS) excels in sensitivity and the detection range of metabolites, nuclear magnetic resonance (NMR), despite being nondestructive, less labor-intensive in terms of sample preparation, and inherently quantitative, has certain limitations (Marshall and Powers 2017). In addition, NMR provided structural information on metabolites that MS could not due to the various sample preparation techniques and derivatization used in particular immunoassays (Lacy et al. 2014). NMR research amalgamates with other "omics" technologies such as genomics, transcriptomics, and proteomics, creating more possibilities and significance in metabolomics (Gowda et al. 2008). Researchers can appreciate regulated cellular functions in their complexity by studying the interconnectedness at various levels within metabolic pathways using a collection of analytical techniques (Walsh et al. 2018). Such helps discover critical biomarkers, explains and disperses metabolomic changes accompanying a certain disease entity, and ultimately leads to precise strategies to alter particular biochemical activities and processes (Zhang et al. 2015).

3 Integrating Metabolomics with Other Omics Techniques

3.1 Overview of Multi-omics Integration

The concept of multi-omics has revolutionized our understanding of systems biology. Multi-omics integration is defined as the data integration of multiple high-throughput omics technologies to examine more complex biological systems (Mirza et al. 2019). Concurrently studying various omics datasets, such as genomic, transcriptomic, proteomic, or metabolic, related to one or more biological pathways is known as multi-omics integration (Bersanelli et al. 2016). In systems biology, the advantage of multi-omics integration lies in the fact that this approach allows to define the stability of cellular processes and their connectedness with other processes (Bersanelli et al. 2016). To better understand the principles governing the functioning of biological systems, researchers try to link the information received at several levels of the organization. This is very helpful for the examination of complex diseases when genes, molecules, and biochemistry work together to determine disease diagnosis and manifestation (Cho et al. 2012).

The integration of metabolomics data with other omics techniques has been one of the most difficult parts of multi-omics studies (Pinu et al. 2019). Metabolomics is the branch of life sciences that deals with the study of cells' metabolites at a given time. More specifically, it connects gene information with protein function in a specific biological state as shown in Fig. 2 (Yıldız-Ozturk and Askin-Uzel 2022). However, there are some challenges, including the integration of other omics data with such inherent complexity and, in practice, context dependencies (Bersanelli et al. 2016). This challenge is further compounded by the fact that the data integration with metabolomics typically requires collaboration across numerous disciplines as shown in Fig. 2 (Pinu et al. 2019). As a result, any comprehensive analysis of metabolism aims to integrate data from genomics, transcriptomics, proteomics, and other relevant domains (Wanichthanarak et al. 2015). Each of these dimensions is subject to changes over time, across organs, or even cell types (Walsh et al. 2018).

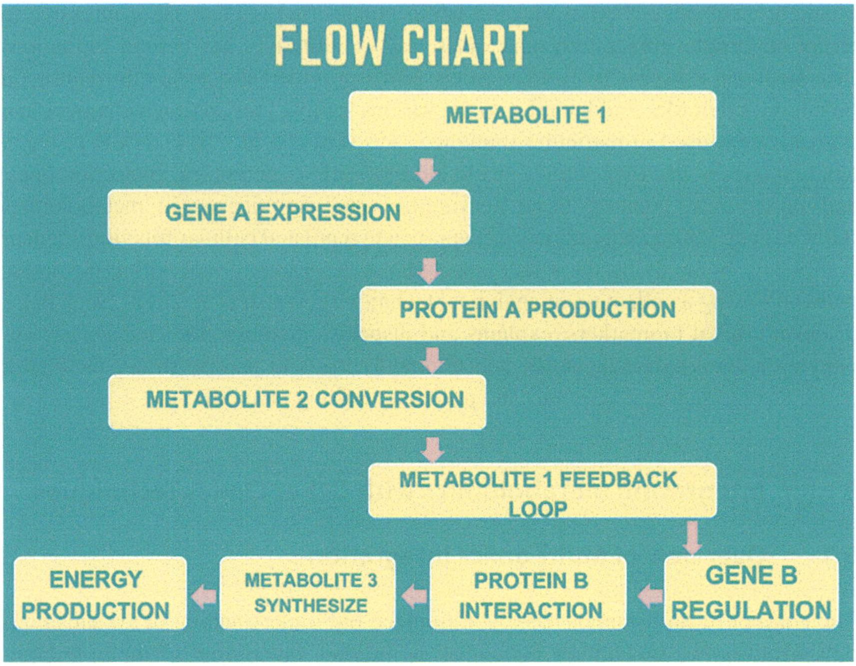

Fig. 2 This metabolic pathway diagram shows a biological system's relationships among proteins, genes, and metabolites

Standardization issues and differences in experimental design, sample preparation, or use of platforms for data analytics add even more complexity, which hinders data integration and interpretation (Fouché and Zinovyev 2023). These issues call for developing advanced and effective computational or statistical tools to analyze the disparate and voluminous data available from multi-omics studies (Rotroff and Motsinger-Reif 2016). Despite the complexity involved, a comprehensive multi-omics approach offers tremendous advantages for interpreting biological systems (Mirza et al. 2019). Leveraging data across different scales and system perspectives enables researchers to better elucidate the biological processes underlying gene expression and metabolic pathways, from the cellular to the organismal level (Diaz-Beltran et al. 2013). The integrated multi-omics knowledge can aid in improving targeted treatment strategies, developing new medicines, and identifying new markers for disease diagnosis and prognosis (Sun and Hu 2016).

3.2 Case Studies of Integrative Omics in Disease Research

Integrated omics is one of the fields that has progressed most in the last years by allowing a thorough understanding of complex disorders and providing direction for targeted treatments (Hasin et al. 2017). This has been made possible largely by

Table 3 Key findings from selected case studies demonstrating the impact of integrative omics approaches

Case study	Key findings	Impact	Reference
Cancer genomics study	Integrated genomics and transcriptomics identified novel driver mutations and therapeutic targets in cancer	Improved cancer treatment strategies through targeted therapies	Asrina (2024)
Neurodegenerative disease study	Proteomics and metabolomics revealed altered pathways associated with neurodegeneration, leading to potential biomarkers	Identification of new biomarkers for early diagnosis of neurodegenerative diseases	Schumacher-Schuh et al. (2022)
Metabolic disorder study	Multi-omics integration uncovered the genetic basis and metabolic disruptions contributing to diabetes	Enhanced understanding of metabolic disorders for personalized medicine approaches	Dar et al. (2023)
Microbiome-host interaction study	Combining metagenomics and metabolomics provided insight into microbiome-host interactions influencing gut health	Better understanding of gut microbiome's role in health, leading to probiotics development	Mathuria et al. (2024)
Cardiovascular disease study	Genomics, proteomics, and metabolomics highlighted molecular mechanisms contributing to heart disease, aiding drug development	Advanced precision medicine approaches for cardiovascular diseases	Banerjee et al. (2021)

metabolomics, which facilitates the understanding of disease mechanisms and finds suitable intervention strategies (Le Gouellec et al. 2023). An example of this type of research is using integrated omics approaches to study cancer (Buescher and Driggers 2016). By integrating the metabolomic, genomic, and transcriptomic data, researchers have traced the chain of metabolic changes associated with the growth and spread of cancers as shown in Table 3 (Cantafio et al. 2019). For example, altered lipid metabolism in some subtypes of breast cancer has been found, and therapies that specifically target these metabolic aberrations have been generated (Mishra and Ambs 2015). In the same way, great details taken into the metabolic processes of cancer cells have enabled the discovery of new marker molecules for follow-up and early disease detection. Hence, more precision management of cancer is designed (Kang et al. 2018).

Integrative omics has also made a considerable contribution to investigating neurological disorders (Puchades-Carrasco and Pineda-Lucena 2017). The incorporation of metabolomic, proteomic, and genomic datasets has assisted scientists in understanding the complex biochemical disturbances that are the root causes of Alzheimer's and Parkinson's disease (Ibáñez et al. 2015). Now, it is possible to provide clinical guidelines with incredible accuracy based on metabolic changes during disease initiation and development (Li and Luo 2022). For instance, by understanding changes in tryptophan metabolism in Alzheimer's disease, it has been possible to formulate new neuroprotection strategies based on metabolism changes (Toledo et al. 2017). In addition to disease study, multi-omics integration has helped elucidate the relationship between gut microbiota and human health (Li and Luo 2022). Utilizing a combination of metabolomics, metagenomics, and metatranscriptomics, researchers have also demonstrated that gut microbial metabolites are important in regulating host processes and disease susceptibility, ranging from inflammatory bowel disease to metabolic syndrome (Fan and Pedersen 2021). Such information has opened avenues for designing effective prebiotics and probiotics and investigating individualized dietary strategies to enhance gut-microbiome interaction health (Vyas and Ranganathan 2012). The integration of metabolomics with other omics techniques has emerged as a valuable, if not essential, approach for comprehending the complexity of biological systems and diseases holistically (Ursell et al. 2014). By using systems biology for their studies, diseases' causes can be thoroughly investigated, and new potential therapeutic targets and biomarkers can be identified, ensuring that after several trials, better treatment options and approaches will be developed (Rai et al. 2018).

3.3 Methodologies for Multi-omics Integration

The incorporation of metabolomics data with other branches of molecular sciences, including genomics, transcriptomics, and proteomics, has emerged as an imperative contemporary trend (Pinu et al. 2019). This integration permits probing deep into the more complex relations and temporal changes within biological entities. Data from multi-omics is so vast and complex that these systems and approaches help perform the requisite computations and analyses to generate meaningful information from such data as shown in Table 2 (Bersanelli et al. 2016). One critical strategy in

Table 2 Comparative analysis of omics techniques

Omics technique	Strengths	Limitations
Genomics	High throughput, provides comprehensive genetic information, stable data	Limited to genetic variations, doesn't reflect actual gene expression or protein activity
Transcriptomics	Measures gene expression in real time, reveals dynamic biological processes	Sensitive to experimental conditions, doesn't directly measure proteins
Proteomics	Directly studies proteins, captures posttranslational modifications	Requires complex sample preparation, highly variable data, expensive
Metabolomics	Reflects functional phenotype, can detect small-molecule metabolites	Highly influenced by environmental factors, lower coverage of metabolites
Epigenomics	Shows gene regulation through epigenetic modifications, dynamic and reversible	Epigenetic changes may not directly correlate with gene or protein changes

multi-omics integration invloves statistical and machine learning-based approaches. These methodologies include techniques such as correlation analysis, network analysis, and supervised and unsupervised learning, which may aim to discover underlying associations, dependencies, or other relationships among the different omics layers (Vahabi and Michailidis 2022). For instance, the correlation of changes in metabolite levels with the alterations in the levels of proteins or gene expression provides a clear overview (Xi and Yu 2021). In contrast, network analysis offers insights into metabolic pathways through established networks (Rosato et al. 2018).

Besides using statistical tools, the development of analytical techniques has also shaped the advancement of multi-omics studies (Haas et al. 2017). In order to aid the management of the multi-omics data, which is naturally heterogeneous and highly dimensional, there has been a need for the creation of bioinformatics computable algorithms or tools that facilitate data normalization, integration, visualization, and other bioinformatics processes. Such tools allow filtering of challenges like incompleteness or missing data, batch effects or uneven sensitivity, and data compatibility challenges, which results in a more reliable and accurate integration of the omics datasets and enhanced biological insight (Fouché and Zinovyev 2023). There are also approaches from systems biology that help build new hypotheses by integrating multi-omics datasets and looking at biological systems rather than single parts. It allows for more data on the multifaceted, intricate cellular interrelations behind the control of multifactorial diseases in order to reveal regulatory mechanisms and interconnections (Fouché and Zinovyev 2023). Achieving this broader vision can lead to the discovery of new biomarkers, more precise and effective therapies, and better-targeted action against disease pathogenesis. In order to understand how this field has evolved, one must consider that the recent progress in multi-omics integration has also stemmed from the availability of advanced computational power and the emergence of numerous large-scale omics repositories made freely accessible to the public (Subramanian et al. 2020). Also, the application of cloud, high performance, and parallel computing has made it possible to address the ever-increasing complexity and data-intensive nature of multi-omics thoughts (Mirza et al. 2019). In addition, the introduction of normalized formats and repositories for the data and measures to share the data has made it easier to do multi-omics studies, enhancing the rate of scientific advancement (Krassowski et al. 2020).

4 Technological Advancements in Multi-omics

4.1 Emerging Trends in Omics Technologies

Metabolomics, as part of omics technologies, has undergone significant changes due to recent advances, opening up new avenues in its evolution (Yang et al. 2019). A key factor in advancing multi-omics is the adoption of high-throughput technologies, which significantly enhance the accuracy and scope of studies conducted, keeping you, our esteemed audience, well-informed and up-to-date (Pan et al.

2023). A remarkable development in this field is the evolution of sophisticated mass spectrometry methods (Pan et al. 2023). These MS-based platforms have become incredibly sensitive to modifications, thereby expanding the range of measurable metabolites. This impressive advancement allows for the integration of MS data with other omics data, such as genomics and proteomics, providing a deeper understanding of biological systems and their interrelations, which we are sure will intrigue you, our esteemed audience (Girolamo et al. 2013). Bioinformatics and computational methods have played a pivotal role in the advancement of metabolomics research, revolutionizing the way data is obtained and processed (Alonso et al. 2015). The use of advanced algorithms and combined computer technologies, such as machine learning, has enabled the analysis of large volumes of data, automating data processing and revealing metabolic data and their composition (Damiani et al. 2020). These technological advancements have significantly improved the accuracy and speed of metabolite profiling, making it easier to study cellular and tissue processes (Patti et al. 2012).

As for multi-omics, high-temporal resolution metabolomics helps dissect biological complexity. The combination of genomic, transcriptomic, proteomic, and metabolomic data allows scientists to gain a deeper insight into cellular functionality, disease development, and response to environmental changes (Blum et al. 2018). For example, the supplementary addition of transcriptomics and genomics to metabolomics has permitted the analysis of new biomarker discovery and gene-metabolite interactions (Walsh et al. 2018). This has been especially helpful in the context of personalized medicine, where different strands of incoming data are merged to help make decisions on the patient's future and the best way to use drug interventions (Zhang et al. 2015). In addition, improvements in sample preparation methods and data collection technologies in the past few years have considerably improved the sensitivity and reproducibility of metabolomics. For example, microfluidic technologies interestingly helped analyze shallow volumes of samples, and improvements in chromatographic separations led to higher resolution and peak capacities in metabolite profiling (Yang et al. 2019). Further advances in omics technologies and the accumulation of multi-omics datasets will certainly contribute to deepening the understanding of biological complexity in the future (Yang et al. 2019). Using these combined methods, researchers will be able to discover new metabolic pathways that have not been identified before, delineate the mechanisms of diseases, and improve and target treatment approaches (Berg et al. 2020).

4.2 Cutting-Edge Technologies in Metabolomics

The sector of metabolomics is rapidly evolving, propelled by the rapid pace of technological changes (Gowda et al. 2008). These advancements have empowered researchers with new capabilities to identify and quantify metabolites (Yang et al. 2019). One of the most exciting developments is the continuous improvement of mass spectrometry instruments and methods (Thomas et al. 2022). Specialized high-resolution MS platforms such as TOF/MS and Orbitrap analyzer have made it possible to detect thousands of metabolites in biological samples with unprecedented sensitivity and accuracy. The adoption of tandem mass spectrometry (MS/MS) has further enhanced metabolite identification by elucidating metabolite structure (Ren et al. 2018). In addition to the above MS improvements, metabolomics has benefited from advancements in nuclear magnetic resonance spectroscopy, a potential alternative method for metabolite analysis in MS (Marshall and Powers 2017). Introducing cryogenic probes and using higher-field magnets have improved the NMR data regarding signal-to-noise ratio and spectral resolution to allow the analysis of metabolites at much lower concentrations (Moco 2022). Moreover, the combination of NMR and various sophisticated data processing and statistical analysis approaches has made identifying and quantifying metabolites easier, especially in complex biological systems (Miao et al. 2015).

In addition to improving these core analytical platforms, the field of metabolomics has also experienced new developments regarding the introduction and enhancement of the techniques for the extraction, separation, and probably even identification of metabolites (Qiu et al. 2023b). For instance, supercritical fluid extraction, liquid-liquid extraction, and solid phase extraction have been used and refined to enhance the recovery or cleanup of metabolites while circumventing the obvious limitations brought about by metabolites' diversity and concentration levels in biological matrices (Khaw KooiYeong et al. 2017). With the advent and the incorporation of metabolomics into other omics branches, including genomics, transcriptomics, and proteomics among others, the horizons of this field are expanding (Yang et al. 2019). This integration has the potential to provide a clearer picture of biological systems by combining data from largely independent sources such as networks and temporal regulation (Cai et al. 2022). The emergence of computational approaches to multi-omics datasets further enhances this integration, enabling the identification of metabolic pathways, therapeutic mechanisms, and disease biomarkers (Mirza et al. 2019).

5 Data Analysis and Bioinformatics

5.1 Role of Bioinformatics in Multi-omics Integration

The term bioinformatics is devoted to the incorporation and scrutiny of multi-omics data, which is valuable, especially for metabolomics (Rai et al. 2018). With the increasing amounts and complexity of omics data, the tools and platforms for bioinformatics are now essential for storing, analyzing, and extracting relevant information from these huge datasets (Mirza et al. 2019). The integration of other omics, particularly genomics, transcriptomics, proteomics, and metabolomics, with constituent biomolecular data, enhances the understanding of biological systems and the processes involved therein (Mirza et al. 2019). Bioinformatics facilitates the integration and analysis of diverse omics data, revealing relationships, correlations, and patterns that would have been challenging to uncover from individual omics datasets (Misra et al. 2019). In metabolomics, bioinformatics domains are applied for data processing and identifying metabolites and the pathways in which metabolites participate (Pinu et al. 2019). Numerous software packages, also known as tools, including XCMS, MetaboAnalyst, or HMDB, are essential in the processing and analysis of metabolomics data, such as peak detection retention time alignment and compound identification. These bioinformatics tools leverage extensive metabolite databases and complex algorithms to facilitate the understanding of metabolic data and elucidate its relationship with disease and normal physiological processes (Sugimoto et al. 2012).

Besides, bioinformatics databases and tools such as Galaxy and Cytoscape provide step-by-step processes for picturing the relationships in various multi-omics datasets (Buescher and Driggers 2016). These platforms enable researchers to integrate and investigate diverse omics datasets, delve into the molecular landscape, and critically examine the networks and pathways within living systems (Zhou et al. 2019). The comprehension of such highly complex data structures is integral in the formulation of hypotheses, presentation of results, and guidance of further work (Haas et al. 2017). Furthermore, the significance of bioinformatics in analyzing and presenting multi-omics results must be acknowledged. On the other hand, bioinformatics tools employ sophisticated statistical methods, machine learning, and data mining to assist research endeavors (Yadav 2015). Such an increased understanding of biological processes holds a better chance of effectively managing the development of strategies aimed at diagnosis, prognostication, and even treatment (Cruz and Wishart 2006). Bioinformatics tools and platforms are shaping the future of research, where the processing, analysis, and interpretation of huge amounts of current omics data have a bearing, most importantly, on the quality of comprehension of biological systems and their mechanisms (Manzoni et al. 2018).

5.2 Data Processing and Analysis Workflows

Bioinformatics is helpful in processing and interpreting multi-omics data, especially in commercializing multi-omics technologies in metabolomics. These workflows are critical to coping with the increasing volume and complexity of omics data, allowing scholars to understand and comprehend biology (Chicco et al. 2023). Applying workflows for analyzing metabolomics data follows a similar order in most studies. First, the raw data obtained from various analytical platforms such as mass spectrometry or NMR is subjected to data processing, which usually entails peak picking, RT correction, and signal intensity normalization. This pre-processing step helps ensure the data is in the required format and appropriately scaled to the following analysis type (Cannataro et al. 2005). Further statistical and data mining methods are utilized to determine the dataset's critical metabolomic features and patterns. Normal unilabiate or multivariate statistical methods such as t-test, ANOVA, or supervised/unsupervised machine learning methods are carried out. These methods help find and analyze known and unknown differentially expressed metabolite patterns, sample clustering according to the metabolic pattern, and correlation of metabolites in multitarget pathways or other omics data (Rosato et al. 2018).

To better understand metabolic pathways and biological processes, metabolite features are commonly assigned to the respective metabolite databases and pathway analysis tools (Pinu et al. 2019). This allows scientists to assimilate the data generated from the different metabolomics analyses within the rest of cellular metabolism, making assumptions and explaining the observations (Johnson et al. 2015). Achieving a comprehensive understanding of the biological system targeted in the research is challenging, particularly when combining multi-omics data such as genomics, transcriptomics, and proteomics (Pinu et al. 2019). Without such integration, there is no chance to analyze networks, their regulatory elements, and biological interactions in detail. A multitude of distinct data categories can now be managed, represented, and scrutinized by researchers with the help of specialized bioinformatic tools and platforms such as Galaxy and Cytoscape, thus disclosing various existing intricate relations (Cline et al. 2007). Bioinformatics plays an undeniably crucial role in processing, analyzing, and integrating multi-omics data. Workflows and tools devised in bioinformatics have the twin advantages of managing vast amounts and a variety of data and efficiency in the accuracy and reproducibility of the research work (Clement et al. 2018). Advanced statistical models, machine learning, and rich metabolic data collection may help discover phenomena, propose ideas, and induce progress in biology using bioinformatics techniques (Clement et al. 2018).

5.3 Comparative Analysis of Methodologies

There has been much progress in multi-omics data analysis in recent years, particularly in metabolomics studies. Specialized strategies have been adopted to gain functional interpretation from the immense data generated by these advanced techniques. Hence, this comparative analysis aims to determine the quality of these methods and provide a discourse about their future trend in multi-omics research (Haas et al. 2017). One of the key aspects of multi-omics integration is the use of univariate and multivariate analysis in data analysis (Haas et al. 2017). While methods like ANOVA, principal component analysis, and partial least squares discriminant analysis have been instrumental in establishing the significance of experimental conditions within the context of various data, they often need to capture the complexity of biological systems and their interactions. This underscores the need for further research and innovation in this field (Sinha and Namdev 2020). However, in response to these challenges, more advanced data integration and network-based approaches are emerging. Techniques like weighted correlation network analysis and integrative network analysis are opening new doors for scientists, enabling them to explore the interactions between different types of omics data and unearth novel features and biological insights (Yan et al. 2018). These developments instill optimism about the future of multi-omics research, promising a deeper understanding of metabolic networks and their regulation (Hasin et al. 2017).

In addition, at this time, multi-omics data analysis is facilitated by using machine learning and deep learning algorithms. These methods are more successful in feature selection, predictive modelling, and pattern recognition from the data (Cuperlovic-Culf 2018). Although the field of multi-omics is expanding, it has become essential to develop efficient and easy-to-interpret data analysis approaches. Such novel methods as Bayesian networks, structural equation modelling, and graph-based methods have shown their efficiency in dealing with biological systems' various dynamic and hierarchical structures (Markowetz and Spang 2007). These techniques allow inferring causality, modelling interactions, and including background knowledge in the analysis, facilitating the comprehension of biological processes and the mechanisms governing those (Acerbi et al. 2014).

6 Applications in Health and Disease Research

6.1 Integrative Omics in Personalized Medicine

Rapid developments within the personalized medicine field have increasingly benefited from adopting a multi-omics approach. In this context, the focus has increasingly shifted towards the use of metabolomics, which allows for the measurement of actual cellular activities through its metabolomic profiling (De Tullio et al. 2019). Pharmacogenomics, transcriptomics, and proteomics aid in the personalization and progression of treatment through therapeutic relief with stronger therapeutics and better choice of treatments (Aboud and Weiss 2013). Integration of the omics information has also led to the development of novel disease-related molecular features and classifiers that can help understand the risk of disease, how it develops or progresses, and the response to a given treatment (Chen and Snyder 2013). Changes in the levels of metabolites over time are what are captured by omics methods, providing insights into metabolic activity. Such a holistic look at the human metabolism allows for demystifying disease states, early disease processes, or disease precursors and choosing relevant treatment (Johnson and Gonzalez 2012).

One of the few examples illustrating the impact of integrative omics in connection with personalized medicine is cancer management. By the combinatorial approaches of genomic, transcriptomic, and metabolomic data, researchers have formulated precise and accurate methods for diagnosis, treatment response prediction, and follow-up of the disease (Cantafio et al. 2019). As such, targeting specific metabolic changes attributed to various cancer subtypes would allow clinicians to choose effective therapeutic options for that specific cancer subtype and improve the outcome by minimizing the side effects, thereby reassuring the audience about the safety of personalized medicine (Ogrodzinski et al. 2017). Similarly, one exposure to integrative omics is personalized risk assessment and risk-preventing strategies for cardiovascular disease (Chen and Snyder 2013). Incorporating the metabolomic data with genetic and clinical data has led to the discovery of new mechanical factors associated with cardiovascular endpoints, such as myocardial infarction and temporal arterial stroke. Such information helps reduce the patient's risk by administering tailored lifestyle changes, specific antidrug therapies, and other preventative measures (Hicks et al. 2018).

The management of diseases has also included the use of integrative omics approaches to neurodegenerative diseases, which arise through interactions between genetic, environmental, and metabolic risk factors (Hara et al. 2019). Strategies such as decision-making characterized by metabolomic profiling can be adopted, which would enable the recognition of the altered metabolic processes that occur in a patient and the creation of treatments targeted at those processes that are important for the particular patient being treated (Qiu et al. 2023a). The fusion of omics data, including the integration of metabolomics, has made significant strides toward personalized medicine. As a result, when these approaches are combined, healthcare delivery can be more precisely tailored to meet the individual needs and characteristics of patients. This paves the way for improved patient outcomes, instilling a sense of optimism and hope for the future of personalized medicine (Evers et al. 2012) (Table 3).

6.2 Metabolomics in Disease Mechanisms and Diagnostic Tools

Metabolomics, with its focus on identifying and quantifying small molecules, has emerged as a powerful tool for understanding the mechanisms of various disease processes. The ability to identify related biomarkers for disease diagnosis is a testament to the potential of metabolomics (Castelli et al. 2022). This technology allows us to understand how different disease processes, including chronic and rare diseases at the genetic level, come about through quantitative changes in the metabolome in different conditions (Clish 2015). Many researchers have begun integrating metabolomics with genomics, transcriptomics, and proteomics, among other fields, to better understand disease mechanisms. These connections between gene variation, gene expression, and the dynamics of the metabolic network have long been established. Using this complementary approach, one can create diagnostic tests that consider the complex nature of the disease (Wu et al. 2009). One of the opportunities presented by the use of metabolomics in diagnosis is the detection of metabolic changes and biochemical alterations, which often occur well in advance before the onset of clinical symptoms (Johnson and Gonzalez 2012). Diagnostic markers can be developed and used well in advance to prevent diseases with great success. In addition, the fact that metabolomic assays are highly sensitive and specific presents prospects for their use in screening and monitoring individuals at risk of developing certain conditions, thereby ensuring that appropriate interventions are implemented early (Zhang et al. 2015).

In this respect, metabolomics has emerged as a new tool for the diagnosis of breast, prostate, and colorectal cancer. Following these metabolic profiles, specific groups of metabolites are expressed, which can ultimately be useful for the early detection and follow-up of the disease (Wang et al. 2018). Metabolomics has also provided clues to novel biomarkers of degenerative diseases of the brain, particularly Alzheimer's and Parkinson's diseases, which would allow more aggressive and tailored treatment (Ibáñez et al. 2015). In addition, multivariate analysis can also be harnessed to develop new diagnostic tools based on multi-omics data, which may boost the quality of such a tool (Yu et al. 2022). In genetic approaches to diseases, the use of different omics technologies can elucidate the integration of dysregulation across multiple aspects of the disease (Chen et al. 2023). Further examination of matched healing with mechanistic aspects of disease pathogenesis can also facilitate the development of effective therapies that account for each patient's metabolic and molecular differences (Naviaux 2019). The use of metabolomics in disease pathways and diagnosis is a pivotal aspect of the precision medicine endeavor. The development of metabolomics technology aids in its effective implementation and integration with other omics technologies, disease forecasting, and targeted therapy (Long et al. 2020). Each scientific revolution is unlocking more of metabolomics' potential, from early disease diagnosis to precision treatment, improved treatment efficacy, and patient management. This progress is a testament to the power of metabolomics in the realization of the precision medicine concept (Clish 2015).

6.3 Therapeutic Strategies and Targeted Therapies

The combination of metabolomics and other multi-omics strategies has been crucial in the progress made in targeted therapies for complex diseases. Analysis done at the genomic, transcriptomic, proteomic, and metabolomic levels has provided researchers with the basis for further development by helping locate new therapeutic targets and estimate how different individuals will respond to particular treatments (Cai et al. 2022). Metabolomics sequencing has also helped psychologists because it is the only "omics" that assesses the system's biochemical workings. Because metabolites are products of molecular processes within the cells, they are adequate indicators of bodily function or disease (Beger et al. 2016). In particular, metabolites are therapeutic strategies against any metabolic disorders brought about by disease (Gowda et al. 2008). Additionally, combining metabolomics with other omics domains like genomics and proteomics helps better understand the intricate molecular networks involved in disease causation. This multidimensional approach helps delineate the contributing, interlinked, and temporally changing factors that characterize the disease processes (Hasin et al. 2017). This detailed characterization can then guide the design of such treatments that seek to inhibit or switch on the molecular pathways and immune cells responsible for the disease (Labonte et al. 2014).

The usefulness of the multi-omics integration in the creation of new forms of treatment has been evidenced in the development of new therapies across various fields of medicine, offering a beacon of hope for patients (Cantafio et al. 2019). In oncology, multidimensional approaches utilizing genomic, transcriptomic, and metabolomic systems biology study emergent features of various cancer types associated with critical metabolic reprogramming, defined in targeted precision therapies of cancer-related metabolic dysregulations (Kang et al. 2018). In the same context, the use of multi-omics data has played a pivotal role in identifying new biomarkers and drug targets in the field of neurodegenerative diseases (Gómez-Cebrián et al. 2019). By studying such integrated networks of genes, transcripts, and metabolites, researchers have been able to create specific therapies against disease from biological processes, not only prohibiting signs of the ailment. This information keeps the audience informed about the latest advancements in the field of medicine (Chan and Loscalzo 2012). It also embraces the creation of novel prophylactic strategies targeted at specific disease pathophysiological processes. It has also paved the way for systems and precision medicine, where treatment modalities are designed based on the metabolic footprint or the molecular signatures of the individual patient (Ginsburg and Phillips 2018).

7 Future Directions and Challenges

7.1 Emerging Trends and Future Perspectives

The area of metabolomics has progressed rapidly in the past few years, and the arrival of several omics integrations has created hope in the comprehension of biological systems. With metabolomics research heading in a favorable direction, several novel concepts and trends are poised for adoption (Yang et al. 2019). A prominent trend in the field is the increasing adoption of a multi-omics approach, which involves the concurrent examination of metabolomics alongside other omics disciplines, such as genomics, transcriptomics, and proteomics (Pinu et al. 2019). As a result, this enhances researchers comprehension of how genes, transcripts, proteins, and metabolites interact in the biological system, enhancing such integration. This is essential for discovering new disease markers, more precise descriptions of complicated metabolic pathways, and devising better and more innovative modes of treatment (Zitnik et al. 2019). The availability of advanced measuring instruments such as mass spectrophotometers and nuclear magnetic resonance (NMR) has provided better detection capacity, selectivity, and throughput to functional metabolomics. With such technological advancements, neglecting further improvements in computational and bioinformatics tools is even more inappropriate (Gowda et al. 2008). Higher-quality and more thorough metabolomics datasets have been achieved in this era. Consequently, new metabolites have been found, and the metabolic networks have been investigated more deeply (Aretz and Meierhofer 2016).

Metabolomics is expected to be further integrated with other omics techniques to give rise to multi-omics platforms, where different types of biological data can be processed and analyzed comprehensively. This union will be vital in resolving intricate biological problems, decoding how diseases start and progress, and developing better solutions (Haas et al. 2017). Furthermore, miniaturized, portable, and easy-to-use analytical devices, judiciously engineered, will enhance the usability of metabolomics technologies. Hence, their use in conventional settings, individual therapy, and environmental assessments will increase (Yang et al. 2019). Advances in metabolomic probing will also be realized through cooperation and multidisciplinary collaborations. When researchers from different fields such as chemistry, biology, bioinformatics, and medicine collaborate, integrative omics approaches can be more effectively developed and applied. (Beger et al. 2016). Such silo-breaking will stimulate the field with new creative ideas and valuable gold mines of complementary skill sets. Improving the data integration process, understanding how the data is analyzed, and how the results are applied and served in real-life clinical practices will be transformational in these kinds of partnerships (Bennink 2020).

7.2 Challenges and Opportunities in Systems Biology

Integrating multi-omics presents a significant challenge for systems biology. In particular, metabolomics lacks a consistent approach for correlating and interpreting the overwhelming amount of data generated by various analytical platforms (Haas et al. 2017). Despite efforts in metabolomics, effective integration of the complexity and dynamic nature of biological information still needs to be available mainly due to the challenges associated with identifying and quantifying biological metabolites (Pinu et al. 2019). However, there is more than hope that the understanding of biological systems will be helped by metabolomics. The metabolome is the end product of cellular activity, thus offering a molecular snapshot of an organism's functional state by integrating feedback between genetic, transcriptomic, and proteomic activity (Gowda et al. 2008). In the context of cells, metabolome analysis enables a comprehensive understanding of cellular metabolism and biochemical processes from a broad systems-level perspective (Walsh et al. 2018).

The multi-omics approach has the potential to revolutionize the study of complex diseases, offering a promising avenue for research. Chronic and multifactorial diseases, such as cancer, cardiovascular diseases, and neurodegenerative diseases, exhibit alterations at a network level that cannot be adequately captured by assessing a single molecular component in isolation (Barabási et al. 2011). The constructive application of these different omics methods should enable researchers to elucidate the mechanisms underlying disease progression, leading to the discovery of new biomarkers and the creation of more effective, tailored treatment options (Olivier et al. 2019). However, to fully realize the promises of multi-omics integration in systems biology, we must address fundamental challenges. These include the documentation and integration of rigorous data collection challenges and analytical methods, the establishment of databases with relevant and appropriate data on metabolites and pathways, and the further development of systems biology approaches in silico (Pinu et al. 2019). To overcome these challenges, interdisciplinary teams of basic scientists, clinical researchers, and computer scientists must work together (Rai et al. 2018).

7.3 Recommendations for Future Research

Integrative omics include a wide range of industrial and academic endeavors and continue to proliferate. These developments comprise the use of novel equipment or strategies (Amer and Baidoo 2021). Researchers are presented with ongoing opportunities when novel technologies and unconventional solutions emerge, enabling better analysis and understanding of complex phenomena (Dreyer et al. 2017). Time integration of data from multi-omics approaches is one area that deserves further investigation. Therefore, as omics data becomes more diverse in volume, the need for advanced software and statistical models to manage and analyze such key variables and heterogeneous data still arises (Mirza et al. 2019). Leveraging high-throughput technologies to enhance the development of more robust data integration procedures, machine learning methods, and practice that uncover relationships and interactions between and within the molecular layers of genomics, transcriptomics, proteomics, and metabolomics is a fascinating pursuit (Fouché and Zinovyev 2023). In addition to planning for redundancy as a mitigation measure, it is crucial to consider factors such as standardizing data formats, metadata, and analytical processes. By doing so, we can facilitate the sharing and utilization of data among different research teams (Robinson et al. 2019). The creation of publicly accessible databases and the implementation of regulations regarding their use can further promote transparency and collaboration in the field of integrative omics and metabolomics (Pinu et al. 2019).

Interdisciplinary teaming will likely be fundamental to improving integrative omics and metabolomics. The solution is to involve experts from different fields, e.g., biologists, chemists, computer scientists, and mathematicians, who would emphasize their facet of problems for researchers to seize the opportunity of positive-sum games (Cavill et al. 2016). Thus, teams' efforts should be directed toward creating new ways of designing experiments and computational techniques and developing effective strategies considering the complex relationships among various biological constituents and their functions (Risi and Togelius 2015). Another crucial direction for future research in integrative omics and metabolomics is the application of these techniques in personalized and precision medicine. Integrating clinical data and patient-specific factors with diverse omics data can enable the development of personalized biomarkers and targeted therapeutic approaches. This personalized approach has the potential to significantly increase the efficacy of therapy, particularly in disease control, treatment monitoring, and progression prediction (Manzoni et al. 2018). The desired progression in integrated omics and metabolomics will be through advancement in new tools and methodology development enhanced within collaborations emerging from various spheres, embracing most especially the focus of these methodologies on solving essential problems in integrated and targeted medicine (Pinu et al. 2019).

8 Conclusion

Integrating metabolomics with other omics technologies, including genomics, transcriptomics, and proteomics, marks an essential improvement in biological sciences by facilitating detailed characterization of cellular activity, disease processes and activities, and drug targets. In this review, we focus on metabolic profiling in the context of multi-omics approaches, which is essential for a comprehensive systems biology understanding of the targeted cells. Dramatic improvements in metabolic data have happened, mainly due to technologies like mass spectrometry and NMR, which have led to more accurate diagnostic approaches and treatment strategies. Despite these advances, many issues can be encountered in implementing multi-omics approaches, mainly related to the diverse and complex nature of the data collected. Besides, the lack of a consistent method and approach in studies hampers the transferability and uptake of research efforts. To fill these gaps, we need much more reliable bioinformatics technologies and common work structures to enhance the synthesis of omics data, leading to a better understanding of biological processes. In addition, future studies should extend investigations of epithelial-mesenchymal transition (EMT) and other dynamic interactions of omics layers to different diseases and, most importantly, to highly complex cancer diseases. Also, multi-omics data may be improved even more by incorporating AI and machine learning, which may lead to the discovery of novel therapeutic targets and biomarker identification. This chapter, while recognizing the limitations of its focus, especially regarding recent trends and approaches, appreciates the role of integrative omics in precision medicine. As this field progresses, ongoing advances and collaborative efforts will be important for harnessing the maximum benefit from multi-omics research for the purpose of scientific exploration as well as clinical applications.

References

Aboud OA, Weiss RH (2013) New opportunities from the cancer metabolome. Clin Chem 59(1):138–146

Acerbi E, Zelante T, Narang V, Stella F (2014) Gene network inference using continuous time Bayesian networks: a comparative study and application to Th17 cell differentiation. BMC Bioinf 15:1–27

Alonso A, Marsal S, Julià A (2015) Analytical methods in untargeted metabolomics: state of the art in 2015. Front Bioeng Biotechnol 3:23

Amer B, Baidoo EE (2021) Omics-driven biotechnology for industrial applications. Front Bioeng Biotechnol 9:613307

Aretz I, Meierhofer D (2016) Advantages and pitfalls of mass spectrometry based metabolome profiling in systems biology. Int J Mol Sci 17(5):632

Asrina A (2024) Precision medicine approaches in oncology: current trends and future directions. Adv Healthc Res 2(1):36–48

Banerjee S, Prabhu Basrur N, Rai PS (2021) Omics technologies in personalized combination therapy for cardiovascular diseases: challenges and opportunities. Pers Med 18(6):595–611

Beger RD, Dunn W, Schmidt MA, Gross SS, Kirwan JA, Cascante M et al (2016) Metabolomics enables precision medicine: "a white paper, community perspective". Metabolomics 12:1–15

Bennink H (2020) Understanding and managing responsible innovation. Philos Manag 19(3):317–348

Berg JA, Zhou Y, Ouyang Y, Waller TC, Cluntun AA, Conway ME et al (2020). Network-aware reaction pattern recognition reveals regulatory signatures of mitochondrial dysfunction. BioRxiv, 2020-06

Bersanelli M, Mosca E, Remondini D, Giampieri E, Sala C, Castellani G, Milanesi L (2016) Methods for the integration of multi-omics data: mathematical aspects. BMC Bioinf 17:167–177

Bertholf RL (2006) Gas chromatography and mass spectrometry in clinical chemistry. In: Encyclopedia of analytical chemistry: applications, theory and instrumentation. Wiley, Chichester

Blum BC, Mousavi F, Emili A (2018) Single-platform 'multi-omic' profiling: unified mass spectrometry and computational workflows for integrative proteomics–metabolomics analysis. Mol Omics 14(5):307–319

Buescher JM, Driggers EM (2016) Integration of omics: more than the sum of its parts. Cancer Metab 4:1–8

Bueschl C, Krska R, Kluger B, Schuhmacher R (2013) Isotopic labeling-assisted metabolomics using LC–MS. Anal Bioanal Chem 405:27–33

Cai Z, Poulos RC, Liu J, Zhong Q (2022) Machine learning for multi-omics data integration in cancer. iScience 25(2):103798

Cannataro M, Guzzi PH, Mazza T, Tradigo G, Veltri P (2005, June) On the preprocessing of mass spectrometry proteomics data. In: Italian workshop on neural nets. Springer, Berlin/Heidelberg, pp 127–131

Cantafio MEG, Grillone K, Caracciolo D, Scionti F, Arbitrio M, Barbieri V, Di Martino MT (2019) From single level analysis to multi-omics integrative approaches: a powerful strategy towards the precision oncology. In: The road from nanomedicine to precision medicine. Jenny Stanford Publishing, New York, pp 829–869

Capati A, Ijare OB, Bezabeh T (2017) Diagnostic applications of nuclear magnetic resonance–based urinary metabolomics. Magn Reson Insights 10. https://doi.org/10.1177/1178623X17694346

Castelli FA, Rosati G, Moguet C, Fuentes C, Marrugo-Ramírez J, Lefebvre T et al (2022) Metabolomics for personalized medicine: the input of analytical chemistry from biomarker discovery to point-of-care tests. Anal Bioanal Chem 414(2):759–789

Cavill R, Jennen D, Kleinjans J, Briedé JJ (2016) Transcriptomic and metabolomic data integration. Brief Bioinform 17(5):891–901

Cesare Marincola F, Mannina L (2020) Special issue on "NMR-based metabolomics and its applications volume 2". Meta 10(2):45

Chan SY, Loscalzo J (2012) The emerging paradigm of network medicine in the study of human disease. Circ Res 111(3):359–374

Chen R, Snyder M (2013) Promise of personalized omics to precision medicine. Wiley Interdiscip Rev Syst Biol Med 5(1):73–82

Chen C, Wang J, Pan D, Wang X, Xu Y, Yan J et al (2023) Applications of multi-omics analysis in human diseases. MedComm 4(4):e315

Chicco D, Cumbo F, Angione C (2023) Ten quick tips for avoiding pitfalls in multi-omics data integration analyses. PLoS Comput Biol 19(7):e1011224

Cho DY, Kim YA, Przytycka TM (2012) Chapter 5: network biology approach to complex diseases. PLoS Comput Biol 8(12):e1002820

Clement L, Emeric D, Laurent M, David L, Eivind H, Kristian V (2018) A data-supported history of bioinformatics tools. arXiv preprint arXiv:1807.06808

Cline MS, Smoot M, Cerami E, Kuchinsky A, Landys N, Workman C et al (2007) Integration of biological networks and gene expression data using Cytoscape. Nat Protoc 2(10):2366–2382

Clish CB (2015) Metabolomics: an emerging but powerful tool for precision medicine. Mol Case Stud 1(1):a000588

Cruz JA, Wishart DS (2006) Applications of machine learning in cancer prediction and prognosis. Cancer Informat 2:117693510600200030

Cuperlovic-Culf M (2018) Machine learning methods for analysis of metabolic data and metabolic pathway modeling. Meta 8(1):4

Damiani C, Gaglio D, Sacco E, Alberghina L, Vanoni M (2020) Systems metabolomics: from metabolomic snapshots to design principles. Curr Opin Biotechnol 63:190–199

Dar MA, Arafah A, Bhat KA, Khan A, Khan MS, Ali A et al (2023) Multiomics technologies: role in disease biomarker discoveries and therapeutics. Brief Funct Genomics 22(2):76–96

De Castro F, Benedetti M, Del Coco L, Fanizzi FP (2019) NMR-based metabolomics in metal-based drug research. Molecules 24(12):2240

De Tullio P, Leenders J, de Ceniga MV, Chakfé N, Kolh PH (2019) Metabolomics as an innovative tool for a personalised approach to vascular disease. Eur J Vasc Endovasc Surg 57(3):329–330

Dettmer K, Aronov PA, Hammock BD (2007) Mass spectrometry-based metabolomics. Mass Spectrom Rev 26(1):51–78

Diaz-Beltran L, Cano C, Wall DP, Esteban FJ (2013) Systems biology as a comparative approach to understand complex gene expression in neurological diseases. Behav Sci 3(2):253–272

Draisma HH, Pool R, Kobl M, Jansen R, Petersen AK, Vaarhorst AA et al (2015) Genome-wide association study identifies novel genetic variants contributing to variation in blood metabolite levels. Nat Commun 6(1):7208

Dreyer M, Chefneux L, Goldberg A, Heimburg JV, Patrignani N, Schofield M, Shilling C (2017) Responsible innovation: a complementary view from industry with proposals for bridging different perspectives. Sustain For 9(10):1719

Eicher T, Kinnebrew G, Patt A, Spencer K, Ying K, Ma Q et al (2020) Metabolomics and multi-omics integration: a survey of computational methods and resources. Meta 10(5):202

Emwas AH, Roy R, McKay RT, Tenori L, Saccenti E, Gowda GN et al (2019) NMR spectroscopy for metabolomics research. Meta 9(7):123

Evers AW, Rovers MM, Kremer JA, Veltman JA, Schalken JA, Bloem BR, Van Gool AJ (2012) An integrated framework of personalized medicine: from individual genomes to participatory health care. Croat Med J 53(4):301–303

Fan Y, Pedersen O (2021) Gut microbiota in human metabolic health and disease. Nat Rev Microbiol 19(1):55–71

Fang Z, Zhang H, Guo J, Guo J (2024) Overview of therapeutic drug monitoring and clinical practice. Talanta 266:124996

Fouché A, Zinovyev A (2023) Omics data integration in computational biology viewed through the prism of machine learning paradigms. Front Bioinf 3:1191961

Garcia A, Barbas C (2011) Gas chromatography-mass spectrometry (GC-MS)-based metabolomics. In: Metabolic profiling: methods and protocols. Humana Press, New York, pp 191–204

Ginsburg GS, Phillips KA (2018) Precision medicine: from science to value. Health Aff 37(5):694–701

Girolamo FD, Lante I, Muraca M, Putignani L (2013) The role of mass spectrometry in the "omics" era. Curr Org Chem 17(23):2891–2905

Gómez-Cebrián N, Rojas-Benedicto A, Albors-Vaquer A, López-Guerrero JA, Pineda-Lucena A, Puchades-Carrasco L (2019) Metabolomics contributions to the discovery of prostate cancer biomarkers. Meta 9(3):48

Gowda GN, Zhang S, Gu H, Asiago V, Shanaiah N, Raftery D (2008) Metabolomics-based methods for early disease diagnostics. Expert Rev Mol Diagn 8(5):617–633

Gruber B, David F, Sandra P (2020) Capillary gas chromatography-mass spectrometry: current trends and perspectives. TrAC Trends Anal Chem 124:115475

Guo Z, Huang S, Wang J, Feng YL (2020) Recent advances in non-targeted screening analysis using liquid chromatography-high resolution mass spectrometry to explore new biomarkers for human exposure. Talanta 219:121339

Haas R, Zelezniak A, Iacovacci J, Kamrad S, Townsend S, Ralser M (2017) Designing and interpreting 'multi-omic' experiments that may change our understanding of biology. Curr Opin Syst Biol 6:37–45

Han J, Li Q, Chen Y, Yang Y (2021) Recent metabolomics analysis in tumor metabolism reprogramming. Front Mol Biosci 8:763902

Hara Y, McKeehan N, Fillit HM (2019) Translating the biology of aging into novel therapeutics for Alzheimer disease. Neurology 92(2):84–93

Hasin Y, Seldin M, Lusis A (2017) Multi-omics approaches to disease. Genome Biol 18:1–15

Hemida MHS (2021) Portable liquid chromatography for on-site analysis. Doctoral dissertation, University of Tasmania

Hicks KA, Mahaffey KW, Mehran R, Nissen SE, Wiviott SD, Dunn B et al (2018) 2017 cardiovascular and stroke endpoint definitions for clinical trials. Circulation 137(9):961–972

Huang R (2021) Application of liquid chromatography-mass spectrometry-based metabolomics in cell culture, drug study, and diseases. Int J Biosci Biochem Bioinf 11(2):34–39

Ibáñez C, Cifuentes A, Simó C (2015) Recent advances and applications of metabolomics to investigate neurodegenerative diseases. Int Rev Neurobiol 122:95–132

Johnson CH, Gonzalez FJ (2012) Challenges and opportunities of metabolomics. J Cell Physiol 227(8):2975–2981

Jousse C, Pujos-Guillot E (2013) Exploring metabolome with GC/MS. In: Advances in botanical research, vol 67. Academic Press, pp 303–329

Kang YP, Ward NP, DeNicola GM (2018) Recent advances in cancer metabolism: a technological perspective. Exp Mol Med 50(4):1–16

Khaw KooiYeong KK, Parat MO, Shaw PN, Falconer JR (2017) Solvent supercritical fluid technologies to extract bioactive compounds from natural sources: a review, vol 22

Kim SJ, Kim SH, Kim JH, Hwang S, Yoo HJ (2016) Understanding metabolomics in biomedical research. Endocrinol Metab 31(1):7–16

Koch WF (1989) Ion chromatography and the certification of standard reference materials. J Chromatogr Sci 27(8):418–420

Krassowski M, Das V, Sahu SK, Misra BB (2020) State of the field in multi-omics research: from computational needs to data mining and sharing. Front Genet 11:610798

Labonte AC, Tosello-Trampont AC, Hahn YS (2014) The role of macrophage polarization in infectious and inflammatory diseases. Mol Cells 37(4):275–285

Lacy P, McKay RT, Finkel M, Karnovsky A, Woehler S, Lewis MJ et al (2014) Signal intensities derived from different NMR probes and parameters contribute to variations in quantification of metabolites. PLoS One 9(1):e85732

Le Gouellec A, Plazy C, Toussaint B (2023) What clinical metabolomics will bring to the medicine of tomorrow. Front Anal Sci 3:1142606

Li Y, Luo Y (2022) Metabolomics of aging and Alzheimer's disease: from single-omics to multi-omics. arXiv preprint arXiv:2212.09870

Long NP, Nghi TD, Kang YP, Anh NH, Kim HM, Park SK, Kwon SW (2020) Toward a standardized strategy of clinical metabolomics for the advancement of precision medicine. Meta 10(2):51

Manzoni C, Kia DA, Vandrovcova J, Hardy J, Wood NW, Lewis PA, Ferrari R (2018) Genome, transcriptome and proteome: the rise of omics data and their integration in biomedical sciences. Brief Bioinform 19(2):286–302

Markowetz F, Spang R (2007) Inferring cellular networks–a review. BMC bioinf 8:1–17

Marshall DD, Powers R (2017) Beyond the paradigm: combining mass spectrometry and nuclear magnetic resonance for metabolomics. Prog Nucl Magn Reson Spectrosc 100:1–16

Mathuria A, Chaudhary A, Sharma H, Mani I (2024) Multi-omics in gut microbiome. In: Multi-omics analysis of the human microbiome: from technology to clinical applications. Singapore, Springer Nature, pp 181–213

Miao Z, Jin M, Liu X, Guo W, Jin X, Liu H, Wang Y (2015) The application of HPLC and microprobe NMR spectroscopy in the identification of metabolites in complex biological matrices. Anal Bioanal Chem 407:3405–3416

Mirza B, Wang W, Wang J, Choi H, Chung NC, Ping P (2019) Machine learning and integrative analysis of biomedical big data. Genes 10(2):87

Mishra P, Ambs S (2015) Metabolic signatures of human breast cancer. Mol Cell Oncol 2(3):e992217

Misra BB, Langefeld C, Olivier M, Cox LA (2019) Integrated omics: tools, advances and future approaches. J Mol Endocrinol 62(1):R21–R45

Moco S (2022) Studying metabolism by NMR-based metabolomics. Front Mol Biosci 9:882487

Muthubharathi BC, Gowripriya T, Balamurugan K (2021) Metabolomics: small molecules that matter more. Mol Omics 17(2):210–229

Nagana Gowda GA, Gowda YN, Raftery D (2015) Expanding the limits of human blood metabolite quantitation using NMR spectroscopy. Anal Chem 87(1):706–715

Nandania J, Peddinti G, Pessia A, Kokkonen M, Velagapudi V (2018) Validation and automation of a high-throughput multitargeted method for semiquantification of endogenous metabolites from different biological matrices using tandem mass spectrometry. Meta 8(3):44

Naviaux RK (2019) Metabolic features and regulation of the healing cycle—a new model for chronic disease pathogenesis and treatment. Mitochondrion 46:278–297

Ogrodzinski MP, Bernard JJ, Lunt SY (2017) Deciphering metabolic rewiring in breast cancer subtypes. Transl Res 189:105–122

Olivier M, Asmis R, Hawkins GA, Howard TD, Cox LA (2019) The need for multi-omics biomarker signatures in precision medicine. Int J Mol Sci 20(19):4781

Pan L, Qin P, Rong P, Zeng X, Liu D, Peng S (2023, December) PACS: prediction and analysis of cancer subtypes from multi-omics data based on a multi-head attention mechanism model. In: 2023 IEEE international conference on bioinformatics and biomedicine (BIBM). IEEE, pp 904–909

Patti GJ, Yanes O, Siuzdak G (2012) Metabolomics: the apogee of the omics trilogy. Nat Rev Mol Cell Biol 13(4):263–269

Pinu FR, Beale DJ, Paten AM, Kouremenos K, Swarup S, Schirra HJ, Wishart D (2019) Systems biology and multi-omics integration: viewpoints from the metabolomics research community. Meta 9(4):76

Puchades-Carrasco L, Pineda-Lucena A (2017) Metabolomics applications in precision medicine: an oncological perspective. Curr Top Med Chem 17(24):2740–2751

Qiu S, Cai Y, Yao H, Lin C, Xie Y, Tang S, Zhang A (2023a) Small molecule metabolites: discovery of biomarkers and therapeutic targets. Signal Transduct Target Ther 8(1):132

Qiu S, Guo S, Yang Q, Xie Y, Tang S, Zhang A (2023b) Innovation in identifying metabolites from complex metabolome—highlights of recent analytical platforms and protocols. Front Chem 11:1129717

Rai S, Raj U, Varadwaj PK (2018) Systems biology: a powerful tool for drug development. Curr Top Med Chem 18(20):1745–1754

Ren JL, Zhang AH, Kong L, Wang XJ (2018) Advances in mass spectrometry-based metabolomics for investigation of metabolites. RSC Adv 8(40):22335–22350

Risi S, Togelius J (2015) Neuroevolution in games: state of the art and open challenges. IEEE Trans Comput Intell AI Games 9(1):25–41

Robinson NH, Hamman J, Abernathey R (2019) Seven principles for effective scientific big-Data-Systems. arXiv preprint arXiv:1908.03356

Rosato A, Tenori L, Cascante M, De Atauri Carulla PR, Martins dos Santos VA, Saccenti E (2018) From correlation to causation: analysis of metabolomics data using systems biology approaches. Metabolomics 14:1–20

Rotroff DM, Motsinger-Reif AA (2016) Embracing integrative multiomics approaches. Int J Genomics 2016(1):1715985

Saurina J, Sentellas S (2017) Strategies for metabolite profiling based on liquid chromatography. J Chromatogr B 1044:103–111

Schumacher-Schuh A, Bieger A, Borelli WV, Portley MK, Awad PS, Bandres-Ciga S (2022) Advances in proteomic and metabolomic profiling of neurodegenerative diseases. Front Neurol 12:792227

Sinha AK, Namdev N (2020, March) Computational approach of tumor growth in human body with a significant technique the rough set. In: IOP conference series: materials science and engineering, vol. 798, no. 1. IOP Publishing, p 012038

Subramanian I, Verma S, Kumar S, Jere A, Anamika K (2020) Multi-omics data integration, interpretation, and its application. Bioinf Biol Insights 14:1177932219899051

Sugimoto M, Kawakami M, Robert M, Soga T, Tomita M (2012) Bioinformatics tools for mass spectroscopy-based metabolomic data processing and analysis. Curr Bioinforma 7(1):96–108

Sun YV, Hu YJ (2016) Integrative analysis of multi-omics data for discovery and functional studies of complex human diseases. Adv Genet 93:147–190

Thomas SN, French D, Jannetto PJ, Rappold BA, Clarke WA (2022) Liquid chromatography–tandem mass spectrometry for clinical diagnostics. Nat Rev Methods Primers 2(1):96

Toledo JB, Arnold M, Kastenmüller G, Chang R, Baillie RA, Han X et al (2017) Metabolic network failures in Alzheimer's disease: a biochemical road map. Alzheimers Dement 13(9):965–984

Ursell LK, Haiser HJ, Van Treuren W, Garg N, Reddivari L, Vanamala J et al (2014) The intestinal metabolome: an intersection between microbiota and host. Gastroenterology 146(6):1470–1476

Vahabi N, Michailidis G (2022) Unsupervised multi-omics data integration methods: a comprehensive review. Front Genet 13:854752

Vyas U, Ranganathan N (2012) Probiotics, prebiotics, and synbiotics: gut and beyond. Gastroenterol Res Pract 2012(1):872716

Walsh CT, Tu BP, Tang Y (2018) Eight kinetically stable but thermodynamically activated molecules that power cell metabolism. Chem Rev 118(4):1460–1494

Wang L, Liu X, Yang Q (2018) Application of metabolomics in cancer research: as a powerful tool to screen biomarker for diagnosis, monitoring and prognosis of cancer. Biom J 4(3):12

Wanichthanarak K, Fahrmann JF, Grapov D (2015) Genomic, proteomic, and metabolomic data integration strategies. Biomark Insights 10:BMI-S29511

Wishart DS (2019) NMR metabolomics: a look ahead. J Magn Reson 306:155–161

Wu S, Lusis AJ, Drake TA (2009) A systems-based framework for understanding complex metabolic and cardiovascular disorders. J Lipid Res 50:S358–S363

Xi J, Yu Z (2021) Unsupervised learning models for unlabeled genomic, transcriptomic & proteomic data. Front Genet 12:781698

Yadav D (2015) Relevance of bioinformatics in the era of omics driven research. J Next Gener Sequ Appl 2(1):e102

Yan J, Risacher SL, Shen L, Saykin AJ (2018) Network approaches to systems biology analysis of complex disease: integrative methods for multi-omics data. Brief Bioinform 19(6):1370–1381

Yang Q, Zhang AH, Miao JH, Sun H, Han Y, Yan GL et al (2019) Metabolomics biotechnology, applications, and future trends: a systematic review. RSC Adv 9(64):37245–37257

Yıldız-Ozturk E, Askin-Uzel R (2022, October) Advanced bioengineering applications with ecotechnology approach for sustainable development. In: International conference on sustainable, circular management and environmental engineering. Springer Nature, Cham, pp 182–208

Yu HY, Myoung S, Ahn S (2021) Recent applications of benchtop nuclear magnetic resonance spectroscopy. Magnetochemistry 7(9):121

Yu J, Kong Z, Zhan L, Shen L, He L (2022) Tensor-based multi-modality feature selection and regression for Alzheimer's disease diagnosis. Comp Sci Inf Technol 12(18):123

Zhang A, Sun H, Yan G, Wang P, Wang X (2015) Metabolomics for biomarker discovery: moving to the clinic. Biomed Res Int 2015(1):354671

Zhang XW, Li QH, Dou JJ (2020) Mass spectrometry-based metabolomics in health and medical science: a systematic review. RSC Adv 10(6):3092–3104

Zhou Y, Zhou B, Pache L, Chang M, Khodabakhshi AH, Tanaseichuk O et al (2019) Metascape provides a biologist-oriented resource for the analysis of systems-level datasets. Nat Commun 10(1):1523

Zitnik M, Nguyen F, Wang B, Leskovec J, Goldenberg A, Hoffman MM (2019) Machine learning for integrating data in biology and medicine: principles, practice, and opportunities. Inf Fus 50:71–91

Metabolic Modeling and Flux Analysis: Intersection with Other Omics Techniques

J. Sushmitha, C. J. Tharun Kumar, Kumar Nupur Hrishikeshan, Tejveer Singh, T. Kavya, and T. Vinutha

Abstract

The emergence of systems, synthetic biology, and meta-omics approaches has advanced the exploration of intricate cellular processes in biological systems. A living system's flux map reflects an integrated functional phenotype resulting from numerous levels of biological organization and regulation, such as the genome, transcriptome, and proteome. Constraint-based models (CBM) employ metabolic reaction network models of metabolism that operate in a steady state, ensuring that reaction rates (fluxes) and amounts of metabolic intermediates remain constant. Constraint-based models (CBM) employ metabolic reaction network models of metabolism that operate in a steady state, maintaining constant reaction rates (fluxes) and stable levels of metabolic intermediates, while providing estimated (metabolic flux analysis, MFA) or anticipated (flux balance analysis, FBA) fluxes through the network in vivo that cannot be computed directly. These fluxes can provide insight into basic biology and are being effectively utilized to drive metabolic engineering efforts. Furthermore, the rapidly evolving discipline of machine learning (ML) and its specialized branch, deep learning (DL), provide critical computational frameworks for decoding complicated and diversified biological data. Despite this, the emergence of these multidisciplinary methodological frameworks is largely independent, limiting the concatenation of biological information from many disciplines. As a result, we

J. Sushmitha (✉) · T. Singh · T. Vinutha
Division of Biochemistry, ICAR- Indian Agricultural Research Institute, New Delhi, Delhi, India

C. J. Tharun Kumar · K. N. Hrishikeshan
Division of Molecular Biology and Biotechnology, ICAR- Indian Agricultural Research Institute, New Delhi, Delhi, India

T. Kavya
Division of Microbiology, ICAR- Indian Agricultural Research Institute, New Delhi, Delhi, India

S. Kaur et al. (eds.), *Microbial Metabolomics*,
https://doi.org/10.1007/978-981-96-4824-5_5

have proposed the possibility of combining multidisciplinary tools and methodologies from many domains, such as CBM, omics, and machine learning, to investigate biochemical phenomena transcending conventional biology dogma. Genome-scale metabolic models (GEMs) remain effective tools used to comprehend metabolism at the systems level. GEMs, in their most basic form, do not take into account cellular regulation. It is worth noting that several strategies are focused on directly integrating omics data into GEMs in order to improve model accuracy. The GEM reconstruction techniques and regulatory systems that control metabolism are also emphasized.

Keywords

Flux · Constraint-based modeling · Machine learning · Metabolic modeling · Systems biology

1 Introduction

Cellular metabolism is a fundamental biological process used by all living organisms to generate and expend energy for growth. Metabolic fluxes provide a quantitative readout of cellular function and, thus, contribute to the understanding of cell growth and maintenance as well as the cell's responses to environmental changes (Sauer 2006). Here, the term "metabolic flux" is used as a synonym for all reaction and transport rates in living cells (the fluxome). The end product of a fluxome study, the flux map, reveals how metabolism actually works in the cells' environment and is, thus, a unique analytical tool for all fields of metabolic research (Stephanopoulos 1999). The monitoring of stable isotope labeling profiles emerged as a key technology providing highly informative flux indicators (Zamboni 2011). By quantifying fluxes under various perturbed conditions, e.g., after generic manipulation of pathway components or their regulation, it is possible to identify bottlenecks in metabolic networks (Stephanopoulos 1999; Cheah et al. 2020), quantify metabolic control (Niederberger et al. 1992; Stephanopoulos and Vallino 1991; Vallino and Stephanopoulos 2000), and even establish predictive kinetic models of metabolism (Foster et al. 2019), which all greatly aid in the design of new strategies to improve desired system outputs. Stable isotope tracing techniques based on ^{13}C tracers are now one of the most accurate and widely used techniques in metabolic engineering to investigate metabolic network models, enabling highly accurate determination of metabolic fluxes, validation of metabolic model reconstructions, and discovery of unusual pathways and enzyme functions in less-characterized organisms (Xiong et al. 2015; Gonzalez and Antoniewicz 2017; Long et al. 2017; Tang et al. 2007a, b). In recent decades, genome-scale metabolic models (GEMs) have emerged as a powerful tool, enabling the elucidation of complex systems-level metabolism. GEMs are computational representations of metabolic networks accounting for the entirety of metabolic activity encoded in the genome for a given organism (Price et al. 2004). Principally, they involve a set of mass-balanced metabolic reactions and metabolites represented in a stoichiometric matrix. GEMs also include gene-protein-reaction (GPR) associations describing the relationship between thousands of genes, proteins, and reactions (Thiele and Palsson 2010). Constraint-based modeling (CBM) is the standard framework for reconstructing and analyzing GEMs, primarily through the addition of model constraints (Lewis et al. 2012). The most basic form of CBM is flux balance analysis (FBA) (Orth et al. 2010). Data generated by approaches such as genomics, transcriptomics, proteomics, epigenomics, metabolomics, pharmacogenomics, fluxomics, or phenomics constitutes most of the big data in biology and medicine (Alyass et al. 2015). GEMs quantitatively define a relationship between genotype and phenotype by contextualizing multi-omics data. GEMs reconstruction can be approached by machine learning, a subclass of artificial intelligence in which statistical tools train computers to learn and adapt to the available data. The existing information-based training allowed machines to make automated decisions without explicit instructions. ML can be applied to increase the predictive capability of the reconstructed GEMs. ML-based method called automated metabolic model ensemble-driven uncertainty elimination using statistical learning (AMMEDEUS) can be used to integrate manual curation stages to the automated GEM reconstruction pipeline, enhancing the quality of the resulting GEM (Medlock and Papin 2020).

2 Models for Flux Analysis

One of the prime reasons for the intricacy of in vivo flux measurements is the complex nature of the cellular environment. Therefore, modeling approaches are essential for estimating or predicting these fluxes, allowing researchers to infer the rates of metabolic reactions within living organisms based on experimental data and computational simulations (Kaste et al. 2023). Metabolic models are mathematical representations of the confluence of biochemical reactions within living organisms. These models record interactions among enzymes, metabolites, and pathways, allowing researchers to simulate and analyze cellular metabolism under the conditions of their interest. Network visualization approaches, such as metabolic models, are particularly important in studies on the functioning of cells under environmental and physiological constraints and the representation of fluxes between nodes of the network. Through a detailed description of the pathways, the cellular model allows one to predict the usage of resources, environmental changes, and shifts in metabolic fluxes, which are the rates of substances changing into different chemical products, through a network (Zeng et al. 2021). A notable strength of metabolic models is that they predict the distribution of metabolic fluxes in response to genetic or environmental perturbations. For instance, the model involves knocking out or overexpressing certain genes. Cells may exhibit varying growth rates or produce more of a specific product, offering a basic understanding of the interactions occurring with other genes or pathways. Researches can predict the effects and simulate the resultant changes by knocking out or overexpressing genes in the model (Segre et al. 2002). This feature is especially useful in metabolic engineering, where the objective is to maximize the production of biofuels, medicines, and other important substances from microbial strains (Bailey 1991). These models are used in flux analysis, especially using methods such as flux balance analysis (FBA), to examine the movement of metabolites along various pathways. This method is essential for comprehending the physiology of cells, streamlining industrial bioprocesses, and designing metabolic pathways for the synthesis of medicines, biofuels, and other important substances.

Furthermore, metabolic models are increasingly integrated with multi-omics data (genomic, transcriptomic, proteomic, and metabolomic data) to provide a more comprehensive understanding of cellular behavior (Maan et al. 2023). This integration enhances the predictive power of the models, enabling more accurate simulations of complex biological processes.

Metabolic models are categorized as stoichiometric models, kinetic models, constraint-based models, and dynamic models. These models differ in complexity and nature of biological processes captured by them. Constraint-based methods can be used to infer and predict metabolic flux distributions and complex phenotypes using stoichiometric and thermodynamic data. These models focus on the conservation of mass and energy, ensuring that the input and output of metabolites in each reaction are balanced (Sher et al. 2024). The advanced versions are kinetic models and dynamic models. Kinetic models integrate biochemical network stoichiometry, mechanistic reaction rate laws, kinetic parameters, and enzyme concentrations and simulate metabolite concentration alterations over time. Dynamic models capture the temporal changes in metabolite concentrations, shedding light on the behavior of metabolic networks in relation to time (Ahn and Antoniewicz 2012). These models are crucial for studying non-steady-state processes, such as metabolic shifts during cell growth, responses to environmental changes, or oscillatory behaviors in metabolism.

3 Constraint-Based Models: FBA, MFA, and ^{13}C Isotopes-MFA

3.1 Flux Balance Analysis (FBA) Models

Flux balance analysis (FBA) is one of the pioneer methods of computational metabolic modeling that is used in metabolic engineering to analyze the capabilities of biological systems by examining their metabolic network models. Stoichiometric representation of metabolic reactions forms the basis for this model, which is used to predict the flow of metabolites through a network under specific constraints (Orth et al. 2010).

3.1.1 Principle of Flux Balance Analysis

At the core of FBA is the metabolic network model, which is mathematically represented by a stoichiometry matrix (*S*). This matrix contains the stoichiometric coefficients for all known metabolic reactions and transport processes occurring within a cell. Each column of the matrix corresponds to a reaction, while each row corresponds to a metabolite. It operates under the assumption of steady state for intracellular metabolites. This means that the concentrations of intracellular metabolites remain relatively constant over time. Consequently, the sum of all fluxes producing a metabolite must equal the sum of all fluxes consuming that metabolite, leading to the equation

$$\left[S \times v = 0\right]$$

Here, S is the stoichiometric matrix. The flux vector, denoted as v, represents the rate of each reaction in the network. The goal of FBA is to determine the values of these fluxes that satisfy the network constraints while optimizing a given objective function, typically the maximization of biomass production or the yield of a desired metabolic product. This is formulated as a linear programming problem, where the objective function is maximized subject to the stoichiometric constraints imposed by the S matrix. FBA employs linear programming techniques to solve the optimization problem. The constraints include not only the stoichiometric equations but also bounds on the fluxes based on biological and thermodynamic considerations [$v_{\min} \leq v \leq v_{\max}$]. FBA requires an objective function, which is a linear combination of the fluxes to be optimized. The objective function typically represents a biological goal, such as maximizing biomass production

$$Maximize\left(\text{or } Minimize\right) Z = c^{\mathrm{T}} \times v$$

where c is a vector of coefficients that define the contribution of each flux to the objective function Z. In a biomass maximization scenario, the coefficients correspond to the stoichiometric coefficients of the metabolites involved in biomass synthesis.

3.1.2 Incorporating FBA for Flux Analysis: Methodology

To reconstruct the metabolic network, identify pathways in the organism using genome annotations and databases such as KEGG or MetaCyc. Define each reaction with its stoichiometry, specifying substrates, products, and their coefficients. Compartmentalize the reactions according to its compartments, like mitochondria or chloroplast. Construct the stoichiometric matrix (S matrix) with row-wise metabolites and column-wise reactions. S_{ij} matrix element denotes the stoichiometric coefficient of metabolite M_i in reaction R_j, with negative for substrate and positive for product.

Apply the constraints assuming the steady state, i.e., constant metabolite concentration. This leads to the core equation $S \times v = 0$. Physiological limits and reaction irreversibility are captured by setting flux bounds $v_{\min} \leq v \leq v_{\max}$. Environmental constraints such as nutrient uptake are also included.

Define the objective function Z that depends on biological contexts such as biomass maximization, ATP production, and byproduct secretion. FBA is formulated as a linear programming problem to maximize (or minimize) the objective function

$$Maximize\left(\text{or } Minimize\right)Z = c^{\mathrm{T}} \times v;\ \text{Subject to}: S \times v = 0; v_{\min} \leq v \leq v_{\max}$$

Use linear programming solvers such as CPLEX, Gurobi, or GLPK to determine pathway activity. In analyzing results from FBA, flux distribution interpretation helps determine how metabolites flow through the network, identifying active and inactive pathways. Sensitivity analysis evaluates the robustness of fluxes to changes in constraints or objectives. Flux variability analysis (FVA) examines the range of fluxes that achieve the optimal objective, revealing alternative pathways. Knockout simulations involve setting a reaction's flux to zero, rerunning FBA, and identifying critical genes or reactions for growth or metabolite production. These methods provide valuable insights into the metabolic network's behavior and adaptability under various conditions.

3.1.3 Applications of FBA

- Predicting maximum yields.
- Growth rate predictions.
- Gene essentiality and metabolic engineering.
- It is computationally efficient and can analyze large-scale metabolic networks, making it suitable for genome-scale models.
- Filling gaps in genome-scale models.
- The method provides valuable predictions about metabolic behavior under various conditions, which can guide experimental designs and metabolic engineering efforts.
- Integration with other approaches like dynamic modeling and constraint-based optimization, to enhance its predictive capabilities.

3.2 Metabolic Flux Analysis and ^{13}C Isotopes

3.2.1 MFA (Metabolic Flux Analysis)

The FBA-based models cannot directly quantify the flux, but instead predict the flux distribution in the metabolic network. MFA (metabolic flux analysis) is also a constraint-based model that can quantify the fluxes within a network based on stoichiometric models and known input and output rates (such as nutrient uptake and product formation). The fluxes are determined by solving a system of linear equations based on experimentally measured mass balances around each metabolite.

Principle and Methodology of Metabolic Flux Analysis

The principle of MFA is based on the mass balance of metabolites within a defined metabolic network. By quantifying the input (nutrient uptake) and output (product formation) rates, MFA can determine the distribution of fluxes through various metabolic pathways (Antoniewicz 2015). Metabolic fluxes under the assumption of a metabolic steady state for intracellular metabolites are determined by coupling stoichiometric constraints ($S \times v = 0$) with measured external metabolic rates ($R \times v = r$), where R represents the measurement matrix. The combined system is then solved using least squares regression (Eq. 4) to minimize the squared residuals between measured and predicted external rates.

$$Minimize\ SSR = \sum_{i=1}^{n} \frac{\left(r_i - r_{m,i}\right)^2}{\sigma_i^2}$$

r_i: The predicted flux or response from the model for observation *i*.
$r_{m,i}$: The measured flux or response for observation *i*.
σ_i^2 : The variance of the measurement errors. It accounts for the uncertainty in the measurements, and if each measurement has the same variance, this term becomes a constant factor. If the measurements have different variances, this would be used instead, making the sum a weighted SSR.

This approach enables the estimation of metabolic fluxes in both fully determined and overdetermined systems, where the latter have redundant sets of external rate measurements. The primary advantage of MFA lies in its simplicity and accessibility, as it requires only basic linear algebra and relies on relatively reliable measurements of extracellular metabolites.

3.2.2 ^{13}C-Metabolic Flux Analysis (^{13}C-MFA)

C-Metabolic flux analysis (^{13}C-MFA) is an advanced version of MFA, a state-of-the-art technique in metabolic engineering and systems biology used to map and quantify carbon flow through cellular metabolic networks (Zamboni 2011). This method uses the stable and nonradioactive isotope of carbon, carbon-13 (^{13}C), as a tracer, to study the metabolic pathways of living organisms. The distribution of isotopomers and isotopologues within the cellular metabolites provides critical information about the flow of carbon through different pathways. By analyzing ^{13}C-labeled substrates in the pathway and their distribution across various metabolic intermediates, fluxes of different metabolic reactions can be estimated within cells. This is typically done using high-precision analytical techniques like mass spectrometry (MS) (Wang et al. 2021) or nuclear magnetic resonance (NMR) spectroscopy (Roscher et al. 2014) providing high accuracy. Additionally, isotopic equilibrium must be achieved so that the ^{13}C distribution among isotopomers corresponds to steady-state fluxes (Antoniewicz et al. 2006).

^{13}C-MFA particularly allows for direct measurement of metabolic fluxes under physiological conditions. Unlike traditional flux analysis methods, which often rely on external measurements like substrate uptake and product secretion rates, ^{13}C-MFA provides a more detailed and precise picture of intracellular metabolic activity. This in vivo approach is crucial because metabolic fluxes can vary significantly depending on the cellular environment, genetic background, and external conditions. This technique has been extensively used in microbial systems, plant cells, and mammalian cells to study metabolism in various contexts, including industrial biotechnology, medical research, and environmental studies.

3.2.2.1 Methodology of ^{13}C-Metabolic Flux Analysis

The selection of ^{13}C-labeled substrates, such as ^{13}C-glucose, that participate in the metabolic pathways of interest is involved. Cells are cultured under controlled conditions that simulate the desired physiological state. The labeled substrate is introduced, allowing ^{13}C to integrate into metabolites as the cells metabolize it, eventually reaching an isotopic steady state. Samples are collected at various time points to track the distribution of ^{13}C in metabolites, and metabolic activity is quickly halted by quenching to preserve the labeling.

Intracellular metabolites are extracted using appropriate solvents, followed by isotopic labeling analysis through mass spectrometry (MS) or nuclear magnetic resonance (NMR) spectroscopy. MS measures the mass-to-charge ratio of isotopomers, while NMR reveals positional information about ^{13}C atoms. A metabolic network is constructed, detailing all relevant reactions and metabolites, and a stoichiometric matrix (S) is created. The matrix describes metabolite balances at steady state.

Next, isotopomer balance equations are formulated to explain the distribution of ^{13}C isotopomers across metabolites. This is represented by the equation $x = A \times v$, where A is the isotopomer mapping matrix. To estimate fluxes, an initial guess is made, and nonlinear optimization is used to minimize the difference between experimental and predicted isotopomer fractions, expressed as

$$Minimize \sum_{i=1}^{n} \left(x_i^{\mathrm{exp}} - x_i^{sim}(v) \right)^2$$

where x_i^{exp} represents the experimentally measured isotopomer fraction and $x_i^{\mathrm{sim}}(v)$ is the simulated fraction based on the fluxes v. The estimated fluxes are then validated by ensuring consistency with known biological constraints, and results are compared with independent data, such as uptake or secretion rates, for further validation. Finally, the methodology is documented, and if the model does not fit well, assumptions are re-evaluated, or additional data is collected to improve accuracy.

4 Non-Steady State Assuming Models (Dynamic Models)

Dynamic models emerged as a natural evolution from constraint-based models such as FBA, MFA, and ^{13}C-MFA, due to the need to capture the time-dependent behavior of metabolic networks. While constraint-based models are powerful tools for analyzing metabolic networks at steady state, they have limitations when it comes to understanding how these systems behave over time, especially under changing environmental conditions or during transitions between different metabolic states or growth phases.

To analyze metabolic dynamics, two constraint-based approaches, FBA and MFA, have been adapted to incorporate the effects of transient changes in cellular growth and metabolism. These include dynamic metabolic flux analysis (DMFA) and dynamic flux balance analysis (DFBA), which are extensions of MFA and FBA, respectively, and are advanced modeling approaches developed to study the temporal behavior of metabolic networks (Mahadevan et al. 2002; Leighty and Antoniewicz 2011). These dynamic models assume constant metabolite concentrations and fluxes. DMFA and DFBA overcome the limitations of these static models by allowing the analysis of metabolic changes over time.

Instead of assuming that intracellular metabolite concentrations remain constant over time, DMFA accounts for their accumulation or depletion (Ahn and Antoniewicz 2012). This is achieved by modeling the time-dependent changes in metabolite concentrations and reaction fluxes using a set of ordinary differential equations (ODEs). It often incorporates kinetic data, enabling a more detailed and accurate depiction of how metabolic processes respond to changes in the environment or internal cellular conditions. The ODEs are derived from the stoichiometric relationships within the metabolic network and the kinetics of the reactions involved. Contrastingly, DFBA combines linear programming (for FBA) with ODEs to model extracellular metabolite dynamics. The complexity arises mainly from the need to solve an optimization problem at each time step. DFBA and DMFA differ primarily in focus and assumptions. DFBA models the dynamics of extracellular metabolite concentrations and biomass growth, assuming intracellular fluxes are in quasi-steady state and optimizing them iteratively over time. In contrast, DMFA focuses on the time-dependent behavior of intracellular metabolites and fluxes, using differential equations without assuming steady-state conditions. DFBA is typically used for bioprocess optimization, where external conditions change, while DMFA is applied to understand intracellular metabolic dynamics, providing a detailed view of how metabolism evolves. ^{13}C dynamic metabolic flux analysis (^{13}C-DMFA) is an advanced method that combines the principles of dynamic metabolic flux analysis (DMFA) with ^{13}C isotopic labeling techniques (Ma et al. 2014).

5 Genome-Scale Metabolic Model (GEM) Reconstruction

The above-discussed models are confined to a set of reactions or pathways with steady or kinetic environments. The metabolism is defined by genome scale of a particular organism. The levels at which the metabolism is regulated are genetic, transcriptional, biochemical, and physiological. This requires a holistic view of metabolic models at the genome scale. GEMs are the integrated mathematical equations that comprehend all the metabolic changes occurring in an organism. These models are built in silico by combining the knowledge of all the metabolites with the genome. This is the bottom-up approach, which links genotype and phenotype using the connectivity among gene-protein-metabolite-reaction associations. Microorganisms dominate the number of genome-scale models available (Tarzi et al. 2024). GEMs are well established in model organisms, even in plants (Arnold and Nikoloski 2014) and animals (Robinson et al. 2020).

GEM operates based on one of the oldest flux analysis models, FBA, based on balancing fluxes of metabolites through linear programming in combination with key objective functions. Recent advances in sequencing methods, data accumulation at multilevel, and understandings led to comprehensive context-specific GEMs (Cho et al. 2019). These models integrate multilevel omics data to enhance the analysis specific to tissue or cell at system level. The in silico use of GEM reconstruction and analysis expedites the biological discoveries and metabolic understandings. Some of the major tools and software are referred in Table 1.

6 Machine Learning for GEM Reconstruction and Analysis (Machine Learning-Based Tools)

6.1 Machine Learning Techniques in GEM Reconstruction

Machine learning (ML) is fundamentally transforming genome-scale metabolic model (GEM) reconstruction and analysis. Traditionally, GEM reconstruction relied heavily on manual curation, making it time-consuming and prone to inaccuracies due to the complexity of metabolic networks. GEMs, which describe cellular metabolism using stoichiometric matrices of biochemical reactions, require comprehensive data on metabolic pathways, gene-protein-reaction (GPR) associations, and flux distributions. ML models have begun to automate several key steps in this process, significantly enhancing the speed, accuracy, and scalability of GEM construction (Turanli et al. 2024).

Deep learning models such as DeepGEM and AutoGEM are key tools that utilize neural networks to predict missing reactions in metabolic pathways, which is crucial for species with incomplete metabolic annotations (CLOSEgaps Framework 2021). These deep learning approaches analyze vast datasets, make plausible predictions about missing metabolic interactions, and subsequently enrich GEM reconstructions. By learning from experimental data, such models enable the filling of gaps in reaction networks and improve model completeness (Liu et al. 2024).

GEMINI, a tool that leverages machine learning algorithms, focuses on refining metabolic pathways by classifying enzyme functions and optimizing reaction networks. It automates the annotation of enzyme functions, addressing a major challenge in GEM development where incomplete or inaccurate enzyme information could skew metabolic predictions. The introduction of machine learning in GEM construction has improved not only reaction prediction but also the refinement of enzyme classification systems, enhancing applications such as metabolic engineering, precision medicine, and synthetic biology (PLOS Computational Biology 2024).

6.2 Supervised and Unsupervised Learning in GEMs

Supervised learning algorithms are widely adopted in GEMs for tasks like enzyme function prediction and metabolic flux estimation. These models are trained on labeled data from known metabolic networks, allowing them to predict enzyme activity in lesser-known organisms or conditions where empirical data is scarce. For instance, support vector machines (SVMs) and random forests (RF) have been

Table 1 Comprehensive list of in silico tools and software used in GEM reconstruction and analysis

Purpose	Tool or software	Link or source
GEM reconstruction tools	CarveMe	https://github.com/cdanielmachado/carveme
	ModelSEED	https://modelseed.org/
	KBase	https://kbase.us/
	RAVEN	https://github.com/SysBioChalmers/RAVEN
	merlin	https://merlin-sysbio.org/
	MetaNetX	https://www.metanetx.org/
	Pathway Tools	https://bioinformatics.ai.sri.com/ptools/
Gap-filling tools	FastGapFilling	http://thielelab.eu.
	MEMOTE	https://memote.io/
	GrowMatch	Kumar and Maranas (2009)
Metabolic network analysis tools	The COBRA Toolbox (COnstraint-Based Reconstruction and Analysis Toolbox)	https://opencobra.github.io/cobratoolbox/stable/
	cobrapy	https://opencobra.github.io/cobrapy/
	Escher	https://escher.github.io/
	FAME (Flux Analysis and Modeling Environment)	http://f-a-m-e.org/
	FASIMU	https://code.google.com/archive/p/fasimu/
	GEMSiRV	https://www.sysbio.se/GEMSiRV/
Flux balance analysis (FBA) and related tools	FBAwMC (Flux Balance Analysis with Molecular Crowding)	Beg et al. (2007)
	efmtool	https://csb.ethz.ch/tools/software/efmtool.html
	FASTCORMICS	https://github.com/bioasp/FASTCORMICS
	OptFlux	http://www.optflux.org/

(continued)

Table 1 (continued)

Purpose	Tool or software	Link or source
Context-specific GEM tools	INIT (Integrative Network Inference for Tissues)	Agren et al. (2012)
	iMAT (Integrative Metabolic Analysis Tool)	http://imat.cs.tau.ac.il/
	MBA (Model Building Algorithm)	Jerby et al. (2010)
	tINIT	Agren et al. (2014)
	FASTCORE	https://pypi.org/project/fastcore/
Metabolic engineering tools	OptKnock	http://gcrg.ucsd.edu/Downloads/OptKnock/
	OptGene	https://sourceforge.net/projects/optgene/
	RobustKnock	Tepper and Shlomi (2010)
	OptFlux	http://www.optflux.org/
Visualization and pathway mapping tools	Escher	https://escher.github.io/
	CellDesigner	http://www.celldesigner.org/https://metexplore.toulouse.inrae.fr/metexplore2/
	MetExplore	https://metexplore.toulouse.inrae.fr/metexplore-portal/
	Cytoscape	https://cytoscape.org/
Model testing and validation tools	MEMOTE	https://memote.io/
	SBML Test Suite	https://sbml.org/software/sbml-test-suite/
Reaction and pathway databases	KEGG (Kyoto Encyclopedia of Genes and Genomes)	https://www.genome.jp/kegg/
	MetaCyc	https://metacyc.org/
	BRENDA	https://www.brenda-enzymes.org/
	BioCyc	https://biocyc.org/
Metabolite and reaction prediction tools	RetroPath	https://www.myexperiment.org/workflows/4987.html

effectively used to predict gene knockouts and their effects on metabolic networks, thereby identifying critical enzymes under specific conditions (Hardikkumar 2024).

In contrast, unsupervised learning models, such as k-means clustering and principal component analysis (PCA), are employed to discover hidden patterns within metabolic networks without labeled data. These algorithms group reactions or enzymes based on similarity metrics, enabling researchers to uncover previously unknown interactions or regulatory mechanisms within the metabolic system (Singh et al. 2024; Ghahramani 2024). This approach is especially useful during early GEM reconstruction stages, where experimental data might be sparse.

The application of unsupervised learning extends to gap-filling in GEMs. AMMEDEUS, for example, combines supervised and unsupervised learning techniques to identify and fill reaction gaps in draft GEMs. This tool uses feature selection methods, including PCA and random forests, to refine predictions by targeting critical metabolic reactions essential for network integrity (Singh et al. 2024).

6.3 Deep Learning for Model Accuracy and Optimization

Deep learning models, particularly neural networks such as convolutional neural networks (CNNs) and recurrent neural networks (RNNs), are revolutionizing GEM reconstruction by simulating dynamic environmental responses of metabolic networks. These models are adept at predicting condition-specific models that reflect changes in metabolic processes under various environmental stimuli or genetic perturbations. For example, CNNs have been applied in enzyme function classification and reaction prediction tasks, while RNNs are used to model time-dependent changes in gene expression and metabolic flux (Angione et al. 2024; Dias et al. 2024).

The optimization of flux distributions is another crucial area where deep learning excels. Models like DeepMetabolism utilize multilayer perceptrons (MLPs) to predict flux distributions across metabolic networks, enabling accurate simulations of cellular metabolism under varying conditions. These predictive capabilities are particularly relevant in precision medicine and metabolic engineering, where understanding cellular responses to environmental changes is critical for targeted therapies or optimizing metabolic production (Turanli et al. 2024; Gurgen et al. 2021).

Reinforcement learning models have also been applied to optimize GEMs by iteratively learning from metabolic network simulations. These models fine-tune flux predictions by exploring different metabolic scenarios and selecting those that optimize specific objectives such as growth rate, biomass yield, or metabolite production. Such approaches significantly improve GEM performance in predicting metabolic outputs, making them valuable tools in industrial biotechnology and personalized healthcare (Canzar et al. 2024).

7 Machine Learning Integration with Big Bio-Data and Omics

7.1 Handling Big Bio-Data with ML

The rapid development of high-throughput technologies has generated vast amounts of biological data, particularly from multi-omics fields such as genomics, transcriptomics, proteomics, metabolomics, and fluxomics. These technologies provide an unprecedented level of insight into cellular mechanisms but also introduce significant challenges in data processing and analysis due to the sheer scale and complexity of the data involved. Machine learning (ML), particularly deep learning models, has emerged as an essential tool for handling this complexity. Deep learning is highly effective at processing and interpreting large, high-dimensional omics datasets, enabling the discovery of hidden patterns and relationships that traditional analytical methods often miss (Chung and Ping 2019; Zeng et al. 2024).

The ability of ML to integrate multiple omics datasets is crucial for developing a comprehensive understanding of cellular functions and metabolic interactions. By integrating data from various omics platforms—such as combining genomic and transcriptomic data—ML models can uncover novel gene-metabolite relationships, providing insights into metabolic networks. This capability is critical for refining genome-scale metabolic models (GEMs), where accurate metabolic predictions are essential for applications in drug discovery, personalized medicine, and metabolic engineering (Ballard et al., 2024; Neo and Peipei 2019).

Deep learning models like autoencoders and neural networks are particularly well-suited for analyzing large, multi-omics datasets. For instance, convolutional neural networks (CNNs) have been used to process transcriptomic and proteomic data, while recurrent neural networks (RNNs) are employed to predict dynamic changes in metabolic fluxes over time. These models enable researchers to integrate cross-omics data to identify complex interactions, thereby improving the predictive power of GEMs (Ballard et al., 2024; Zeng et al. 2024).

7.2 Data Integration and Omics Analysis

One of the critical tasks in handling big bio-data is the integration of diverse omics datasets to develop a holistic view of biological systems. ML-driven frameworks such as Multi-Omics Factor Analysis (MOFA) have been designed to integrate genomics, transcriptomics, proteomics, and metabolomics data, offering insights into how different biological layers interact (Argelaguet et al. 2020). This integration is essential for constructing GEMs that accurately reflect cellular metabolism. MOFA combines various types of omics data to reveal underlying patterns and relationships, improving the interpretation of metabolic behaviors under different conditions (Argelaguet et al. 2020). Dimensionality reduction techniques like principal component analysis (PCA) and t-distributed stochastic neighbor embedding (t-SNE) are frequently used to simplify large-scale omics datasets while preserving the most critical relationships within the data (García Sarmina et al. 2023). These methods are crucial for reducing computational complexity and enhancing the interpretability of the data, allowing ML algorithms to identify key features and relationships across omics datasets. PCA has been widely used in transcriptomic and proteomic studies to reduce the dimensionality of high-throughput data, enabling more accurate predictions in GEM reconstructions (Ballard et al. 2024).

Moreover, the integration of multi-omics data often requires addressing missing data points, which is a common challenge in large-scale biological datasets. ML-based imputation algorithms, such as k-nearest neighbors (k-NN) and random forests, have proven effective at filling in missing values in omics datasets, ensuring that incomplete data does not hinder the GEM reconstruction process (Ballard et al. 2024). These imputation techniques allow for more comprehensive and accurate multi-omics data integration, further refining the predictions made by GEMs.

7.3 Deep Learning and Multi-omics Fusion

Deep learning techniques, including CNNs and RNNs, have been increasingly applied to fuse multi-omics data, providing a unified framework for understanding cellular metabolism (Ballard et al. 2024). By integrating data from various omics layers, these models can capture intricate biological interactions that might otherwise be missed. For example, deep learning frameworks can integrate genomic, transcriptomic, and metabolomic data to predict how changes in gene expression influence metabolic fluxes, helping refine GEMs to simulate metabolic behaviors under different environmental or genetic conditions (Ballard et al. 2024).

The challenge of missing or incomplete datasets is also addressed through ML-based data fusion. Techniques like imputation models and generative adversarial networks (GANs) are used to predict and fill in missing omics data, ensuring that multi-omics datasets remain comprehensive (Ballard et al. 2024). These approaches not only improve the accuracy of GEM reconstructions but also allow for the simulation of metabolic responses to environmental and genetic changes. Additionally, graph-based machine learning models, such as graph neural networks (GNNs), are applied to integrate regulatory networks derived from multi-omics data, providing insights into complex gene-protein interactions and metabolic pathways (Ballard et al. 2024). By leveraging the power of deep learning for multi-omics fusion, researchers can build more accurate and dynamic GEMs that better reflect the complexity of biological systems. These integrated models are essential for applications in precision medicine, where personalized treatment strategies depend on an in-depth understanding of the interactions between various biological layers (Ballard et al. 2024).

Machine learning has become an indispensable tool for integrating big bio data and multi-omics data into GEMs. By handling the vast complexity of omics datasets, ML enables more accurate and comprehensive metabolic models, facilitating breakthroughs in systems biology, biotechnology, and medical research. The fusion of multi-omics data through deep learning models is particularly valuable for understanding cellular metabolism, offering enhanced predictive power and the ability to simulate complex biological interactions (Ballard et al. 2024). These advancements are reshaping the field of metabolic modeling, providing new opportunities for therapeutic discovery, metabolic engineering, and personalized medicine.

7.4 Challenges of Omics Data Integration in GEM

Integrating multi-omics data into genome-scale metabolic models (GEMs) presents several challenges. The diverse nature of omics data introduces complexities in quality, integration, and interpretation due to noise, missing data, and biases (Flores et al. 2023). Proper annotation is difficult as omics datasets often use different identifiers for genes, proteins, and metabolites, complicating integration into GEMs (Nanda and Ghosh 2021). Additionally, the high-dimensional nature of multi-omics data creates computational hurdles, especially with large-scale models (Wang et al. 2014). Advances in omics, computational methods, and validation are needed to overcome these issues.

A key challenge in integrating multi-omics data into GEMs is the risk of model overfitting, where the model becomes overly complex and too tailored to the specific dataset used for its construction (Schinn et al. 2021). This reduces the model's generalizability and reliability when applied to new data or conditions, as it captures noise or random fluctuations rather than true biological patterns. To address overfitting in context-specific models, strategies like incorporating prior biological knowledge and using cross-validation can help ensure the model accurately reflects the biological system. Additionally, machine learning approaches have been introduced to mitigate overfitting risks (Oyetunde et al. 2019).

8 Multi-omics Knowledge for GEM Reconstruction: Genomics, Transcriptomics, Epigenomics, Proteomics, Metabolomics, Fluxomics

8.1 Genomics and Transcriptomics in GEMs

Genomic data serves as the foundational layer for genome-scale metabolic model (GEM) reconstruction. It provides essential information about the metabolic genes that encode enzymes involved in various biochemical pathways (Argelaguet et al. 2020). The integration of genomic data enables the identification of gene-protein-reaction (GPR) associations, a critical step in defining the stoichiometric relationships within metabolic networks. For example, the mapping of GPRs allows for the prediction of gene knockouts and their impacts on metabolic fluxes, facilitating the development of more precise GEMs (Ballard et al. 2024). Transcriptomics data adds another crucial dimension by reflecting the dynamic changes in gene expression under different environmental or physiological conditions. By integrating transcriptomic profiles into GEMs, machine learning (ML) models can predict condition-specific metabolic activity (Ballard et al. 2024). Supervised and unsupervised learning algorithms have been implemented to classify and interpret transcriptomic data, allowing the accurate estimation of reaction fluxes from gene expression levels (García Sarmina et al. 2023). For instance, transcriptome-based GEMs enable the modeling of metabolic behaviors in cancer cells by integrating gene expression data from tumor tissues (Argelaguet et al. 2020). Recent advancements in ML tools like convolutional neural networks (CNNs) have been particularly effective in processing RNA-seq data to predict gene expression patterns (Ballard et al. 2024). Such methods are useful for identifying how changes in gene expression influence metabolic fluxes, allowing GEMs to more accurately simulate metabolic responses under various conditions.

8.2 Epigenomics and Regulatory Mechanisms

Epigenomics, which encompasses DNA methylation, histone modifications, and other regulatory modifications, provides vital insights into the control of gene expression without altering the underlying DNA sequence. Epigenomic data reveals how environmental factors and developmental cues influence gene accessibility and transcriptional activity, thus modulating metabolic pathways (García Sarmina et al. 2023). ML models, including random forests and neural networks, have been successfully applied to analyze epigenomic data, identifying key regulatory elements that impact metabolic function (Ballard et al. 2024). For instance, machine learning-based approaches have been used to predict the transcription factor binding sites and chromatin accessibility associated with specific metabolic genes (Argelaguet et al. 2020). Integrating this epigenomic information into GEMs enables researchers to explore how regulatory changes, such as histone acetylation or DNA methylation, influence enzyme activity and metabolic fluxes.

8.3 Proteomics and Metabolomics

Proteomic data, which quantifies the abundance of enzymes and other proteins, is indispensable for understanding the functional state of metabolic networks (Ballard et al. 2024). Since enzyme concentrations directly affect reaction rates, integrating proteomics into GEMs provides a more accurate representation of the active metabolic state of a cell or organism (García Sarmina et al. 2023). Proteomics data can be processed using ML algorithms to predict the functional activity of enzymes, enabling better simulations of metabolic fluxes in GEMs. Metabolomics complements proteomics by providing information on metabolite concentrations, which are crucial for quantifying metabolic reactions. Metabolomics data can be analyzed using ML methods such as Bayesian networks, support vector machines (SVMs), and decision trees to estimate the distribution of metabolic fluxes across pathways (Ballard et al. 2024). For example, machine learning has been used to identify patterns in metabolomic datasets that correspond to specific metabolic states, which can be integrated into GEMs to predict metabolic shifts in response to environmental changes (Argelaguet et al. 2020).

8.4 Fluxomics and Metabolic Flux Estimation

Fluxomics focuses on measuring and predicting the flow of metabolites through metabolic pathways, providing a direct quantification of metabolic activities. Fluxomics is critical for identifying bottlenecks in metabolic networks and optimizing pathway performance, especially in the context of metabolic engineering and synthetic biology (Ballard et al. 2024). Traditional fluxomics approaches, such as ^{13}C-metabolic flux analysis (MFA), have been enhanced by ML-based predictive models, which estimate metabolic fluxes based on multi-omics data (García Sarmina et al. 2023). These models, including deep learning frameworks and SVMs, allow researchers to predict metabolic responses under different conditions, such as nutrient limitations or genetic modifications (Ballard et al. 2024). For instance, machine learning has been applied to predict metabolic fluxes in bacteria and yeast, facilitating the optimization of industrial fermentation processes (Argelaguet et al. 2020). ML-based flux estimation tools such as SUMOFLUX and k-nearest neighbor (k-NN) have been used to predict flux partitions in parallel pathways. This capability is particularly useful in bioengineering, where optimizing specific fluxes can lead to increased production of desired compounds, such as biofuels or pharmaceuticals (Ballard et al. 2024). Furthermore, integrating fluxomics data into GEMs allows for a more comprehensive understanding of how metabolic fluxes are distributed across entire networks, enabling the design of more efficient and productive metabolic systems.

The integration of multi-omics data—genomics, transcriptomics, epigenomics, proteomics, metabolomics, and fluxomics—into GEMs has revolutionized the field of metabolic modeling. By utilizing ML tools to process and interpret these diverse data types, researchers can develop highly accurate and predictive GEMs (Ballard et al. 2024). These models provide a robust framework for simulating complex metabolic systems, offering new insights into cellular metabolism and paving the way for advanced applications in biotechnology, medicine, and environmental science (Argelaguet et al. 2020).

9 Conclusions and Future Perspectives

Integrating genomic, transcriptomic, proteomic, and metabolomic data is crucial for a comprehensive understanding of biological systems. Knowledge of metabolic networks can be derived through various methods, such as metabolic flux analysis (MFA), flux balance analysis (FBA), or monitoring of ^{13}C-labeled metabolites. Genome-scale metabolic models (GEMs) have become essential for modeling metabolism from a systems-level perspective, enabling the study of metabolic adaptations to environmental changes. Context-specific GEMs driven by omics data have helped uncover unique biochemical properties of cells, tissues, or communities. However, the increasing complexity and multidimensionality of high-throughput omics data present new challenges for existing methodologies in systems and synthetic biology (Hu et al. 2022; Roy et al. 2021). Next-generation GEMs that integrate multiple regulatory mechanisms offer the potential to reveal complex cellular interactions underlying human health and disease. Additionally, constraint-based simulations are often used to predict metabolic phenotypes, and incorporating omics data can enhance the predictive accuracy of these models. Machine learning will play a key role in integrating genomic, transcriptomic, proteomic, and metabolomic data, providing a comprehensive view of cellular metabolism under specific conditions (Li et al. 2022). This will help address current challenges and expand the use of multi-omics data in GEM reconstruction and analysis. ML algorithms will improve context-specific model reconstruction by efficiently integrating diverse omics data, reducing noise, and optimizing parameters, ultimately offering deeper insights into complex biological systems. In conclusion, while GEMs have significantly advanced computational modeling of metabolic systems, further progress can be made by integrating omics data and machine learning approaches.

References

Agren R, Bordel S, Mardinoglu A, Pornputtapong N, Nookaew I, Nielsen J (2012) Reconstruction of genome-scale active metabolic networks for 69 human cell types and 16 cancer types using INIT. PLoS Comput Biol 8(5):e1002518. https://doi.org/10.1371/journal.pcbi.1002518

Agren R, Mardinoglu A, Asplund A, Kampf C, Uhlen M, Nielsen J (2014) Identification of anti-cancer drugs for hepatocellular carcinoma through personalized genome-scale metabolic modeling. Mol Syst Biol 10(3):721. https://doi.org/10.1002/msb.145122

Ahn WS, Antoniewicz MR (2012) Towards dynamic metabolic flux analysis in CHO cell cultures. Biotechnol J 7(1):61–74. https://doi.org/10.1002/biot.201100052

Alyass A, Turcotte M, Meyre D (2015) From big data analysis to personalized medicine for all: challenges and opportunities. BMC Med Genom 8:1–12. https://doi.org/10.1186/s12920-015-0108-y

Angione C, Sullivan N, Zampieri G (2024) Machine learning applications in metabolic modeling. PLoS Comput Biol 15:e1007084. https://doi.org/10.1371/journal.pcbi.1007084

Antoniewicz MR (2015) Methods and advances in metabolic flux analysis: a mini-review. J Ind Microbiol Biotechnol 42(3):317–325. https://doi.org/10.1007/s10295-015-1585-x

Antoniewicz MR, Kelleher JK, Stephanopoulos G (2006) Determination of confidence intervals of metabolic fluxes estimated from stable isotope measurements. Metab Eng 8(4):324–337. https://doi.org/10.1016/j.ymben.2006.01.004

Argelaguet R, Velten B, Arnol D et al (2020) MOFA+: a statistical framework for comprehensive integration of multi-modal single-cell data. Genome Biol 21(1):111. https://genomebiology.biomedcentral.com/articles/10.1186/s13059-020-02015-1

Arnold A, Nikoloski Z (2014) Bottom-up metabolic reconstruction of Arabidopsis and its application to determining the metabolic costs of enzyme production. Plant Physiol 165(3):1380–1391. https://doi.org/10.1104/pp.114.235358

Bailey JE (1991) Toward a science of metabolic engineering. Science (New York, N.Y.) 252(5013):1668–1675. https://doi.org/10.1126/science.2047876

Ballard T, Wang Z, Li W, Shen L, Long Q (2024) Deep learning-based approaches for multi-omics data integration and prediction. BioData Mining 17(1):38. https://biodatamining.biomedcentral.com/articles/10.1186/s13040-024-00391-z

Beg QK, Vazquez A, Ernst J, de Menezes MA, Bar-Joseph Z, Barabási A-L, Oltvai ZN (2007) Intracellular crowding defines the mode and sequence of substrate uptake by Escherichia coli and constrains its metabolic activity. Proc Natl Acad Sci USA 104(31):12663–12668. https://doi.org/10.1073/pnas.0609845104

Canzar S, Brunk E, Hanson AD (2024) Reinforcement learning techniques in genome-scale metabolic modeling. Oxford University Press Journals

Cheah YE, Xu Y, Sacco SA, Babele PK, Zheng AO, Johnson CH, Young JD (2020) Systematic identification and elimination of flux bottlenecks in the aldehyde production pathway of Synechococcus elongatus PCC 7942. Metab Eng 60:56–65. https://doi.org/10.1016/j.ymben.2020.03.007

Cho JS, Gu C, Han TH, Ryu JY, Lee SY (2019) Reconstruction of context-specific genome-scale metabolic models using multiomics data to study metabolic rewiring. Curr Opin Syst Biol 15:1–11. https://doi.org/10.1016/j.coisb.2019.02.009

Chung NC, Ping P (2019) Machine learning and integrative analysis of biomedical big data. MDPI Genes 10(2):87. https://doi.org/10.3390/genes10020087

CLOSEgaps Framework (2021) Automated annotating and curating gaps with hypergraph-integrated deep learning methods. Research Square

Dias R, Pereira F, Oliveira J (2024) Deep learning models for metabolic flux prediction. Oxford University Press Bioinformatics

Flores JE, Claborne DM, Weller ZD, Webb-Robertson BJM, Waters KM, Bramer LM (2023) Missing data in multi-omics integration: Recent advances through artificial intelligence. Front Artif Intell 6:1098308. https://doi.org/10.3389/frai.2023.1098308

Foster CJ, Gopalakrishnan S, Antoniewicz MR, Maranas CD (2019) From Escherichia coli mutant 13C labeling data to a core kinetic model: A kinetic model parameterization pipeline. PLoS Comput Biol 15(9):e1007319. https://doi.org/10.1371/journal.pcbi.1007319

García Sarmina B, Sun G-H, Dong S-H (2023) Principal component analysis and t-distributed stochastic neighbor embedding analysis in the study of quantum approximate optimization algorithm entangled and non-entangled mixing operators. Entropy 25(11):1499. https://www.mdpi.com/1099-4300/25/11/1499

Ghahramani Z (2024) Unsupervised learning techniques for metabolic analysis. SpringerLink

Gonzalez JE, Antoniewicz MR (2017) Tracing metabolism from lignocellulosic biomass and gaseous substrates to products with stable-isotopes. Curr Opin Biotechnol 43:86–95. https://doi.org/10.1016/j.copbio.2016.10.002

Gurgen F, Celik D, Ozcan A (2021) Leveraging deep learning for GEM optimization. Research Square

Hardikkumar H (2024) Support vector machines and principal component analysis tutorial. Analytics Vidhya

Hu B, Canon S, Eloe-Fadrosh EA, Anubhav, Babinski M, Corilo Y, ... Chain PS (2022) Challenges in bioinformatics workflows for processing microbiome omics data at scale. Front Bioinform 1:826370. https://doi.org/10.3389/fbinf.2021.826370

Jerby L, Shlomi T, Ruppin E (2010) Computational reconstruction of tissue-specific metabolic models: application to human liver metabolism. Mol Syst Biol 6(1):401. https://doi.org/10.1038/msb.2010.56

Kaste JAM, Green A, Shachar-Hill Y (2023) Integrative teaching of metabolic modeling and flux analysis with interactive python modules. Biochem Mol Biol Educ 51(6):653–661. https://doi.org/10.1002/bmb.21777

Kumar VS, Maranas CD (2009) GrowMatch: an automated method for reconciling in silico/in vivo growth predictions. PLoS Comput Biol 5(3):e1000308. https://doi.org/10.1371/journal.pcbi.1000308

Leighty RW, Antoniewicz MR (2011) Dynamic metabolic flux analysis (DMFA): a framework for determining fluxes at metabolic non-steady state. Metab Eng 13(6):745–755. https://doi.org/10.1016/j.ymben.2011.09.010

Lewis NE, Nagarajan H, Palsson BO (2012) Constraining the metabolic genotype–phenotype relationship using a phylogeny of in silico methods. Nat Rev Microbiol 10(4):291–305. https://doi.org/10.1038/nrmicro2737

Li R, Li L, Xu Y, Yang J (2022) Machine learning meets omics: applications and perspectives. Brief Bioinform 23(1):bbab460. https://doi.org/10.1093/bib/bbab460

Liu X, Yang H, Ai C, Dong R, Ding Y, Yuan Q, Tang J, Guo F (2024) A generalizable framework for unlocking missing reactions in genome-scale metabolic networks using deep learning. https://doi.org/10.48550/arXiv.2409.13259

Long CP, Au J, Sandoval NR, Gebreselassie NA, Antoniewicz MR (2017) Enzyme I facilitates reverse flux from pyruvate to phosphoenolpyruvate in Escherichia coli. Nat Commun 8(1):14316. https://doi.org/10.1038/ncomms14316

Ma F, Jazmin LJ, Young JD, Allen DK (2014) Isotopically nonstationary 13C flux analysis of changes in Arabidopsis thaliana leaf metabolism due to high light acclimation. Proc Natl Acad Sci USA 111(47):16967–16972. https://doi.org/10.1073/pnas.1319485111

Maan K, Baghel R, Dhariwal S, Sharma A, Bakhshi R, Rana P (2023) Metabolomics and transcriptomics based multi-omics integration reveals radiation-induced altered pathway networking and underlying mechanism. npj Syst Biol Appl 9(1):42. https://doi.org/10.1038/s41540-023-00305-5

Mahadevan R, Edwards JS, Doyle FJ III (2002) Dynamic flux balance analysis of diauxic growth in Escherichia coli. Biophys J 83(3):1331–1340. https://doi.org/10.1016/s0006-3495(02)73903-9

Medlock GL, Papin JA (2020) Guiding the refinement of biochemical knowledgebases with ensembles of metabolic networks and machine learning. Cell Syst 10(1):109–119. https://doi.org/10.1016/j.cels.2019.11.006

Nanda P, Ghosh A (2021) Genome scale-differential flux analysis reveals deregulation of lung cell metabolism on SARS-CoV-2 infection. PLoS Comput Biol 17(4):e1008860. https://doi.org/10.1371/journal.pcbi.1008860

Neo C, Peipei P (2019) Integrating multi-omics datasets for comprehensive data-driven research. MDPI Genes 10(2):87

Niederberger P, Prasad R, Miozzari G, Kacser H (1992) A strategy for increasing an in vivo flux by genetic manipulations. The tryptophan system of yeast. Biochem J 287(2):473–479. https://doi.org/10.1042/bj2870473

Orth JD, Thiele I, Palsson BØ (2010) What is flux balance analysis? Nat Biotechnol 28(3):245–248. https://doi.org/10.1038/nbt.1614

Oyetunde T, Liu D, Martin HG, Tang YJ (2019) Machine learning framework for assessment of microbial factory performance. PloS ONE 14(1):e0210558. https://doi.org/10.1371/journal.pone.0210558

Price ND, Reed JL, Palsson BØ (2004) Genome-scale models of microbial cells: evaluating the consequences of constraints. Nature reviews microbiology 2(11):886–897

PLOS Computational Biology (2024) Machine and deep learning meet genome-scale metabolic modeling. PLoS Comput Biol 15:e1007084. https://doi.org/10.1371/journal.pcbi.1007084

Robinson JL, Kocabaş P, Wang H, Cholley P-E, Cook D, Nilsson A et al (2020) An atlas of human metabolism. Sci Signal 13(624):eaaz1482. https://doi.org/10.1126/scisignal.aaz1482

Roscher A, Troufflard S, Taghki AI (2014) In vivo NMR for ^{13}C metabolic flux analysis. Methods Mol Biol (Clifton, N.J.) 1090:143–152. https://doi.org/10.1007/978-1-62703-688-7_10

Roy S, Radivojevic T, Forrer M, Marti JM, Jonnalagadda V, Backman T, ... Garcia Martin H (2021) Multiomics data collection, visualization, and utilization for guiding metabolic engineering. Front Bioeng Biotechnol 9:612893. https://doi.org/10.3389/fbioe.2021.612893

Sauer U (2006) Metabolic networks in motion: 13C-based flux analysis. Mol Syst Biol 2(1):62. https://doi.org/10.1038/msb4100109

Schinn SM, Morrison C, Wei W, Zhang L, Lewis NE (2021) Systematic evaluation of parameters for genome-scale metabolic models of cultured mammalian cells. Metab Eng 66:21–30. https://doi.org/10.1016/j.ymben.2021.03.013
Segre D, Vitkup D, Church GM (2002) Analysis of optimality in natural and perturbed metabolic networks. Proc Natl Acad Sci 99(23):15112–15117
Sher D, Segrè D, Follows MJ (2024) Quantitative principles of microbial metabolism shared across scales. Nat Microbiol 9(8):1940–1953. https://doi.org/10.1038/s41564-024-01764-0
Singh A, Patel M, Zhao R (2024) Supervised and unsupervised models in biological networks. SpringerLink
Stephanopoulos G (1999) Metabolic fluxes and metabolic engineering. Metab Eng 1(1):1–11. https://doi.org/10.1006/mben.1998.0101
Stephanopoulos G, Vallino JJ (1991) Network rigidity and metabolic engineering in metabolite overproduction. Science 252(5013):1675–1681. https://doi.org/10.1126/science.1904627
Tang YJ, Chakraborty R, Martín HG, Chu J, Hazen TC, Keasling JD (2007a) Flux analysis of central metabolic pathways in Geobacter metallireducens during reduction of soluble Fe (III)-nitrilotriacetic acid. Appl Environ Microbiol 73(12):3859–3864. https://doi.org/10.1128/AEM.02986-06
Tang Y, Pingitore F, Mukhopadhyay A, Phan R, Hazen TC, Keasling JD (2007b) Pathway confirmation and flux analysis of central metabolic pathways in Desulfovibrio vulgaris Hildenborough using gas chromatography-mass spectrometry and Fourier transform-ion cyclotron resonance mass spectrometry. J Bacteriol 189(3):940–949. https://doi.org/10.1128/jb.00948-06
Tarzi C, Zampieri G, Sullivan N, Angione C (2024) Emerging methods for genome-scale metabolic modeling of microbial communities. Trends Endocrinol Metab 35(6):533–548. https://doi.org/10.1016/j.tem.2024.02.018
Tepper N, Shlomi T (2010) Predicting metabolic engineering knockout strategies for chemical production: accounting for competing pathways. Bioinformatics (Oxford, England) 26(4):536–543. https://doi.org/10.1093/bioinformatics/btp704
Thiele I, Palsson BØ (2010) A protocol for generating a high-quality genome-scale metabolic reconstruction. Nat Protoc 5(1):93–121. https://doi.org/10.1038/nprot.2009.203
Turanli B, Gulfidan G et al (2024) Genome-scale metabolic models in translational medicine: The current status and potential of machine learning. Molecular Omics 20:234–247. https://doi.org/10.1039/D3MO00152K
Vallino JJ, Stephanopoulos G (2000) Metabolic flux distributions in Corynebacterium glutamicum during growth and lysine overproduction. Biotechnol Bioeng 67(6):872–885. https://doi.org/10.1002/(SICI)1097-0290(20000320)67:6<872::AID-BIT21>3.0.CO;2-X
Wang X, Zhang C, Wang M, Lu W (2014) Genome-scale metabolic network reconstruction of Saccharopolyspora spinosa for spinosad production improvement. Microb Cell Factories 13:1–7. https://doi.org/10.1186/1475-2859-13-41
Wang Y, Hui S, Wondisford FE, Su X (2021) Utilizing tandem mass spectrometry for metabolic flux analysis. Lab Investig 101(4):423–429. https://doi.org/10.1038/s41374-020-00488-z
Xiong W, Lee TC, Rommelfanger S, Gjersing E, Cano M, Maness PC et al (2015) Phosphoketolase pathway contributes to carbon metabolism in cyanobacteria. Nat Plants 2(1):1–8. https://doi.org/10.1038/nplants.2015.187
Zamboni N (2011) 13C metabolic flux analysis in complex systems. Curr Opin Biotechnol 22(1):103–108. https://doi.org/10.1016/j.copbio.2010.08.009
Zeng H, Rohani R, Huang WE, Yang A (2021) Understanding and mathematical modelling of cellular resource allocation in microorganisms: a comparative synthesis. BMC Bioinformatics 22(1):467. https://doi.org/10.1186/s12859-021-04382-3
Zeng J, Li H, Wang Y (2024) Advances in deep learning for multi-omics data integration. Bioinformatics 23(1):549–560. https://doi.org/10.1093/bioinformatics/bbab460

Proteome Engineering: An Evolutionary Perspective of Biochemical Systems

Mukul Machhindra Barwant, Balwant Singh, Mohit, Shalagha Sharma, Varsha S. Patil, Suresh Babu Naidu Krishna, and Venugopal Krishnan

Abstract

The rapid advancements in omics technologies, including proteomics, genomics, and transcriptomics, have revolutionized the field of metabolic engineering by offering unprecedented insights into cellular mechanisms and metabolic pathways. Synthetic biology tools, when integrated with these omics approaches, enable the rational design and optimization of microbial and plant systems for the efficient production of value-added products such as biofuels, pharmaceuti-

M. M. Barwant (✉)
Department of Botany, Sanjivani Arts Commerce and Science College Kopargoan, Ahilynagar, Maharashtra, India

B. Singh
Department of Botany, K.S. Saket PG College Ayodhya, Dr. Ram Manohar Lohia Avadh University Ayodhya, Faizabad, Uttar Pradesh, India

Mohit
Department of Prosthodontics, Faculty of Dental Science, King George Medical University, Lucknow, UP, India

Center for Advance Research, Faculty of Medicine, King George Medical University, Lucknow, UP, India

S. Sharma
School of Biological Science, Shobhit Deemed to be University, Meerut, Uttar Pradesh, India

V. S. Patil
Department of Botany, Sanjivani Arts, Commerce and Science College Rahata, Ahmednagar, Maharashtra, India

S. B. N. Krishna
Faculty of Health Sciences, Durban University of Technology, Durban, South Africa

V. Krishnan
HIFU Lab, Department of Radiology, UT Southwestern Medical Center, Dallas, TX, USA

S. Kaur et al. (eds.), *Microbial Metabolomics*,
https://doi.org/10.1007/978-981-96-4824-5_6

cals, and specialty chemicals. Genomics provides a comprehensive understanding of the genetic architecture, while transcriptomics reveals dynamic changes in gene expression in response to environmental stimuli or genetic modifications. Proteomics, in turn, offers a detailed view of the functional protein landscape, enabling the identification of bottlenecks and the fine-tuning of metabolic pathways. This chapter highlights the synergistic application of these omics platforms to enhance synthetic biology projects, emphasizing case studies of successful metabolic engineering endeavors. Furthermore, it discusses cutting-edge tools, including CRISPR-Cas systems, computational modelling, and machine learning, which are reshaping the design-build-test-learn (DBTL) cycle in synthetic biology. Challenges such as data integration, system complexity, and scalability are also addressed, along with potential solutions. By exploring the interplay between omics technologies and synthetic biology, this chapter underscores the transformative potential of a systems biology approach in overcoming metabolic engineering bottlenecks, paving the way for sustainable and economically viable biotechnological solutions.

Keywords

Biotechnology · Metabolomics · Metabolic pathway · Proteomics · Sustainability

1 Introduction

Metabolic engineering is at the forefront of modern biotechnology, offering transformative solutions across a spectrum of industries, from pharmaceuticals and biofuels to agriculture and environmental management. By systematically redesigning an organism's metabolic pathways, metabolic engineering enables optimized production of valuable compounds, degradation of environmental pollutants, and synthesis of novel biomolecules (Subramanian et al. 2020). Central to the success of these endeavors is the integration of synthetic biological tools with comprehensive biological data derived from proteomics, genomics, and transcriptomics. This chapter delves into the pivotal roles that these "omics" disciplines play in harnessing synthetic biology for effective metabolic engineering projects (Zhao et al. 2020).

Metabolic engineering has emerged as a multidisciplinary field that combines principles from biology, engineering, and computational sciences to modify cellular processes for desired outcomes (Zhang et al. 2017a, b). Traditional approaches to metabolic engineering rely heavily on trial-and-error techniques, which are time-consuming and inefficient. The advent of synthetic biology has revolutionized this landscape by providing a toolkit for standardized genetic parts, modular systems, and precise gene editing technologies that facilitate the rational design and construction of complex biological systems (Chistoserdova 2010; Baidoo and Teixeira 2019).

This broad range of applications in metabolic engineering highlights its importance. Therapeutic chemicals and medications are produced by modified microorganisms in the pharmaceutical sector (Roberts et al. 2012). Metabolic engineering helps the energy industry create biofuels, which provide sustainable substitutes for fossil fuels. Furthermore, engineered bacteria are essential for bioremediation, as they break down pollutants in the environment and reduce pollution. There is great potential to solve some of the most important problems of our day if organisms can be modified to carry out particular jobs very effectively (VerBerkmoes et al. 2009a, b).

Designing and building new biological systems or reprogramming existing ones for particular uses, synthetic biology has become a transformative field by combining concepts from biology, engineering, and computer sciences. Metabolic engineering is a crucial use in synthetic biology, with the goal of maximizing biological processes to produce valuable substances such as specialty chemicals, medicines, and biofuels (Keasling 2010). Despite tremendous progress, there are still difficulties in comprehending and modifying the intricacy of metabolic networks, necessitating a system-level strategy.

Omics technologies, such as genomics, proteomics, and transcriptomics, have become indispensable for driving innovation in synthetic biology and metabolic engineering. These high-throughput approaches provide comprehensive datasets that elucidate the molecular mechanisms governing cellular functions and metabolic pathways. Advances in genomics, such as next-generation sequencing (NGS), provide insights into the genetic architecture of organisms, including the identification of genes, regulatory elements, and genome-wide interactions (Fig. 1) (Mardis 2017).

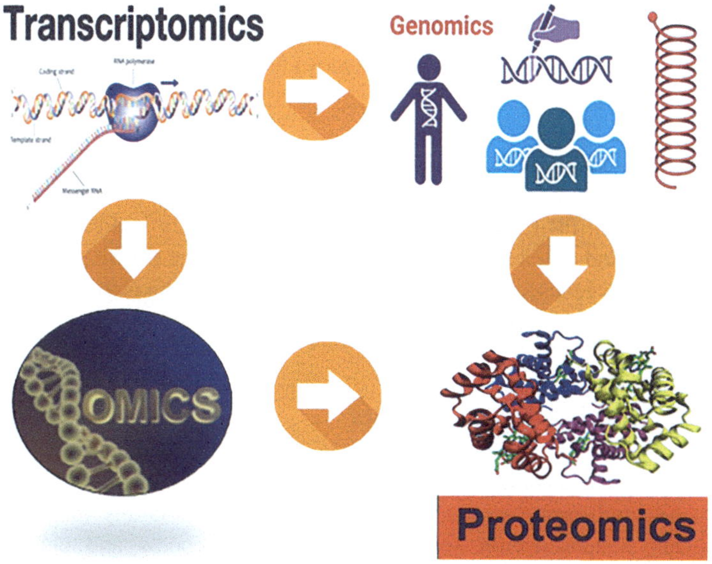

Fig. 1 Omics technologies or advancement in technology

Proteomics complements genomics by focusing on the functional outputs of genes and proteins. Proteins are the primary effectors of cellular metabolism, and their abundance, modifications, and interactions directly affect metabolic flux. Advances in mass spectrometry-based proteomics have enabled the detailed mapping of proteomes, providing critical data for metabolic pathway reconstruction (Aebersold and Mann 2016). Similarly, transcriptomics, which analyzes RNA expression profiles, offers a dynamic view of gene regulation under different conditions, facilitating the identification of regulatory bottlenecks in metabolic pathways (Wang et al. 2009).

By integrating omics technologies with synthetic biology tools, researchers can achieve a system-level understanding of metabolism, enabling more effective pathway engineering. For example, genome editing technologies, such as CRISPR-Cas systems, combined with omics data, have been used to fine-tune metabolic pathways with high precision (Doudna and Charpentier 2014). Furthermore, computational modelling driven by omics datasets allows for the simulation and prediction of metabolic fluxes, guiding rational engineering strategies (Lewis et al. 2012).

Synthetic biology provides a suite of tools that are instrumental in the design, construction, and optimization of metabolic pathways. These tools encompass a variety of techniques and technologies, including:

1.1 Genetic Circuits and Pathway Design

Synthetic biology enables the construction of genetic circuits that regulate gene expression in response to specific stimuli. By precisely designing and assembling metabolic pathways, scientists can control the flux of metabolites and optimize the production of target compounds (Chistoserdova 2010).

1.2 Gene Editing Technologies

Tools such as CRISPR-Cas9 have revolutionized the ability to edit genomes with high specificity and efficiency. These technologies facilitate the insertion, deletion, or modification of genes, enabling the fine-tuning of metabolic pathways to enhance the desired outcomes (Urgen and Mann 2011).

1.3 Standardized Genetic Parts

The development of standardized biological parts, such as promoters, ribosome-binding sites, and terminators, allows for the modular assembly of genetic constructs. This standardization simplifies the design process and enhances the predictability and reproducibility of the engineered systems (Roberts et al. 2012).

1.4 Computational Modelling and Design

Computational tools and models can aid in predicting the behavior of engineered metabolic pathways, optimizing gene expression levels, and identifying potential bottlenecks. These models are essential for guiding experimental design and improving the efficiency of metabolic engineering projects (Pagani et al. 2012).

2 Omics Technologies/Advances in Omics Technologies

Omics technologies, including genomics, transcriptomics, proteomics, and metabolomics, have emerged as essential tools in systems biology to enable a comprehensive understanding of cellular processes at multiple levels. These high-throughput approaches generate large datasets that serve as a foundation for designing and implementing synthetic biological solutions in metabolic engineering. Each omics technology provides a unique perspective, collectively forming a holistic view of the cellular systems.

Although synthetic biology tools provide the means to engineer biological systems with unprecedented precision, their effective utilization hinges on a deep understanding of the underlying biological processes. Proteomics, genomics, and transcriptomics play a crucial role.

To effectively engineer metabolism, it is imperative to have a comprehensive understanding of an organism's biological systems at multiple levels. Proteomics, genomics, and transcriptomics offer complementary perspectives that inform the design and optimization of metabolic engineering strategies. Figure 1 represents the following advancement in omics:

2.1 Genomics

Genomics involves the study of an organism's complete DNA set, encompassing all genes and regulatory elements. Advances in genomic sequencing technologies have enabled rapid and accurate determination of genomic sequences, facilitating the identification of genes involved in specific metabolic pathways. Genomic data provides a blueprint for understanding an organism's potential capabilities and limitations, informing the selection of targets for genetic modification (Roberts et al. 2012; Chekmeneva et al. 2017).

Genomics involves the study of an organism's entire genome and offers insights into its genetic blueprint. Advances in sequencing technologies, particularly next-generation sequencing (NGS), have transformed genomics into a cornerstone of modern biology. Genomics facilitates the identification of genes, regulatory elements, and structural variations that are critical for metabolic function (Mardis 2017). Additionally, genome editing tools such as CRISPR-Cas9 are often guided by genomic data to target specific loci for pathway optimization (Doudna and Charpentier 2014).

Comparative genomics enables researchers to identify conserved genes and pathways across species, thereby offering insights into the evolutionary adaptations of metabolic systems. For example, genomics has been pivotal in engineering microbial platforms for biofuel production by uncovering pathways in naturally efficient organisms (Keasling 2010).

2.2 Transcriptomics

Transcriptomics focuses on a complete set of RNA molecules expressed by an organism under specific conditions. Technologies such as RNA-Seq provide a high-resolution view of gene expression dynamics, revealing regulatory networks and identifying key genes that respond to environmental or metabolic stimuli (Wang et al. 2009; Vavricka et al. 2020).

In metabolic engineering, transcriptomics is instrumental in identifying rate-limiting steps within biosynthetic pathways. For example, transcriptomic analysis can reveal overexpressed or under-expressed genes in engineered strains, allowing researchers to fine-tune expression levels to improve metabolite production (Choi et al. 2010).

2.3 Proteomics

Proteomics involves large-scale study of proteins, including their expression levels, structures, functions, and interactions. Proteins are functional executors of cellular processes, and their abundance and activity are directly linked to metabolic functions. Proteomic analyses provide insights into the actual protein expression levels, posttranslational modifications, and protein-protein interactions that govern metabolic pathways. Understanding the proteomic landscape is essential for identifying potential bottlenecks, optimizing enzyme activities, and ensuring proper functioning of engineered pathways (Basher Edward 2021).

Together, these "omics" disciplines provide a holistic view of the biological system, enabling a systems biology approach to metabolic engineering. By integrating data from genomics, transcriptomics, and proteomics, researchers can gain a comprehensive understanding of an organism's biology, identify key targets for modification, and design more effective engineering strategies (VerBerkmoes et al. 2009a, b).

By analyzing all the proteins that an organism produces, proteomics sheds light on the key participants in cellular metabolism. The proteome is dynamic and changes with the environment, tissues, and developmental stages, in contrast to the static genome. The gold standard for describing protein abundance, changes, and interactions is mass spectrometry (MS) proteomics (Aebersold and Mann 2016).

Proteomics bridges the gap between genotypes and phenotypes by directly linking genetic information to biochemical activity. In synthetic biology, proteomic data guide the design of synthetic pathways by identifying enzymes with desirable properties or by uncovering bottlenecks in metabolic flux (Dulermo et al. 2015).

2.4 Integration of Omics Technologies

Although individual omics platforms provide valuable insights, their integration is critical for addressing the complexity of metabolic networks. Multi-omics approaches combine genomic, transcriptomic, and proteomic data to create comprehensive models of cellular behavior, enabling predictive metabolic engineering (Lewis et al. 2012). For instance, integrated omics studies have been employed to optimize the production of secondary metabolites by correlating gene expression with protein activity and metabolite accumulation (Zhang et al. 2017a, b).

Omics integration also enhances the precision of synthetic biology tools such as metabolic flux analysis and computational pathway design, fostering more efficient and predictable engineering outcomes. The integration of proteomic, genomic, and transcriptomic data with synthetic biological tools represents a paradigm shift in metabolic engineering. This integrative approach facilitates a more rational and informed design process, thereby enhancing the efficiency and success rate of engineering projects. Several key aspects underscore the importance of the integration.

2.4.1 Identification of Engineering Targets

"Omics" data enable the identification of genes, transcripts, and proteins that are critical for desired metabolic functions. Genomic analyses can reveal gene clusters involved in specific pathways, transcriptomic studies can identify genes that are differentially expressed under certain conditions, and proteomic data can identify enzymes present at suboptimal levels. This comprehensive identification process ensures that engineering efforts are directed toward the most impactful targets (Subramanian et al. 2020).

2.4.2 Understanding System Dynamics

Metabolic pathways are dynamic and interconnected, with complex regulatory mechanisms governing their behavior. Transcriptomics and proteomics provide temporal and quantitative data on gene expression and protein levels, respectively, allowing researchers to understand how a system responds to genetic modifications or environmental changes. This understanding is crucial for predicting the effects of engineering interventions and designing strategies to maintain system stability and robustness (Zhang et al. 2017a, b).

2.4.3 Optimization of Pathway Performance

"Omics" data inform the optimization of metabolic pathways by identifying rate-limiting steps, potential feedback inhibition points, and interactions between pathways. For example, proteomic analyses can reveal whether certain enzymes are present at limiting concentrations, suggesting targets for overexpression. Transcriptomic data can indicate whether genes are appropriately regulated in response to pathway flux, guiding the design of regulatory circuits to achieve optimal expression levels (Baidoo and Teixeira 2019).

2.4.4 Detection of Unintended Consequences

Engineering metabolic pathways can lead to unintended perturbations in cellular systems. "Omics" analyses provide a means to monitor these changes comprehensively. Genomic stability can be assessed to ensure that no off-target mutations occur. Transcriptomic and proteomic profiles can reveal unexpected shifts in gene expression or protein abundance, allowing the identification and mitigation of potential issues before compromising the project (VerBerkmoes et al. 2009a, b).

2.4.5 Facilitation of Iterative Design-Build-Test-Learn Cycles

The integration of "omics" data supports an iterative approach to metabolic engineering, where design, construction, testing, and learning are conducted in cycles. "Omics" analyses provide the data necessary to assess the outcomes of engineering interventions, learn from the results, and refine the design in subsequent iterations. This iterative process enhances the efficiency and effectiveness of metabolic engineering projects, leading to more robust and optimized engineered systems (Roberts et al. 2012; Zhao et al. 2020).

Proteomics, genomics, and transcriptomics have been successfully integrated with synthetic biology methods in metabolic engineering projects, as numerous studies have shown. For example, thorough "omics" analyses have been beneficial in the engineering of *Escherichia coli* for the generation of biofuels. Proteomic results indicated enzyme levels that require optimization, transcriptome analysis showed regulatory responses to pathway alterations, and genomic investigations identified important genes involved in fatty acid metabolism (Roberts et al. 2012; Zhang et al. 2017a, b). Researchers have been able to greatly increase the yields of biofuel production by combining this knowledge with synthetic biology technologies, including synthetic regulatory circuits and CRISPR-based gene editing (Vavricka et al. 2020).

Similarly, in the pharmaceutical industry, the production of complex natural products often requires engineering of multiple biosynthetic pathways. Genomic sequencing of producer organisms identifies biosynthetic gene clusters, transcriptomic analyses elucidate the conditions under which these clusters are expressed, and proteomic studies ensure that the necessary enzymes are present and active (Zhao et al. 2020). Synthetic biology tools facilitate the assembly and optimization of these pathways in host organisms, leading to the efficient and scalable production of valuable pharmaceuticals.

The convergence of proteomics, genomics, and transcriptomics with synthetic biology tools represents powerful synergy in the field of metabolic engineering. By leveraging comprehensive biological data, researchers can design more precise and effective engineering interventions, optimize metabolic pathways, and achieve higher yields of desired products (Baidoo and Teixeira 2019). This integrative approach not only enhances the success rate of metabolic engineering projects but also expands the horizons of what is achievable through biotechnology. As our understanding of biological systems continues to deepen and our technological capabilities advance, the role of proteomics, genomics, and transcriptomics in the utilization of synthetic biology tools will undoubtedly become even more pivotal, driving innovation and enabling the realization of new possibilities in metabolic engineering (Roberts et al. 2012).

3 Integration of Genomics, Transcriptomics, and Proteomics

Genotype must predict accurately to phenotype: biological engineering necessitates testing and validation of modified or synthesized genomes (the study of the genomics: "how much genome"). Genome engineering could be properly analyzed, although based on transcriptome, the complete set of RNA transcripts expressed from a genome, otherwise known as the study of transcriptomics. In addition, proteomics and metabolomics have gained significant attention because of the provision of metabolic information related to both functions and phenotypes (Baidoo 2019). Four decades ago, scientists came to realize that biochemical information flows in biological systems are not strictly one-way from the genome to metabolome but a complex interplay of the "omics" (Roberts et al. 2012). Therefore, a multi-omics approach is required to elucidate the chemical structure, function, development, adaptation, and evolution of biological systems for a deeper understanding of the principles of life (Baidoo 2019).

Genes consist of DNA composed of four fundamental nucleotides: guanine, adenine, cytosine, and thymine. In contrast, proteins are constructed from a combination of 32 amino acids, and their metabolites exhibit greater variability in their chemical structures (Wang et al. 2010). It is analytically less challenging to perform genomics and transcriptomics than metabolomics and proteomics (Aizat et al. 2018).

Research over the last few decades has demonstrated that transcriptomics and genomics alone cannot fully describe complex biological systems because genetic information can raise more questions than it can provide answers. For example, while genomics can measure genotypes (define genes and their interactions), it is unable to explain phenotypes. Therefore, to promote a better understanding of biological systems, the focus shifts to the application of additional "omics" approaches, like as proteomics and metabolomics, which can close the gap between genetic potential and final phenotype (Smith and Figeys 2006; Wilmes et al. 2015). Although proteomics (translation) and transcriptomics (transcription) offer insights into gene expression, the latter establishes a direct correlation between the genotype and phenotype. A snapshot of a cell's true metabolic and physiological status is provided by the metabolome, which offers phenotypic information and an immediate reaction to genetic and/or environmental perturbations (Tang 2011).

However, metabolomics alone cannot measure changes at the gene level and relate them to phenotypes or outwardly visible traits of organisms, which are first influenced by the genotype (Fiehn 2001). Therefore, "omics" data must be merged to completely understand the molecular composition of an organism and find new substances and pathways.

By sequencing the target genome, genomic and transcriptomic approaches aim to explore inter-individual differences at the germline and somatic levels. First-generation Sanger sequencing (Moniruzzaman et al. 2020), second-generation massively parallel sequencing (also called next-generation sequencing, or NGS) (Chen et al. 2021), DNA microarray technology (Thomas et al. 2020), and third-generation long-read sequencing (TGS) (Ou et al. 2020) have made it possible to sequence the entire genome or exome in sufficient detail to describe the mutational landscape of a particular sample.

DNA microarray technology was first developed by Schena et al. (1995), in which samples were labeled with fluorescent dyes for detection following hybridization and thousands of probes were glued to a surface (Kinaret et al. 2020). There are two types of DNA microarrays: two-channel and one-channel arrays. The two most common commercial arrays are Agilent (Painter et al. 2012) and Affymetrix GeneChip (Graham et al. 2011).

The first generation of DNA sequencing or Sanger sequencing was performed in 1977 (Sanger et al. 1977). Its foundation is the way DNA polymerase selectively incorporates chain-terminating di-deoxynucleotides during in vitro DNA replication. Owing to its high precision per base (~99.999%) and comparatively long read length (up to ~1000 bp) (Shendure and Ji 2008), Sanger sequencing has dominated this field for approximately 30 years and has been utilized to make several significant achievements, including the completion of the Human Genome Project (Frederick Sanger 2004) (Table 1).

Proteomic studies of proteins and peptides produced by cells at different stages of their life cycles and development, as well as in biological systems under certain growth conditions, are known as proteomics. Proteomics can also shed light on the temporal dynamics of posttranslational modification or protein expression levels (VerBerkmoes et al. 2009a, b).

Protein engineering is the process of modifying the protein structure through targeted or random mutagenesis to provide optimal cofactor utilization, reduced substrate competition, greater catalytic rates, better substrate sensitivity, and less product inhibition (Pleiss 2011). Several rounds of directed evolution are frequently involved in engineering processes (Cobb et al. 2012). These changes have the potential to significantly increase the yield and titer of metabolically generated molecules (Table 2).

4 Recent Trends in Synthetic Biology

To create, build, and reengineer biological organisms, the interdisciplinary area of synthetic biology integrates concepts from computer science, engineering, and biology (Garner 2021). According to Roberts et al. (2013), these endeavors include the development of new biological components, tools, and systems as well as the modification of pre-existing natural biological systems for particular uses. Since its inception in the early 2000s, this area has been greatly influenced by developments in systems biology, genetic engineering, and molecular biology (Chuang et al. 2010). The capacity to alter and control biological systems has been transformed by

Table 1 Genomics and transcriptomics analysis techniques

Technique	Applications	References
Metagenomics	Actively employed in agricultural research to investigate different bacteria that can promote the cycling of macro- and micronutrients and the release of vital enzymes that improve crop productivity, as well as to comprehend the microbial communities in the soil system	Durot et al. (2009), Cupples (2005)
Oxford Nanopore Technologies (ONT)	Identify alterations in DNA and RNA. To use long-range information technologies, such as synthetic long reads, Hi-C, and 10x Genomics. To create new algorithms that streamline the labor-intensive assembly and polishing process	Somerville et al. (2018)
Fluorescence-activated cell sorting (FACS)	It makes it possible to sort individual cells and enrich particular cells to produce high-resolution transcriptional and gene expression maps	Kambale et al. (2019)
Metabolic pathway analysis	Thermodynamic favorability of each step, reaction processes, and substrate binding selectivity are some of the reasons that make pathway analysis useful for comparing several approaches to a desired output	Marcheschi et al. (2012)
Quantum mechanical (QM) modelling	Examine whether the intended or anticipated enzymatic reactions are likely to happen, what the reaction's potential energy barriers are, and what the chemical transition state might resemble	Acevedo and Jorgensen (2010)
Rosetta, a molecular modelling program	With the use of this tool, it is possible to analyze billions of protein mutants by performing site-saturation mutagenesis at several sites in a protein at once	Privett et al. (2012)
Transcription factor engineering for complex phenotypes	Multiple genes can have their expression altered at the exact same time, which might result in drastically distinct cell phenotypes. For one to accomplish this, mutagenesis of the TF protein—which affects numerous distant genes—is crucial	Lanza and Alper (2011)

the development of reliable tools and approaches made possible by this interdisciplinary convergence (Koide et al. 2009).

A cornerstone of synthetic biology is the concept of standardization, particularly through the creation of standardized biological components known as BioBricks (Radde et al. 2024). These DNA sequences, which are engineered to perform specific functions, can be assembled in a modular fashion to construct complex genetic circuits and systems with predictable behaviors (Shakiba et al. 2021). The BioBrick standard, introduced by Tom Knight in 2003, establishes a framework for the interchangeable use of genetic components, thereby promoting the sharing and collaborative development of biological systems within the scientific community (Shetty et al. 2008). This standardization has been instrumental in transforming biology from a descriptive science into a more quantitative and engineering-driven discipline. The field also emphasizes the application of engineering principles, such as abstraction, modularity, and iterative design processes, particularly the design-build-test (DBT) cycle (Katz et al. 2021).

Table 2 Proteomics tools used in metabolic pathway analysis

Proteomics tool	Description	Application in metabolic pathways	References
Mass spectrometry (MS)	Analytical technique used for protein identification and quantification	Identification of pathway-specific enzymes and quantification of their abundance in engineered systems	Smith et al. (2021)
Tandem mass tag (TMT)	Isobaric labeling for comparative proteomic analysis	Comparative analysis of metabolic shifts in response to genetic modifications or environmental conditions	Johnson et al. (2020)
2D gel electrophoresis	Separation of proteins based on their isoelectric point and molecular weight	Mapping and analysis of pathway-specific proteins in microbial systems	Müller et al. (2019)
SWATH-MS	Data-independent acquisition method for comprehensive proteomic profiling	Time-resolved analysis of metabolic pathways under varying conditions	Zhang et al. (2022)
Liquid chromatography-MS	Combination of liquid chromatography with MS for detailed protein analysis	Identification and quantification of low-abundance enzymes in metabolic pathways	Patel and Gupta (2021)
Protein-Protein interaction networks	Bioinformatics tools to study protein interactions	Mapping of enzyme-enzyme interactions critical for pathway regulation and optimization	Wang et al. (2021)
iTRAQ	Isobaric tags for relative and absolute quantification	Quantitative proteomics for detecting pathway bottlenecks in synthetic biology projects	Chen et al. (2020)

Modularity allows the recombination of standardized parts in various configurations, enabling the creation of diverse functional outputs from a limited set of components (Melo et al. 2016). The DBT (design-build-test) cycle is a systematic approach in which designs are iteratively built and tested, with each iteration refining the system based on the experimental feedback. This cycle is crucial for optimizing synthetic constructs and ensuring that they function as intended within the complex milieu of living cells (Katz et al. 2021). The advent of advanced DNA synthesis and sequencing technologies has led to exponential growth in synthetic biology. The ability to synthesize long, custom de novo DNA sequences has enabled the creation of entirely new genomes, as exemplified by the synthesis of the *Mycoplasma mycoides* genome (Ma et al. 2012). This groundbreaking achievement demonstrates the feasibility of constructing life forms with tailored genetic content, opening new avenues for biological research and biotechnology (Hughes and

Ellington 2017). Moreover, these technological advancements have significantly reduced the cost and time required for DNA synthesis, making it accessible for academic and industrial research.

Synthetic biology has far-reaching implications across multiple domains, including medicine, agriculture, and environmental sustainability (Douglas et al. 2013). In medicine, synthetic biology has revolutionized the production of pharmaceuticals by engineering microorganisms to produce therapeutic compounds. Notable examples include the microbial synthesis of insulin and artemisinin, an antimalarial drug, both of which have had significant impacts on global health (Yan et al. 2023). In agriculture, synthetic biology offers the potential to engineer crops with enhanced traits such as drought resistance, improved nutritional content, and resistance to pests and diseases. This could play a critical role in addressing the food security challenges posed by the growing global population and climate change (Zaidi et al. 2020). Synthetic biology is being harnessed to develop bio-based solutions for pollution mitigation and renewable energy production. Engineered microbes are being explored for their ability to degrade environmental pollutants and produce biofuels from renewable biomass, thereby contributing to a more sustainable and circular economy (Aminian-Dehkordi et al. 2023).

As synthetic biology continues to advance, it also raises important ethical, social, and safety considerations. The ability to create and modify life forms introduces unprecedented risks, including the potential for unintended ecological impacts, biosecurity concerns, and ethical implications of creating synthetic organisms (Ou and Guo 2023). These concerns necessitate the development of responsible governance frameworks, public engagement, and robust regulatory mechanisms to ensure that the benefits of synthetic biology are maximized while minimizing potential harm (Trump et al. 2020). The ongoing discourse among scientists, policymakers, and the public is crucial for addressing these challenges and guiding the responsible development of synthetic biology. Synthetic biology represents a transformative approach for understanding and manipulating life at the molecular level. By integrating principles from multiple disciplines, this field has enabled the development of innovative tools and technologies that are driving progress in medicine, agriculture, and environmental sustainability. However, the rapid pace of advancement underscores the need for careful consideration of ethical and safety implications, ensuring that synthetic biology can be harnessed for the greater good.

5 Omics in Metabolic Engineering

Systems biology powered by "omics" is a fundamental part of biotechnology-based approaches for understanding and analyzing metabolic networks, such as metabolic engineering and synthetic biology. By assessing the information flow from one "omics" level to another, the integration of "omics" omics data creates a link between the genotype and phenotype (Subramanian et al. 2020). Furthermore, the incorporation of "omics" approaches is essential to address unresolved biological questions that accelerate our understanding of the system as a whole and improve the use of systems metabolic engineering tools in commercial settings (Zhao et al. 2020).

Proteins and metabolites, which can originate from tissues, cells, and organs, constitute the majority of bioproducts produced by biomanufacturing processes. Natural or genetically modified biological systems, metabolic engineering, synthetic biology, and protein engineering are sources of these bioproducts (Zhang et al. 2017a, b). Synthetic biology and metabolic engineering can benefit from systems biology methods, which in turn use "omics" technology to describe and understand metabolic networks. Such high-end information obtained from omics-driven systems biology research can be used to drive progress toward metabolic engineering and synthetic biology technologies. This will make it easier to manipulate complex biological systems in the most reliable ways toward creating consistent industrial biomanufacturing platforms (Roberts et al. 2012). Biological engineering requires the prediction of phenotypes based on genotypes to be highly accurate. Hence, the evaluation of genome engineering must involve the testing and validation of changed and constructed genomes (genomics), as well as the exploration of the transcriptome expressed from the genome (transcriptomics). Proteomics and metabolomics have been in great focus as they have been capable of providing metabolic information that is pertinent to both function and phenotype (Baidoo and Teixeira Benites 2019).

In living things, the flow of biochemical information involves several interactions between the different "omes" rather than only going from the genome to the metabolome (Roberts et al. 2012). Therefore, for a deeper understanding of the principles of life, a multi-omics approach is required to reveal the chemical structure, function, growth, adaptability, and evolution of biological systems. A challenge in traditional genomics is that not all microorganisms can be grown in the laboratory for analysis. Moreover, strains that are separated may exhibit different behaviors when grown in a laboratory compared to their behavior in the wild. Therefore, the development of cultivation-independent methods for studying microbial communities is urgently required (VerBerkmoes et al. 2009a, b). Bypassing cultivation and metagenomics can analyze genetic material in a variety of samples from various organisms and entire microbial communities to solve problems and help discover new genes, enzymes, and metabolic pathways. Sequence-based screening and function-based screening are two types of metagenomic techniques used to identify novel natural genes and chemicals in environmental materials (Chistoserdova 2010).

The establishment of desired and expected traits is crucial in synthetic biology; hence, the accurate and adaptable control of gene expression is crucial. Recent developments in genome editing and genomics have also had a significant impact on biological engineering, providing access to cutting-edge tools for altering and developing natural systems. Additionally, advancements in the production of novel compounds from scratch and the targeting of genes in real organisms have enabled the efficient assessment of model-based theories. Additionally, high-throughput DNA sequencing and extensive bimolecular modelling of metabolic and signaling networks in both natural and modified strains are made possible by genomics (Pagani et al. 2012).

The study of proteins and peptides produced by cells in many biological systems and at different stages of development under growth conditions is the focus of proteomics. According to VerBerkmoes et al. (2009a, b), proteomics is also used to detect posttranslational modifications or temporal variations in protein expression levels. Proteomics data analysis tools are frequently used for bioinformatics-based protein identification and quantification as well as bioinformatics-based proteomics data processing. To link the MS data gathered from peptide/protein analysis to their base peptides/proteins and within silico projected intact masses and fragmentation patterns, several algorithm-based programs, including Tandem, Skyline, UniProt, UniNovo, and SWPepNovo, were created (Urgen Cox and Mann 2011).

The detection sensitivity and dynamic range of metabolites in tissues and biofluids during human research have been significantly improved by developments in nano-electrospray-ionization and direct-infusion nano-electrospray high-resolution mass spectrometry (Chekmeneva et al. 2017). Multivariate methods, such as principal component analysis (PCA) and partial least squares (PLS) analysis, are used to examine vast volumes of metabolic profile data (i.e., reveal clustering based on characteristics). Furthermore, to interpret metabolomics data, solve some of the most challenging biological puzzles, and determine the optimal conditions for biological systems, advanced pathway analysis methods are needed. These techniques allow systems biologists to leverage metabolomics data as a resource to support the iterative cyclical cycles of hypothesis creation and testing (Vavricka et al. 2020). New developments in metabolomics tools have been made to achieve this goal, including WebSpecmine (Cardoso et al. 2019), MetaboAnalyst 4.0 (Chong et al. 2018), and SECIM (Kirpich et al. 2018).

To improve desired phenotypic features like color, taste, drought tolerance, pesticide resistance, insecticide resistance, etc., "omics" technologies are also being used in agricultural biotechnology. In addition to producing food crops with higher nutritional content, "omics" has a major influence on raising agricultural output, quality, and consistency (Van Emon 2019). "Omics" technologies have been crucial to the development of crops and the metabolic engineering of biofuels and bioproducts. Proteomics and metabolomics are frequently used to examine altered metabolic pathways in microbial hosts. It seems that the development of next-generation DNA and mRNA vaccines against viruses such as SARS-CoV-2 depends on data from genome sequencing initiatives. However, transcriptome, proteomic, and metabolomic data are being used to guide the development of COVID-19 medicinal medicines. The rise of transcriptomics, proteomics, and metabolomics in the field of genomics shows the development of robust "omics" technologies and methodologies, even though the number of biotechnology publications in this field has increased steadily over the past 20 years (Basher and Edward 2021).

6 Challenges and Future Directions

With the growing number and variety of molecular measurements, many new omics technologies will probably have an impact on how omics-based tests are developed in the future. A view of the transcriptome with greater resolution is now possible thanks to developments in RNA sequencing technology. These innovative methods may make it easier to create more cutting-edge molecular diagnostics. It may be possible to develop omics-based tests in the future that use RNA editing events, tiny noncoding RNAs, or alternative splice variants that were not detectable with previous hybridization-based technologies such as microarrays. As a result, it is now possible to find uncommon or yet undiscovered mutations that may have important experimental implications, through the sequencing of DNA. Next-generation sequencing methods hold great promise for the identification of whole DNA and RNA sequences, as well as for the high-throughput identification of epigenetic and posttranscriptional alterations to DNA and RNA, respectively.

Significant progress has been made in tissue expression research and antibody-based protein identification (Ayoglu et al. 2011). Of the 20,300 gene-coded proteins, antibody data for more than 12,000 are available in the Human Protein Atlas. The Protein Atlas can help with experiment planning, and the Human Proteome Project will integrate it with mass spectrometry data to further improve it (Legrain et al. 2011). The lack of high-affinity capture agents with high sensitivity and specificity for specific proteins is a major barrier to the successful application of broad proteomic techniques. In very complicated mixes, such as blood, where the quantities of various proteins vary by more than 10 orders of magnitude, this issue becomes more challenging.

Emerging technologies will enable omics measurements on individual cells, including transcriptomics and proteomics (Tang et al. 2011). Such in-depth molecular analyses offer profound understanding of the biology underlying tissues and have the potential to be a strong foundation for the development of omics-based tests. Nevertheless, because of the heterogeneity of cell states, the measurements become more variable as their resolution rises (Ma et al. 2011). Therefore, there is a probability that new omics technologies will raise the risk of overfitting the computational model to the data, even though they have a great deal of promise to improve the development of omics-based testing. It is expected that methods that integrate several omics data kinds into a single clinical test will become increasingly important as the number of simultaneous capacities that may be created continues to increase. Increasing the number of genomic and transcriptomic measurements is simple because DNA and RNA have complementary binding partners. However, it is more challenging to increase the number of protein measurements because high-affinity capture agents are required, as previously discussed in this section.

To increase the identification of disease-related mutations, datasets from genome-wide association revisions, in which a sample of cases and a sample of controls are taken from a large population and genotyped and each mutation found is assessed for association with the phenotype of interest, can also be analyzed within the framework of biological pathways (Segre et al. 2010).

These investigations could be highly helpful in the creation of omics-based diagnostics; in fact, these approaches are already being applied in clinical settings to identify mutated genes that cause unknown and rare hereditary disease states. Advancements in multi-omics will depend heavily on the creation of increasingly complex bioinformatics platforms and tools. From proteomics and metabolomics to transcriptomics and genomics, these tools must facilitate the integration and analysis of various omics datasets. Cloud computing and improvements in processing and storage capacity will facilitate the processing of massive volumes of multi-omics data. Multidisciplinary cooperation in multi-omics is expected to grow in the future by combining knowledge from clinical research, systems biology, bioinformatics, and computational biology. Working together is crucial for solving intricate biological puzzles and creating thorough disease models. Multidisciplinary research is also necessary to translate the findings of multi-omics research into clinical practice. Collaboration between data scientists and physicians, for instance, can assist in closing the gap between complicated multi-omics data analysis and practical healthcare applications, ensuring that the insights obtained are both actionable and clinically meaningful.

Multi-omics has an optimistic future ahead of it thanks to several new trends, interdisciplinary methods, and technology developments that will likely spur precision medicine's next wave of innovation. Realizing the full potential of multi-omics research will require addressing existing obstacles and investigating their possible uses. Sustained funding in this expanse combined with interdisciplinary cooperation will guarantee that multi-omics remains at the forefront of science and medicine.

7 Conclusion

The integration of proteomics, genomics, and transcriptomics with synthetic biology tools represents a pivotal advancement in metabolic engineering, offering unparalleled opportunities for engineering organisms for enhanced productivity, novel compound production, and improved resilience to environmental stressors. This study underscores the critical role of these omics technologies in overcoming the complexities associated with metabolic pathways and achieving scalable and sustainable metabolic engineering solutions.

Proteomics, genomics, and transcriptomics provide complementary data layers that, when integrated, offer a holistic view of cellular functions and regulatory mechanisms. Genomics and transcriptomics are fundamental for mapping the genetic architecture of organisms, identifying key genes and regulatory elements, and elucidating gene expression patterns under different conditions. Proteomics, however, reveals the actual expression levels and modifications of proteins, providing insights into posttranslational modifications and functional interactions within cellular networks. The integration of omics data into synthetic biology tools enhances the precision and efficiency of metabolic engineering strategies, leading to optimized strain designs and improved biosynthetic pathways.

In conclusion, the synergistic application of proteomics, genomics, and transcriptomics with synthetic biology tools has immense potential for revolutionizing metabolic engineering. Overcoming current challenges and advancing technological capabilities will require sustained research efforts, collaborative networks, and integration of ethical considerations. By addressing these issues, researchers can pave the way for sustainable production of valuable bio-based products, advancing the fields of industrial biotechnology and synthetic biology.

References

Acevedo O, Jorgensen WL (2010) Advances in quantum and molecular mechanical (QM/MM) simulations for organic and enzymatic reactions. Acc Chem Res 43(1):142–151. https://doi.org/10.1021/ar900171c

Aebersold R, Mann M (2016) Mass-spectrometry-based proteomics. Nature 537(7620):347–355. https://doi.org/10.1038/nature19949

Aizat WM, Ismail I, Noor NM (2018) Recent development in omics studies. In: Advances in experimental medicine and biology. Springer International Publishing, pp 1–9

Amer B, Baidoo EEK (2021) Omics-driven biotechnology for industrial applications front. Bioeng Biotechnol 9. https://doi.org/10.3389/fbioe.2021.613307

Aminian-Dehkordi J, Rahimi S, Golzar-Ahmadi M et al (2023) Synthetic biology tools for environmental protection. Biotechnol Adv 68:108239. https://doi.org/10.1016/j.biotechadv.2023.108239

Ayoglu B, Haggmark A, Neiman M, Igel U, Uhlen M, Schwenk JM, Nilsson P (2011) Systematic antibody and antigen-based proteomic profiling with microarrays. Expert Rev Mol Diagn 11(2):219–234

Baidoo EEK (2019) Microbial metabolomics: a general overview. Methods Mol Biol (Clifton, N.J.) 1859:1–8. https://doi.org/10.1007/978-1-4939-8757-3_1

Baidoo EEK, Teixeira Benites V (2019) Mass spectrometry-based microbial metabolomics: techniques, analysis, and applications. Methods Mol Biol 1859:11–69. https://doi.org/10.1007/978-1-4939-8757-3_2

Cardoso S, Afonso T, Maraschin M, Rocha M (2019) WebSpecmine: a website for metabolomics data analysis and mining. Meta 9:237. https://doi.org/10.3390/metabo9100237

Chekmeneva E, Dos Santos Correia G, Chan Q, Wijeyesekera A, Tin A, Young JH et al (2017) Optimization and application of direct infusion nanoelectrospray HRMS method for large-scale urinary metabolic phenotyping in molecular epidemiology. J Proteome Res 16:1646–1658. https://doi.org/10.1021/acs.jproteome.6b01003

Chen L, Zhao J, Yang G (2020) iTRAQ-based quantitative proteomics reveals critical enzymes in metabolic pathways. J Proteome 232:103708. https://doi.org/10.1016/j.jprot.2020.103708

Chen X, Dong Y, Huang Y, Fan J, Yang M, Zhang J (2021) Whole-genome resequencing using next-generation and Nanopore sequencing for molecular characterization of T-DNA integration in transgenic poplar 741. BMC Genomics 22(1). https://doi.org/10.1186/s12864-021-07625-y

Chistoserdova L (2010) Recent progress and new challenges in metagenomics for biotechnology. Biotechnol Lett 32:1351–1359. https://doi.org/10.1007/s10529-010-0306-9

Choi KR, Jang WD, Yang D, Cho JS, Park SJ (2010) Systems metabolic engineering strategies: integrating systems biology with synthetic biology. Nature 468(7324):67–72. https://doi.org/10.1038/nature09450

Chong J, Soufan O, Li C, Caraus I, Li S, Bourque G et al (2018) MetaboAnalyst 4.0: towards more transparent and integrative metabolomics analysis. Nucleic Acids Res 46:W486–W494. https://doi.org/10.1093/nar/gky310

Chuang HY, Hofree M, Ideker T (2010) A decade of systems biology. Annu Rev Cell Dev Biol 26:721–744. https://doi.org/10.1146/annurev-cellbio-100109-104122

Cobb RE, Si T, Zhao H (2012) Directed evolution: an evolving and enabling synthetic biology tool. Curr Opin Chem Biol 16(3–4):285–291. https://doi.org/10.1016/j.cbpa.2012.05.186

Cupples AM (2005) Principles and applications of soil microbiology, second edition. J Environ Qual 34(2):731–732. https://doi.org/10.2134/jeq2005.0731dup

Doudna JA, Charpentier E (2014) The new frontier of genome engineering with CRISPR-Cas9. Science 346(6213):1258096. https://doi.org/10.1126/science.1258096

Douglas T, Powell R, Savulescu J (2013) Is the creation of artificial life morally significant? Stud Hist Phil Biol Biomed Sci 44(4 Pt B):688–696. https://doi.org/10.1016/j.shpsc.2013.05.016

Dulermo T, Nicaud JM, Sterckx YGJ (2015) Proteomics in metabolic engineering: enabling tools for pathway optimization. Metab Eng 32:103–111. https://doi.org/10.1016/j.ymben.2015.01.004

Durot M, Bourguignon P-Y, Schachter V (2009) Genome-scale models of bacterial metabolism: reconstruction and applications. FEMS Microbiol Rev 33(1):164–190. https://doi.org/10.1111/j.1574-6976.2008.00146.x

Fiehn O (2001) Combining genomics, metabolome analysis, and biochemical modelling to understand metabolic networks. Comp Funct Genoms 2(3):155–168. https://doi.org/10.1002/cfg.82

Garner KL (2021) Principles of synthetic biology. Essays Biochem 65(5):791–811. https://doi.org/10.1042/EBC20200059

Graham NS, May ST, Daniel ZCTR, Emmerson ZF, Brameld JM, Parr T (2011) Use of the Affymetrix human GeneChip array and genomic DNA hybridisation probe selection to study ovine transcriptomes. Anim Int J Anim Biosci 5(6):861–866. https://doi.org/10.1017/s1751731110002533

Hughes RA, Ellington AD (2017) Synthetic DNA synthesis and assembly: putting the synthetic in synthetic biology. Cold Spring Harb Perspect Biol 9(1):a023812. Published 2017 Jan 3. https://doi.org/10.1101/cshperspect.a023812

Johnson S, Green R, Patel K (2020) TMT labeling in metabolic pathway proteomics. Mol Cell Proteomics 19(12):1234–1245. https://doi.org/10.1074/mcp.TIR120.002128

Kambale R, Rahman H, Omics S (2019) Fluorescence activated cell sorting single-cell omics approaches in plants. Elsevier, London

Katz N, Tripto E, Granik N et al (2021) Overcoming the design, build, test bottleneck for synthesis of nonrepetitive protein-RNA cassettes. Nat Commun 12(1):1576. Published 2021 Mar 11. https://doi.org/10.1038/s41467-021-21578-6

Keasling JD (2010) Manufacturing molecules through metabolic engineering. Nat Biotechnol 27(12):987–993. https://doi.org/10.1038/nbt.1591

Kinaret PAS, Serra A, Federico A, Kohonen P, Nymark P, Liampa I, Ha MK, Choi J-S, Jagiello K, Sanabria N, Melagraki G, Cattelani L, Fratello M, Sarimveis H, Afantitis A, Yoon T-H, Gulumian M, Grafström R, Puzyn T, Greco D (2020) Transcriptomics in toxicogenomics, part I: experimental design, technologies, publicly available data, and regulatory aspects. Nanomaterials (Basel, Switzerland) 10(4):750. https://doi.org/10.3390/nano10040750

Kirpich AS, Ibarra M, Moskalenko O, Fear JM, Gerken J, Mi X et al (2018) SECIMTools: a suite of metabolomics data analysis tools. BMC Bioinformatics 19:151. https://doi.org/10.1186/s12859-018-2134-1

Koide T, Pang WL, Baliga NS (2009) The role of predictive modelling in rationally re-engineering biological systems. Nat Rev Microbiol 7(4):297–305. https://doi.org/10.1038/nrmicro2107

Lanza AM, Alper HS (2011) Global strain engineering by mutant transcription factors. Methods Mol Biol (Clifton, NJ) 765:253–274. https://doi.org/10.1007/978-1-61779-197-0_15

Legrain P, Aebersold R, Archakov A, Bairoch A, Bala K, Beretta L, Bergeron J, Borchers CH, Corthals GL, Costello CE, Deutsch EW, Domon B, Hancock W, He F, Hochstrasser D, Marko-Varga G, Salekdeh GH, Sechi S, Snyder M, Srivastava S, Uhlen M, Wu CH, Yamamoto T, Paik

YK, Omenn GS et al (2011) The human proteome project: current state and future direction. (009993). Mol Cell Proteomics 10(7):M111
Lewis NE, Nagarajan H, Palsson BO (2012) Constraining the metabolic genotype–phenotype relationship using a phylogeny of in silico methods. Nat Rev Microbiol 10(4):291–305. https://doi.org/10.1038/nrmicro2763
Ma C, Fan R, Ahmad H, Shi Q, Comin-Anduix B, Chodon T, Koya RC, Liu CC, Kwong GA, Radu CG, Ribas A, Heath JR et al (2011) A clinical microchip for evaluation of single immune cells reveals high functional heterogeneity in phenotypically similar T cells. Nat Med 17(6):738–743
Ma S, Tang N, Tian J (2012) DNA synthesis, assembly and applications in synthetic biology. Curr Opin Chem Biol 16(3–4):260–267. https://doi.org/10.1016/j.cbpa.2012.05.001
Marcheschi RJ, Li H, Zhang K, Noey EL, Kim S, Chaubey A, Houk KN, Liao JC (2012) A synthetic recursive "+1" pathway for carbon chain elongation. ACS Chem Biol 7(4):689–697. https://doi.org/10.1021/cb200313e
Mardis ER (2017) DNA sequencing technologies: 2006–2016. Nat Protoc 12(2):365–368. https://doi.org/10.1146/annurev-genom-083115-022413
Melo D, Porto A, Cheverud JM, Marroig G (2016) Modularity: genes, development and evolution. Annu Rev Ecol Evol Syst 47:463–486. https://doi.org/10.1146/annurev-ecolsys-121415-032409
Moniruzzaman M, Hossain MU, Islam MN, Rahman MH, Ahmed I, Rahman TA, Bhattacharjee A, Amin MR, Rashed A, Keya CA, Das KC, Salimullah M (2020) Coding-complete genome sequence of SARS-CoV-2 isolate from Bangladesh by Sanger sequencing. Microbiol Resour Announce 9(28). https://doi.org/10.1128/mra.00626-20
Müller H, Fischer R, Meier D (2019) 2D-gel electrophoresis in microbial proteomics: applications in pathway analysis. Proteomics 19(9):e1900003. https://doi.org/10.1002/pmic.201900003
Ou Y, Guo S (2023) Safety risks and ethical governance of biomedical applications of synthetic biology. Front Bioeng Biotechnol 11:1292029. Published 2023 Oct 24. https://doi.org/10.3389/fbioe.2023.1292029
Ou Y-J, Ren Q-Q, Fang S-T, Wu J-G, Jiang Y-X, Chen Y-R, Zhong Y, Wang D-D, Zhang G-X (2020) Complete genome insights into Lactococcus petauri CF11 isolated from a healthy human gut using second- and third-generation sequencing. Front Genet 11. https://doi.org/10.3389/fgene.2020.00119
Pagani I, Liolios K, Jansson J, Chen IMA, Smirnova T, Nosrat B et al (2012) The Genomes OnLine Database (GOLD) v.4: status of genomic and metagenomic projects and their associated metadata. Nucleic Acids Res 40:571–579. https://doi.org/10.1093/nar/gkr1100
Painter HJ, Altenhofen LM, Kafsack BFC, Llinás M (2012) Whole-genome analysis of Plasmodium spp. utilizing a new Agilent technologies DNA microarray platform. In: Methods in molecular biology. Humana Press, pp 213–219
Patel D, Gupta N (2021) Liquid chromatography-mass spectrometry in proteomics. TrAC Trends Anal Chem 134:116136. https://doi.org/10.1016/j.trac.2021.116136
Pleiss J (2011) Protein design in metabolic engineering and synthetic biology. Curr Opin Biotechnol 22(5):611–617. https://doi.org/10.1016/j.copbio.2011.03.004
Privett HK, Kiss G, Lee TM, Blomberg R, Chica RA, Thomas LM, Hilvert D, Houk KN, Mayo SL (2012) Iterative approach to computational enzyme design. Proc Natl Acad Sci USA 109(10):3790–3795. https://doi.org/10.1073/pnas.1118082108
Radde N, Mortensen GA, Bhat D et al (2024) Measuring the burden of hundreds of BioBricks defines an evolutionary limit on constructability in synthetic biology. Nat Commun 15(1):6242. Published 2024 Jul 24. https://doi.org/10.1038/s41467-024-50639-9
Roberts LD, Souza AL, Gerszten RE, Clish CB (2012) Targeted metabolomics. Curr Protoc Mol Biol 1:1–24. https://doi.org/10.1002/0471142727.mb3002s98
Roberts MAJ, Cranenburgh RM, Stevens MP, Oyston PCF (2013) Synthetic biology: biology by design. Microbiology (Reading) 159(Pt 7):1219–1220. https://doi.org/10.1099/mic.0.069724-0
Sanger F (2004) Determination of nucleotide sequences in DNA. Biosci Rep 24(4–5):237–253. https://doi.org/10.1007/s10540-005-2733-8
Sanger F, Nicklen S, Coulson AR (1977) DNA sequencing with chain-terminating inhibitors. Proc Natl Acad Sci USA 74(12):5463–5467. https://doi.org/10.1073/pnas.74.12.5463

Schena M, Shalon D, Davis RW, Brown PO (1995) Quantitative monitoring of gene expression patterns with a complementary DNA microarray. Science (New York, NY) 270(5235):467–470. https://doi.org/10.1126/science.270.5235.467

Segre AV, Groop L, Mootha VK, Daly MJ, Altshuler D (2010) Common inherited variation in mitochondrial genes is not enriched for associations with type 2 diabetes or related glycemic traits. PLoS Genet 6(8):e1001058

Shakiba N, Jones RD, Weiss R, Del Vecchio D (2021) Context-aware synthetic biology by controller design: engineering the mammalian cell. Cell Syst 12(6):561–592. https://doi.org/10.1016/j.cels.2021.05.011

Shendure J, Ji H (2008) Next-generation DNA sequencing. Nat Biotechnol 26(10):1135–1145. https://doi.org/10.1038/nbt1486

Shetty RP, Endy D, Knight TF (2008) Jr. Engineering BioBrick vectors from BioBrick parts. J Biol Eng 2:5. Published 2008 Apr 14. https://doi.org/10.1186/1754-1611-2-5

Smith JC, Figeys D (2006) Proteomics technology in systems biology. Mol BioSyst 2(8):364. https://doi.org/10.1039/b606798k

Smith A, Roberts D, Clark P (2021) Mass spectrometry for pathway-specific enzyme analysis. Curr Opin Biotechnol 67:122–130. https://doi.org/10.1016/j.copbio.2021.01.005

Somerville V, Lutz S, Schmid M, Frei D, Moser A, Irmler S, Frey JE, Ahrens CH (2018) Long read-based de novo assembly of low complex metagenome samples results in finished genomes and reveals insights into strain diversity and an active phage system. In bioRxiv. https://doi.org/10.1101/476747

Subramanian I, Verma S, Kumar S, Jere A, Anamika K (2020) Multi-omics data integration, interpretation, and its application. Bioinform Biol Insights 14:7–9. https://doi.org/10.1177/1177932219899051

Tang J (2011) Microbial metabolomics. Curr Genomics 12(6):391–403. https://doi.org/10.2174/138920211797248619

Tang F, Lao K, Surani MA (2011) Development and applications of single-cell transcriptome analysis. Nat Methods 8(4 Suppl):S6–S11

Thomas R, Pontius JU, Borst LB, Breen M (2020) Development of a genome-wide oligonucleotide microarray platform for detection of DNA copy number aberrations in feline cancers. Veterin Sci 7(3):88. https://doi.org/10.3390/vetsci7030088

Trump BD, Galaitsi SE, Appleton E et al (2020) Building biosecurity for synthetic biology. Mol Syst Biol 16(7):e9723. https://doi.org/10.15252/msb.20209723

Urgen Cox J, Mann M (2011) Quantitative, high-resolution proteomics for data-driven systems biology. Annu Rev Biochem 80:273–299. https://doi.org/10.1146/annurev-biochem-061308-093216

Van Emon JM (2019) Omics in fruit nutrition: concepts and application. Elsevier Inc, Amsterdam

Vavricka CJ, Hasunuma T, Kondo A (2020) Dynamic metabolomics for engineering biology: accelerating learning cycles for bioproduction. Trends Biotechnol 38:68–82. https://doi.org/10.1016/j.tibtech.2019.07.009

VerBerkmoes NC, Denef VJ, Hettich RL, Banfield JF (2009a) Systems biology: functional analysis of natural microbial consortia using community proteomics. Nat Rev Microbiol 7:196–205. https://doi.org/10.1038/nrmicro2080

VerBerkmoes NC, Denef VJ, Hettich RL, Banfield JF (2009b) Functional analysis of natural microbial consortia using community proteomics. Nat Rev Microbiol 7(3):196–205. https://doi.org/10.1038/nrmicro2080

Wang Z, Gerstein M, Snyder M (2009) RNA-Seq: a revolutionary tool for transcriptomics. Nat Rev Genet 10(1):57–63. https://doi.org/10.1038/nrg2484

Wang JH, Byun J, Pennathur S (2010) Analytical approaches to metabolomics and applications to systems biology. Semin Nephrol 30(5):500–511. https://doi.org/10.1016/j.semnephrol.2010.07.007

Wang Y, Li Q, Sun J (2021) Network-based analysis of protein-protein interactions in metabolic engineering. Nucleic Acids Res 49(D1):D324–D330. https://doi.org/10.1093/nar/gkab847

Wilmes P, Heintz-Buschart A, Bond PL (2015) A decade of metaproteomics: where we stand and what the future holds. Proteomics 15(20):3409–3417. https://doi.org/10.1002/pmic.20150018

Yan X, Liu X, Zhao C, Chen GQ (2023) Applications of synthetic biology in medical and pharmaceutical fields. Signal Transduct Target Ther 8(1):199. Published 2023 May 11. https://doi.org/10.1038/s41392-023-01440-5

Zaidi SS, Mahas A, Vanderschuren H, Mahfouz MM (2020) Engineering crops of the future: CRISPR approaches to develop climate-resilient and disease-resistant plants. Genome Biol 21(1):289. Published 2020 Nov 30. https://doi.org/10.1186/s13059-020-02204-y

Zhang X, Liu J, Cheng L (2017a) Multi-omics integration for optimizing secondary metabolite biosynthesis in microorganisms. Curr Opin Biotechnol 46:38–44. https://doi.org/10.1016/j.copbio.2017.07.002

Zhang YHP, Sun J, Ma Y (2017b) Biomanufacturing: history and perspective. J Ind Microbiol Biotechnol 44:773–784. https://doi.org/10.1007/s10295-016-1863-2

Zhang X, Wu Y, Chen J (2022) SWATH-MS for time-resolved metabolic pathway analysis. Nat Biotechnol 40:152–160. https://doi.org/10.1038/s41587-022-01234-7

Zhao J, Wang G, Chu J, Zhuang Y (2020) Harnessing microbial metabolomics for industrial applications. World J Microbiol Biotechnol 36:1. https://doi.org/10.1007/s11274-019-2775-x

Uncovering Pathogen Metabolism: A Key to Infectious Disease Mitigation

C. V. Babu, B. N. Ramesh, B. N. Ganavi, and Gatta Vishnu Vyas

Abstract

Understanding pathogen metabolism is crucial for combating infectious diseases effectively. This chapter explores the dynamic world of pathogen metabolism and how it affects infectious disease mitigation. Through advanced tools and technologies, such as omics approaches, researchers gain insights into the diverse metabolic pathways utilized by pathogens. Exploiting metabolic dependencies emerges as a promising strategy for therapeutic intervention, with antimetabolite therapy offering a targeted approach to disrupt pathogen growth. Additionally, the interplay between pathogen and host metabolism unveils novel targets for intervention, including host metabolic defenses against pathogen propagation. Metabolism-based diagnostics leverage unique metabolic signatures to enable rapid and precise disease detection, while vaccine development targeting metabolic antigens holds promise for long-term protection. Despite significant progress, challenges remain, including the adaptive nature of pathogen metabolism and the complexity of pathogen-host metabolic interactions. By addressing these challenges and advancing our understanding of pathogen metabolism, we pave the way for innovative strategies to mitigate infectious diseases and safeguard public health.

C. V. Babu · B. N. Ganavi · G. V. Vyas
Department of Plant Biotechnology, University of Agricultural Sciences, GKVK, Bengaluru, India

B. N. Ramesh (✉)
ICAR-AICRP on Post-Harvest Engineering and Technology, University of Agricultural Sciences, GKVK, Bengaluru, Karnataka, India
e-mail: rameshbn20@uasbangalore.edu.in

S. Kaur et al. (eds.), *Microbial Metabolomics*,
https://doi.org/10.1007/978-981-96-4824-5_7

Keywords

Metabolomics · Pathogen metabolism · Therapeutic targets · Host-pathogen

1 Introduction to Pathogen Metabolism

Pathogen metabolism encompasses the complex biochemical processes essential for pathogenic organisms to sustain life, including nutrient uptake, energy production, and synthesis of biomolecules crucial for their growth and survival (Salway 2017). These processes are intricate and tightly regulated, allowing pathogens to adapt to various host environments and effectively evade host immune responses (Nicholson et al. 2007). Understanding these metabolic pathways in detail is crucial as it reveals potential vulnerabilities within the pathogen that can be strategically targeted (Eisenreich et al. 2017). For example, by elucidating how pathogens acquire and utilize nutrients, researchers can identify critical points for intervention that could disrupt these essential functions. Furthermore, insights into energy production pathways and biomolecule synthesis can result in the development of unique antimicrobial therapies designed to impair these processes and reduce pathogen viability (Nathwani and Wood 1993). This comprehensive understanding not only enhances our ability to design effective treatments but also aids in the creation of precise diagnostic tools capable of detecting and monitoring infections more accurately. Ultimately, the strategic targeting of these metabolic pathways represents a promising approach to combating infectious diseases and addressing the growing challenge of antimicrobial resistance (Salway 2017; Nicholson et al. 2007; Eisenreich et al. 2017; Nathwani and Wood 1993).

1.1 Dynamic World and Importance

Pathogens inhabit diverse and often hostile environments, requiring them to adapt their metabolism dynamically to survive and propagate effectively (Roemhild et al. 2022; Martínez and Rojo 2011). For example, the cause of tuberculosis, *Mycobacterium tuberculosis*, demonstrates remarkable metabolic plasticity by switching between aerobic respiration and anaerobic fermentation based on the availability of oxygen in its host microenvironment (Weiner et al. 2012; Taylor et al. 2019). This metabolic flexibility not only allows pathogens to persist within different host tissues but also enhances their virulence by facilitating adaptation to varying nutrient availabilities and immune pressures (Roemhild et al. 2022; Martínez and Rojo 2011). Understanding dynamic metabolic adaptations of pathogens is crucial for developing effective therapeutic strategies against infections, as these adaptations allow pathogens to thrive in various host environments and evade immune responses (Troha and Ayers 2020). By elucidating how pathogens modify their metabolism in response to environmental changes, researchers can identify specific metabolic pathways as potential targets for novel antimicrobial treatments (Wand and de Carvalho 2023). For instance, competition for glucose between host immune cells and pathogens highlights the metabolic warfare that occurs during infections, with pathogens often upregulating glycolytic pathways to enhance their virulence and survival (Nouwen, and Everts 2020). Furthermore, insights gained from metabolomic profiling can inform the design of interventions that disrupt essential metabolic pathways, potentially leading to innovative therapeutic strategies that improve disease control and prevention efforts (Garcia et al. 2023).

1.2 Recent Developments

Recent studies have unveiled the intricate metabolic adaptations that pathogens employ to thrive in diverse ecological niches and evade host immune responses (Shah et al. 2020; Ortega-Prieto and Postic 2019). *Candida albicans*, a commensal fungus capable of becoming pathogenic, adjusts its metabolism to survive within the hostile environment of the host immunological system. This includes switching from oxidative phosphorylation to fermentation to meet energy demands under stress conditions (Rodaki et al. 2009; Rodriguez 2020). Such metabolic versatility not only enhances the pathogen's survival but also poses challenges for treatment strategies aimed at disrupting its adaptive metabolic pathways (Vylkova et al. 2011).

Understanding the metabolic adaptations of pathogens like *Candida albicans* is crucial for infectious disease mitigation, as these adaptations enable pathogens to adjust their metabolic processes in response to environmental changes (Sharma et al. 2023). By comprehending these mechanisms, researchers can develop targeted therapeutic strategies that disrupt adaptive pathways, such as identifying key enzymes involved in metabolic switching (Paul et al. 2024). Inhibiting these enzymes can hinder the pathogen's ability to adapt and survive under hostile conditions, ultimately enhancing treatment efficacy (Troha and Ayres 2020). Additionally, this knowledge can facilitate the design of combination therapies that concurrently target multiple metabolic pathways, thereby lowering the likelihood of resistance development (Sharma et al. 2023). Ultimately, insights into pathogen metabolism can inform the creation of more effective and sustainable approaches to infectious disease control and prevention (Paul et al. 2024).

1.3 Evasion of Host Immune Responses

Pathogens employ sophisticated strategies to evade immune detection, often involving metabolic alterations that affect their surface antigens and immune recognition (Seeger et al. 2023; Juneja and Tasleem 2024). For instance, certain bacterial pathogens modify their cell wall components and surface proteins through metabolic pathways to avoid detection and clearance by host immune cells, contributing to chronic infections and treatment resistance.

Understanding the metabolic alterations of pathogens is crucial for mitigating infectious diseases, as these alterations enable pathogens to modify their surface antigens and evade immune detection. By unraveling the specific metabolic pathways that pathogens use to achieve this, researchers able to recognize novel targets for therapeutic intervention. For instance, drugs designed to inhibit these metabolic pathways could prevent pathogens from modifying their surface proteins, making them more susceptible to immune recognition and clearance (Paul et al. 2024). Additionally, vaccines could be developed to target the unmodified, native forms of surface antigens, boosting the immune system's ability to recognize and combat the pathogen (Sharma et al. 2023). Furthermore, insights into these evasion strategies can inform the design of diagnostic tools that detect the specific metabolic markers associated with immune evasion, allowing for earlier and more accurate identification of infections (Troha and Ayres 2020). This can lead to timely and targeted treatment, reducing the duration and severity of infections and decreasing the likelihood of chronic disease and resistance development. Ultimately, a deeper understanding of pathogen metabolism and immune evasion mechanisms can drive the development of innovative strategies to attack infectious diseases more effectively (Paul et al. 2024).

1.4 Interplay Between Pathogen and Host Metabolism

The interplay between pathogen and host metabolism is complex and multifaceted, influencing development of the disease and treatment outcomes (Shah et al. 2020; Ortega-Prieto and Postic 2019). Intracellular pathogens like *Toxoplasma gondii* exemplify this interaction by hijacking host metabolic pathways, such as lipid metabolism as depicted in Table 1, to sustain their intracellular growth and

Table 1 Key metabolic pathways of pathogens and their significance

Pathogen	Key metabolic pathway	Significance to pathogen	References
Mycobacterium tuberculosis	Aerobic and anaerobic respiration	Adaptation to hypoxic environments	Shi et al. (2005)
Escherichia coli	Glycolysis and TCA cycle	Versatility in nutrient utilization	Shimizu (2013)
Plasmodium falciparum	Glycolysis	Energy production in anaerobic conditions	Kita et al. (2002)
Candida albicans	Glycolysis and fatty acid oxidation	Adaptation to immune response	Pellon et al. (2022)
Toxoplasma gondii	Lipid metabolism	Dependency on host lipids	Nolan et al. (2017a)
Pseudomonas aeruginosa	Diverse carbon metabolism	Thriving in diverse environments	Rojo (2010)

This table presents the primary metabolic pathways utilized by various pathogens and their significance. Understanding these pathways can aid researchers in developing therapeutics targeting these pathogens replication (Hossain 2023; Kamali and Sarvtin 2023; Jiao et al. 2024). Targeting these metabolic dependencies offers promising avenues for therapeutic interventions aimed at disrupting pathogen survival and enhancing host immune responses against infections.

2 Tools and Technologies for Studying Pathogen Metabolism

2.1 Omics Technologies and Integration

Genomics Next-generation sequencing (NGS) has revolutionized pathogen genomics by enabling rapid and comprehensive sequencing, significantly enhancing our understanding of pathogen biology and public health responses (Ferreira et al. 2024; Salway 2017). This technology facilitates the swift identification of metabolic pathways crucial for pathogen survival, essential for developing targeted therapeutic interventions (Ferreira et al. 2024). For instance, NGS has been pivotal in mapping the genomes of pathogens such as *Mycobacterium tuberculosis*, providing profound insights into genetic variations that inform treatment strategies and public health measures, particularly concerning emerging infectious diseases and antimicrobial resistance (Gavaghan et al. 2000a, Diray et al. 2020). New advancements in long-read sequencing technologies, like PacBio and Oxford Nanopore, have further enhanced our ability to assemble complete pathogen genomes, bolstering genome assembly accuracy and allowing researchers to discern intricate genomic features, including structural variations and complex genomic regions (Ferreira et al. 2024; Salway 2017). This advanced capability has been crucial in studying pathogens like *Listeria monocytogenes*, where understanding genetic diversity has critical implications for outbreak investigations and public health responses (Diray et al. 2020; Ferreira et al. 2024). The integration of NGS into public health frameworks has transformed disease surveillance, enabling real-time tracking of pathogen strains during outbreaks and facilitating timely responses and control measures (Salway 2017). The SARS-CoV-2's genome was sequenced during the COVID-19 pandemic, allowing health officials to track variants such as Delta and Omicron and informing vaccine development and public health responses. The UK Health Security Agency (UKHSA) has utilized NGS to sequence over one million genomes, significantly enhancing outbreak investigations and providing crucial data for vaccine development and epidemiological studies (Salway 2017). Furthermore, NGS has been instrumental in identifying genetic markers associated with virulence and resistance, as seen in *Escherichia coli* strains linked to foodborne outbreaks, aiding in the development of rapid diagnostic tools (Diray et al. 2020; Gavaghan et al. 2000a, b). By identifying specific genetic markers, NGS enables the creation of precise and efficient diagnostic assays, which are critical for timely public health interventions and the management of infectious diseases

(Ferreira et al. 2024). The ability to quickly and accurately identify pathogenic strains and their resistance profiles through NGS has profound implications for improving treatment outcomes and curbing the spread of resistant infections (Diray et al. 2020).

Transcriptomics The investigation of pathogen gene expression profiles under various circumstances has been transformed by RNA sequencing (RNA-seq), which has also shown active metabolic pathways during infection (Gavaghan et al. 2000a, b; Diray et al. 2020). By providing a comprehensive view of the transcriptome, researchers can find genes that exhibit differential expression in response to host interactions, antimicrobial treatments, and environmental changes using RNA-seq. Understanding the intricate regulatory networks that viruses employ to adapt and flourish within their hosts has been made possible thanks in large part to this technology. For instance, investigations into *Salmonella enterica* have pinpointed upregulated metabolic genes that may serve as promising targets for novel antimicrobial therapies (Salway 2017). Studies have shown that during infection, *Salmonella* upregulates genes involved in amino acid metabolism, stress response, and virulence, revealing potential vulnerabilities that could be exploited for therapeutic purposes (Ferreira et al. 2024; Salway 2017). Moreover, the identification of non-coding RNAs and regulatory elements has been made easier by RNA-seq that play crucial roles in pathogen biology, further expanding the potential targets for drug development. In addition to *Salmonella enterica*, RNA-seq has been applied to a wide range of pathogens, including *Mycobacterium tuberculosis*, *Plasmodium falciparum*, and *Candida albicans*, shedding light on their unique metabolic adaptations and survival strategies (Gavaghan et al. 2000a, b; Diray et al. 2020). For *Mycobacterium tuberculosis*, RNA-seq has identified genes involved in lipid metabolism and cell wall synthesis that are essential for persistence in the host, offering new avenues for drug targeting (Gavaghan et al. 2000a, b). In *Plasmodium falciparum*, the causing agent of malaria, RNA-seq has revealed stage-specific gene expression patterns that are critical for its life cycle, enabling the identification of novel antimalarial targets (Diray et al. 2020). Furthermore, RNA-seq data can be combined with other omics technologies, for example, proteomics and metabolomics, to provide a holistic understanding of pathogen biology and host-pathogen interactions (Gavaghan et al. 2000a, b; Diray et al. 2020). This integrative approach enhances the ability to recognize key metabolic pathways and regulatory networks that can be targeted by new therapeutic agents, ultimately improving the effectiveness of antimicrobial treatments.

Overall, RNA-seq has not only deepened our understanding of pathogen metabolism and gene regulation but has also opened up new possibilities for the development of innovative antimicrobial therapies. By continuously advancing RNA-seq technology and its applications, researchers are well-positioned to tackle the challenges posed by infectious diseases and improve global health outcomes (Gavaghan et al. 2000a, b; Diray et al. 2020).

Proteomics Proteomics, particularly mass spectrometry-based approaches, offers detailed insights into pathogen protein composition, providing a comprehensive analysis of the enzymes involved in various metabolic pathways and their regulatory mechanisms (Gavaghan et al. 2000a, b; Diray et al. 2020). With the aid of this technology, it is possible to identify and measure the proteins present in a pathogen, providing insight into their functions related to adaptability, pathogenicity, and metabolism.

Mass spectrometry-based proteomics has significantly advanced our understanding of pathogen biology by enabling the discovery of novel biomarkers and drug targets. For example, recent studies on *Plasmodium falciparum*, the causative agent of malaria, have utilized proteomics to identify critical metabolic enzymes involved in glycolysis and the pentose phosphate pathway (Ferreira et al. 2024; Salway 2017). These pathways are crucial for the parasite's survival and proliferation within the host, making their enzymes promising targets for antimalarial drug development.

Proteomics has revealed that enzymes such as aldolase and enolase, which are key players in glycolysis, are highly expressed in *Plasmodium falciparum* and play crucial roles in the parasite's energy metabolism (Ferreira et al. 2024). Similarly, enzymes in the pentose phosphate pathway, like glucose-6-phosphate dehydrogenase, are upregulated to support nucleotide and lipid biosynthesis, further highlighting their potential as drug targets (Salway 2017). By focusing on these enzymes, researchers aim to develop inhibitors that can disrupt essential metabolic processes, thereby reducing the parasite's ability to thrive and cause disease. In addition to identifying targets, proteomics also provides insights into the post-translational modifications (PTMs) of pathogen proteins, such as phosphorylation, ubiquitination, and acetylation (Gavaghan et al. 2000a, b). These modifications can influence protein function, interactions, and stability, offering additional layers of regulation that can be targeted for therapeutic intervention. For instance, understanding how PTMs affect enzyme activity in *Plasmodium falciparum* can help in designing drugs that specifically disrupt these regulatory processes. Moreover, proteomics can be integrated with other omics technologies, such as genomics and metabolomics, in order to offer a holistic view of pathogen metabolism and host-pathogen interactions (Diray et al. 2020). This integrative approach enables researchers to correlate protein expression with genetic data and metabolic profiles, helping to create a deeper comprehension of pathogen biology and the identification of new therapeutic strategies. Overall, mass spectrometry-based proteomics is a powerful tool for advancing our knowledge of pathogen protein composition and function. By elucidating the roles of key enzymes and regulatory mechanisms, proteomics contributes to the evolution of targeted therapies that can address the specific metabolic needs of pathogens and improve treatment outcomes (Gavaghan et al. 2000a, b; Diray et al. 2020).

Metabolomics Metabolomics techniques, comprising nuclear magnetic resonance (NMR) spectroscopy and mass spectrometry (MS), provide a powerful means of analyzing pathogen metabolites, offering critical insights into active metabolic pathways and possible therapeutic targets (Gavaghan et al. 2000a, b; Diray et al. 2020). By profiling the small molecules involved in various metabolic processes, these techniques enable researchers to gain a comprehensive understanding of pathogen biology, determine biomarkers of disease, and uncover novel drug targets. NMR spectroscopy is a nondestructive technique that provides detailed information about the chemical environment and concentration of metabolites within a sample (Gavaghan et al. 2000a, b). It allows for the identification and quantification of metabolites in complex biological systems without the need for extensive sample preparation. For example, NMR has been used to study the metabolic profiles of *Mycobacterium tuberculosis*, revealing key metabolic changes associated with drug resistance and bacterial survival (Ferreira et al. 2024). By examining the metabolite profiles of drug-resistant strains, researchers have identified distinctive metabolic signatures that could inform the development of new treatment strategies. Mass spectrometry (MS), however, offers high sensitivity and precision in the detection and quantification of metabolites (Diray et al. 2020). It enables the measurement of a wide range of metabolites with high accuracy, making it an essential tool for metabolomics studies. MS-based metabolomics has been employed to investigate the metabolic adaptations of pathogens in response to environmental changes and drug treatments. For instance, studies on *Mycobacterium tuberculosis* have utilized MS to identify alterations in lipid and amino acid metabolism that are associated with resistance to antitubercular drugs (Ferreira et al. 2024; Salway 2017). These findings have provided beneficial insights into the mechanisms of drug resistance and have guided the development of novel therapeutic approaches. In addition to identifying metabolic signatures associated with drug resistance, metabolomics has been instrumental in exploring pathogen-host interactions (Gavaghan et al. 2000a, b). By comparing the metabolite profiles of infected versus non-infected hosts, researchers can uncover changes in host metabolism that result from pathogen infection. This information is crucial for understanding how pathogens manipulate host metabolic pathways to their advantage and for identifying potential targets for therapeutic intervention. Metabolomics also allows for the integration of information from different omics technologies, such as genomics and proteomics, to provide a holistic view of pathogen biology (Diray et al. 2020). By combining metabolomics data with information about gene expression and protein function, researchers can build comprehensive models of metabolic networks and identify key regulatory nodes that could be targeted by new drugs.

Overall, metabolomics techniques like NMR and MS are invaluable for advancing our understanding of pathogen metabolism and identifying novel therapeutic targets. By revealing the dynamic metabolic changes that occur during infection and drug resistance, these techniques contribute to the enhancement of more effective targeted treatment strategies, ultimately improving patient outcomes and combating infectious diseases more effectively (Gavaghan et al. 2000a, b; Diray et al. 2020; Ferreira et al. 2024; Salway 2017).

Integration Integrating data from genomics, transcriptomics, proteomics, and metabolomics gives a holistic view of pathogen metabolism (Gavaghan et al. 2000a, b; Diray et al. 2020). This integrative approach facilitates the development of detailed metabolic models, predicting the impact of metabolic interventions and informing

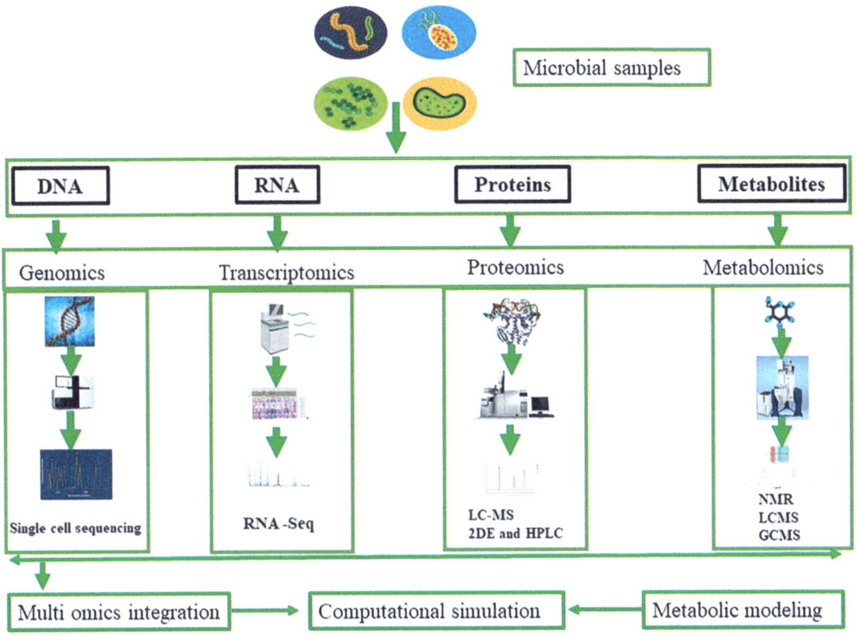

Fig. 1 Workflow of omics integration in pathogen metabolism studies. This figure shows how multi-omics methods—genomics, transcriptomics, proteomics, and metabolomics—are used to analyze pathogen metabolism. It focuses on computational simulation and metabolic modeling for therapeutic treatments and incorporates data from proteins, RNA, DNA, and metabolites to understand metabolic pathways

the design of effective antimicrobial therapies (Ferreira et al. 2024; Salway 2017) as described in Fig. 1.

2.2 Computational Modeling and Bioinformatics

The integration of omics data requires robust computational tools to analyze and interpret the vast amounts of data generated. Bioinformatics and systems biology play a crucial role in this process as presented in Table 2.

Metabolic Network Reconstruction This involves creating a comprehensive model of a pathogen's metabolic network based on genomic, transcriptomic, proteomic, and metabolomic information. These models can anticipate the growth and metabolic fluxes of pathogens under various conditions.

The expanded metabolic reconstruction of *Helicobacter pylori*, known as *iIT341 GSM/GPR*, significantly enhances our understanding of this human gastric pathogen, which infects nearly half of the global population and is linked to numerous gastric diseases, including peptic ulcers and gastric cancer (Thiele et al. 2005). This comprehensive model integrates 341 metabolic genes, 476 intracellular reactions,

Table 2 Computational tools for metabolic analysis

Tool	Description	Application	References
COBRA Toolbox	Framework for constraint-based reconstruction and analysis	Metabolic network reconstruction	Ng et al. (2022)
KEGG PATHWAY	Database of metabolic pathways	Pathway mapping and analysis	Kanehisa et al. (2022)
MetaboAnalyst	Web-based tool for metabolomics data analysis	Data visualization and statistical analysis	Dong and Malitsky (2024)
BioCyc	Gathering of pathway/genome databases	Pathway reconstruction and analysis	Caspi et al. (2014)
MetaCyc	Enzyme and metabolic pathway database	Understanding metabolic diversity	Caspi et al. (2014)
ModelSEED	Automated model reconstruction	Metabolism models at the genome size	Faria et al. (2018)

This table provides an overview of various computational tools used in metabolic analysis. These tools are essential for reconstructing, mapping, and analyzing metabolic networks and pathways, facilitating a deeper understanding of metabolic processes in various organisms

78 exchange reactions, and 485 metabolites, providing a detailed framework for studying *H. pylori* metabolism (Thiele et al. 2005; Dutta et al. 2006). Notably, the model introduces novel features such as gene-protein-reaction associations and charge-balanced reactions, which improve the accuracy of metabolic predictions (Thiele et al. 2005).

Through in silico deletion studies, the researchers identified 128 essential and 75 conditionally crucial metabolic genes, which are critical for the bacterium's survival in the acidic gastric environment (Thiele et al. 2005; Dutta et al. 2006). Additionally, the study revealed 47 synthetic lethal mutants, indicating potential targets for new therapeutic strategies aimed at disrupting *H. pylori* metabolism without harming beneficial gut microbiota (Thiele et al. 2005). This targeted approach could result in the creation of innovative antibiotics or treatments specifically inhibiting *H. pylori*, thereby reducing infection rates and associated health complications. Furthermore, understanding the metabolic pathways of *H. pylori* can inform public health strategies to manage infections and mitigate the risk of related metabolic disorders, ultimately improving patient outcomes in populations at risk for such diseases (Thiele et al. 2005; Dutta et al. 2006).

Flux Balance Analysis (FBA) FBA is a mathematical strategy used to analyze the flow of metabolites throughout a metabolic network. It helps in identifying key metabolic pathways and potential drug targets by predicting the effect of gene deletions on pathogen growth.

Recent studies utilizing genome-scale metabolic models (GEMs) of *Pseudomonas aeruginosa* have provided valuable insights into the pathogen's metabolic pathways and potential targets for antimicrobial therapy. The *iSD1509* model, which comprises 1509 genes, is the largest GEM developed for *P. aeruginosa* to date (Dahal et al. 2023). These models have demonstrated high predictive accuracy for growth, substrate utilization, and gene essentiality, making them powerful tools for studying this opportunistic pathogen (Dahal et al. 2023).

Flux balance analysis (FBA) studies conducted with these GEMs have been instrumental in identifying essential genes that are implicated in critical metabolic pathways, like amino acid, lipid, and cofactor biosynthesis. These essential genes represent potential targets for antimicrobial therapy, offering new avenues for drug development (Dahal et al. 2023). For instance, the *iSD1509* model has highlighted the crucial function the ubiquinone-9 biosynthesis pathway in the anaerobic growth of *P. aeruginosa*, indicating a potential target for disrupting the pathogen's metabolism under low-oxygen conditions. Additionally, these models elucidate mechanisms behind the excess production of drug susceptibility biomarkers in *P. aeruginosa* mutants, providing insights that could inform the design of more effective treatments (Dahal et al. 2023; Oberhardt et al. 2008; Oberhardt et al. 2010; Payne et al. 2021).

Overall, the integration of GEMs and FBA has significantly improved our understanding of *P. aeruginosa* metabolism, elucidating the roles of various virulence factors, such as phenazines, which are critical to the pathogen's capability to cause disease (Dahal et al. 2023). By leveraging these models, researchers can systematically explore the metabolic vulnerabilities of *P. aeruginosa*, establishing the way for the development of novel antimicrobial strategies aimed at mitigating infections caused by this formidable pathogen.

Machine Learning Machine learning (ML) algorithms are increasingly being utilized to identify patterns in omics data that may not be evident through traditional analytical methods. These algorithms are particularly effective in predicting how pathogens respond to varying environmental conditions and treatments. Recent advancements in ML have enabled the prediction of metabolic pathway alterations in antibiotic-resistant bacteria, providing critical insights into potential therapeutic strategy. By integrating and analyzing diverse omics data, including genomics, proteomics, and metabolomics, ML facilitates the discovery of new biomarkers and enhances disease prediction and patient stratification. The strength of ML lies in its capacity to manage complexity and heterogeneity of omics data, which often involves a vast number of variables with relatively few samples. Techniques such as dimensionality reduction and models designed for limited sample sizes are employed to address these challenges (Arjmand et al. 2022; Harris. 2023).

By examining metabolic pathway alterations in antibiotic-resistant bacteria, ML can offer valuable insights into potential therapeutic strategies, aiding in predicting how pathogens will respond to different treatments. This contributes to the development of more effective antibiotics. However, among the primary difficulties in applying ML to omics data is the integration of data from diverse sources, which requires advanced computational analyses and sophisticated ML techniques (Arjmand et al. 2022; Harris. 2023). Despite these challenges, the integration of omics data with ML possesses great promise for advancing precision medicine and discovering new biomarkers.

3 Identifying Metabolic Pathways of Pathogens

3.1 Diversity and Adaptability

Pathogens exhibit a wide range of metabolic pathways that enable them to adapt to various environments. For instance, *Escherichia coli* can utilize different carbon sources, while *Mycobacterium tuberculosis* has specialized pathways for nitrogen assimilation. The ability to adapt metabolically is vital for pathogen survival and virulence as presented in Table 3 and Fig. 2.

Glycolysis and TCA Cycle Pathogens exhibit remarkable metabolic flexibility, primarily relying on glycolysis and the TCA cycle for the generation of energy. This flexibility allows them to thrive in both aerobic and anaerobic conditions, as noted by Hofer (2014) and Papp-Wallace et al. (2011). The adaptability of these metabolic pathways is a key aspect of pathogen survival, enabling them to efficiently generate energy regardless of oxygen availability (Hofer 2014). Under aerobic conditions, pathogens maximize ATP production through oxidative phosphorylation, while in oxygen-scarce environments, they switch to anaerobic metabolism, relying on glycolysis and fermentation to produce energy. This metabolic versatility allows pathogens to survive and proliferate in various host tissues with fluctuating oxygen levels or in external oxygen-deprived environments (Karn et al. 2017). Additionally, the ability to switch between aerobic and anaerobic metabolism helps pathogens evade host immune responses; for instance, in hypoxic regions within a host, such as necrotic tissue, anaerobic metabolism becomes crucial for their survival and virulence (Hofer 2014). This metabolic flexibility also enhances resistance to certain treatments, as some antibiotics are less effective under anaerobic conditions. Furthermore, pathogens exploit various other metabolic pathways to support their growth and replication, altering their use of nutrients by switching between differ-

Table 3 Examples of pathogen metabolic pathways

Pathogen	Metabolic pathway	Adaptability feature	References
Escherichia coli	Glycolysis, TCA cycle	Aerobic and anaerobic metabolism	Passalacqua et al. (2016)
Mycobacterium tuberculosis	Nitrogen assimilation	Survival in low-nitrogen environments	Krysenko et al. (2022)
Plasmodium falciparum	Glycolysis	Energy production in low-oxygen conditions	Penkler et al. (2015)
Candida albicans	Glycolysis and fatty acid oxidation	Adaptation to immune response	Burgain et al. (2020)
Toxoplasma gondii	Lipid metabolism	Dependence on host lipids	Bisanz et al. (2006)
Pseudomonas aeruginosa	Diverse carbon metabolism	Thriving in diverse environments	La Rosa et al. (2019)

This table lists examples of key metabolic pathways in various pathogens, along with features that highlight their adaptability. Understanding these pathways can help in developing strategies to combat these pathogens by targeting their metabolic processes

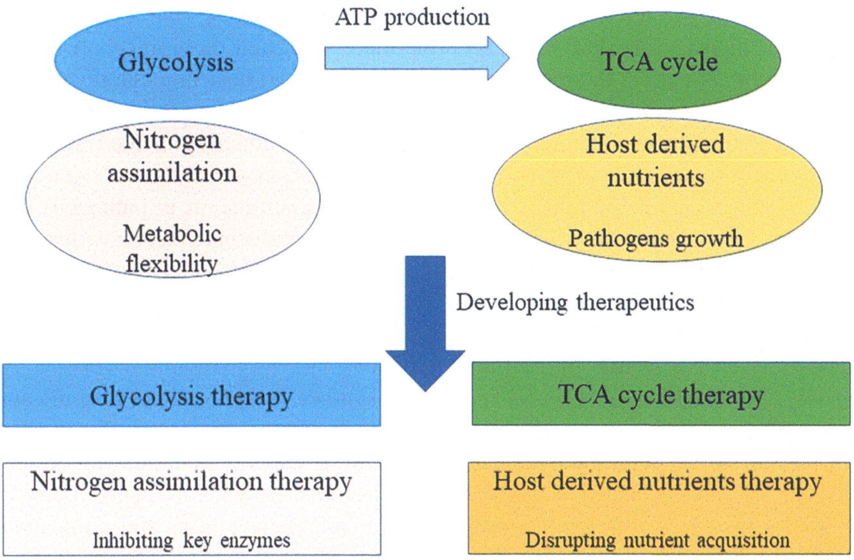

Fig. 2 This figure illustrates the pathogen metabolic pathways and corresponding therapeutic targets, emphasizing metabolic flexibility for pathogen survival. Therapeutic strategies aim to disrupt key metabolic functions

ent carbon sources depending on availability (Karn et al. 2017). This adaptability extends to the control of metabolic enzymes and transport systems, allowing pathogens to fine-tune their metabolism in response to environmental signals (Papp-Wallace et al. 2011).

Nitrogen Assimilation Many pathogens, such as *Mycobacterium tuberculosis*, have developed specialized pathways to assimilate nitrogen from the host environment, which is crucial for their survival and pathogenicity (Gouzy et al. 2014). The ability to utilize different nitrogen sources can significantly influence pathogen survival and virulence in various niches (Gouzy et al. 2014). For instance, *M. tuberculosis* can assimilate nitrogen from host-derived compounds, enabling it to thrive in the nutrient-limited environments it often encounters within the host. This capacity to metabolize various nitrogen sources not only supports the pathogen's growth but also enhances its ability to establish and maintain infections in diverse tissue environments. Additionally, a recent study on the nitrogen metabolism of *Plasmodium falciparum* revealed that this parasite can utilize host-derived amino acids, underscoring the potential of targeting nitrogen assimilation pathways in antimalarial therapy (Allary et al. 2007). *P. falciparum*, the causative agent of malaria, depends heavily on the amino acids obtained from host to fuel its rapid development and replication cycles within red blood cells. Disrupting these nitrogen assimilation pathways could critically impair the pathogens' ability to synthesize essential cellular components, thereby reducing their pathogenicity and survival. This approach could be particularly useful in developing antimalarial therapies by hindering the parasite's ability to utilize host-derived amino acids, which are vital for its growth and replication (Allary et al. 2007). Furthermore, targeting these metabolic pathways might also affect the pathogen's ability to adapt to host immune responses, potentially making them more susceptible to treatment. By interfering with these fundamental processes, it may be possible to develop treatments that are less prone to resistance, as they strike at core biological functions essential for pathogen survival. Thus, understanding and disrupting nitrogen assimilation in pathogens not only provides significant therapeutic advantages in combating infectious diseases but also opens new avenues for research into the metabolic dependencies that underpin pathogenic success.

Host-Derived Nutrients Intracellular pathogens such as *Toxoplasma gondii* rely heavily on host-derived lipids for membrane synthesis, representing a significant metabolic dependency and potential target for therapeutic intervention. *T. gondii* is an obligate intracellular parasite that resides within a specialized parasitophorous vacuole (PV) inside host cells. Within this environment, it depends on host-derived lipids, particularly cholesterol, to support its membrane synthesis and replication (Coppens 2006). The parasite has evolved sophisticated mechanisms to acquire these lipids across the PV membrane (PVM), illustrating its metabolic reliance on the host for several essential nutrients. Disrupting the pathways involved in the acquisition or utilization of host-derived lipids could impair the parasite's ability to synthesize membranes, leading to growth inhibition or death. Additionally, *T. gondii* relies on various other host-derived nutrients, like amino acids, sugars, and vitamins, using specialized mechanisms including transporters and endocytic pathways to acquire these essential nutrients (Blume et al. 2015; Nolan et al. 2017a, b). Targeting these nutrient acquisition pathways offers a promising strategy for developing new antimicrobial therapies against *T. gondii* and other intracellular pathogens with similar dependencies on host-derived nutrients. By impairing the parasite's ability to access and utilize these vital resources, therapeutic interventions can effectively hinder the growth and proliferation of the pathogen within the host (Coppens 2013; Nolan et al. 2017a, b).

3.2 Metabolic Interactions and Dependencies

Symbiotic Relationships Some pathogens assemble symbiotic relationships with other microbes to fulfil their metabolic needs as depicted in Table 4. For example, in the human gut, certain pathogenic bacteria can rely on metabolites produced by commensal bacteria.

A study has elucidated the symbiotic relationship between the pathogenic bacterium *Clostridium difficile* and the commensal gut microbiota, revealing that *C. difficile* thrives in the gut by utilizing metabolites produced by other gut microbiota members. Specifically, *C. difficile* leverages succinate produced by commensal bacteria as a nutrient source to support its growth and virulence. This metabolic interdependence allows *C. difficile* to outcompete resident gut microbiota, establishing

Table 4 Host-pathogen metabolic interactions

Pathogen	Interaction type	Impact on host metabolism	References
Clostridium difficile	Symbiosis	Utilization of commensal metabolites	Theriot and Young (2015)
Chlamydia trachomatis	Host manipulation	Hijacking host nutrient pathways	Wenbo et al. (2024)
Helicobacter pylori	Host manipulation	Alteration of glucose metabolism	Martín-Núñez et al. (2019)
Salmonella enterica	Host manipulation	Induction of inflammatory metabolites	Anderson and Kendall (2017)
Streptococcus pneumoniae	Symbiosis	Nutrient acquisition from commensal bacteria	Krishnamurthy and Kyd (2014)
Staphylococcus aureus	Host manipulation	Modification of lipid metabolism	Teoh et al. (2021)

This table highlights various types of metabolic interactions between pathogens and their hosts, along with the impact on host metabolism. Understanding these interactions can help in developing strategies to disrupt these processes and combat infections

an infection within the host (Curtis and Sperandio 2011; Gast et al. 2009). Pathogens like *C. difficile* have evolved to exploit the metabolic by-products of other microbes in their environment, obtaining essential nutrients and energy sources they cannot synthesize independently. This interdependence confers a significant advantage to the pathogen, facilitating its proliferation in the host (Garcia and Gerardo 2014; Clay 2014). Understanding these symbiotic relationships is essential for developing effective antimicrobial therapies. Disrupting the metabolic interactions between pathogens and their symbiotic partners could hinder the pathogen's ability to colonize and cause disease. Therapies targeting the gut microbiome, such as fecal microbiota transplantation or the use of specific probiotic strains, may help restore microbiome balance and limit the growth of opportunistic pathogens like *C. difficile* (Clay 2014).

Host Interactions Intracellular pathogens such as *Chlamydia trachomatis* and *Helicobacter pylori* adeptly exploit host cell metabolism to procure essential nutrients crucial for their growth and replication. *Chlamydia trachomatis,* for instance, utilizes host lipid and glucose metabolism, diverting these resources to facilitate its development within the host cell environment. Meanwhile, *Helicobacter pylori* has been found to induce significant alterations in host glucose metabolism, specifically influencing glycolysis and gluconeogenesis pathways. These modifications create a metabolic milieu favoring the pathogen's survival and persistence within the gastric environment (Eisenreich et al. 2017). Beyond nutrient acquisition, pathogens employ metabolic manipulation as a strategy for evading or suppressing the host's immune responses. By modulating host amino acid metabolism, certain pathogens can inhibit T cell activation and proliferation, thereby ensuring their sustained presence within the host (Omsland et al. 2012; Seeger et al. 2023). A comprehensive understanding of these intricate metabolic interactions is crucial for devising targeted antimicrobial therapies aimed at disrupting pathogen survival strategies and improving host immune responses against intracellular infections (Juneja and Tasleem 2024; Yu et al. 2024).

4 Exploiting Metabolic Dependencies for Therapeutic Intervention

Pathogens heavily rely on specific metabolic pathways for their survival and growth, presenting potential targets for therapeutic intervention (Eisenreich et al. 2017), as presented in Table 5. Understanding and targeting these critical metabolic dependencies have led to the development of various strategies aimed at disrupting pathogen viability and pathogenicity.

4.1 Targeting Essential Pathways and Antimetabolites

Antimetabolites are key agents in these strategies, designed to interfere with pathogens' normal metabolic processes by mimicking natural substrates (Nathwani and Wood 1993). These compounds inhibit essential enzymes, thereby blocking crucial metabolic pathways necessary for the synthesis of biomolecules essential for pathogen survival.

Sulfonamides represent a class of antimetabolites that disrupt folate biosynthesis, a vital pathway for nucleotide synthesis in bacteria. By mimicking para-aminobenzoic acid (PABA), sulfonamides inhibit the formation of dihydrofolic acid, essential for DNA synthesis. Recent studies highlight the efficacy of combining sulfonamides with trimethoprim, another folate pathway inhibitor, in treating multidrug-resistant bacterial infections (Eliopoulos and Huovinen 2001).

Table 5 Examples of metabolic inhibitors in antimicrobial therapy

Inhibitor	Target pathway	Pathogen	Mechanism of action	References
Sulfonamides	Folate biosynthesis	Bacteria (e.g., *E. coli*)	Inhibit dihydropteroate synthase	Capasso and Supuran (2019)
Methotrexate	Folate biosynthesis	Bacteria, parasites	Inhibits dihydrofolate reductase	Aragaw et al. (2024)
Isoniazid	Mycolic acid synthesis	*M. tuberculosis*	Inhibits enoyl-ACP reductase	Prasad et al. (2021)
Trimethoprim	Folate biosynthesis	Bacteria (e.g., *E. coli*)	Inhibits dihydrofolate reductase	Wróbel et al. (2020)
Fluconazole	Ergosterol biosynthesis	Fungi (e.g., *C. albicans*)	Inhibits lanosterol 14α-demethylase	Sagatova et al. (2015)
Pyrimethamine	Folate biosynthesis	Parasites (e.g., *P. falciparum*)	Inhibits dihydrofolate reductase	Yadav et al. (2024)

This table lists various metabolic inhibitors used in antimicrobial therapy, detailing their target pathways, the pathogens they affect, and their mechanisms of action. These inhibitors are crucial in disrupting the metabolic processes essential for pathogen survival and proliferation

Similarly, methotrexate, a potent antimetabolite that inhibits dihydrofolate reductase, has demonstrated significant efficacy in combating bacterial and parasitic infections. This drug's mechanism of action involves disrupting folate metabolism, which is essential for nucleotide synthesis in both *Mycobacterium tuberculosis* and malaria parasites. For instance, methotrexate has been effectively utilized in the treatment of *Mycobacterium tuberculosis*, where its inhibition of dihydrofolate reductase interferes with the bacterium's folate metabolism, thus impeding its growth and replication (Shamshad et al. 2022; Sharma and Chauhan 2012). Similarly, in the context of malaria, methotrexate's role in targeting folate metabolism proves beneficial. Malaria parasites rely heavily on folate pathways for their development, and methotrexate's ability to inhibit dihydrofolate reductase disrupts these crucial processes, leading to reduced parasite viability and proliferation. Consequently, methotrexate's application extends beyond traditional cancer treatment, offering a valuable strategy against these challenging infectious diseases (Shamshad et al. 2022; Sharma and Chauhan 2012).

Isoniazid, another potent antimetabolite, plays a vital role in combating *Mycobacterium tuberculosis* by inhibiting mycolic acid synthesis, which is crucial for the integrity of the bacterial cell wall. This disruption of mycolic acid biosynthesis compromises the structural integrity of the cell wall, ultimately leading to bacterial cell death and effective intervention of tuberculosis (Vilchèze and Jacobs 2007). The efficacy of isoniazid is significantly enhanced when used in combination with other antitubercular drugs, as this multidrug approach not only improves therapeutic outcomes but also minimizes the risk of drug resistance development. Moreover, there is emerging evidence suggesting potential applications of isoniazid in treating infections caused by *Staphylococcus aureus* (Vilchèze and Jacobs 2019). This is due to its ability to disrupt cell wall synthesis, a critical factor in the pathogenicity of *Staphylococcus aureus*. By targeting similar pathways involved in cell wall biosynthesis, isoniazid may offer a novel therapeutic strategy against this challenging pathogen, expanding its utility beyond tuberculosis and into the treatment of staphylococcal infections. Consequently, the continued exploration of isoniazid's applications could provide valuable insights into novel treatment strategies for both established and emerging bacterial infections (Vilchèze and Jacobs 2007, 2019).

Combining inhibitors that target different metabolic pathways further enhances therapeutic efficacy and reduces the likelihood of resistance development. By employing a multi-target approach, it becomes possible to comprehensively disrupt pathogen metabolic processes, thereby improving treatment outcomes. For instance, synergistic effects have been observed when combining folate pathway inhibitors with those targeting peptidoglycan biosynthesis. This combination strategy not only amplifies the overall disruption of crucial bacterial metabolic pathways but also limits the pathogen's ability to adapt and develop resistance. The integration of these different inhibitors results in a more effective antimicrobial strategy by attacking multiple essential functions within the pathogen. Such combinatorial approaches highlight the importance of leveraging diverse therapeutic mechanisms to tackle bacterial infections more robustly and are a key area of interest in the development of novel treatments (Worthington and Melander 2013; Farha et al. 2013).

By exploiting these metabolic dependencies, researchers can develop more effective antimicrobial therapies with reduced susceptibility to resistance development. This approach holds promise in tackling the challenge of multidrug-resistant infections and improving patient outcomes. However, the design and optimization of antimetabolites must carefully consider specificity and minimize off-target effects on host cells (Nathwani and Wood 1993).

5 Metabolism-Based Diagnostics and Vaccine Development

Metabolic biomarkers have emerged as crucial tools for diagnosing infections by providing unique insights into pathogen-specific metabolic signatures (Shah et al. 2020; Ortega-Prieto and Postic 2019). These biomarkers, detectable in biological fluids such as blood or urine, offer the potential to not only identify the presence of infection but also differentiate between different pathogens (Farha et al. 2013). Developed analytical techniques such as mass spectrometry and nuclear magnetic resonance (NMR) spectroscopy play a crucial role in this field, allowing for comprehensive profiling of the metabolome and yielding a detailed snapshot of small molecule compositions in samples (Shah et al. 2020; Farha et al. 2013), as depicted in Fig. 3. Recent studies exemplify the application of metabolic biomarkers in tuberculosis (TB) diagnosis, showcasing specific blood metabolites as reliable indicators of *Mycobacterium tuberculosis* infection (Weiner et al. 2012). Such findings underscore the potential of metabolomics in early disease detection and monitoring treatment responses. Beyond diagnostics, identifying metabolic enzymes and pathways essential for pathogen survival presents promising targets for vaccine development. Vaccines designed to target these metabolic antigens can stimulate immune responses that disrupt pathogen metabolism, aiding in pathogen clearance and offering novel avenues for combating infectious diseases (Ortega-Prieto and Postic 2019; Salway 2017). Understanding the metabolic alterations induced by vaccination is critical for assessing vaccine efficacy, safety, and potential adverse effects (Diray et al. 2020). This knowledge can inform the development of biomarkers indicative of vaccine-induced immunity, facilitating more precise evaluations of vaccine responses (Diray et al. 2020; Ortega-Prieto).

Advancements in diagnostic technologies continue to leverage pathogen-specific metabolites for the development of rapid, noninvasive detection methods (Shah et al. 2020; Farha et al. 2015), as represented in Table 6. Breath tests, for instance, capitalize on detecting volatile organic compounds produced by pathogens, offering a promising approach for prompt diagnosis and surveillance of infections. These metabolite-based tools represent significant strides toward enhancing diagnostic accuracy and efficiency in clinical settings, thereby improving patient outcomes and reducing the burden of infectious diseases. By enabling quicker and more precise identification of pathogens, these advancements facilitate timely intervention and appropriate treatment, minimizing complications and transmission risks. Additionally, the noninvasive nature of these diagnostic methods enhances patient

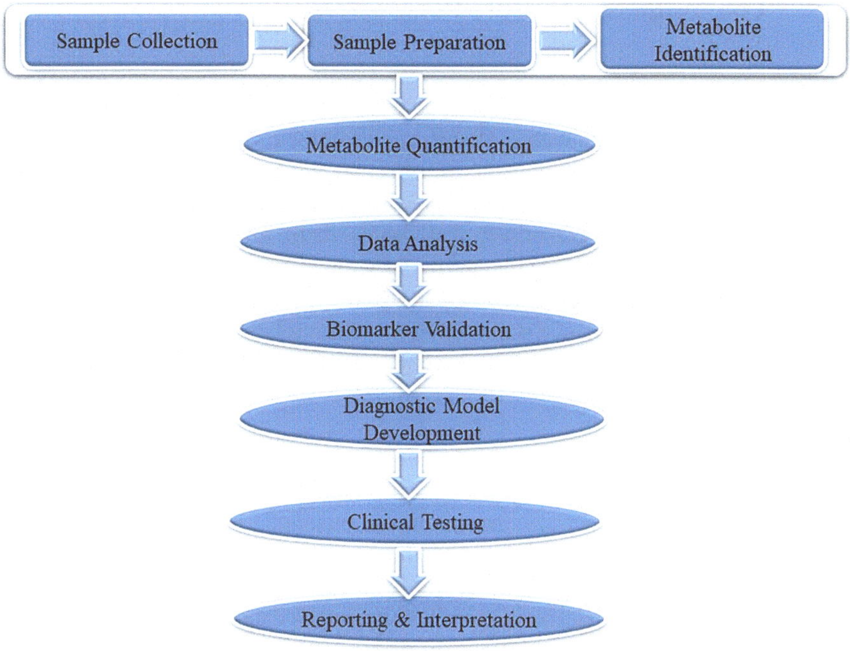

Fig. 3 Metabolite-based diagnostic workflow. This figure illustrates the metabolite-based diagnostic workflow for therapeutic invention. This workflow is designed to identify and validate biomarkers, supporting the development of targeted therapeutic interventions

Table 6 Examples of metabolic biomarkers for infectious diseases

Disease	Pathogen	Biomarker	Diagnostic method	References
Bacterial infections	Various bacteria	Lactate, succinate	NMR, mass spectrometry	Yu et al. (2023)
Fungal infections	*Candida* species	Mannan, β-glucan	Enzyme immunoassay, PCR	Malhotra et al. (2023)
Viral infections	*Hepatitis C virus*	Viral RNA	PCR, next-generation sequencing	Thomson et al. (2016)
Parasitic infections	*Trypanosoma* species	Trypanothione	Mass spectrometry, PCR	Stauning et al. (2023)

This table presents examples of metabolic biomarkers associated with various infectious diseases, identifying the pathogens involved and the diagnostic methods used to detect these biomarkers. For the early identification, diagnosis, and surveillance of infectious illnesses, these biomarkers are essential

compliance and comfort, further contributing to their widespread adoption in clinical practice. As research continues to refine these technologies, the integration of metabolite-based diagnostics is poised to revolutionize the landscape of infectious disease management, paving the way for more effective and personalized healthcare solutions (Shah et al. 2020; Farha et al. 2013).

6 Challenges and Future Perspectives

Adaptation Pathogens exhibit a remarkable capacity to adapt their metabolic strategies in response to environmental changes and therapeutic pressures, posing ongoing challenges in the development of effective antimicrobial treatments. This adaptability allows pathogens to quickly evolve alternative metabolic pathways or develop compensatory mechanisms that enable them to survive and proliferate despite metabolic inhibition (Baquero and Levin 2021; Stokes et al. 2019). Understanding the dynamic nature of pathogen metabolism and the evolutionary processes leading to resistance is crucial for devising robust and durable antimicrobial strategies that can withstand microbial adaptation (Roemhild et al. 2022; Martínez and Rojo 2011).

Complex Interactions The metabolic interactions between pathogens and their hosts are intricate and multifaceted yet still incompletely understood (Nicholson et al. 2007). Delving deeper into these complex metabolic networks and signalling pathways is essential to pinpointing new therapeutic targets and developing interventions that disrupt pathogen survival within host environments (Ortega-Prieto and Postic 2019; Salway 2017). Integration of multi-omics data and sophisticated computational modeling approaches offers promising avenues for unraveling these host-pathogen metabolic dynamics (Shah et al. 2020; Gavaghan et al. 2000a, b).

Resistance Development Resistance development against metabolic inhibitors remains a formidable obstacle in antimicrobial therapy (Roemhild et al. 2022; Martínez and Rojo 2011). Pathogens employ diverse strategies to acquire resistance, including genetic mutations in metabolic enzymes and activation of alternative metabolic pathways that bypass antimetabolite effects (Baquero and Levin 2021; Stokes et al. 2019). Overcoming resistance involves strategic combinations of therapies targeting multiple metabolic pathways simultaneously and the development of novel inhibitors with unique mechanisms of action (Worthington and Melander 2013; Farha et al. 2013). Recent studies underscore the prevalence of mutations in metabolic enzymes as a common mechanism by which bacteria evade antimetabolite therapies, underscoring the critical importance of targeting essential metabolic processes (Weiner et al. 2012; de Villiers and Du Loots 2013).

Technological Advancements Technological advancements in omics technologies, computational modeling, and bioinformatics continue to revolutionize our understanding of pathogen metabolism (Shah et al. 2020; Gavaghan et al. 2000a, b). These cutting-edge tools enable precise targeting of pathogen-specific metabolic pathways and help to establish the growth of innovative diagnostic tools and therapeutic methods that exploit vulnerabilities in pathogen metabolism. Integrating these technological breakthroughs with a comprehensive understanding of host-pathogen metabolic interactions holds promise for translating metabolic-based interventions into effective clinical practices, thereby addressing the growing challenges posed by antimicrobial resistance and infectious diseases (Ortega-Prieto and Postic 2019; Salway 2017).

7 Conclusion

In conclusion, the metabolic adaptability of pathogens showcases their evolutionary ingenuity and survival capabilities. By continuing to unravel the complexities of pathogen metabolism, we can discover new ways to combat infectious diseases. Integrating metabolic research with clinical strategies promises more potent and targeted therapies, which can enhance patient outcomes and reduce the global burden of infectious diseases. Future research should explore the metabolic pathways that allow a wider range of pathogens to adapt and develop innovative strategies to disrupt these pathways. This proactive approach can help us stay ahead in the ongoing battle against infectious diseases and antimicrobial resistance. Understanding pathogen metabolism is crucial for addressing future health challenges and emphasizes the significance of ongoing research and innovation in infectious disease management. Investing in this research will provide immediate benefits through new therapeutic options and deepen our understanding of the microbial world. Therefore, the scientific community must continue to prioritize research on pathogen metabolism to develop robust, effective, and sustainable strategies for fighting infectious diseases.

References

Allary M, Lu JZ, Zhu L, Prigge ST (2007) Scavenging of the cofactor lipoate is essential for the survival of the malaria parasite *Plasmodium falciparum*. Mol Microbiol 63(5):1331–1344

Anderson CJ, Kendall MM (2017) Salmonella enterica serovar Typhimurium strategies for host adaptation. Front Microbiol 8:1983

Aragaw WW, Negatu DA, Bungard CJ, Dartois VA, Marrouni AE, Nickbarg EB, Olsen DB, Warrass R, Dick T (2024) Pharmacological validation of dihydrofolate reductase as a drug target in Mycobacterium abscessus. Antimicrob Agents Chemother 68(1):e00717–e00723

Arjmand B, Hamidpour SK, Tayanloo-Beik A, Goodarzi P, Aghayan HR, Adibi H, Larijani B (2022) Machine learning: a new prospect in multi-omics data analysis of cancer. Front Genet 13:824451

Baquero F, Levin BR (2021) Proximate and ultimate causes of the bactericidal action of antibiotics. Nat Rev Microbiol 19(2):123–132

Bisanz C, Bastien O, Grando D, Jouhet J, Maréchal E, Cesbron-Delauw MF (2006) Toxoplasma gondii acyl-lipid metabolism: de novo synthesis from apicoplast-generated fatty acids versus scavenging of host cell precursors. Biochem J 394(1):197–205

Blume M, Seeber F et al (2015) Host-derived lipids control cholesteryl ester synthesis and storage in *Toxoplasma gondii*. J Biol Chem 290(6):3010–3021

Burgain A, Tebbji F, Khemiri I, Sellam A (2020) Metabolic reprogramming in the opportunistic yeast Candida albicans in response to hypoxia. mSphere 5(1):10–1128

Capasso C, Supuran CT (2019) Dihydropteroate synthase (sulfonamides) and dihydrofolate reductase inhibitors. In: Bacterial resistance to antibiotics–From molecules to man. Wiley, pp 163–172

Caspi R, Altman T, Billington R, Dreher K, Foerster H, Fulcher CA, Holland TA, Keseler IM, Kothari A, Kubo A, Krummenacker M (2014) The MetaCyc database of metabolic pathways and enzymes and the BioCyc collection of Pathway/Genome Databases. Nucleic Acids Res 42(D1):D459–D471

Clay K (2014) Defensive symbiosis: a microbial perspective. Funct Ecol 28(2):293–298

Coppens I (2006) Contribution of host lipids to *Toxoplasma* pathogenesis. Cell Microbiol 8(1):1–9

Coppens I (2013) Targeting lipid biosynthesis and salvage in apicomplexan parasites for improved therapies. Nat Rev Microbiol 11:823–835

Curtis MM, Sperandio V (2011) A complex relationship: the interaction among symbiotic microbes, invading pathogens, and their mammalian host. Mucosal Immunol 4(2):133–138

Dahal S, Renz A, Dräger A, Yang L (2023) Genome-scale model of Pseudomonas aeruginosa metabolism unveils virulence and drug potentiation. Commun Biol 6(1):165

de Villiers L, Du Loots T (2013) Using metabolomics for elucidating the mechanisms related to tuberculosis treatment failure. Curr Metabol 1(4):306–317

Diray-Arce J, Conti MG, Petrova B, Kanarek N, Angelidou A, Levy O (2020) Integrative metabolomics to identify molecular signatures of responses to vaccines and infections. Meta 10(12):492

Dong Y, Malitsky S (2024) MetaboReport: from metabolomics data analysis to comprehensive reporting. Bioinformatics 40:btae373

Dutta A, Singh SK, Ghosh P, Mukherjee R, Mitter S, Bandyopadhyay D (2006) In silico identification of potential therapeutic targets in the human pathogen *Helicobacter pylori*. In Silico Biol 6(1–2):43–47

Eisenreich W, Rudel T, Heesemann J, Goebel W (2017) To eat and to be eaten: mutual metabolic adaptations of immune cells and intracellular bacterial pathogens upon infection. Front Cell Infect Microbiol 7:316

Eliopoulos GM, Huovinen P (2001) Resistance to trimethoprim-sulfamethoxazole. Clin Infect Dis 32(11):1608–1614

Farha MA, Verschoor CP, Bowdish D, Brown ED (2013) Collapsing the proton motive force to identify synergistic combinations against *Staphylococcus aureus*. Chem Biol 20(9):1168–1178

Faria JP, Rocha M, Rocha I, Henry CS (2018) Methods for automated genome-scale metabolic model reconstruction. Biochem Soc Trans 46(4):931–936

Ferreira R, Fardilha M, Rocha H, Coimbra MA, Oliveira PF (2024) Glycolysis: breaking paradigms since Meyerhof's contribution. In: Glycolysis. Academic Press, pp 3–14

Garcia JR, Gerardo NM (2014) The symbiont side of symbiosis: do microbes really benefit? Front Microbiol 5:510

Gast RJ, Sanders RW, Caron DA (2009) Ecological strategies of protists and their symbiotic relationships with prokaryotic microbes. Trends Microbiol 17(12):563–569

Gavaghan CL, Holmes E, Lenz E, Wilson ID, Nicholson JK (2000a) An NMR-based metabonomic approach to investigate the biochemical consequences of genetic strain differences: application to the C57BL10J and Alpk: ApfCD mouse. FEBS Lett 484(3):169–174

Gavaghan CL, Nicholson JK, Connor SC, Wilson ID, Wright B, Holmes E (2000b) Directly coupled high-performance liquid chromatography and nuclear magnetic resonance spectroscopic with chemometric studies on metabolic variation in Sprague–Dawley rats. Anal Biochem 291(2):245–252

Gouzy A, Poquet Y, Neyrolles O (2014) Nitrogen metabolism in *Mycobacterium tuberculosis* physiology and virulence. Nat Rev Microbiol 12(11):729–737

Harris M (2023) Machine learning and artificial intelligence for pathogen identification and antibiotic resistance detection: advancing diagnostics for urinary tract infections. BioMed 3(2):246–255

Hofer U (2014) The value of vaccines. Nat Rev Microbiol 12(1):6–6

Hossain S (2023) Exploring the role of protein homeostasis in regulating *Candida albicans* Morphogenesis (Doctoral dissertation, University of Toronto (Canada)

Jiao W, Li M, Lei T, Liu X, Zhang J, Hu J, Zhang X, Liu J, Shi S, Pan H, Zhang Y (2024) The APSES transcription factor SsStuA regulating cell wall integrity is essential for Sclerotia formation and pathogenicity in Sclerotinia sclerotiorum. J Fungi 10(4):238

Juneja S, Tasleem M (2024) Role of *Helicobacter pylori* virulence factor CagA in gastric cancer. Life Sci Res Commun 1(1):14–17

Kamali M, Sarvtin MT (2023) Insights into *candida albicans*: a new perspective on pathogenic factors and regulatory mechanisms. Int J Med Lab 10:91

Kanehisa M, Sato Y, Kawashima M (2022) KEGG mapping tools for uncovering hidden features in biological data. Protein Sci 31(1):47–53

Karn SK, Fang G, Duan J (2017) Bacillus sp. acting as dual role for corrosion induction and corrosion inhibition with carbon steel (CS). Front Microbiol 8:2038

Kita K, Hirawake H, Miyadera H, Amino H, Takeo S (2002) Role of complex II in anaerobic respiration of the parasite mitochondria from Ascaris suum and Plasmodium falciparum. Biochim Biophys Acta Bioenerg 1553(1–2):123–139

Krishnamurthy A, Kyd J (2014) The roles of epithelial cell contact, respiratory bacterial interactions and phosphorylcholine in promoting biofilm formation by Streptococcus pneumoniae and nontypeable Haemophilus influenzae. Microbes Infect 16(8):640–647

Krysenko S, Wohlleben W (2022) Polyamine and ethanolamine metabolism in bacteria as an important component of nitrogen assimilation for survival and pathogenicity. Med Sci 10(3):40

La Rosa R, Johansen HK, Molin S (2019) Adapting to the airways: metabolic requirements of Pseudomonas aeruginosa during the infection of cystic fibrosis patients. Meta 9(10):234

Lopez-Garcia AV, AbuOun M, Nunez-Garcia J, Nale JY, Gaylov EE, Phothaworn P, Sukjoi C, Thiennimitr P, Malik DJ, Korbsrisate S, Clokie MR (2023) Pathogen genomics and phage-based solutions for accurately identifying and controlling *Salmonella* pathogens. Front Microbiol 14:1166615

Malhotra S, Ranjan V, Suman C, Patil S, Malhotra A, Bhatia NK (2023) Advanced microbiological diagnostic techniques in fungal infections of the central nervous system. In: Viral and fungal infections of the central nervous system: a microbiological perspective. Springer Nature, Singapore, pp 419–463

Martínez JL, Rojo F (2011) Metabolic regulation of antibiotic resistance. FEMS Microbiol Rev 35(5):768–789

Martín-Núñez GM, Cornejo-Pareja I, Coin-Aragüez L, Roca-Rodríguez MDM, Muñoz-Garach A, Clemente-Postigo M, Cardona F, Moreno-Indias I, Tinahones FJ (2019) H. pylori eradication with antibiotic treatment causes changes in glucose homeostasis related to modifications in the gut microbiota. PLoS One 14(3):e0213548

Nathwani D, Wood MJ (1993) Penicillins: a current review of their clinical pharmacology and therapeutic use. Drugs 45:866–894

Ng RH, Lee JW, Baloni P, Diener C, Heath JR, Su Y (2022) Constraint-based reconstruction and analyses of metabolic models: open-source python tools and applications to cancer. Front Oncol 12:914594

Nicholson JK, Holmes E, Lindon JC (2007) Metabonomic and metabolomics techniques and their applications in mammalian systems. In: The handbook of metabonomics and metabolomics, pp 1–34

Nolan SJ, Romano JD, Coppens I (2017a) Host lipid droplets: an important source of lipids salvaged by the intracellular parasite Toxoplasma gondii. PLoS Pathog 13(6):e1006362

Nolan SJ, Romano JD, Kline JT, Coppens I (2017b) Novel approaches to combating *Toxoplasma gondii* infection: targeting parasite and host cell metabolism. Trends Parasitol 33(6):479–497

Nouwen LV, Everts B (2020) Pathogens MenTORing macrophages and dendritic cells: manipulation of mTOR and cellular metabolism to promote immune escape. Cells 9(1):161

Oberhardt MA, Puchałka J, Fryer KE, Martins dos Santos VA, Papin JA (2008) Genome-scale metabolic network analysis of the opportunistic pathogen Pseudomonas aeruginosa PAO1. J Bacteriol 190(8):2790

Oberhardt MA, Goldberg JB, Hogardt M, Papin JA (2010) Metabolic network analysis of Pseudomonas aeruginosa during chronic cystic fibrosis lung infection. J Bacteriol 192(20):5534–5548

Omsland A, Sager J, Nair V, Sturdevant DE, Hackstadt T (2012) Developmental stage-specific metabolic and transcriptional activity of *Chlamydia trachomatis* in an axenic medium. Proc Natl Acad Sci 109(48):19781–19785

Ortega-Prieto P, Postic C (2019) Carbohydrate sensing through the transcription factor ChREBP. Front Genet 10:472

Papp-Wallace KM, Endimiani A, Taracila MA, Bonomo RA (2011) Carbapenems: past, present, and future. Antimicrob Agents Chemother 55(11):4943–4960

Passalacqua KD, Charbonneau ME, O'riordan MX (2016) Bacterial metabolism shapes the host–pathogen interface. In: Virulence mechanisms of bacterial pathogens, pp 15–41

Paul S, Todd OA, Eichelberger KR, Tkaczyk C, Sellman BR, Noverr MC, Cassat JE, Fidel PL Jr, Peters BM (2024) A fungal metabolic regulator underlies infectious synergism during *Candida albicans-Staphylococcus aureus* intra-abdominal co-infection. Nat Commun 15(1):5746

Payne DD, Renz A, Dunphy LJ, Lewis T, Dräger A, Papin JA (2021) An updated genome-scale metabolic network reconstruction of Pseudomonas aeruginosa PA14 to characterize mucin-driven shifts in bacterial metabolism. npj Syst Biol Appl 7(1):37

Pellon A, Begum N, Sadeghi Nasab SD, Harzandi A, Shoaie S, Moyes DL (2022) Role of cellular metabolism during Candida-host interactions. Pathogens 11(2):184

Penkler G, Du Toit F, Adams W, Rautenbach M, Palm DC, Van Niekerk DD, Snoep JL (2015) Construction and validation of a detailed kinetic model of glycolysis in Plasmodium falciparum. FEBS J 282(8):1481–1511

Prasad MS, Bhole RP, Khedekar PB, Chikhale RV (2021) Mycobacterium enoyl acyl carrier protein reductase (InhA): a key target for antitubercular drug discovery. Bioorg Chem 115:105242

Rodaki A, Bohovych IM, Enjalbert B, Young T, Odds FC, Gow NA, Brown AJ (2009) Glucose promotes stress resistance in the fungal pathogen *Candida albicans*. Mol Biol Cell 20(22):4845–4855

Rodriguez Sánchez A (2020) Optimization of RNA-extraction for the study of human host: fungal pathogen interactions (Doctoral dissertation, Ghent University)

Roemhild R, Bollenbach T, Andersson DI (2022) The physiology and genetics of bacterial responses to antibiotic combinations. Nat Rev Microbiol 20(8):478–490

Rojo F (2010) Carbon catabolite repression in Pseudomonas: optimizing metabolic versatility and interactions with the environment. FEMS Microbiol Rev 34(5):658–684

Sagatova AA, Keniya MV, Wilson RK, Monk BC, Tyndall JD (2015) Structural insights into binding of the antifungal drug fluconazole to Saccharomyces cerevisiae lanosterol 14α-demethylase. Antimicrob Agents Chemother 59(8):4982–4989

Salway JG (2017) Metabolism at a Glance. Wiley, Germany

Seeger AY, Zaidi F, Alhayek S, Jones RM, Zohair H, Holland RL, Kim IJ, Blanke SR (2023) Host cell sensing and restoration of mitochondrial function and metabolism within *Helicobacter pylori* VacA intoxicated cells. MBio 14(5):e02117–e02123

Shah RM, McKenzie EJ, Rosin MT, Jadhav SR, Gondalia SV, Rosendale D, Beale DJ (2020) An integrated multi-disciplinary perspective for addressing challenges of the human gut microbiome. Meta 10(3):94

Shamshad H, Bakri R, Mirza AZ (2022) Dihydrofolate reductase, thymidylate synthase, and serine hydroxy methyltransferase: successful targets against some infectious diseases. Mol Biol Rep 49(7):6659–6691

Sharma M, Chauhan PM (2012) Dihydrofolate reductase as a therapeutic target for infectious diseases: opportunities and challenges. Future Med Chem 4(10):1335–1365

Sharma G, Agarwal S, Verma K, Bhardwaj R, Mathur V (2023) Therapeutic compounds from medicinal plant endophytes: molecular and metabolic adaptations. J Appl Microbiol 134(4):lxad074

Shi L, Sohaskey CD, Kana BD, Dawes S, North RJ, Mizrahi V, Gennaro ML (2005) Changes in energy metabolism of Mycobacterium tuberculosis in mouse lung and under in vitro conditions affecting aerobic respiration. Proc Natl Acad Sci 102(43):15629–15634
Shimizu K (2013) Metabolic regulation of a bacterial cell system with emphasis on Escherichia coli metabolism. Int Scholar Res Notice 1(2013):645983
Stauning MA, Jensen CS, Staalsøe T, Kurtzhals JA (2023) Detection and quantification of Plasmodium falciparum in human blood by matrix-assisted laser desorption/ionization time-of-flight mass spectrometry: a proof of concept study. Malar J 22(1):285
Stokes JM, Lopatkin AJ, Lobritz MA, Collins JJ (2019) Bacterial metabolism and antibiotic efficacy. Cell Metab 30(2):251–259
Taylor R, Al-Mrabeh A, Sattar N (2019) Understanding the mechanisms of reversal of type 2 diabetes. Lancet Diab Endocrinol 7(9):726–736
Teoh WP, Chen X, Laczkovich I, Alonzo F III (2021) Staphylococcus aureus adapts to the host nutritional landscape to overcome tissue-specific branched-chain fatty acid requirement. Proc Natl Acad Sci 118(13):e2022720118
Theriot CM, Young VB (2015) Interactions between the gastrointestinal microbiome and Clostridium difficile. Ann Rev Microbiol 69(1):445–461
Thiele I, Vo TD, Price ND, Palsson BØ (2005) Expanded metabolic reconstruction of *Helicobacter pylori (i IT341 GSM/GPR):* an Insilco genome-scale characterization of single-and double-deletion mutants. J Bacteriol 187(16):5818–5830
Thomson E, Ip CL, Badhan A, Christiansen MT, Adamson W, Ansari MA, Bibby D, Breuer J, Brown A, Bowden R, Bryant J (2016) Comparison of next-generation sequencing technologies for comprehensive assessment of full-length hepatitis C viral genomes. J Clin Microbiol 54(10):2470–2484
Troha K, Ayres JS (2020) Metabolic adaptations to infections at the organismal level. Trends Immunol 41(2):113–125
Vilchèze C, Jacobs WR Jr (2007) The mechanism of isoniazid killing: clarity through the scope of genetics. Ann Rev Microbiol 61(1):35–50
Vilchèze C, Jacobs WR Jr (2019) The isoniazid paradigm of killing, resistance, and persistence in *Mycobacterium tuberculosis*. J Mol Biol 431(18):3450–3461
Vylkova S, Carman AJ, Danhof HA, Collette JR, Zhou H, Lorenz MC (2011) The fungal pathogen *Candida albicans* autoinduces hyphal morphogenesis by raising extracellular pH. MBio 2(3):10–1128
Wang H, de Carvalho LPS (2023) Metabolomic profiling reveals bacterial metabolic adaptation strategies and new metabolites. Curr Opin Chem Biol 74:102287
Weiner J 3rd, Parida SK, Maertzdorf J, Black GF, Repsilber D, Telaar A, Mohney RP, Arndt-Sullivan C, Ganoza CA, Faé KC, Walzl G (2012) Correction: biomarkers of inflammation, immunosuppression and stress with active disease are revealed by metabolomic profiling of tuberculosis patients. PLoS One 7(12):e40221
Wenbo L, Yewei Y, Hui Z, Zhongyu L (2024) Hijacking host cell vesicular transport: new insights into the nutrient acquisition mechanism of Chlamydia. Virulence 15(1):2351234
Worthington RJ, Melander C (2013) Combination approaches to combat multidrug-resistant bacteria. Trends Biotechnol 31(3):177–184
Wróbel A, Arciszewska K, Maliszewski D, Drozdowska D (2020) Trimethoprim and other non-classical antifolates an excellent template for searching modifications of dihydrofolate reductase enzyme inhibitors. J Antibiot 73(1):5–27
Yadav GP, Arukha AP, Kothapalli Y, Singh US (2024) Antimalarial drugs: discovery, mechanism of action, and drug resistance. In: Falciparum Malaria. Academic Press, pp 89–112
Yu Y, Jiang XX, Li JC (2023) Biomarker discovery for tuberculosis using metabolomics. Front Mol Biosci 10:1099654
Yu D, Huang W, Sheng M, Zhang S, Pan H, Ren F, Luo L, Zhou J, Huang D, Tang L (2024) Angiotensin-(1-7) modulates the Warburg effect to alleviate inflammation in LPS-induced macrophages and septic mice. J Inflamm Res 17:469–485

Integrating Metabolomics with Next-Generation Approaches for Mitigating Emerging Pathogen

T. Kavya, Merugu Shashank Goud, K. K. Chetan, K. S. Rishika, K. Aswini, J. Sushmitha, J. H. Ashwini, Mallanna Malagatti, Rashmitha, and Praveen S. Patted

Abstract

Climate change and ecological invasions are altering temperature patterns and weather conditions, creating environments favorable for endemic pests, microbial pathogens, and soil-borne pathogens, particularly those with broad host ranges, which affect plant health and, consequently, environmentally sustainable crop production. Developing superior cultivars resistant to abiotic and biotic stresses through plant breeding faces numerous challenges. Plants have evolved primary defense mechanisms that produce secondary metabolites like terpenes, flavonoids, alkaloids, saponins, and glucosinolates upon pathogen attack, playing crucial roles in microbial defense infection. In recent years, pathologists have increasingly applied techniques in the omics field, such as genomic, proteomics, transcriptomics, volatilomics, spectranomics, and metabolomics to gain deeper insights into these interactions. The comprehensive analysis of small molecules

T. Kavya (✉) · M. S. Goud · K. Aswini
Division of Microbiology, ICAR-Indian Agricultural Research Institute, New Delhi, India

K. K. Chetan · J. H. Ashwini
Division of Plant Pathology, ICAR-Indian Agricultural Research Institute, New Delhi, India

K. S. Rishika
ICAR-National Institute for Plant Biotechnology (NIPB) LBS Centre, Pusa Campus, New Delhi, India

J. Sushmitha
Division of Biochemistry, ICAR-Indian Agricultural Research Institute, New Delhi, India

M. Malagatti · Rashmitha
Division of Seed Science and Technology, ICAR-Indian Agricultural Research Institute, New Delhi, India

P. S. Patted
ICAR-CPRI, Regional Station, Ooty, Tamil Nadu, India

S. Kaur et al. (eds.), *Microbial Metabolomics*,
https://doi.org/10.1007/978-981-96-4824-5_8

or metabolites in cells, tissues, or organisms has become a valuable approach for investigating plant metabolites produced in response to pathogen attacks. In recent years, the emergence of next-generation sequencing, particularly whole genome sequencing (WGS), has transformed our understanding of plant-pathogen interactions, providing in-depth insights into the intricate biochemical networks that govern plant defense mechanisms. By integrating metabolomics with the NGS technologies which helps to sequence many samples within a short time will enable the researchers understand the specific metabolites involved in plant defense but also facilitate the identification of novel biomarkers and pathways that can be targeted for crop improvement. This chapter discusses the critical role of metabolomics in conjunction with next-generation approaches, highlighting its potential to transform plant pathology and crop protection research.

1 Introduction

Agricultural production presents a formidable challenge in meeting the nutritional needs of the projected 10 billion population anticipated by 2050 (Singh 2019). Researchers emphasize the necessity of achieving a balance between crop diversity and quality by augmenting yields to sustainably address the burgeoning food demand (Singh et al. 2024). However, it is critical to diagnose fungal and oomycete diseases quickly and accurately in order to reduce the use of chemical treatments and production losses, as these pathogens cause significant losses in forestry and agriculture (Comtet et al. 2015). When it comes to non-culturable infections, for example, diagnostic techniques relying on culturing the pathogen and direct morphological observation are laborious and frequently unreliable. Although molecular methods, like DNA hybridization-based techniques, PCR-based protocols, and DNA sequence analysis, are more accurate, they can occasionally produce false-positive results and/or fail to detect all species present in a complex community or unknown species. These methods use specific oligonucleotide primers or probes. Even while Sanger sequencing of PCR products is required to acquire DNA analysis of certain marker genes, such ITS and TEF for fungi, it may be done in a few days and yield high-resolution taxonomic data—a trait that is frequently observed in noncoding sections of eukaryotes—but it is also time-consuming.

In recent decades, several promising omics technologies have emerged. Through changes in DNA, transcript levels, proteins, metabolites, and mineral nutrients against a backdrop of environmental and physiological stress responses, these omics-based approaches have shown themselves to be useful for investigating the genetic and molecular basis of crop development (Muthamilarasan et al. 2019). Various omics techniques have been employed to uncover the molecular biological aspects associated with plant systems. These techniques include genomics, mutagenomics, transcriptomics, proteomics, metabolomics, phenomics, and ionomics (Houle et al. 2010; Talukdar and Sinjushin 2015; Yang et al. 2021). The harnessing of genetic resources through genomics and metabolomics methodologies can facilitate the development of resilient plant varieties capable of withstanding future stressors (Mmbando 2024). The advent of next-generation sequencing (NGS) technologies has established a crucial framework for elucidating the genetic potential of various plant species and advancing our comprehension of the domestication processes, thereby revealing the genetic attributes of wild-type plants for application in crop enhancement. NGS forums have the capability to generate genomic sequences utilizing libraries that are formulated from fragmented RNA/DNA/amplicon with attached adapters. This methodology surpasses traditional vector-based cloning techniques, thus effectively reducing or eliminating the contaminants that often arise from cloned DNA sequences in genomic sequencing endeavors (Park 2009; Hurd and Nelson 2009). Notably, the sequencing protocols adhere to a standardized workflow, regardless of the specific sequencing forum involved: these steps include (i) the construction of libraries utilizing nucleic acids, (ii) the operation of sequencing machinery to accumulate sequenced data, and (iii) the ultimate interpretation of data via bioinformatics tools or software. The library construction phase in NGS is a crucial component that delineates sequencing technologies based on the chemical characteristics of (i) synthesis reaction systems, (ii) single-molecule long-read capabilities, and (iii) ligation chemistry (Unamba et al. 2015; Anandhakumar et al. 2015). Advancements in NGS technologies have facilitated a plethora of research platforms that provide extensive genomic data concerning various plant species through the elucidation of more intricate genomes, such as those of pea, barley, maize, cotton (allotetraploid), wheat (hexaploid), and sugarcane (octoploid).

2 Benefits of Integration of Metabolomics and NGS

Integrating metabolomics with next-generation sequencing (NGS) offers a powerful approach to combating plant pathogens by providing a comprehensive understanding of plant-pathogen interactions at both genetic and metabolic levels. NGS enables the identification of plant resistance genes, pathogen virulence factors, and genetic variations associated with susceptibility or immunity. Metabolomics complements this by capturing the dynamic changes in plant metabolites during infection, shedding light on defense mechanisms such as the production of phytoalexins, signaling molecules, and other secondary metabolites critical for resistance (Scossa et al. 2018). Together, these tools enable the discovery of biomarkers for early pathogen detection and the development of disease-resistant crops through targeted breeding or genetic engineering. This integration also aids in monitoring pathogen evolution, including the emergence of new virulent strains, by linking genetic adaptations with metabolic outcomes. Additionally, it facilitates a deeper understanding of plant microbiome interactions, as NGS profiles microbial communities, while metabolomics elucidates their functional contributions. Ultimately, this combined approach enhances precision in pathogen control strategies, improves crop resilience, and supports sustainable agricultural practices (Fig. 1).

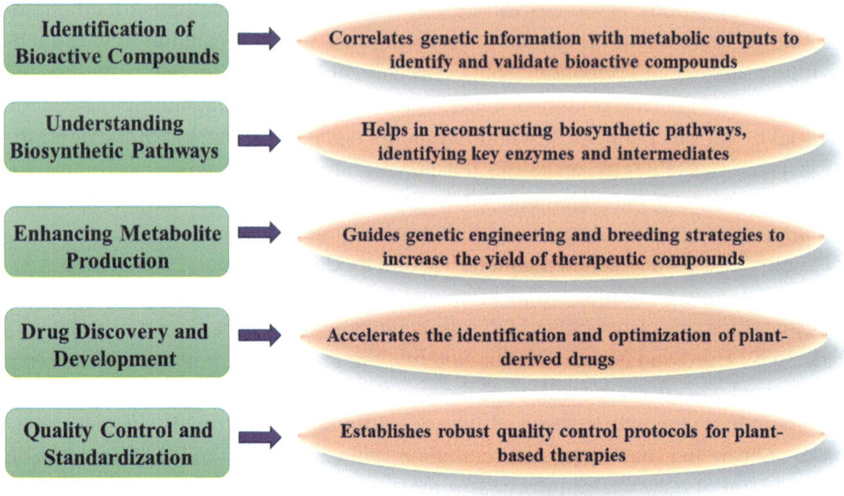

Fig. 1 Some of the key benefits of integrating metabolomics with next-generation sequencing (NGS)

3 Important Defensive Metabolites That Are Produced by Plant Under Pathogen Attack

Phytopathogens such as bacteria, fungi, protozoa, nematodes, and viruses infect the plants at initial stage of seed germination to the storage of the harvested produce. To overcome the infection by the phytopathogens, plants have developed different mechanisms by which plants produce different types of defensive metabolites to cope with the pathogens. These compounds, often produced in response to biotic stress, include phenolics, alkaloids, terpenoids, and various secondary metabolites. They not only provide a direct line of defense by inhibiting pathogen growth and reproduction but also function in signalling pathways that activate systemic acquired resistance, enhancing the plant's ability to resist future attacks. Studying the importance of these defensive metabolites is crucial not only for comprehending plant-pathogen interactions but also for their potential applications in sustainable agriculture. Based on their chemical nature and mechanism of action, these defense metabolites are classified into different categories. They are secondary metabolites, proteins, volatile compounds, and phytohormones (Fig. 2).

3.1 Secondary Metabolites

Secondary metabolites represent a wide array of organic compounds that primarily serve as defense agents, playing a crucial role in a plant's interaction with the pathogens (Zaynab et al. 2018). Secondary metabolites can be constitutively produced and stored in plant tissues or synthesized de novo when a threat is detected,

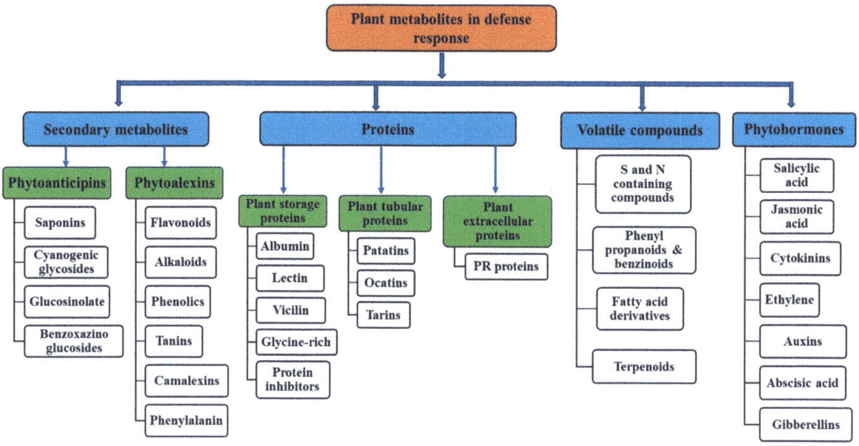

Fig. 2 Metabolites involved in plant defense mechanisms

providing both passive and active defense mechanisms. Defensive metabolites that are continuously produced and stored in plant tissues are known as phytoanticipins (VanEtten et al. 1994; Kliebenstein and Kvitko 2023), while those that are synthesized a new in response to infection are referred to as phytoalexins.

3.1.1 Phytoanticipins

Phytoanticipins are defense metabolites which are constitutively produced in the cells. Phytoanticipins require significant amounts of carbon and energy for their production, which can result in a fitness cost under natural conditions. Several such kind of molecules has been identified in plant systems. Based on their nature, these phytoanticipins are categorized into different classes such as saponins, cyanogenic glucosides, glucosiniolates, and benzoxazino glucosides (Table 1).

3.1.2 Phytoalexins

Phytoalexins are low molecular weight antimicrobial compounds synthesized by plants in response to biotic and abiotic stresses These metabolites, generated de novo in response to pathogen invasion, are collectively referred to as phytoalexins. Based on their nature, phytoalexins are categorized into different classes (Table 2).

3.2 Proteins

Plants have developed a complex defense system to safeguard against various pathogens, including bacteria, fungi, and viruses. Central to this defense mechanism are proteins that play crucial roles in recognizing, responding to, and combating pathogenic threats. These proteins are classified into storage proteins, tubular proteins, and extracellular proteins. Through the concerted action of these proteins,

Table 1 Some of the examples of phytoanticipins and their importance

Saponins	Saponins are plant glycosides named for their soap-like characteristics. The primary mechanism by which saponins act against pathogens is by forming pores in their membranes, resulting in membrane disruption and loss of membrane integrity (Sreij et al. 2019). For example, avenacins, components of an active glycolysis pathway that accumulate in the epidermal cells of root tips, exhibit potent antifungal properties, providing broad resistance against soil-borne pathogens (Osbourn 1996)
Cyanogenic glycosides	Cyanogenic glucosides are natural phytotoxins and nitrogen-containing compounds found in various parts of plants. Cyanogenic glycosides can inhibit the hyphal growth of certain fungi and have been reported to provide protection against fungal attacks, like the rice blast fungus *Magnaporthe grisea* (Seo et al. 2011)
Glucosinolates	Glucosinolates (GSLs) are secondary metabolites present in Brassicaceae species that contribute to the plant's defense against pathogens. Upon tissue damage caused by a pathogen, the myrosinase enzyme breaks down glucosinolates into various products, which demonstrate antimicrobial properties against a broad spectrum of fungi and bacteria in vitro. It has been reported that glucosinolates are toxic to the fungal pathogen like *Sclerotinia sclerotiorum* and bacteria *Xanthomonas compestris* pv. compestris (Madloo et al. 2019)
Benzoxazino glucosides	Benzoxazino glucosides represent a unique category of defensive metabolites found in the root exudates of the Gramineae family. Benzoxazino glucosides have shown significant potential in protecting plants from both biotic and abiotic stresses, drawing considerable interest in recent years (Mwendwa et al. 2021). Notably, DIMBOA (2,4-dihydroxy-7-methoxy-1,4-benzoxazin-3-one) and its degradation product MBOA (6-methoxy-benzoxazolin-2-one) are two of the most thoroughly researched benzoxazino glucosides. It has been reported that benzoxazino glucosides are very efficient in controlling the pathogen *Fusarium oxysporum* f. sp. Fabae (Ma et al. 2024)

plants can mount effective defenses, minimizing damage and ensuring survival despite ongoing pathogen challenges.

3.2.1 Storage Proteins

Storage proteins play crucial roles in plant persistence by serving as molecular reserves vital for growth. Additionally, these proteins contribute to defense mechanisms owing to their insecticidal and antimicrobial properties. Plant albumin proteins are a diverse group of soluble proteins that are primarily found in the aqueous portions of seeds, leaves, and other plant tissues. Unlike storage proteins such as glutenins and prolamins, which are primarily involved in nutrient storage, albumins play a variety of roles that support plant health and development. The antifungal activity of seed proteins from *Passiflora* sp. has been studied that albumin-like proteins in *Passiflora edulis f. flavicarpa* have shown effectiveness against the phytopathogens *Fusarium oxysporum*, *Colletotrichum musae*, and *Colletotrichum lindemuthianum*, as well as the yeast *Saccharomyces cerevisiae* (Agizzio et al. 2003). Vicilins, which are part of the cupin protein superfamily, is characterized by their spherical shape and capability to form trimers or hexamers. They play a

Table 2 Different types of phytoalexins and their importance

Flavonoids	Flavonoids are a diverse class of secondary metabolites found abundantly in plants, known for their biological significance and chemical diversity. They are effective against the pathogenic fungi such as *Fusarium* spp., *Botrytis cinerea* (gray mold), and *Colletotrichum* spp. (Jha and Mohamed 2022)
Alkaloids	Alkaloids represent one of the largest groups of plant secondary metabolites, serving as defense compounds that effectively deter pathogens and predators through their toxic properties. Zhao et al. (1998) reported that furoquinoline alkaloids are effective against the phytopathogenic fungus *Cladosporium cucumerinum*. Defense compounds like pyrrolizidine alkaloids influence the selection processes that shape the composition of the soil-borne microbial community in the rhizosphere (Kowalchuk et al. 2006)
Phenolics	Phenolic compounds are secondary metabolites that originate from the phenylpropanoid, shikimate, and pentose phosphate pathways in plants. As one of the most significant groups of phytochemicals, they have significant physiological and morphological functions. They contribute to plant growth and reproduction and provide defense against pathogens. Structurally, phenolic compounds comprise an aromatic ring, bearing one or more hydroxyl substituents, and range from simple phenolic molecules to highly polymerized compounds (Bravo 1998). Phenolic compounds have been reported to enhance plant defense mechanisms, reducing the incidence of Fusarium wilt in tomatoes caused by the fungus *Fusarium oxysporum* (Orr and Nelson 2018)
Tannins	Tannins are water-soluble polyphenolic compounds known for their capacity to precipitate proteins and include various types of phenolic substances. They are broadly categorized into two main groups: condensed tannins (CTs) and hydrolyzable tannins (HTs) (Peter Constabel et al. 2014). Tannins are some of the most commonly found secondary metabolites in plants, although their distribution is not uniform across all species and commonly present in woody plants and forest trees. Studies have reported that the production of tannins in plants is linked to defense mechanisms against various pathogens, including *Aspergillus* and *Xylella fastidiosa* in grapes (Wallis and Chen 2012) and *Colletotrichum fragariae* in strawberries
Camalexins	Camalexin is one of the most widely examined pathogen-inducible secondary metabolites with antimicrobial properties. Unlike indole glucosinolates (IGs), this tryptophan-derived compound is not widely found within the Brassicaceae family and seems to be limited to species within the Camelinae tribe (Bednarek 2012). It has been reported the production of camalexin reduced the pathogen *Botrytis cinerea* in *Arabidopsis thaliana* plants (Kachroo and Kachroo 2009).
Phenylpropanoids	Phenylpropanoids are a category of compounds that originate from the aromatic amino acid phenylalanine (Phe). Their metabolism starts either through decarboxylation, catalyzed by phenylacetaldehyde synthase, or through deamination of phenylalanine, catalyzed by phenylalanine ammonia lyase (Gonda et al. 2018). These phenylpropanoids exhibit antifungal properties and act as signaling molecules in plant-pathogen interactions (Vanholme et al. 2019). It has been shown that increased levels of phenylalanine reduced the pathogenic attack of *Botrytis cinerea* in chrysanthemum plants (Kumar et al. 2020)

protective role in plants by defending against a variety of threats, including bacteria, fungi, nematodes, insects, and other pests. The effectiveness of vicilins' antifungal activity can vary depending on the source material. Vicilins were extracted from *Vigna unguiculata*, which demonstrated inhibitory effects against the yeast *Saccharomyces cerevisiae* (Gil-Salido et al. 2024). Additionally, this vicilin interfered with the development of several fungi, including *Colletotrichum musae*, *Neurospora crassa*, *Phytophthora capsici*, *Fusarium oxysporum*, *Fusarium solani*, and *Ustilago maydis*. It was particularly effective in impacting the spore germination processes of these fungi (de Souza Candido et al. 2011). Glycine-rich proteins are versatile molecules that serve essential functions in maintaining plant resilience and adaptability, serving as both structural components and dynamic participants in the plant's reaction to environmental challenges. A glycine-rich, proline-rich protein from *Sorghum bicolor* has been reported to function as an antimicrobial protein, preventing infestation by *Rhodococcus fascians*. Protease inhibitors typically represent about 1–10 percent of the total protein content in seeds and tubers. Protease inhibitors discovered in *Carica papaya* latex, referred as papaya protease inhibitors (PPIs), show resilience to proteolytic degradation. Additionally, it has been reported that these inhibitors possess antimicrobial activity against Gram-positive bacteria (de Souza Candido et al. 2011).

3.2.2 Plant Tubular Proteins

Plant tubular proteins are a category of structural proteins that play a critical role in maintaining the architecture and function of plant cells. Patatins are a group of glycoproteins primarily found in the tubers of potatoes (*Solanum tuberosum*) and other members of the Solanaceae family. Beyond their function as nutrient reservoirs, patatins possess enzymatic activities, particularly lipid acyl hydrolase activity, which contributes to plant defense against various biotic stresses. A patatin isoform in potato tubers has been identified with β-1,3-glucanase activity and the presence of Glu-40, both of which have potential antifungal properties (Tonon et al. 2001). Ocatins are a class of proteins found in certain plant species, notably in the yam (*Dioscorea*) family. These proteins are recognized for their role in plant defense mechanisms against various pathogens and pests. As part of the plant's innate immune system, ocatins contribute to both direct and indirect defense strategies, enhancing the plant's ability to survive in hostile environments. The primary role of these proteins is to store nutrients for the plant, but they also exhibit antimicrobial properties against a range of phytopathogenic bacteria, including *Pseudomonas aureofaciens*, *Serratia marcescens*, *Agrobacterium tumefaciens*, and *Agrobacterium radiobacter*. Additionally, they have antifungal properties effective against fungi such as *Phytophthora cinnamomi*, *Nectria haematococca*, *Fusarium oxysporum*, and *Rhizoctonia solani* (de Souza Candido et al. 2011). Tarins are a unique group of lectin proteins found in certain plant species, particularly in the taro plant and *Taraxacum* genus (dandelions). These proteins are known for their ability to bind specifically to carbohydrates, which play a key role in the plant defense mechanisms against diverse biotic stressors, including pathogens and herbivores. As part of the plant's innate immune system, tarins contribute to both direct defense actions and signaling pathways that strengthen the plant's overall resistance to external threats.

3.2.3 Plant Extracellular Proteins

Plant extracellular proteins are a diverse group of proteins released into the extracellular space, or apoplast, surrounding plant cells. These proteins play a key role in the plant's protection against biotic stressors, such as pathogens and herbivores, also in response to environmental stresses (Chowdhary and Tank 2023). Positioned at the front lines of interaction with the environment, extracellular proteins are essential for detecting threats, initiating defense signaling, and executing protective responses.

3.2.4 Pathogenesis-Related Proteins (PR Proteins)

Pathogenesis-related (PR) proteins are a category of proteins that plants produce when they encounter pathogen attack or environmental stress. PR proteins are typically induced as part of the plant's systemic acquired resistance (SAR) and local defense mechanisms, providing both direct antimicrobial effects and supporting other aspects of the plant's immune response. Most of the PR proteins with demonstrated antibacterial or antifungal activity in vitro exhibit chitinase, glucanase activity, or the capacity to bind chitin. In fungi, the mechanism of action of these PR proteins involves the breakdown of structural polysaccharides in the fungal cell wall, which affects fungal growth and inhibits its development. Tomato plants that expressed tobacco class I chitinase and β-1,3-glucanase transgenes have shown to enhance tolerance levels to infection by *Fusarium oxysporum* f. sp. lycopersici (Jongedijk et al. 1995).

3.3 Plant Volatile Compounds

Plant volatile compounds constitute a varied group of organic molecules that evaporate into the air, serving as crucial signaling agents in the plant's defense response. Plant sulfur- and nitrogen-containing volatile compounds are key players in the plant defense response, contributing to a range of protective mechanisms against biological and environmental stressors (Bezerra et al. 2021). Nitriles and oximes are volatile compounds obtained from amino acids and are released by plants following herbivore damage. Many toxic nitrogen-containing volatiles are generated through the breakdown of non-volatile glycosylates through hydrolysis. For instance, glucosinolates found in the Brassicaceae family and cyanogenic glycosides present in various plants, such as cassava and rubber, are examples of such compounds. Consequently, these volatile compounds act as both physical and chemical barriers, providing a defense mechanism against herbivores and other threats (Chowdhary and Tank 2023). Phenylpropanoids and benzenoids are key classes of volatile compounds in plants that play significant roles in defense responses. These compounds, derived from the phenylpropanoid pathway, are crucial for plant protection against a variety of biotic and abiotic stressors. Phenylpropanoids absorb UV-B radiation, offering plants protection against UV light. Among benzenoids, methyl salicylate (MeSA) is particularly notable. Its release is commonly triggered by herbivore or pathogen attacks and plays a crucial role in both physical and chemical defenses against these threats (Chowdhary and Tank 2023). Fatty acid-derivative volatile compounds are crucial elements in the plant defense response, and fatty acid derivatives are synthesized in plants in response to herbivory or invasion by necrotrophic pathogens. They are derived from C18 unsaturated fatty acids. These volatiles play diverse functions in safeguarding plants against biological and environmental stressors by mediating interactions with pests, pathogens, and environmental conditions. Terpenoids are a varied group of volatile compounds that originate from the 5-carbon isoprene unit. They play a pivotal role in plant defense mechanisms, serving as key players in the plant's response to biotic and abiotic stresses. It has been reported that plant producing terpenes exhibit defense against the black spot disease caused by the fungal pathogen *Alternaria tenuissima* in the ornamental plant *Chrysanthemum morifolium* (He et al. 2023).

3.4 Phytohormones

Phytohormones, also known as plant hormones, are crucial signaling molecules that control various aspects of plant growth, development, and stress responses. Phytohormones play a pivotal role in regulating how plants perceive, respond to, and manage biotic and abiotic stresses. These hormones enable plants to mount effective defense responses against pathogens, herbivores, and environmental challenges, thereby improving their survival and resilience. Some of the phytohormones are listed in Table 3.

4 Metabolomics as a Tool to Study the Plant-Pathogen Interactions

The term "metabolome" pertains to the comprehensive examination of secondary metabolites in plants that govern various cellular processes within living organisms. Metabolomics is recognized as a rapidly advancing field of plant sciences and system biology, which encompasses the study of extensive array of antimicrobial metabolites and secondary metabolites such as proteins, amino acids, and organic acids which are synthesized through metabolic pathways within plant systems or metabolic signals that reorganize plant physiology and help adapt to under abiotic and biotic stress conditions (Xavier et al. 2017; Wang et al. 2017). Their metabolism starts either through decarboxylation, catalyzed by phenylacetaldehyde synthase, or through deamination of phenylalanine, catalyzed by phenylalanine ammonia-lyase. Research in metabolomics enables us better understanding of the functional roles of the metabolites in relation to their growth and physiological development in response to the ongoing climate change or genetic disturbances (Turner et al. 2016). Metabolic alterations are directly indicative of the postgenomic modifications

Table 3 Phytohormones involved in plant defense mechanisms

Jasmonic acid	These compounds are crucial for regulating various physiological processes related to stress responses, including the triggering of defense genes in response to herbivores and pathogens. Jasmonic acid (JA) signaling has been reported to be essential for resistance to *Peronospora parasitica* and *Pseudomonas syringae* in *Arabidopsis* plants (Shah 2003)
Ethylene	This phytohormone regulates numerous aspects of the plant life cycle, including flower development, root hair formation, root initiation, seed germination, fruit ripening, sex determination, and senescence (Mathur et al. 2023; Khan et al. 2024). Additionally, ethylene plays a crucial role in the plant's response to both abiotic stresses, such as drought and flooding, and biotic stresses, including pathogen and herbivore attacks. It has been studied that treating plants with ethylene boosts their resistance to the fungus *Botrytis cinerea* (Diaz et al. 2002)
Salicylic acid	Plants synthesize salicylic acid (SA) in reaction to challenges from various phytopathogens, which is crucial for establishing both local and systemic acquired resistance (SAR). Salicylic acid and jasmonic acid (JA) signaling has been reported to be essential for resistance to *Pseudomonas syringae* and *Peronospora parasitica* in *Arabidopsis* plants (Shah 2003)
Gibberellins	Gibberellins interact with other phytohormones which are typically linked to plant immune responses. It has been reported that GA is involved in the defense against Fusarium head blight causing fungus *Fusarium graminearum* in wheat (Buhrow et al. 2016)
Abscisic acid	ABA is recognized for its role in regulating plant responses to environmental stresses, and recent studies have also highlighted its involvement in plant defense mechanisms against pathogens (Pigolev et al. 2023). It has been reported that regulatory capacity of ABA in soya bean has shown the effective control over the growth of *Phytophthora sojae* (Cahill et al. 1993)
Auxin	Auxin signaling is crucial for a diverse range of plant growth and developmental processes, from embryogenesis to senescence. These processes are largely initiated and mediated by the expression of genes regulated by auxin (Song et al. 2023). It has been observed that auxin levels increase during infection with *Pseudomonas syringae*, aiding in the plant's ability to overcome the infestation (Park et al. 2007)

occurring in plants and aid in elucidating the plant phenome or phenotype. Consequently, metabolites are increasingly recognized as crucial and dependable instruments for deciphering mechanisms of stress tolerance or resistance in plants (Pilu et al. 2009). Numerous pathogens, such as the wide variety of bacteria, fungi, protozoa, mollicutes, viruses, and viroids, are incidental disease agents (Jayawardena et al. 2021; Gougherty and Davies 2021). Plants use their extensive and intricate chemical machinery as a primary defense mechanism against infections and abiotic stressors. Plant-stress or plant-pathogen interactions determine the processes underlying the metabolic responses of plants to stress. In the face of diverse biotic stresses, the critical function of metabolites in cereal crops, including rice, maize, and barley, has been established (Balmer et al. 2013). Given the complexity of the metabolome and the diverse physicochemical properties of metabolites, multiple analytical techniques are necessary. Gas chromatography-mass spectrometry (GC-MS), mass spectrometry (MS), liquid chromatography-mass spectrometry (LC-MS), and nuclear magnetic resonance (NMR) are the most commonly used platforms. These methods facilitate substantial data generation and advanced chemometric analysis, providing fundamental insights into metabolites. Compared to the NMR, mass spectrometry has advantage due to its higher sensitivity in finding the metabolites. Recent advancements in ionization technologies, including air pressure chemical ionization (APCI), electrospray ionization (ESI), and MALDI-TOF, have enhanced the accuracy of mass spectrometry (Manickam et al. 2023). Despite NMR's large sample requirements and lower sensitivity, it remains advantageous over MS in its ability to identify physical properties of binding sites, ligands, on proteins, direct binding to target proteins, and detect protein-ligand complex structures. Advent of GC X GC-TOF-MS has also helped in higher sample throughput and co-eluting peak separation which enabled in the identification of underivatized compounds. Recently, mass spectrometry (MS) has been combined with field asymmetric waveform ion mobility spectrometry (FAIMS), an electrophoretic technique based on ion mobility. The FAIMS method is used to detect biological samples, including volatile chemicals produced during bacterial growth. However, because of the significant diversity in metabolite polarity and volatility, the heterogeneous mixtures of pathogen and plant cells, and the rapid turnover rates of metabolites, it is difficult to understand the biochemical responses of plants to various stresses.

This necessitates both targeted and untargeted metabolomics methods, along with a range of protein assays, enzymatic assays, and transcriptomic and proteomic technologies. The targeted metabolomics involves the study of particular categories of metabolites with focused selectivity and sensitivity, while untargeted metabolomics aims to analyze all detectable metabolites, including unknown compounds. The metabolomics approach has been used in barley; about 194 resistance-related (RR) metabolites were found whose abundances are higher in resistant than in susceptible genotypes belonging to flavonoid, phenylpropanoid, fatty acid, and terpenoid metabolic pathways (Bollina et al. 2010). By employing GC-MS, the metabolic profiles of three different rice cultivars have shown many metabolites or biomarkers that are resistant to the gall midge biotype 1 (GMB1) pathogen (Agarrwal et al. 2014). Similarly, in reaction to *Xanthomonas oryzae* pv. *oryzae*, many metabolites were found in the rice crop (Sana et al. 2010). An additional investigation revealed that the application of GC-MS may detect the build-up of several metabolites in rice and barley crops as a defense against *Magnaporthe oryzae* (Parker et al. 2009). In the meanwhile, wheat crops have been shown to produce phenolic and phenylpropanoid metabolites in response to biotic stress (Gunnaiah et al. 2012). ^{1}H NMR metabolomics investigation, complemented with physicochemical and analyte-specific analyses of citrus plants infected with huanglongbing disease caused by nonculturable bacterium "*Candidatus* Liberibacter asiaticus" (Las), revealed significant differences in the concentrations of sugars, amino and organic acids, limonin glucoside, and limonin between healthy and infected citrus plants (Slisz et al. 2012). The analysis of the metabolome strawberry plants using GC-MS between the control and infected plants identified 20 metabolites as a tentative bio-marker of *Colletotrichum theobromicola* infection which helps in early diagnosis of disease and controls in further spread of disease (Dai et al. 2019). The untargeted metabolomics study of wheat-resistant genotype and susceptible genotype infected with

Fusarium head blight (FHB) caused by *F. graminearum* showed the existence of differential amino acid metabolic pathways. And also results showed that exogenous application of amino acids and *F. graminearum* inoculation assay showed that proline (Pro) and alanine (Ala) increased wheat resistance to FHB, while cysteine (Cys) aggravated the susceptibility (Zhao et al. 2021). Untargeted metabolomics analysis of tomato leaf sample infected with *Alternaria solani* versus control plants showed a significant difference in leaf metabolites, and also it was observed an upregulation and downregulation of many biomarker metabolites. Downregulated metabolite biomarkers were mapped with pathways specifically known for plant defense, suggesting their prominent role in pathogen resistance (Singh et al. 2023). Thus, metabolomics study in plant pathogen interaction strengthens plant defenses to effectively resist pathogenic attack before infection (Tegeder 2014).

5 Metabolomics with NGS Technologies as a Powerful Tool for Mitigation of Pathogens

The field of genomics has experienced significant advancements in sequencing technologies over the years, leading to the widespread use of high-throughput sequencing or next-generation sequencing (NGS), to generate data across various disciplines. NGS platforms can generate genome sequences using libraries constructed from fragmented, adapter-attached RNA, DNA, or amplicons. This method is superior to the traditional vector-based cloning approach, as it reduces or eliminates impurities caused by cloned DNA sequences in genome sequencing projects. It can also be used to obtain general data by deciphering millions of different DNA sequences at the same time (Petersen et al. 2019), making it possible to qualitatively investigate multiple types of genetic alterations. NGS is parallel sequencing, which can be used to simultaneously assess multiple genes, and also advancements in NGS technologies have provided enriched genomic information for various plant species by decoding more complex genomes to generate long reads instead of short reads with limited decoding capabilities (Montenegro et al. 2017). Using diagnostic procedures, NGS technology may also be used to identify crops certified, insect vectors that harbor infections, and quarantine programs. Thus, progress in NGS has also reduced time and costs, improved genome assemblies, and made it more feasible to manage whole-genome sequencing (WGS) tasks using NGS technologies. Library construction in next-generation sequencing (NGS) is a crucial step that determines the sequencing technologies based on the chemical composition of (i) the synthesis reaction system, (ii) single-molecule long reads, and (iii) ligation chemistry. In this context, we will briefly explore short-read sequencing, commonly known as second-generation sequencing (SGS), and long-read sequencing, referred to as third-generation sequencing (TGS). These platforms generate short reads of DNA or RNA, making sequencing faster and more affordable than first-generation methods like Sanger sequencing. The most prominent SGS (second-generation sequencing) and TGS (third-generation sequencing) technologies are showed in Fig. 3.

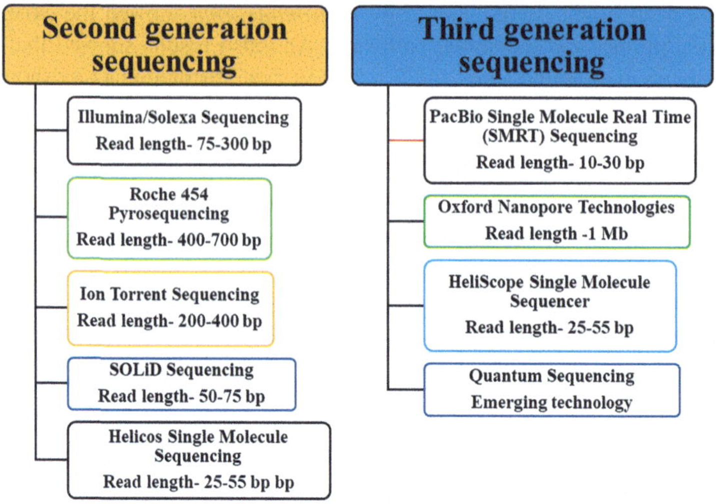

Fig. 3 Figure showing second-generation sequencing and third-generation sequencing platform

Plant metabolites, such as flavonoids, saponins, cyanogenic glycosides, terpenes, benzoxazinoids, and terpenoids, can now be detected and are formed in response to pathogen infection in a variety of crops, including rice, maize, rye, barley, oat, millet, sorghum, etc. (Diretto et al. 2010; Turner et al. 2016; Yang et al. 2019). The field of next-generation sequencing technology has shown great promise in comprehending the molecular underpinnings of cellular responses to biotic and abiotic challenges in crops, as well as the control of gene expression (Abdelrahman et al. 2018). Further, these tools can help the scientist to improve the plant yield by determining the unique treasure of genome information about the desirable traits which can help in genetic manipulation to develop high-quality cultivars that enhance crop sustainability. NGS technologies have enabled the discovery of many novel metabolites that impact plant nutrients, taste, and tolerance to biotic and abiotic stresses. In summary, DNA from plants producing metabolites of interest is extracted and sent to an NGS core facility. Currently, around 80% of DNA sequencing is performed using Illumina® technology, which produces reads between 120 and 300 bp. However, newer technologies like PacBio® have emerged, offering longer reads of up to 15 kb, enhancing the sequencing process with greater sensitivity and specificity and reduced error rates. NGS-based genome-wide genotyping utilizing techniques such as restriction site-associated DNA sequencing (RAD-seq), reduced-representation libraries (RRLs), complexity reduction of polymorphic sequences (CRoPS), and genotyping-by-sequencing (GBS) helps in SNP discovery and genotyping (Arafa et al. 2019). In plant pathology and microbiology, NGS-based methods have significantly contributed to the identification of novel molecular markers linked to resistance genes (Devran et al. 2015). For example, RAD-Seq technology was used to identify a resistance locus to an Egyptian isolate of *P. infestans* from *S. habrochaites* and also in identification of SNP markers from a tomato wild relative, *Solanum pimpinellifolium* (Arafa and Shirasawa 2018; Chen et al. 2014); GBS is also used to assess the genetic variability within four clonal lineages (US-8, US-11, US-23, and US-24) of *P. infestans*, revealing the discovery of 3774, 4363, 5070, and 4353 SNPs, respectively (Hansen et al. 2016). The pathogenic oomycete fungi *Phytophthora capsici* is the causal agent of root, stem, and foliar blights of pepper. A novel resistance gene, *PhR10*, was recently detected through NGS technology on the Illumina HiSeq 2500 platform. This race-specific gene can be utilized in breeding programs for developing varieties resistant to *Phytophthora* root rot through marker-assisted selection (Xu et al. 2016). Recently NGS technologies also emerged as promising tools for breeding, providing insights into the regulatory mechanisms and cellular responses to environmental stimuli, including both biotic and abiotic stresses. Understanding the changes in the numerous metabolites such as proline; glycine-butane; tryptophan; phenolic; organic acids; sulfur-responsible metabolites including methionine, cysteine, and glutathione; as well as phytohormones associated with critical stages or different time points during attack or infection by the pathogen allows for precise genetic manipulation by targeting specific transcripts to develop a resistant variety against the particular disease (Yang et al. 2018). For example, maize chlorotic mottle virus (MCMV) is a major threat to corn production worldwide. A real-time TaqMan reverse transcription PCR targeting coat protein gene was used in early diagnosis of MCMV viral diseases (Liu et al. 2016). NGS methods enable the identification of millions of SNPs within a genome, many of which are linked to secondary metabolites. GWAS combines phenotype and single-nucleotide polymorphism (SNP) data from natural populations to study the genetic basis of heritable phenotypes, providing valuable information for gene hunting, understanding of biological processes, and plant breeding. Matsuda et al. (2015) conducted genome-wide association studies to investigate the natural variation of rice secondary metabolites. Using GWAS and the metabolome data which sequenced a total of 175 rice accessions successfully revealed 323 associations between 143 single nucleotide polymorphisms (SNPs) and 89 metabolites. Once genes containing SNPs with direct effects on metabolites are identified, they can be edited using CRISPR/Cas9 technology to produce novel metabolic compounds, either in increased or decreased abundances (Khan et al. 2017). This technology has been widely regarded by researchers as a cost-effective, safe, and efficient tool for human cancer therapies, including lung and blood cancers (Sachdeva et al. 2015). As a result, it is considered highly safe for use in crop gene editing, offering potential for improving tolerance to biotic and abiotic stresses, enhancing nutritional quality, and creating new metabolites (Kharabian-Masouleh and Henry 2015). Nevertheless, metabolomics and NGS have given rise to a foundation for predicting an organism's early metabolic network based on its genome sequence (Weckwerth 2011). To create strategies for crop development, the information was combined using the genome sequencing technique (NGS) and measurement of metabolites (via MS) (Pandey et al. 2016). Therefore, the global decline in plant yield can be stabilized by implementing a combination approach based on the study of metabolomics and genomics.

6 Conclusion

Integrating metabolomics with next-generation approaches offers a powerful paradigm for mitigating the impacts of emerging plant pathogens. Metabolomics, with its ability to comprehensively profile biochemical changes in response to pathogen attacks, provides invaluable insights into the mechanisms of plant defense. When combined with next-generation sequencing (NGS), which deciphers the genetic and genomic intricacies of plant-pathogen interactions, this integrated approach enables the identification of novel resistance genes, defense-related metabolites, and biomarkers for early disease detection.

This synergy facilitates the development of disease-resistant crop varieties through targeted breeding and genetic engineering while enhancing our understanding of pathogen evolution and host adaptability. Furthermore, the integration of metabolomics and NGS contributes to the identification of microbial community dynamics, offering new avenues for managing plant microbiomes and improving resilience to biotic stresses.

In the face of rising threats posed by climate change, global population growth, and pathogen diversity, the combined use of metabolomics and next-generation technologies provides a robust framework for sustainable agricultural practices. This integration not only strengthens pathogen mitigation strategies but also promotes crop productivity and food security, marking a significant step forward in modern plant pathology and crop science.

References

Abdelrahman M, El-Sayed MA, Hashem A, Abd Allah EF, Alqarawi AA, Burritt DJ, Tran LSP (2018) Metabolomics and transcriptomics in legumes under phosphate deficiency in relation to nitrogen fixation by root nodules. Front Plant Sci 9:922

Agarrwal R, Bentur JS, Nair S (2014) Gas chromatography mass spectrometry based metabolic profiling reveals biomarkers involved in rice-gall midge interactions. J Integr Plant Biol 56(9):837–848

Agizzio AP, Carvalho AO, Suzanna de Fátima FR, Machado OL, Alves EW, Okorokov LA, Gomes VM (2003) A 2S albumin-homologous protein from passion fruit seeds inhibits the fungal growth and acidification of the medium by Fusarium oxysporum. Arch Biochem Biophys 416(2):188–195

Anandhakumar C, Kizaki S, Bando T, Pandian GN, Sugiyama H (2015) Advancing small-molecule-based chemical biology with next-generation sequencing technologies. Chembiochem 16(1):20–38

Arafa RA, Shirasawa K (2018) Technical review of molecular markers and next-generation sequencing technology to manage plant pathogenic oomycetes. Afr J Biotechnol 17(12):369–379

Arafa RA, Rakha MT, Kamel SM, Soliman NEK, Moussa OM, Shirasawa K (2019) Identification of resistance locus and characterization of effector genes toward breeding for resistance to *Phytophthora infestans*. In: VI International Symposium on Tomato Diseases: Managing Tomato Diseases in the Face of Globalization and Climate Change 1316, pp 9–14

Balmer D, Flors V, Glauser G, Mauch-Mani B (2013) Metabolomics of cereals under biotic stress: current knowledge and techniques. Front Plant Sci 4:82

Bednarek P (2012) Sulfur-containing secondary metabolites from *Arabidopsis thaliana* and other Brassicaceae with function in plant immunity. Chembiochem 13(13):1846–1859

Bezerra RHS, Sousa-Souto L, Santana AEG, Ambrogi BG (2021) Indirect plant defenses: volatile organic compounds and extrafloral nectar. Arthropod Plant Interact 15(4):467–489

Bollina V, Kumaraswamy GK, Kushalappa AC, Choo TM, Dion Y, Rioux S et al (2010) Mass spectrometry-based metabolomics application to identify quantitative resistance-related metabolites in barley against Fusarium head blight. Mol Plant Pathol 11(6):769–782

Bravo L (1998) Polyphenols: chemistry, dietary sources, metabolism, and nutritional significance. Nutr Rev 56(11):317–333

Buhrow LM, Cram D, Tulpan D, Foroud NA, Loewen MC (2016) Exogenous abscisic acid and gibberellic acid elicit opposing effects on Fusarium graminearum infection in wheat. Phytopathology 106(9):986–996

Cahill DM, Morris PF, Ward EWB (1993) Influence of metalaxyl on abscisic acid levels in soybean hypocotyls infected with *Phytophthora sojae*. Physiol Mol Plant Pathol 42(2):109–121

Chen AL, Liu CY, Chen CH, Wang JF, Liao YC, Chang CH et al (2014) Reassessment of QTLs for late blight resistance in the tomato accession L3708 using a restriction site associated DNA (RAD) linkage map and highly aggressive isolates of Phytophthora infestans. PLoS One 9(5):e96417

Chowdhary VA, Tank JG (2023) Biomolecules regulating Defense mechanism in plants. Proc Natl Acad Sci India B: Biol Sci 93(1):17–25

Comtet T, Sandionigi A, Viard F, Casiraghi M (2015) DNA (meta) barcoding of biological invasions: a powerful tool to elucidate invasion processes and help managing aliens. Biol Invasions 17:905–922

Dai T, Chang X, Hu Z, Liang L, Sun M, Liu P, Liu X (2019) Untargeted metabolomics based on GC-MS and chemometrics: a new tool for the early diagnosis of strawberry anthracnose caused by Colletotrichum theobromicola. Plant Dis 103(10):2541–2547

de Souza Candido E, Pinto MFS, Pelegrini PB, Lima TB, Silva ON, Pogue R et al (2011) Plant storage proteins with antimicrobial activity: novel insights into plant Defense mechanisms. FASEB J 25(10):3290–3305

Devran Z, Kahveci E, Özkaynak E, Studholme DJ, Tör M (2015) Development of molecular markers tightly linked to Pvr4 gene in pepper using next-generation sequencing. Mol Breed 35:1–9

Diaz J, ten Have A, van Kan JA (2002) The role of ethylene and wound signaling in resistance of tomato to Botrytis cinerea. Plant Physiol 129(3):1341–1351

Diretto G, Al-Babili S, Tavazza R, Scossa F, Papacchioli V, Migliore M et al (2010) Transcriptional-metabolic networks in β-carotene-enriched potato tubers: the long and winding road to the Golden phenotype. Plant Physiol 154(2):899–912

Gil-Salido AA, Rojas-Cabeza JF, Sotelo-Mundo RR, Islas-Osuna MA (2024) Role of vicilin in plant defense. In: Defense-related proteins in plants. Academic Press, pp 379–395

Gonda I, Davidovich-Rikanati R, Bar E, Lev S, Jhirad P, Meshulam Y et al (2018) Differential metabolism of L–phenylalanine in the formation of aromatic volatiles in melon (Cucumis melo L.) fruit. Phytochemistry 148:122–131

Gougherty AV, Davies TJ (2021) Towards a phylogenetic ecology of plant pests and pathogens. Philos Trans R Soc B 376(1837):20200359

Gunnaiah R, Kushalappa AC, Duggavathi R, Fox S, Somers DJ (2012) Integrated metabolo-proteomic approach to decipher the mechanisms by which wheat QTL (Fhb1) contributes to resistance against Fusarium graminearum. PLoS One 7(7):e40695

Hansen ZR, Everts KL, Fry WE, Gevens AJ, Grünwald NJ, Gugino BK et al (2016) Genetic variation within clonal lineages of Phytophthora infestans revealed through genotyping-by-sequencing, and implications for late blight epidemiology. PLoS One 11(11):e0165690

He X, Jiang Y, Chen S, Chen F, Chen F (2023) Terpenoids and their possible role in Defense against a fungal pathogen *Alternaria tenuissima* in *Chrysanthemum morifolium* cultivars. J Plant Growth Regul 42(2):1144–1157

Houle D, Govindaraju DR, Omholt S (2010) Phenomics: the next challenge. Nat Rev Genet 11(12):855–866

Hurd PJ, Nelson CJ (2009) Advantages of next-generation sequencing versus the microarray in epigenetic research. Brief Funct Genomics Proteomics 8(3):174–183
Jayawardena RS, Hyde KD, de Farias ARG, Bhunjun CS, Ferdinandez HS, Manamgoda DS et al (2021) What is a species in fungal plant pathogens? Fungal Divers 109(1):239–266
Jha Y, Mohamed HI (2022) Plant secondary metabolites as a tool to investigate biotic stress tolerance in plants: a review. Gesunde Pflanzen 74(4):771–790
Jongedijk E, Tigelaar H, Van Roekel JS, Bres-Vloemans SA, Dekker I, van den Elzen PJ, Melchers LS (1995) Synergistic activity of chitinases and β-1, 3-glucanases enhances fungal resistance in transgenic tomato plants. Euphytica 85:173–180
Kachroo A, Kachroo P (2009) Fatty acid–derived signals in plant Defense. Annu Rev Phytopathol 47(1):153–176
Khan N, Ali S, Shahid MA, Kharabian-Masouleh A (2017) Advances in detection of stress tolerance in plants through metabolomics approaches. Plant Omics 10(3):153–163
Khan S, Alvi AF, Khan NA (2024) Role of ethylene in the regulation of plant developmental processes. Stress 4(1):28–53
Kharabian-Masouleh A, Henry RJ (2015) Genome editing can boost yield and quality in rice. TropAg2015. 16–18 November, Brisbane, Australia
Kliebenstein DJ, Kvitko BH (2023) Better living through phytochemistry: "Phytoavengins" and reappraising the production-focused dichotomy for defensive phytochemicals. Physiol Mol Plant Pathol 125:101978
Kowalchuk GA, Hol WG, Van Veen JA (2006) Rhizosphere fungal communities are influenced by Senecio jacobaea pyrrolizidine alkaloid content and composition. Soil Biol Biochem 38(9):2852–2859
Kumar V, Hatan E, Bar E, Davidovich-Rikanati R, Doron-Faigenboim A, Spitzer-Rimon B, Oren-Shamir M (2020) Phenylalanine increases *chrysanthemum* flower immunity against *Botrytis cinerea* attack. Plant J 104(1):226–240
Ma G, Hu B, Yang S, Cen Z, Zheng Y, Dong Y (2024) Benzoxazinoids secreted by wheat root weaken the pathogenicity of *Fusarium oxysporum* f. sp. fabae by inhibiting linoleic acid and nucleotide metabolisms. Plant Cell Rep 43(4):109
Madloo P, Lema M, Francisco M, Soengas P (2019) Role of major glucosinolates in the Defense of kale against *Sclerotinia sclerotiorum* and *Xanthomonas campestris* pv. campestris. Phytopathology 109(7):1246–1256
Manickam S, Rajagopalan VR, Kambale R, Rajasekaran R, Kanagarajan S, Muthurajan R (2023) Plant metabolomics: current initiatives and future prospects. Curr Issues Mol Biol 45(11):8894–8906
Mathur M, Khan EA, Prajapat RK, Ahmed HM, Sharma M, Sharma D (2023) Ethylene synthesis and redox homeostasis in plants: recent advancement. In: Gasotransmitters signaling in plant abiotic stress: gasotransmitters in adaptation of plants to abiotic stress. Springer, pp 93–107
Matsuda F, Nakabayashi R, Yang Z, Okazaki Y, Yonemaru JI, Ebana K et al (2015) Metabolome-genome-wide association study dissects genetic architecture for generating natural variation in rice secondary metabolism. Plant J 81(1):13–23
Mmbando GS (2024) Omics: a new, promising technologies for boosting crop yield and stress resilience in African agriculture. Plant Stress 11:100366
Montenegro JD, Golicz AA, Bayer PE, Hurgobin B, Lee H, Chan CKK et al (2017) The pangenome of hexaploid bread wheat. Plant J 90(5):1007–1013
Muthamilarasan M, Singh NK, Prasad M (2019) Multi-omics approaches for strategic improvement of stress tolerance in underutilized crop species: a climate change perspective. Adv Genet 103:1–38
Mwendwa JM, Weston PA, Weidenhamer JD, Fomsgaard IS, Wu H, Gurusinghe S, Weston LA (2021) Metabolic profiling of benzoxazinoids in the roots and rhizosphere of commercial winter wheat genotypes. Plant Soil 466:467–489
Orr R, Nelson PN (2018) Impacts of soil abiotic attributes on Fusarium wilt, focusing on bananas. Appl Soil Ecol 132:20–33

Osbourn AE (1996) Preformed antimicrobial compounds and plant Defense against fungal attack. Plant Cell 8(10):1821

Pandey MK, Roorkiwal M, Singh VK, Ramalingam A, Kudapa H, Thudi M et al (2016) Emerging genomic tools for legume breeding: current status and future prospects. Front Plant Sci 7:455

Park PJ (2009) ChIP–seq: advantages and challenges of a maturing technology. Nat Rev Genet 10(10):669–680

Park JE, Park JY, Kim YS, Staswick PE, Jeon J, Yun J et al (2007) GH3-mediated auxin homeostasis links growth regulation with stress adaptation response in Arabidopsis. J Biol Chem 282(13):10036–10046

Parker D, Beckmann M, Zubair H, Enot DP, Caracuel-Rios Z, Overy DP et al (2009) Metabolomic analysis reveals a common pattern of metabolic re-programming during invasion of three host plant species by *Magnaporthe grisea*. Plant J 59(5):723–737

Peter Constabel C, Yoshida K, Walker V (2014) Diverse ecological roles of plant tannins: plant Defense and beyond. In: Recent advances in polyphenol research, Springer, pp 115–142

Petersen LM, Martin IW, Moschetti WE, Kershaw CM, Tsongalis GJ (2019) Third-generation sequencing in the clinical laboratory: exploring the advantages and challenges of nanopore sequencing. J Clin Microbiol 58(1):10–1128

Pigolev AV, Degtyaryov EA, Miroshnichenko DN, Savchenko TV (2023) Prospects for the application of jasmonates, salicylates, and abscisic acid in agriculture to increase plant stress resistance. Sel'skokhozyaistvennaya Biologiya 58(1):3–22

Pilu R, Panzeri D, Cassani E, Badone FC, Landoni M, Nielsen E (2009) A paramutation phenomenon is involved in the genetics of maize low phytic acid1-241 (lpa1-241) trait. Heredity 102(3):236–245

Sachdeva M, Sachdeva N, Pal M, Gupta N, Khan IA, Majumdar M, Tiwari A (2015) CRISPR/Cas9: molecular tool for gene therapy to target genome and epigenome in the treatment of lung cancer. Cancer Gene Ther 22:509–517. https://doi.org/10.1038/cgt.2015.54

Sana TR, Fischer S, Wohlgemuth G, Katrekar A, Jung KH, Ronald PC, Fiehn O (2010) Metabolomic and transcriptomic analysis of the rice response to the bacterial blight pathogen Xanthomonas oryzae pv. oryzae. Metabolomics 6:451–465

Scossa F, Benina M, Alseekh S, Zhang Y, Fernie AR (2018) The integration of metabolomics and next-generation sequencing data to elucidate the pathways of natural product metabolism in medicinal plants. Planta Med 84(12/13):855–873

Seo S, Mitsuhara I, Feng J, Iwai T, Hasegawa M, Ohashi Y (2011) Cyanide, a coproduct of plant hormone ethylene biosynthesis, contributes to the resistance of rice to blast fungus. Plant Physiol 155(1):502–514

Shah J (2003) The salicylic acid loop in plant Defense. Curr Opin Plant Biol 6(4):365–371

Singh P (2019) Feeding 1.7 billion. In: Presidential Address: Foundation Day and 26th General Body Meeting, National Academy of Agricultural Sciences (NAAS), 5th June

Singh DP, Bisen MS, Prabha R, Maurya S, Yerasu SR, Shukla R et al (2023) Untargeted metabolomics of Alternaria solani-challenged wild tomato species Solanum cheesmaniae revealed key metabolite biomarkers and insight into altered metabolic pathways. Metabolites 13(5):585

Singh A, Sharma V, Grenier M, Singh V, Thakur M (2024) Nourishing the future: introduction to sustainable food systems with concepts and framework. In: Sustainable food systems (Volume I) SFS: framework, sustainable diets, traditional food culture & food production. Cham, Springer Nature Switzerland, pp 3–24

Slisz AM, Breksa AP III, Mishchuk DO, McCollum G, Slupsky CM (2012) Metabolomic analysis of citrus infection by '*Candidatus* Liberibacter' reveals insight into pathogenicity. J Proteome Res 11(8):4223–4230

Song X, Xiong Y, Kong X, Huang G (2023) Roles of auxin response factors in rice development and stress responses. Plant Cell Environ 46(4):1075–1086

Sreij R, Dargel C, Hannappel Y, Jestin J, Prévost S, Dattani R, Hellweg T (2019) Temperature dependent self-organization of DMPC membranes promoted by intermediate amounts of the saponin aescin. Biochim Biophy Acta Biomembr 1861(5):897–906

Talukdar D, Sinjushin A (2015) Cytogenomics and mutagenomics in plant functional biology and breeding. In: PlantOmics: the omics of plant science. Springer, New Delhi, pp 113–156

Tegeder M (2014) Transporters involved in source to sink partitioning of amino acids and ureides: opportunities for crop improvement. J Exp Bot 65:1865–1878. https://doi.org/10.1093/jxb/eru012

Tonon C, Daleo G, Oliva C (2001) An acidic β-1, 3 glucanase from potato tubers appears to be patatin. Plant Physiol Biochem 39(10):849–854

Turner MF, Heuberger AL, Kirkwood JS, Collins CC, Wolfrum EJ, Broeckling CD et al (2016) Non-targeted metabolomics in diverse sorghum breeding lines indicates primary and secondary metabolite profiles are associated with plant biomass accumulation and photosynthesis. Front Plant Sci 7:953

Unamba CI, Nag A, Sharma RK (2015) Next generation sequencing technologies: the doorway to the unexplored genomics of non-model plants. Front Plant Sci 6:1074

VanEtten HD, Mansfield JW, Bailey JA, Farmer EE (1994) Two classes of plant antibiotics: phytoalexins versus "phytoanticipins". Plant Cell 6(9):1191

Vanholme R, De Meester B, Ralph J, Boerjan W (2019) Lignin biosynthesis and its integration into metabolism. Curr Opin Biotechnol 56:230–239

Wallis CM, Chen J (2012) Grapevine phenolic compounds in xylem sap and tissues are significantly altered during infection by *Xylella fastidiosa*. Phytopathology 102(9):816–826

Wang C, Hu S, Gardner C, Lübberstedt T (2017) Emerging avenues for utilization of exotic germplasm. Trends Plant Sci 22(7):624–637

Weckwerth W (2011) Unpredictability of metabolism—the key role of metabolomics science in combination with next-generation genome sequencing. Anal Bioanal Chem 400:1967–1978

Xavier A, Hall B, Hearst AA, Cherkauer KA, Rainey KM (2017) Genetic architecture of phenomic-enabled canopy coverage in Glycine max. Genetics 206(2):1081–1089

Xu X, Chao J, Cheng X, Wang R, Sun B, Wang H et al (2016) Mapping of a novel race specific resistance gene to phytophthora root rot of pepper (Capsicum annuum) using bulked segregant analysis combined with specific length amplified fragment sequencing strategy. PLoS One 11(3):e0151401

Yang L, Wen KS, Ruan X, Zhao YX, Wei F, Wang Q (2018) Response of plant secondary metabolites to environmental factors. Molecules 23(4):762

Yang M, Yang J, Su L, Sun K, Li D, Liu Y et al (2019) Metabolic profile analysis and identification of key metabolites during rice seed germination under low-temperature stress. Plant Sci 289:110282

Yang Y, Saand MA, Huang L, Abdelaal WB, Zhang J, Wu Y et al (2021) Applications of multiomics technologies for crop improvement. Front Plant Sci 12:563953

Zaynab M, Fatima M, Abbas S, Sharif Y, Umair M, Zafar MH, Bahadar K (2018) Role of secondary metabolites in plant Defense against pathogens. Microb Pathog 124:198–202

Zhao W, Wolfender JL, Hostettmann K, Xu R, Qin G (1998) Antifungal alkaloids and limonoid derivatives from *Dictamnus dasycarpus*. Phytochemistry 47(1):7–11

Zhao P, Gu S, Han C, Lu Y, Ma C, Tian J et al (2021) Targeted and untargeted metabolomics profiling of wheat reveals amino acids increase resistance to fusarium head blight. Front Plant Sci 12:762605

Microbial Metabolomics for Nutraceutical Developments and Their Applications

Gopinath Ramalingam, Arundadhi Muthukumar, and Sucila Thangam Ganesan

Abstract

Microbial metabolomics has developed as an important tool in nutraceutical research, offering information about the various bioactive substances produced by microbes and their potential health benefits. This technique uses thorough metabolite profiling to discover novel bioactive chemicals such as polyphenols, flavonoids, alkaloids, and vitamins that have significant health advantages. Furthermore, microbial metabolomics improves production processes by identifying critical metabolic pathways and environmental variables that influence chemical yield and efficacy. The characterization of these bioactive metabolites focuses on their functional features, such as antioxidant, anti-inflammatory, antibacterial, and anticancer activity, emphasizing their therapeutic benefits. This data guides formulation techniques, assuring the stability, bioavailability, and efficacy of finished products. Furthermore, microbial metabolomics plays an important role in quality control by providing accurate methods for validating the consistency and integrity of nutraceutical products. The discipline also investigates personalized nutrition techniques, which use metabolomic data to adapt dietary recommendations based on individual metabolic profiles for the best health outcomes. Beyond nutraceutical uses, microbial metabolomics holds promise for drug discovery and environmental sustainability, potentially leading to advances in pharmaceuticals and improved microbial processes for bioremediation and wastewater treatment. This study investigates the use of microbial metabolomics in nutraceutical research, concentrating on its involvement in immunological modulation, anti-inflammatory effects, and metabolic control, as well as the possibility for generating functional foods and supplements that benefit human health. The implications of personalized nutrition based on microbi-

G. Ramalingam (✉) · A. Muthukumar · S. T. Ganesan
Department of Microbiology-VRDL, Government Theni Medical College, Theni, Tamil Nadu, India

S. Kaur et al. (eds.), *Microbial Metabolomics*,
https://doi.org/10.1007/978-981-96-4824-5_9

ome profiles are also explored, demonstrating how tailored nutraceuticals can improve efficacy and safety.

Keywords

Microbial metabolomics · Nutraceuticals · Bioactive compounds · Microbial fermentation · Personalized nutrition

1 Introduction

In recent years, the field of nutraceutical development has undergone substantial transformations with the introduction of microbial metabolomics, which serves as a breakthrough tool for understanding and utilizing the potential of microbial metabolites (Graham and Ledesma-Amaro 2023). Microorganisms have complex metabolic processes and are known as important producers of bioactive chemicals with significant health benefits. Metabolomics can help researchers understand the diversity and variety of microbial metabolism, revealing significant opportunities for nutraceutical innovation (Galanakis 2019). Microbial metabolomics is the systematic examination of tiny chemicals, known as metabolites, produced by bacteria. These metabolites are vital for energy metabolism, communication, and defense mechanisms, and they contribute to many aspects of microbial physiology and ecology. Recent advances in analytical methodologies and computer technologies have enabled remarkable precision in assessing microbial communities' metabolic fingerprints (Tang 2011). The basic goal of microbial metabolomics is to fully define and understand microorganisms' metabolic patterns. This process entails finding, quantifying, and comprehending the activities of metabolites produced by microbial communities in various environments. Advanced techniques, such as mass spectrometry and nuclear magnetic resonance spectroscopy, help to investigate microbial metabolic pathways and interactions (Ye et al. 2022). Understanding these small chemicals, which are a fast emerging subfield of metabolomics, is crucial for many biological functions such as nutrition absorption, energy production, and cellular communication. Microbial metabolomics insights can greatly promote nutraceutical development by shedding light on the complex interactions between bacteria, their surroundings, and host organisms (Covington et al. 2017). Microbial metabolomics is important in nutraceutical research because it can find bioactive chemicals that may have health benefits. These microbial metabolites provide therapeutic characteristics to a wide range of functional foods, herbal medicines, and dietary supplements, known as nutraceuticals. By studying the metabolomes of various microbes, researchers can identify novel compounds with desirable biological functions, paving the path for the development of novel nutraceutical products (Merlo et al. 2011). Furthermore, microbial metabolomics aids in the optimization of fermentation processes required for the production of probiotics, prebiotics, and other microbially based supplements. Scientists can optimize culture conditions to enhance bioactive metabolite output and efficacy by analyzing changes in metabolite profiles throughout fermentation. This optimization increases not only the stability and reproducibility of manufacturing processes but also the overall quality of nutraceutical goods (Oka and Sartor 2020). In addition to product development, microbial metabolomics helps us understand the mechanisms that underpin the health-promoting effects of probiotics and prebiotics. Researchers can investigate the intricate interplay between the gut microbiota and human health by better understanding the link between microbial metabolites and host physiology. This understanding aids in the development of targeted medicines aiming at modifying the gut microbiota to address a variety of health conditions, including metabolic illnesses and gastrointestinal ailments (Vernocchi et al. 2016). Overall, microbial metabolomics is a potential field for improving nutraceutical production. This interdisciplinary topic explores microorganisms' metabolic capacities and their effects on human health, opening up new pathways for the discovery of bioactive chemicals and the improvement of microbial-derived products. Microbial metabolomics has enormous potential to revolutionize the nutraceutical landscape, providing novel techniques to promote health and well-being through continuing research and development (Kellogg and Kang 2020). This chapter also provides practical insights into optimizing fermentation processes for nutraceutical manufacturing, highlighting the necessity of fine-tuning growth conditions to increase bioactive metabolite output, efficiency, and consistency. Furthermore, it investigates the complex interactions between microbial metabolites and the gut microbiome, offering light on the impact of microbial metabolomics on immunological and gut health. During this investigation, the chapter finds possible uses of microbial metabolomics in microbiome-based therapeutics and customized nutrition, providing useful insights into future paths for nutraceutical research.

2 Bioactive Compounds Unveiled Through Microbial Metabolomics

The discovery of bioactive chemicals via microbial metabolomics has immense promise for the advancement of nutraceuticals and human health. An advanced analytical method called "microbial metabolomics" explores the complex metabolic pathways of microbes and uncovers a wide range of tiny compounds with various biological functions (Mishra et al. 2022). Table 1 outlines the main kinds of bioactive chemicals discovered using microbial metabolomics, including examples, health advantages, and methods of action. Each class of chemical demonstrates the broad potential of microbial metabolites to benefit human health, focusing on their roles in antioxidant activity, inflammation control, and gut health support.

2.1 Exploration of Microbial Metabolites

Microbial metabolomics is the systematic study of the metabolites generated by bacteria in varied settings. This field exposes a vast range of chemicals, including

Table 1 Bioactive compounds unveiled through microbial metabolomics

Compound class	Examples	Health benefits	Mechanism of action
Polyphenols	Resveratrol, curcumin	Antioxidant, anti-inflammatory, cardioprotective	Scavenges free radicals, modulates inflammatory pathways
Flavonoids	Quercetin, catechins	Antioxidant, anticancer, anti-allergic	Inhibits oxidative stress, modulates cell signaling
Alkaloids	Berberine, caffeine	Antimicrobial, anti-inflammatory, metabolic regulation	Interacts with receptors, alters metabolic pathways
Vitamins	B12, folate	Supports energy metabolism, DNA synthesis	Cofactor in enzymatic reactions
Short-chain fatty acids (SCFAs)	Acetate, propionate	Anti-inflammatory, gut health support	Modulates immune response, enhances gut barrier function
Terpenoids	Limonene, linalool	Antimicrobial, anti-inflammatory	Alters membrane permeability, modulates immune pathways
Amino acids	Tryptophan, glutamine	Neurotransmitter precursor, gut health	Precursor for neurotransmitter synthesis, supports gut lining

simple organic acids and complex secondary metabolites (Bartmanski et al. 2023). Advanced analytical techniques like mass spectrometry and nuclear magnetic resonance spectroscopy are required to detect and characterize these chemicals, as well as gain insights into the metabolic capacity of microbial communities (Dunn et al. 2011). These methods allow researchers to investigate the complex metabolic profiles of microbes, facilitating the discovery of new bioactive substances and improving our understanding of microbial activity in a variety of ecological circumstances.

2.2 Nutraceutical Potential of Discovered Compounds

The bioactive compounds discovered through microbial metabolomics have great nutraceutical potential and provide a variety of health benefits. These chemicals can serve as antioxidants, lowering oxidative stress and scavenging free radicals. Furthermore, several of these metabolites have anti-inflammatory properties that assist regulate immune responses and reduce inflammation associated with chronic illnesses (Singh et al. 2024). Certain microbial metabolites have antibacterial properties that prevent infections and promote a healthy gut microbiota (Balakumaran et al. 2022). Furthermore, microbial metabolites can affect a variety of physiological functions in humans, such as neurotransmission, cell signaling, and metabolic processes. Some metabolites, for example, operate as cofactors or precursors in critical metabolic pathways, facilitating energy generation and nutrient absorption. Others control immunological responses, neurotransmitter release, and hormone signaling by interacting with host receptors and enzymes. Researchers can use these bioactive chemicals to create unique nutraceutical products that improve human health and welfare (Al-Obaidi et al. 2021). Functional foods, nutritional supplements, and herbal medications enhanced with microbial metabolites provide tailored treatment for specific health problems such as metabolic syndrome, gastrointestinal disorders, and cardiovascular disease. Furthermore, taking into account individual metabolic profiles and microbiome compositions when developing tailored nutraceutical medicines may result in superior health outcomes and chronic disease prevention (Thakur and Belwal 2022). Microbial metabolomics identifies bioactive chemicals that can be used to make nutraceuticals and improve human health. This area opens up new avenues for identifying novel bioactive chemicals and designing personalized therapies to improve health and well-being by revealing microorganisms' metabolic capabilities and interactions with the host (Sun et al. 2020).

3 Optimization of Nutraceutical Production Processes

The manufacturing procedures used in nutraceuticals are critical to guaranteeing the effectiveness, scalability, and quality of goods that promote health. According to Gupta et al. (2013), optimizing these processes entails putting methods into place to increase yield, boost productivity, and boost efficiency. The metabolomics process flow consists of the following steps: sample pretreatment, sample analysis, raw data preprocessing, statistical analysis, and finally results interpretation (Fig. 1).

3.1 Process Enhancement Strategies

1. Selection of Optimal Production Strains: A crucial aspect of producing nutraceuticals is selecting the manufacturing strains. Manufacturers can increase the productivity of their processes by choosing strains with robust growth characteristics and high metabolic efficiency (Liu et al. 2017). Genetic engineering methods can also be utilized to create strains that are more capable of producing metabolites or that are more resilient to stress (Yuan and Alper 2019).
2. Optimization of Growth Conditions: Microbial growth and metabolite production are significantly influenced by cultivation conditions, which include temperature, pH, oxygenation, and nutrient availability. To improve product yield and quality, these factors must be optimized by methodical testing and process monitoring (Hugenholtz et al. 2002). Modern bioreactor systems with automated control systems and online sensors provide real-time growth condition optimization, giving precise control over fermentation processes (Mitra and Murthy 2022).
3. Substrate Selection and Feed Optimization: The cost-effectiveness and product yield of nutraceutical production are highly influenced by the substrate(s) selected. Reductions in production costs and environmental impact can be achieved by choosing sustainable and affordable substrates, such as industrial

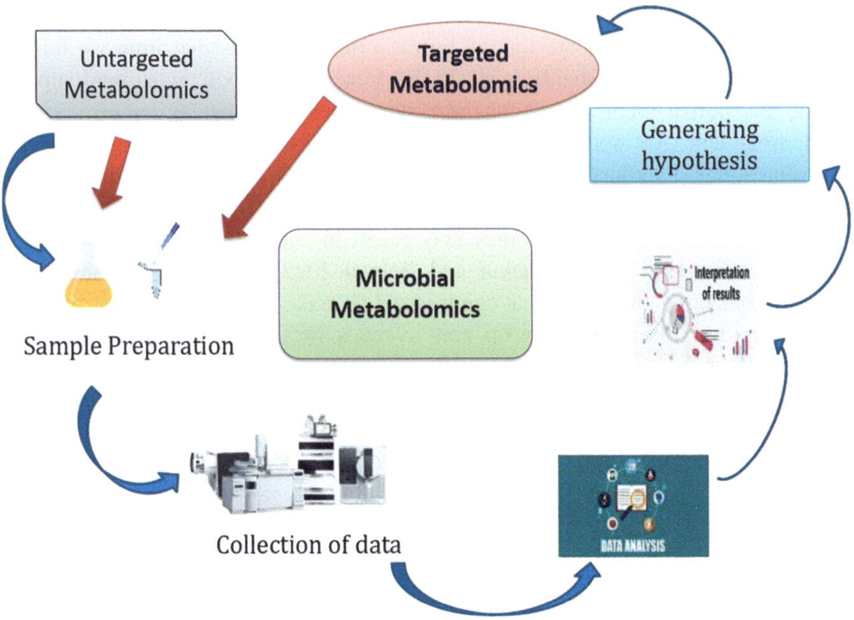

Fig. 1 General metabolomics workflow

byproducts or agricultural wastes. Additionally, larger product yields can be achieved by optimizing feeding strategies, such as fed-batch or continuous feeding procedures, which can improve substrate utilization efficiency and extend the period of fermentation (Conceição et al. 2022).

4. Process Monitoring and Control: Stability and productivity of the process depend on constant monitoring of important parameters such as cell density, metabolite concentrations, and bioreactor conditions. Real-time monitoring of fermentation kinetics and metabolite profiles is made possible by sophisticated analytical techniques such as mass spectrometry, chromatography, and online sensors. Process parameters can be dynamically adjusted to maximize both product yield and quality through the use of feedback control systems that rely on sophisticated algorithms or predictive models (McNeil et al. 2013).

3.2 Maximizing Yield and Efficiency

1. Metabolic Engineering and Synthetic Biology: In the manufacture of nutraceuticals, metabolic engineering and synthetic biology techniques provide effective methods for optimizing product yield and efficiency. Researchers can create microbial strains with higher substrate utilization efficiency and enhanced biosynthetic capacities by changing metabolic pathways, enzyme activity, or regulatory networks (Joana Gil-Chávez et al. 2013). Producing high-value nutraceuticals

with previously unheard-of yields is made possible by the rational design and optimization of biosynthetic pathways by computer modeling and directed evolution (Yuan and Alper 2019).

2. Downstream Processing Optimization: To maximize product production and purity, downstream processing processes such as biomass separation, extraction, purification, and formulation are essential. By choosing the right techniques for each step, such as membrane separation, chromatography, centrifugation, and filtration, the process can be optimized to reduce product loss and guarantee effective target component recovery. In addition, integrating in-line monitoring systems and continuous processing technology lowers total production costs and improves process efficiency (Rodríguez-Sifuentes et al. 2021).
3. Quality Control and Assurance: To guarantee the consistency, safety, and effectiveness of nutraceutical goods, extensive quality control and assurance procedures are required. Product integrity and regulatory compliance are protected by the application of strict quality control procedures, which include stability studies, microbiological screening, and analytical testing. Comprehensive assessment of product quality features and identification of probable contaminants or impurities are made possible by advanced analytical techniques such as mass spectrometry, chromatography, and spectroscopy (Inbathamizh et al. 2022).

Achieving the highest possible product yield, efficiency, and quality requires optimizing the production processes of nutraceuticals. Manufacturers can create higher health benefits and commercial viability in high-quality nutraceuticals by maximizing production and efficiency and using process enhancement measures. The nutraceutical business can address the growing demand for safe, effective, and sustainable health-promoting goods by integrating sophisticated technology and engaging in continuous innovation (Ravishankar and Ambati 2019).

4 Functional Properties and Formulation Development

Nutraceuticals' essential qualities, or functional features, are what make them effective in advancing health and well-being. In order to ensure the delivery of bioactive substances in a form that is safe, efficacious, and bioavailable, formulation development is essential to maximizing these features (Caleja et al. 2017). Figure 2 shows the nutraceuticals: advancement in functional properties and formulation development in microbial metabolomics production.

4.1 Assessing Functional Attributes of Nutraceuticals

1. Bioactivity Assays: The biological effects of nutraceuticals are assessed using bioactivity assays, which offer insights into their possible health advantages. Certain functional characteristics, such as antioxidant activity, anti-inflammatory activity, antibacterial activity, or enzyme inhibition, are measured by these

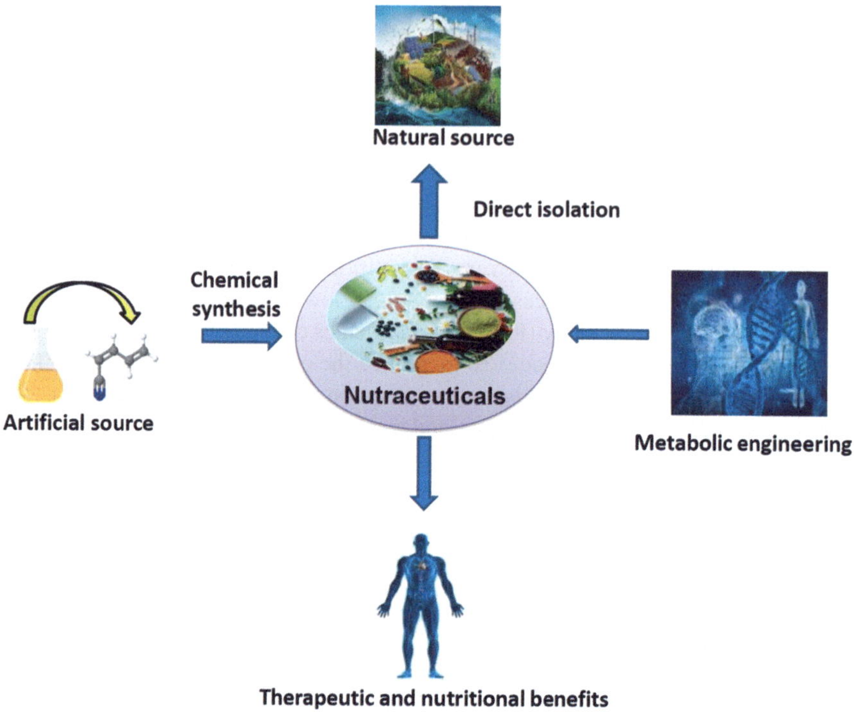

Fig. 2 Nutraceuticals: advancement in formulation development

assays. Common bioassays that evaluate the capacity of nutraceuticals to alter cellular processes and physiological functions include cell-based assays, cytokine assays, and radical scavenging assays (Chanput et al. 2016).

2. Stability Studies: To evaluate the biological, chemical, and physical stability of nutraceutical formulations under varied storage circumstances, stability studies are carried out. These studies assess variables that guarantee the integrity and efficacy of the product over time, including formulation compatibility, shelf-life, and degradation kinetics. To improve stability, formulation compositions are optimized, and breakdown pathways are found using accelerated stability testing and real-time monitoring (Bell 2002).
3. Pharmacokinetic Studies: Pharmacokinetic studies assess the body's absorption, distribution, metabolism, and excretion (ADME) of nutraceuticals, revealing information on their pharmacological and bioavailability characteristics. The amounts of nutraceuticals in blood, tissues, and biological fluids are measured using methods like mass spectrometry, chromatography, and imaging modalities. To evaluate absorption and systemic exposure, pharmacokinetic parameters such as peak plasma concentration (Cmax), duration to achieve peak concentration (Tmax), and area under the curve (AUC) are computed (Mittal et al. 2023).

4.2 Formulation Techniques for Enhanced Bioavailability

1. Nanoencapsulation: To increase the stability, solubility, and bioavailability of nutraceuticals, nanoencapsulation entails encasing them in nanoscale carriers as liposomes, nanoparticles, or nanofibers (Jafari 2017). Nutraceuticals are shielded from deterioration by nanoencapsulation, which also improves their solubility in aqueous solutions and enables targeted distribution to particular tissues or cells. Furthermore, controlled release of nutraceuticals is made possible by nanostructured delivery systems, which increases the duration of the substances' residence in the body and improves therapeutic efficacy (Leena et al. 2020).
2. Self-Emulsifying Drug Delivery Systems (SEDDS): Lipid-based formulations known as SEDDS are intended to improve the absorption and solubility of nutraceuticals that are not very soluble in water. When these formulations are dissolved in aqueous solutions, they generate fine oil-in-water emulsions that increase the surface area accessible for absorption and encourage intestinal permeability. SEDDS eliminate the drawbacks of low water solubility by adding surfactants, cosolvents, and co-solubilizers, hence increasing the oral bioavailability of nutraceuticals (Garg et al. 2016).
3. Solid Dispersion Techniques: To improve the solubility and dissolution rate of nutraceuticals, solid dispersion procedures entail dispersing them inside a hydrophilic carrier matrix. Improved bioavailability of solid dispersions can be achieved by the use of techniques like freeze drying, hot melt extrusion, and spray drying. These formulations improve gastrointestinal transit, decrease particle aggregation, and increase surface area available for dissolution, which increases nutraceutical absorption and systemic exposure (Kulkarni et al. 2022).

Developing nutraceutical products with improved bioavailability and efficacy requires evaluating functional characteristics and refining formulation methods. Researchers can determine formulation strategies to optimize the health-promoting benefits of nutraceuticals and clarify their therapeutic potential by utilizing pharmacokinetic evaluations, stability studies, and bioactivity assays. The development of novel and effective nutraceutical formulations is made possible by the incorporation of sophisticated formulation techniques such as solid dispersion, SEDDS, and nanoencapsulation, which improve the solubility, stability, and absorption of nutraceuticals.

5 Quality Control and Authentication of Nutraceuticals

In order to guarantee the integrity and safety of the products, quality control and authentication are crucial components of the nutraceutical production process. Strict quality control procedures are used throughout the manufacturing process, from screening raw materials to evaluating the final product, to ensure consistency and purity (Nagy et al. 2022). Manufacturers are able to identify impurities, validate potency, and verify compliance with regulatory criteria through the use of sophisticated analytical techniques such as spectroscopy, chromatography, and microbiological assays (Xu et al. 2024). Furthermore, nutraceutical component validity and provenance can be confirmed with the help of useful procedures like isotope analysis, DNA barcoding, and chromatographic fingerprints. Manufacturers can establish consumer confidence, protect public health, and guarantee the effectiveness of nutraceutical goods by adhering to strict quality control and authentication requirements (Nehal et al. 2021).

5.1 Ensuring Product Quality and Purity

In order to protect the efficacy, safety, and integrity of the product, quality control and authentication are crucial components of the manufacture of nutraceuticals (Ghosh and Mukherjee 2019). Bioactive substances produced from natural sources are frequently found in nutraceuticals, which include nutritional supplements, functional foods, and herbal medicines (Vattem and Maitin 2015). Therefore, strict quality control procedures are required to guarantee the purity and quality of the product throughout the manufacturing process.

1. Raw Material Screening: The selection and screening of raw materials used in the manufacturing of nutraceuticals is the first step in quality control. Certificates of analysis, or COAs, are usually needed from suppliers in order to confirm the identification, potency, and purity of raw ingredients. In order to ensure compliance with safety regulations and standards, analytical techniques including mass spectrometry, spectroscopy, and chromatography are used to detect pollutants, adulterants, and microbiological pathogens in raw materials (Minoja and Napoli 2014).
2. In-Process Monitoring: Maintaining consistency and quality in the final product during the manufacturing of nutraceuticals requires constant monitoring of critical process parameters. Real-time monitoring of crucial process parameters, such as temperature, pressure, pH, and mixing efficiency, is made possible by process analytical technologies (PAT), which include spectroscopy, chromatography, and rheology. Manufacturers can prevent batch failures and guarantee product quality by identifying deviations from target specifications and taking corrective action (Workman Jr et al. 2011). This can be achieved by deploying in-process monitoring systems.
3. Finished Product Testing: Nutraceutical formulations are thoroughly analyzed as part of finished product testing to confirm their authenticity, potency, purity, and safety. To determine the concentration of active substances, identify possible impurities, and verify compliance with regulatory standards, analytical techniques like chromatography, spectroscopy, and microbiological assays are utilized (Chahar 2019). In order to assess the shelf-life and degradation kinetics of nutraceutical goods under various storage settings, stability testing is also carried out.

5.2 Authentication Methods for Nutraceuticals

1. Chromatographic Fingerprints: Chromatographic fingerprinting is the process of utilizing chromatography methods like gas chromatography (GC) or high-performance liquid chromatography (HPLC) to analyze the intricate chemical profiles of nutraceutical compositions. Researchers can confirm the validity and quality of nutraceutical goods by comparing the chromatographic fingerprints of authentic samples with reference standards. This allows researchers to detect distinctive peaks that belong to particular substances (Parys et al. 2022).
2. DNA Barcoding: A molecular authentication technique called DNA barcoding entails amplifying and sequencing particular DNA regions found in raw materials or compounds used in nutraceuticals. By cross-referencing DNA sequences with reference databases, scientists may identify plant species and identify any pollutants or adulterants. In the case of herbal supplements and plant extracts, where species replacement and mislabeling are frequent problems, DNA barcoding is especially helpful (Nehal et al. 2021).
3. Isotopic Analysis: Isotopic analysis uses mass spectrometry methods like isotope ratio mass spectrometry (IRMS) to measure the stable isotopic composition of the components in nutraceuticals. Particular geographic areas, environmental factors, and production processes all contribute to the uniqueness of isotopic signatures, which offer important data for product identification and provenance tracking (Zhao et al. 2014). According to Nwachukwu et al. (2024), isotopic analysis can be used to detect adulteration or dilution of nutraceuticals with lower-quality ingredients and to verify that labeling claims regarding regional origin or organic certification are being followed.

To guarantee the quality, purity, and validity of the product, quality control and authentication are essential elements of the nutraceutical production process. Manufacturers may preserve the safety and integrity of nutraceutical products, gain the trust of customers, and adhere to legal requirements by putting strong quality control and authentication procedures in place. The production of safe, dependable, and effective health-promoting products is made easier by the continuous improvement of quality control and authentication capabilities in the nutraceutical business, which is made possible by advancements in analytical techniques and molecular authentication technologies.

6 Personalized Nutrition and Metabolomics

In recent years, there has been a lot of interest in personalized nutrition—a customized approach to dietary recommendations based on individual traits, preferences, and metabolic profiles. By offering insights into a person's metabolic status and reaction to dietary changes, metabolomics is a potent analytical tool that is essential to the advancement of personalized nutrition (LeVatte et al. 2022). Knowing that every person has different nutritional needs and physiological reactions to food components is the fundamental idea behind personalized nutrition. Comprehensive examination of the tiny compounds, or metabolites, found in biological samples like blood, urine, or saliva is made possible by metabolomics. These metabolites show the complex interactions between heredity, diet, lifestyle, and environmental factors and function as indicators of dietary intake, nutrient metabolism, and physiological processes (Nalbantoglu 2019).

Researchers can find metabolic fingerprints linked to particular food patterns, metabolic disorders, or nutritional deficits by studying metabolomics data. Metabolomic profiling, for instance, can identify biomarkers for inflammation, insulin resistance, or dyslipidemia, offering information about a person's metabolic health and risk for disease. Equipped with this data, medical practitioners can formulate customized dietary suggestions based on each person's tastes, metabolic profile, and health objectives (Rezzi et al. 2007). Utilizing metabolomics data, precision nutrition techniques maximize dietary interventions and enhance health outcomes on a personal level. Through the integration of metabolomic data with other omics technologies, including transcriptomics, genomes, and microbiomics, scientists can clarify the fundamental mechanisms that underlie interindividual diversity in responses to food. The creation of customized dietary plans with the goals of preventing chronic illnesses, fostering optimum health, and enhancing general well-being is made possible by this all-encompassing approach (Verma et al. 2018). Furthermore, developments in data analytics and machine learning make it easier to create predictive models that use metabolomics data to provide individualized dietary recommendations. To customize dietary interventions for the best results, these models take into account each person's unique metabolic profile, dietary preferences, lifestyle circumstances, and health objectives. Precision nutrition techniques enable proactive control of metabolic health and illness prevention by leveraging artificial intelligence and big data (O'sullivan et al. 2018). Metabolomics-guided personalized nutrition has enormous potential to improve dietary recommendations and health outcomes. Through the utilization of metabolomics' abundant information, medical practitioners can create customized dietary plans that cater to each patient's unique metabolic requirements, inclinations, and objectives. Individualized nutrition techniques powered by metabolomics data have the potential to change dietary recommendations and encourage individualized health optimization as our understanding of the intricate relationship between diet, metabolism, and health continues to grow.

6.1 Tailoring Nutritional Interventions with Metabolomics Data

Metabolomics measures the amount of tiny molecules, or metabolites, in biological samples like blood, urine, or saliva to provide a complete picture of a person's metabolic condition. These metabolites represent the intricate relationships between heredity, lifestyle, and nutrition and act as indicators of food intake, nutrient metabolism, and physiological processes. Researchers can produce individualized dietary recommendations by identifying metabolic signatures linked to particular food patterns, metabolic illnesses, or nutritional deficiencies through the analysis of metabolomics data (Raparthi et al. 2022).

6.2 Precision Nutrition Approaches

Precision nutrition customizes dietary recommendations and maximizes health outcomes by utilizing customized data, such as genetic, phenotypic, and lifestyle information (Laddu and Hauser 2019). Because metabolomics data reveal a person's distinct metabolic reactions to dietary treatments, they are essential to precision nutrition strategies. Researchers can identify personalized dietary strategies to reduce disease risk and promote optimal health by integrating metabolomics data with other omics technologies like transcriptomics, genomics, and microbiomics. This allows researchers to better understand the underlying mechanisms underlying interindividual variability in dietary responses (Brennan and de Roos 2023).

Furthermore, developments in data analytics and machine learning make it possible to create predictive models that use metabolomics data to provide individualized dietary recommendations. These models can predict an individual's metabolic response to particular foods, nutrients, or dietary patterns by utilizing large-scale metabolomics datasets and sophisticated algorithms. This allows for customized dietary interventions to optimize metabolic health and prevent chronic diseases (Armand et al. 2024). With the use of metabolomics data, customized nutrition strategies have enormous potential to improve individual health outcomes and optimize dietary interventions. Researchers and medical experts may create precision nutrition plans that are suited to each person's own metabolic profile, preferences, and health objectives by utilizing the abundance of data offered by metabolomics. Customized nutrition powered by metabolomics data is set to transform dietary recommendations and advance customized health optimization as our understanding of the intricate interactions between diet, metabolism, and health continues to grow (Singh et al. 2023).

7 Translating Microbial Metabolomics: From Bench to Bedside

By shedding light on the metabolic processes of bacteria and how they affect human health, the rapidly developing science of microbial metabolomics has the potential to completely transform healthcare. However, bridging the gap between scientific discovery and real-world application is necessary for the trip from laboratory research to clinical application (Wehling 2021). Researchers use state-of-the-art analytical methods at the lab bench, like nuclear magnetic resonance spectroscopy and mass spectrometry, to decipher the intricate metabolic patterns of microbes. Through the analysis of the wide range of metabolites that bacteria create, scientists can learn more about the physiology of microbes, ecological relationships, and possible health effects. Microbial metabolomics makes a significant contribution to the discovery of novel biomarkers, treatment targets, and diagnostic instruments for a variety of medical disorders by means of rigorous experimentation and data analysis (Pinu et al. 2019).

Finding and validating biomarkers is the first crucial step in translating microbial metabolomics to the bedside. Microbial metabolomic investigations have yielded biomarkers that show promise in the diagnosis of infectious illnesses, treatment response monitoring, and prognostication. To evaluate their accuracy, sensitivity, specificity, and repeatability, biomarkers must undergo thorough validation in preclinical and clinical settings before moving from biomarker discovery to clinical use (Monteiro et al. 2013).

Moreover, overcoming numerous logistical, technical, and regulatory obstacles is necessary to integrate microbial metabolomics into clinical practice (Anklam et al. 2022). It is imperative to establish standardization in analytical protocols, data interpretation techniques, and quality control systems to guarantee uniformity and repeatability among diverse laboratories and research environments. Furthermore, the predictive ability and clinical significance of microbial metabolomic biomarkers are improved by integrating microbial metabolomic data with other omics datasets, clinical factors, and electronic health records (Ezzelle et al. 2008). The effective application of microbial metabolomics at the bedside also depends on interdisciplinary cooperation and coordination between scientists, physicians, business partners, and government regulators. The development and adoption of microbial metabolomic applications in clinical practice can be facilitated by stakeholders addressing crucial issues like data sharing, technology transfer, intellectual property rights, and regulatory compliance through collaboration and knowledge exchange (Mehta 2022). In the end, there is a great deal of potential for improving patient outcomes, customized treatment, and illness diagnosis and management through the application of microbial metabolomics from the bench to the bedside. Researchers and clinicians can gain new insights into human health and disease by using microbial metabolomics to understand the metabolic language of microorganisms. This can open the door to the development of novel diagnostic tools, therapeutic interventions, and precision medicine strategies catered to the needs of specific patients (Wishart 2016).

7.1 Bridging the Gap Between Research and Application

Several critical processes are required to move microbial metabolomics from the laboratory to clinical practice, including the development, validation, and application of biomarkers in therapeutic settings. Biomarker identification requires methodically evaluating metabolomic data to identify metabolite signatures linked to specific microbial activity or health problems (McIntyre 2005). Advanced analytical techniques, such as mass spectrometry and nuclear magnetic resonance spectroscopy, help to identify and quantify microbial metabolites, providing information on disease causes and prospective therapeutic targets (Agnew et al. 2006). Validating microbial metabolomic biomarkers through rigorous preclinical and clinical investigations is critical for determining their diagnostic, prognostic, and therapeutic value. Validation studies assess biomarkers' clinical relevance, repeatability, sensitivity, and specificity across a wide range of patient populations. Furthermore, translational research aims to improve the use of microbial metabolomics in clinical settings by creating reference databases, standardized techniques, and analytical platforms (Di Minno et al. 2022).

7.2 Clinical Implications and Patient Outcomes

Microbial metabolomics has substantial therapeutic applications in a variety of medical domains, including oncology, infectious diseases, gastrointestinal, and personalized medicine. In the setting of infectious disorders, microbial metabolomics allows for the quick identification and characterization of microbial pathogens, assisting in the selection of appropriate antimicrobial treatments and monitoring therapy responses. In gastroenterology, profiling the gut microbiome using microbial metabolomics provides insights into the pathogenesis of gastrointestinal illnesses and informs the use of probiotic medicines and dietary changes to improve gut health (Rahman and Schellhorn 2023). Microbial metabolomics in oncology offers the ability to predict treatment results and improve patient care. Metabolomic screening of tumor tissues and biofluids identifies metabolic indicators linked to medication resistance, tumor progression, and patient survival. By combining microbial metabolomic data with other omics technologies, doctors can create individualized treatment plans and track therapy responses in real time, improving patient outcomes and cancer care. Furthermore, in the field of personalized medicine, microbial metabolomics permits the creation of tailored medicines based on individual metabolic profiles and microbial fingerprints. Healthcare practitioners can find dietary components, probiotics, or prebiotics that boost health by examining metabolic interactions between the microbiota and the host. Dietary recommendations based on microbial metabolomic data have the potential to improve metabolic health, prevent chronic diseases, and increase general well-being across a wide range of patient populations (Bekri 2016). The application of microbial metabolomics in clinical settings has great potential to enhance patient outcomes and progress precision medicine. Microbial metabolomics facilitates the creation of novel diagnostic instruments, treatment plans, and individualized interventions based on individual metabolic profiles and microbial fingerprints by bridging the gap between research and implementation. Microbial metabolomics has the potential to change patient treatment in a variety of medical specializations and improve clinical practice as our understanding of microbial metabolism advances (Wishart 2016).

8 Future Directions and Emerging Trends

Innovation and integration are critical factors influencing the future of microbial metabolomics and nutraceutical development. The combination of high-throughput screening platforms and multi-omic approaches with microbial metabolomics can provide new information about microbial metabolism and its applications in biotechnology and medicine (Daliri and Lee 2015). Concurrently, advances in plant-based functional meals, personalized approaches, and sophisticated delivery technologies are poised to transform the development of health-promoting products suited to specific requirements and tastes. Embracing these developments enables researchers and industry partners to push the boundaries of microbial metabolomics and nutraceutical development, ultimately improving human health and well-being (Panda et al. 2018). Table 2 summarizes possible future paths and new trends in the development of nutraceuticals and microbial metabolomics.

8.1 Integration of Advanced Analytical Techniques

Future study should concentrate on merging diverse analytical techniques, such as mass spectrometry, nuclear magnetic resonance (NMR), and high-performance liquid chromatography (HPLC), with microbial metabolomics. This multifaceted method will improve the identification and measurement of complex metabolites, offering more information about microbial metabolism and its health implications (Anand and Padmanabhan 2024).

Table 2 Future directions and emerging trends in microbial metabolomics and nutraceutical development

Area	Future directions and emerging trends
Microbial metabolomics	Single-cell metabolomics: The study of microbial metabolites through improvements in single-cell analytical methods
	Spatial metabolomics: Techniques enabling spatial mapping of metabolites within microbial communities
	Dynamic metabolomics: Real-time monitoring and analysis of metabolic fluxes within microbial populations
	Integrating metabolomics data with other omics techniques to gain comprehensive insights is known as integration into systems biology
Nutraceutical development	Customizing nutritional formulas according to each person's unique microbiological and genetic makeup is known as personalized nutrition
	Creating food items enhanced with bioactive substances obtained from microbial metabolites is known as "functional food development"
	Engineering microbial strains to generate certain nutraceuticals with improved bioavailability is known as synthetic biology
	Microbiome modulation: Changing the gut microbiota to get better health outcomes by using microbial metabolites

8.2 Development of Standardized Protocols

Standardized methods for microbial metabolomics research are critical for ensuring repeatability and comparability of results across laboratories. This includes standardizing sample preparation, analytical methodologies, and data processing pipelines to improve collaboration and data sharing across the research community (Dolatmoradi et al. 2022).

8.2.1 Exploration of Host-Microbe Relationships

Future research should focus on the relationships between microbial metabolites and host physiology. Understanding how individual metabolites affect metabolic pathways, immunological responses, and gut health can help design targeted nutraceuticals to address specific health conditions.

8.2.2 High-Throughput Screening for Bioactive Chemicals

Using high-throughput screening technologies will speed up the discovery of new bioactive chemicals from microbial sources. Automated platforms can quickly assess vast metabolite libraries for health-promoting qualities, paving the way for the discovery of new nutraceutical compounds.

Microbial Diversity Investigation Future research should focus on microbial diversity in a variety of contexts, including hitherto unexplored habitats such as harsh environments, traditional fermented foods, and the human microbiome. This could lead to the discovery of novel microbial strains and metabolites with potential health advantages.

Personalized Nutrition and Metabolomics Research should focus on the development of nutraceuticals tailored to individual microbiome profiles, metabolic health, and genetic origins. Using metabolomic data, individualized food recommendations can be developed to improve individual health outcomes.

Sustainable Practices in Nutraceutical Manufacturing It is critical to investigate sustainable practices in nutraceutical manufacturing, such as the use of green chemistry and eco-friendly fermentation processes. Research should look into how microbial metabolomics can help to reduce the environmental impact of nutraceutical manufacturing.

Clinical Studies and Efficacy Studies Future research should emphasize clinical studies to assess the safety and efficacy of nutraceuticals derived from microbial metabolites. This will help to establish their health benefits and support regulatory approvals, easing their adoption into mainstream health practices.

Microbiome Modulation Study Future study should focus on how dietary interventions including microbial metabolites can modulate the gut microbiome for better health outcomes. This includes investigating the role of probiotics, prebiotics, and postbiotics in maintaining a healthy microbiome.

Collaboration with Industry and Regulatory Bodies Future research should encourage collaboration among academics, industry stakeholders, and regulatory bodies. These collaborations can assist convert scientific discoveries into practical applications, ensuring that novel nutraceuticals are developed safely and effectively.

Spatial Metabolomics The imaging and mapping of metabolite distributions inside microbial communities or host tissues is made possible by spatial metabolomics approaches. Researchers can spatially resolve metabolite localization and dynamics in situ by integrating mass spectrometry imaging with microscopy or microfluidics. This allows for insights into microbial localization, metabolic gradients, and microenvironmental interactions (Geier et al. 2020).

8.3 Anticipated Trends in Nutraceutical Development

1. Personalized Nutraceuticals: Personalized nutraceuticals that are suited to each person's unique metabolic profile, genetic make-up, and health requirements are being developed as a result of the growing popularity of precision nutrition and personalized medicine. The identification of biomarker-based signatures for individualized dietary interventions is made possible by developments in biomarker discovery, omics technology, and data analytics (Zeisel 2020). This improves health outcomes and prevents disease.
2. Functional Ingredients from Microbial Sources: It is anticipated that the investigation of microbial metabolomes as a source of bioactive chemicals would provide new functional components with health-promoting qualities (Kumar et al. 2018). The discovery and synthesis of bioactive metabolites for application in nutraceuticals and functional foods are being driven by microbial fermentation techniques, modified microbial strains, and bioprospecting initiatives (Sorour et al. 2017).
3. Sustainable and Plant-Based Nutraceuticals: Nutraceuticals made from plant sources are developing as a result of rising customer demand for environmentally friendly and plant-based goods. Potential health advantages, such as anti-inflammatory, cardioprotective, and neuroprotective effects, are being investigated for plant-based substances rich in phytochemicals, antioxidants, and bioactive compounds (Sharma et al. 2024).

Future developments in the fields of nutraceutical development and microbial metabolomics could result in revolutionary breakthroughs. Our knowledge of microbial ecosystems and their significance to human health will expand with the help of innovations in microbial metabolomics, such as single-cell analysis, multi-omic integration, and spatial metabolomics. Future directions in the development of nutraceuticals include customized nutraceuticals, microbially sourced functional ingredients, and plant-based products that are sustainable. These developments are driven by scientific discoveries and changing customer tastes, which will shape the products that promote health in the next years.

9 Conclusion

Microbial metabolomics is a game-changing method to nutraceutical creation, providing valuable insights into the vast range of bioactive chemicals produced by bacteria. This discipline not only helps to identify novel metabolites like polyphenols, flavonoids, and alkaloids, but it also improves our understanding of their potential health advantages, such as antioxidant, anti-inflammatory, and antibacterial properties. The optimization of fermentation processes, guided by microbial metabolomics, is critical for increasing the output and efficacy of bioactive compounds. Researchers can improve production procedures by understanding metabolic pathways and environmental circumstances, ensuring that nutraceuticals are effective and sustainable. Furthermore, microbial metabolomics plays a critical role in formulation development, enabling the generation of stable and bioavailable products customized to specific health goals. As the need for customized nutrition grows, incorporating metabolomic data into individualized dietary recommendations will increase the therapeutic potential of nutraceuticals. In addition to its use in nutraceuticals, microbial metabolomics has the potential to have larger ramifications in drug discovery and environmental sustainability. Microbial metabolic insights can lead to new solutions to health concerns as well as sustainable food production techniques. As research in this subject advances, the adoption of standardized protocols, collaboration between academia and industry, and the advancement of analytical techniques will become critical. Future research will focus on improving our understanding of host-microbe interactions, generating tailored nutraceuticals, and ensuring product quality and integrity through strong authentication methods.

References

Agnew LJ, Lyon S, Gérard-Marchant P, Collins VB, Lembo AJ, Steenhuis TS, Walter MT (2006) Identifying hydrologically sensitive areas: bridging the gap between science and application. J Environ Manag 78(1):63–76

Al-Obaidi JR, Jambari NN, Ahmad-Kamil EI (2021) Mycopharmaceuticals and nutraceuticals: promising agents to improve human well-being and life quality. J Fungi 7(7):503

Anand S, Padmanabhan P (2024) Role of omics in the microbial characterization. In: Emerging innovative trends in the application of biological processes for industrial wastewater treatment. Elsevier, pp 91–103

Anklam E, Bahl MI, Ball R, Beger RD, Cohen J, Fitzpatrick S, Girard P, Halamoda-Kenzaoui B, Hinton D, Hirose A, Hoeveler A (2022) Emerging technologies and their impact on regulatory science. Exp Biol Med 247(1):1–75

Armand TP, Nfor KA, Kim JI, Kim HC (2024) Applications of artificial intelligence, machine learning, and deep learning in nutrition: a systematic review. Nutrients 16(7):1073

Balakumaran PA, Divakar K, Sindhu R, Pandey A, Binod P (2022) Gut microbes: role in production of nutraceuticals. In: Current developments in biotechnology and bioengineering. Elsevier, pp 273–299

Bartmanski BJ, Rocha M, Zimmermann-Kogadeeva M (2023) Recent advances in data-and knowledge-driven approaches to explore primary microbial metabolism. Curr Opin Chem Biol 75:102324

Bekri S (2016) The role of metabolomics in precision medicine. Expert Rev Precis Med Drug Dev 1(6):517–532

Bell LN (2002) Stability testing of nutraceuticals and functional foods. In: Handbook of nutraceuticals and functional foods. CRC Press, pp 523–538

Brennan L, de Roos B (2023) Role of metabolomics in the delivery of precision nutrition. Redox Biol 65:102808

Caleja C, Ribeiro A, Filomena Barreiro M, Ferreira ICFR (2017) Phenolic compounds as nutraceuticals or functional food ingredients. Curr Pharm Des 23(19):2787–2806

Chahar D (2019) Analytical validation and method development for functional food and nutraceutical manufacturing. In: Nutraceutical and functional food regulations in the United States and around the world. Academic Press, pp 89–96

Chanput W, Krueyos N, Ritthiruangdej P (2016) Anti-oxidative assays as markers for anti-inflammatory activity of flavonoids. Int Immunopharmacol 40:170–175

Conceição AA, Mendes TD, Mendonça S, Quirino BF, Almeida EG, Siqueira FG (2022) Nutraceutical enrichment of animal feed by filamentous fungi fermentation. Fermentation 8(8):402

Covington BC, McLean JA, Bachmann BO (2017) Comparative mass spectrometry-based metabolomics strategies for the investigation of microbial secondary metabolites. Nat Prod Rep 34(1):6–24

Daliri EB, Lee BH (2015) Current trends and future perspectives on functional foods and nutraceuticals. In: Beneficial microorganisms in food and nutraceuticals, pp 221–244

Di Minno A, Gelzo M, Caterino M, Costanzo M, Ruoppolo M, Castaldo G (2022) Challenges in metabolomics-based tests, biomarkers revealed by metabolomic analysis, and the promise of the application of metabolomics in precision medicine. Int J Mol Sci 23(9):5213

Dolatmoradi M, Samarah LZ, Vertes A (2022) Single-cell metabolomics by mass spectrometry: opportunities and challenges. Analysis Sensing 2(1):e202100032

Dunn WB, Broadhurst DI, Atherton HJ, Goodacre R, Griffin JL (2011) Systems level studies of mammalian metabolomes: the roles of mass spectrometry and nuclear magnetic resonance spectroscopy. Chem Soc Rev 40(1):387–426

Ezzelle J, Rodriguez-Chavez IR, Darden JM, Stirewalt M, Kunwar N, Hitchcock R, Walter T, D'souza MP (2008) Guidelines on good clinical laboratory practice: bridging operations between research and clinical research laboratories. J Pharm Biomed Anal 46(1):18–29

Galanakis CM (ed) (2019) Nutraceuticals and natural product pharmaceuticals. Academic Press

Garg V, Gupta R, Kapoor B, Singh SK, Gulati M (2016) Application of self-emulsifying delivery systems for effective delivery of nutraceuticals. In: Emulsions. Academic Press, pp 479–518

Geier B, Sogin EM, Michellod D, Janda M, Kompauer M, Spengler B, Dubilier N, Liebeke M (2020) Spatial metabolomics of in situ host–microbe interactions at the micrometre scale. Nat Microbiol 5(3):498–510

Ghosh D, Mukherjee PK (eds) (2019) Natural medicines: clinical efficacy, safety and quality. CRC Press

Graham AE, Ledesma-Amaro R (2023) The microbial food revolution. Nat Commun 14(1):2231

Gupta S, Jaiswal AK, Abu-Ghannam N (2013) Optimization of fermentation conditions for the utilization of brewing waste to develop a nutraceutical rich liquid product. Ind Crop Prod 44:272–282

Hugenholtz J, Hunik J, Santos H, Smid E (2002) Nutraceutical production by propionibacteria. Lait 82(1):103–112

Inbathamizh L, Prabavathy D, Sudha S (2022) Quality assurance of nutraceuticals and their approval, registration, marketing. In: Handbook of nutraceuticals and natural products: biological, medicinal, and nutritional properties and applications, vol 2, pp 337–360. John Wiley & Sons, Inc.

Jafari SM (ed) (2017) Nanoencapsulation technologies for the food and nutraceutical industries. Academic Press

Joana Gil-Chávez G, Villa JA, Fernando Ayala-Zavala J, Basilio Heredia J, Sepulveda D, Yahia EM, González-Aguilar GA (2013) Technologies for extraction and production of bioactive compounds to be used as nutraceuticals and food ingredients: an overview. Compr Rev Food Sci Food Saf 12(1):5–23

Kellogg J, Kang S (2020) Metabolomics, an essential tool in exploring and harnessing microbial chemical ecology. Phytobiomes J 4(3):195–210

Kulkarni AD, Patel UM, Kandalkar PS, Gorhe AA, Sabale AS, Kedar AK, Jagtap SS, Mahale VA (2022) Solid dispersion techniques for improvement of dissolution and bioavailability of nutraceuticals. In: Nutraceutical delivery systems. Apple Academic Press, pp 279–311

Kumar A, Mosa KA, Ji L, Kage U, Dhokane D, Karre S, Madalageri D, Pathania N (2018) Metabolomics-assisted biotechnological interventions for developing plant-based functional foods and nutraceuticals. Crit Rev Food Sci Nutr 58(11):1791–1807

Laddu D, Hauser M (2019) Addressing the nutritional phenotype through personalized nutrition for chronic disease prevention and management. Prog Cardiovasc Dis 62(1):9–14

Leena MM, Mahalakshmi L, Moses JA, Anandharamakrishnan C (2020) Nanoencapsulation of nutraceutical ingredients. In: Biopolymer-based formulations. Elsevier, pp 311–352

LeVatte M, Keshteli AH, Zarei P, Wishart DS (2022) Applications of metabolomics to precision nutrition. Lifestyle Genom 15(1):1–9

Liu L, Guan N, Li J, Shin HD, Du G, Chen J (2017) Development of GRAS strains for nutraceutical production using systems and synthetic biology approaches: advances and prospects. Crit Rev Biotechnol 37(2):139–150

McIntyre D (2005) Bridging the gap between research and practice. Camb J Educ 35(3):357–382

McNeil B, Archer D, Giavasis I, Harvey L (eds) (2013) Microbial production of food ingredients, enzymes and nutraceuticals. Elsevier

Mehta SS (2022) Commercializing successful biomedical technologies. Cambridge University Press

Merlo ME, Jankevics A, Takano E, Breitling R (2011) Exploring the metabolic state of microorganisms using metabolomics. Bioanalysis 3(21):2443–2458

Minoja AP, Napoli C (2014) NMR screening in the quality control of food and nutraceuticals. Food Res Int 63:126–131

Mishra S, Priyanka, Sharma S (2022) Metabolomic insights into endophyte-derived bioactive compounds. Front Microbiol 13:835931

Mitra S, Murthy GS (2022) Bioreactor control systems in the biopharmaceutical industry: a critical perspective. In: Systems microbiology and biomanufacturing, pp 2:91–112 https://doi.org/10.1007/s43393-021-00048-6. Springer Nature Link

Mittal S, Sawarkar S, Doshi G, Pimple P, Shah J, Bana T (2023) Pharmacokinetics and bioavailability of nutraceuticals. In: Industrial application of functional foods, ingredients and nutraceuticals, pp 725–783. Elsevier

Monteiro MS, Carvalho M, Bastos ML, Guedes de Pinho P (2013) Metabolomics analysis for biomarker discovery: advances and challenges. Curr Med Chem 20(2):257–271

Nagy MM, Wang S, Farag MA (2022) Quality analysis and authentication of nutraceuticals using near IR (NIR) spectroscopy: a comprehensive review of novel trends and applications. Trends Food Sci Technol 123:290–309

Nalbantoglu S (2019) Metabolomics: basic principles and strategies. Mol Med 10(1)

Nehal N, Choudhary B, Nagpure A, Gupta RK (2021) DNA barcoding: a modern age tool for detection of adulteration in food. Crit Rev Biotechnol 41(5):767–791

Nwachukwu SC, Edo GI, Jikah AN, Emakpor OL, Akpoghelie PO, Agbo JJ (2024) Recent advances in the role of mass spectrometry in the analysis of food: a review. J Food Meas Charact:1–6

O'sullivan A, Henrick B, Dixon B, Barile D, Zivkovic A, Smilowitz J, Lemay D, Martin W, German JB, Schaefer SE (2018) 21st century toolkit for optimizing population health through precision nutrition. Crit Rev Food Sci Nutr 58(17):3004–3015
Oka A, Sartor RB (2020) Microbial-based and microbial-targeted therapies for inflammatory bowel diseases. Dig Dis Sci 65(3):757–788
Panda A, Parida AK, Rangani J (2018) Advancement of metabolomics techniques and their applications in plant science: current scenario and future prospective. In: Plant metabolites and regulation under environmental stress. Academic Press, pp 1–36
Parys W, Dołowy M, Pyka-Pająk A (2022) Significance of chromatographic techniques in pharmaceutical analysis. Processes 10(1):172
Pinu FR, Goldansaz SA, Jaine J (2019) Translational metabolomics: current challenges and future opportunities. Metabolites. 9(6):108
Rahman M, Schellhorn HE (2023) Metabolomics of infectious diseases in the era of personalized medicine. Front Mol Biosci 10:1120376
Raparthi M, Maruthi S, Dodda SB, Reddy SR (2022) AI-driven metabolomics for precision nutrition: tailoring dietary recommendations based on individual health profiles. European Econ Lett (EEL) 12(2):172–179
Ravishankar GA, Ambati RR (eds) (2019) Handbook of algal technologies and phytochemicals: volume I. Food, health and nutraceutical applications. CRC Press
Rezzi S, Ramadan Z, Fay LB, Kochhar S (2007) Nutritional metabonomics: applications and perspectives. J Proteome Res 6(2):513–525
Rodríguez-Sifuentes L, Marszalek JE, Hernández-Carbajal G, Chuck-Hernández C (2021) Importance of downstream processing of natural astaxanthin for pharmaceutical application. Front Chem Eng 2:601483
Sharma N, Yeasmen N, Dube L, Orsat V (2024) Rise of plant-based beverages: a consumer-driven perspective. Food Rev Intl:1–27
Singh S, Sarma DK, Verma V, Nagpal R, Kumar M (2023) Unveiling the future of metabolic medicine: omics technologies driving personalized solutions for precision treatment of metabolic disorders. Biochem Biophys Res Commun 682:1
Singh B, Mal G, Sharma R, Gopinath D, Jairath G, Rialch A, He F, Marotta F (2024) Biotransformation of phytochemicals: way forward to gut microbial, nutraceuticals and herbal therapeutics advancements. Funct Food Sci-Online ISSN: 2767–3146. 4(3):96–118
Sorour NM, Tayel AA, Abbas RN, Abonama OM (2017) Microbial biosynthesis of health-promoting food ingredients. In: Food biosynthesis. Academic Press, pp 55–93
Sun W, Chen Z, Hong J, Shi J (2020) Promoting human nutrition and health through plant metabolomics: current status and challenges. Biology 10(1):20
Tang J (2011) Microbial metabolomics. Curr Genomics 12(6):391–403
Thakur M, Belwal T (eds) (2022) Bioactive components: a sustainable system for good health and well-being. Springer Nature
Vattem DA, Maitin V (eds) (2015) Functional foods, nutraceuticals and natural products: concepts and applications.
Verma M, Hontecillas R, Tubau-Juni N, Abedi V, Bassaganya-Riera J (2018) Challenges in personalized nutrition and health. Front Nutr 5:117
Vernocchi P, Del Chierico F, Putignani L (2016) Gut microbiota profiling: metabolomics based approach to unravel compounds affecting human health. Front Microbiol 7:186329
Wehling M (ed) (2021) Principles of translational science in medicine: from bench to bedside . Academic Press, Elsevier Inc:
Wishart DS (2016) Emerging applications of metabolomics in drug discovery and precision medicine. Nat Rev Drug Discov 15(7):473–484
Workman J Jr, Lavine B, Chrisman R, Koch M (2011) Process analytical chemistry. Anal Chem 83(12):4557–4578
Xu C, Yang X, Wang X, Jing L, Zhou Z, Cao Y, Zheng H, Kuo CL, Huang D (2024) Authentication of ten distinctive triterpenoids in Antrodia cinnamomea serves as a crucial aspect for ensuring the quality control of associated nutraceutical products. Curr Res Food Sci 8:100721

Ye D, Li X, Shen J, Xia X (2022) Microbial metabolomics: from novel technologies to diversified applications. TrAC Trends Anal Chem 148:116540
Yuan SF, Alper HS (2019) Metabolic engineering of microbial cell factories for production of nutraceuticals. Microb Cell Factories 18(1):46
Zeisel SH (2020) Precision (personalized) nutrition: understanding metabolic heterogeneity. Annu Rev Food Sci Technol 11:71–92
Zhao Y, Zhang B, Chen G, Chen A, Yang S, Ye Z (2014) Recent developments in application of stable isotope analysis on agro-product authenticity and traceability. Food Chem 145:300–305

Exploring Synergies of Microbial Metabolomics with Other Omics Techniques: Enhancing Drug Discovery and Pharmaceutical Production

Preeti Rajesh, Ram Ajore, and Harsimran Kaur

Abstract

Natural products originating from microorganisms have been an important source of many successful lead molecules in the drug discovery and development process. The technological advances in analytical instrumentation and the development of various omics methodologies have provided new opportunities to accelerate the discoveries of new lead structures of microbial origin. Metabolomics is a rapidly emerging omics-based platform for the identification of diverse microbial metabolites on a large scale. The chapter emphasizes the importance of microbial metabolomics in identifying bioactive compounds and provides a general overview of the standard workflow of mass spectrometry (MS) and nuclear magnetic resonance (NMR) spectroscopy-based metabolomics to identify large numbers of metabolites from cells. In addition, the chapter focuses on the synergistic use of multi-omics approaches—genomics, transcriptomics, proteomics, and metabolomics—in the discovery of novel drug molecules from microbial sources. Special attention is given to how integrative omics techniques improve the identification and characterization of microbial natural products, including how genomics and metabolomics or transcriptomics and metabolomics work together to guide the discovery process. The synergistic use of multiple omics techniques in identifying the microbial targets to treat infectious diseases is also highlighted.

P. Rajesh
Department of Biotechnology, School of Biotechnology and Biosciences, Brainware University, Kolkata, India

R. Ajore
Department of Hematogenomics, Lund University, Lund, Sweden

H. Kaur (✉)
Department of Biotechnology, University Institute of Biotechnology, Chandigarh University, Mohali, Punjab, India

S. Kaur et al. (eds.), *Microbial Metabolomics*,
https://doi.org/10.1007/978-981-96-4824-5_10

Keywords

Metabolomics · Multi-omics · Drug discovery · Genomics · Microbes

1 Introduction

The reformulated ecological theory that combines the universal dominance scaling law with the lognormal model of biodiversity predicts that roughly a trillion microbial species inhabit the Earth (Locey and Lennon 2016). The last decade has seen a rapid surge in our understanding of microbial biodiversity. Recent developments in sequencing technologies have removed the obstacle of uncultivability and opened the door for genome-resolved metagenomics revealing novel information about microbial biodiversity. The discoveries of new species are changing the tree of life. The current state-of-the-art methods employing multi-omics approaches including metagenomics, transcriptomics, proteomics, and metabolomics have added orders of magnitude to the inventory of microbial species and to the repertoire of empirical information that can be applied to biologically relevant trends (Shaffer et al. 2022). While sequencing of the microbial genomes has been instrumental in defining the genetic diversity of microbial communities, metagenome mining and transcriptomic profiling are effective techniques to characterize function in microbial communities (Kobras et al. 2021). Furthermore, the identification of proteins linked to microbial activity, microbial host-pathogen interactions, and antimicrobial-resistant processes is facilitated by microbial proteomics (Kathera 2018). Apart from these omics techniques, metabolomics research has uncovered novel metabolic pathways, providing an understanding of how microbes endure in nearly every sort of terrestrial, aquatic, artificial, and host-related environment. The diverse biochemical compositions and vast range of metabolic variety make these microbes an attractive source of novel therapeutic alternatives, in addition to the evolutionary advantage of changing environmental conditions (Liu et al. 2021). Microbial metabolites are particularly valuable as natural antioxidants, immunosuppressants, enzymes, enzyme inhibitors, vitamins, and antibiotics and as a means of treating illnesses like cancer, anemia, diabetes, obesity, atopic dermatitis, Crohn's disease, and so forth (Abdelghani et al. 2021). These vast arrays of microbe-derived metabolic compounds can have important implications for human health.

In a biological system, the flow of information occurs in general from the genomic DNA (deoxyribonucleic acid) through the RNA (ribonucleic acid) to proteins. The proteins are the workhorses of the cellular processes including the metabolism. The interplay between the tightly regulated metabolic pathways generates a flux of metabolites within a cell, tissue, or organism influencing the phenome. However, it is now understood that the information flow in a biological system is not linear or unidirectional. Many complex networks exist between DNA, RNA, proteins, and metabolites in the cells in an overlapping manner with various intersections and multiple feedback loops. The omics technologies provide invaluable tools for the study of these interactions on a global level and offer an alternative means of investigation to reductionist molecular biology approaches (Roberts et al. 2012). In microbial systems, the metabolites are responsible not only for growth, maintenance, or normal function in a specific physiological state but also for the adaptive responses in various environments. Across the microbial world, the chemical classes and physical properties that constitute metabolites and the dynamic range of metabolite concentrations vary across large orders of magnitude, making it necessary to employ an extensive array of analytical techniques in metabolomics research. Metabolomics methodologies employ both targeted and untargeted approaches to detect and characterize thousands of metabolites produced by microorganisms. While the untargeted strategy focuses on an intended comprehensive analysis of all the measurable analytes in a sample including chemical unknowns, the targeted approach measures defined groups of chemically characterized and biochemically annotated metabolites. Targeted metabolomics provides valuable insights into microbial metabolic processes. Untargeted metabolomics coupled with advanced chemometric techniques offers a comprehensive chemical profile of complex microbial systems. The metabolomics datasets with information on scores of metabolites produced by microorganisms thus serve as a gold mine for the identification of bioactive compounds with therapeutic potential (Bauermeister et al. 2022; Gomes et al. 2023; Mashego et al. 2007). In addition to discovering new lead molecules for drug development, metabolomics studies can help uncover the mechanism of action of various drug molecules and generate their toxicology profiles. Metabolomics studies also provide a comprehensive understanding of the metabolic changes occurring within the microbial pathogens allowing for the identification of unique metabolic targets that can be exploited for therapeutic intervention (Azhar et al. 2023). Even though it is still evolving, metabolomics has emerged as an exciting mechanism-based method that opens up new avenues for drug development, discovers new targets, and presents enormous potential for the discovery of novel natural compounds with unique biological activities, thus revolutionizing drug discovery (Aminov 2022; Krug and Müller 2014).

2 Scope of Metabolomics in Drug Discovery and Pharmaceutical Production

The discovery of new drug molecules, identification of drug targets, and metabolic markers for early disease detection are a few areas of human health where new perspectives have been opened by the synergistic use of multiple omics techniques with metabolomics and the advancements in supporting bioinformatic tools. Recent developments in technology and informatics have allowed researchers to generate and analyze large volumes of postgenomic data that has already shown an unexpected diversity of microbial metabolites and metabolic networks. Combining metabolomics with other "omics" data may thus enable precise characterization of metabolic activities offering a thorough understanding of the altered pathways. The developing field of microbial metabolomics clarifies the mechanisms underlying the microbe-host interaction, facilitating the translation of this knowledge into

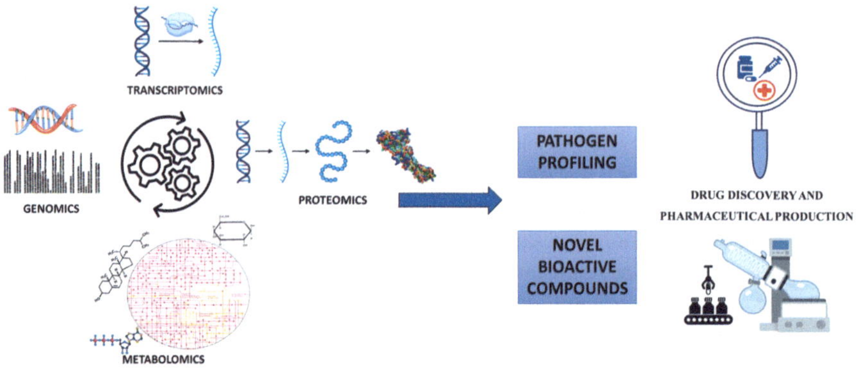

Fig. 1 The integration of various omics sciences, including genomics, transcriptomics, proteomics, and metabolomics, has greatly impacted not only the approach and execution of antimicrobial drug discovery and development but also the industrial application of microbial bioproduction

druggable therapeutic advancements for a range of human diseases as well as molecular diagnostics. It seeks to identify biological alterations in endogenous metabolites because of external and internal stimuli. In addition to accelerating the drug development process by predicting pharmacokinetics, pharmacodynamics, and drug response, metabolomics facilitates the exploration of drug repurposing and drug-drug interactions. The collaborative application of these techniques holds promise for the development of industrial-scale bioproduction processes via Design–Build–Test–Learn (DBTL)-based strain engineering, culture, and model improvement by revealing pathway bottlenecks and metabolic mechanisms (Pang and Hu 2023; Tanaka et al. 2024).

This chapter explores the synergetic role of genomics, transcriptomics, proteomics techniques, and metabolomics in enhancing drug discovery and pharmaceutical production (Fig. 1). The chapter provides a quick overview of the main metabolome research in the discovery of microbial products with pharmaceutical value.

3 Analytical Techniques in Microbial Metabolomics

High-throughput analytical technologies are clustered on a common platform referred as "omics." The important data from biological samples and their metabolites can be collected and analyzed through proteomics, epigenomics, lipidomics, transcriptomics, metabolomics, and nutrigenomics. Diverse array of biological sample metabolites such as polyphenols, organic-inorganic components, amino acids, vitamins, hormones, nucleic acids, alkaloids, biomarkers, and carbohydrates participating in different biological reaction undergoing in cells, tissues, and organs can be studied by metabolomics (Ren et al. 2018). Qualitative and quantitative analysis of various metabolites involved in biological reactions and metabolic processes is attained by metabolomics. The collected data can be further organized to generate useful information in the field of personalized medicine, drug development, and biomarker discoveries. Traditionally the analysis of metabolites was not accomplished due to the physical and chemical complexity of different metabolites indulged in biological reaction. Numerous high-end advanced technologies have been developed to gain a comprehensive understanding of the complex metabolites involved in biological samples (Aggarwal et al. 2023; Lybbert et al. 2020). Analysis of each metabolite necessitates with the collection and preparation of sample, followed by processing under various analytical techniques and finally analysis and compilation of data. Accurate analytical data of metabolites with high precision and high recall values can be achieved by high-resolution techniques. Viz., mass spectroscopy (GC-MS, LC-MS, RPLC-MS, and CE-MS), nuclear magnetic resonance spectroscopy (NMR), high-performance liquid chromatography (HPLC) and ultra-performance liquid chromatography (UPLC) technology, hydrophilic-interaction liquid chromatography (HILIC), and super critical fluid chromatography mass spectrometry (SFC-MS) (Table 1) (Shih et al. 2014).

Mass spectrometry (MS) is one of the most extensively used high-throughput analytic tools of choice in metabolomics. It is noteworthy MS possesses remarkable advanced technology to qualitatively and quantitatively identify infinite range of metabolites produced in biological system by measuring their mass-to-charge ratio. Accordingly, metabolomics could be utilized to get accurate data of the exogenous or endogenous metabolites to further understand the host-microbial interaction for the prognosis of microbial based disease and drug development. The broad range of end products (metabolites) of cellular reactions can be canvassed by mass spectrometry (MS) coupled with separation techniques like gas chromatography (GC-MS), liquid chromatography (LC-MS), and capillary electrophoresis-MS (Bauermeister et al. 2022; Liu et al. 2021).

Gas chromatography-MS (GC-MS) is a metabolomic technique applied to extract the analytical data of the complex microbial metabolites for understanding the synergistic relationship with host. Gas and liquid chromatographies are effective for separation of low molecular weight/volatile and high molecular weight/non-volatile metabolites, respectively, which are further detected by mass spectrometers. Preparation of microbial samples is a prerequisite for reproducibility in the chromatography column along with the removal of residual salts which is accomplished by solid-phase extraction (SPE) method. Samples are run to the column for a known duration to obtain a chromatograph with various metabolite peaks on the basis of interaction of sample and polarity if the column (Segers et al. 2019; Wu et al. 2024).

Furthermore, GC-MS is upgraded with Time-of-Flight (ToF) systems to facilitate the separation of the ultra-complex metabolites in the chromatographic column at a high scan rate so that the retention time is calculated on the basis of chemical interaction of the metabolites with the absorbent of the column. The quantitative GC-MS spectral data is plotted with retention time and mass/charge ratio of the metabolites. Various types of ionization sources, electron ionization mass spectrometry (EI-MS), atmospheric pressure chemical ionization (APCI), and atmospheric pressure photo ionization (APPI), can be employed in GC-MS. Quadruple

Table 1 Comparative analysis of metabolomic analytical techniques

Tool/techniques	Features		Sample		Properties			References
	Advantage	Disadvantages	Type	Volume	Sensitivity	Reproducibility	Resolution	
Gas chromatography-mass spectrometry (GC-MS)	Highly sensitive for volatile metabolites Ability to analyze different groups of metabolites Large linear range	Unable to identify thermolabile compounds Slow analysis complex sample preparation	Cells Biofluids Tissues Organoids	±1–2 μL (microliter)	High	Low	High	Aretz and Meierhofer (2016)
Liquid chromatography-mass spectrometry (LC-MS)	Simple and fast sample preparation Diverse array of metabolites can be identified along with thermolabile compounds	Expensive Slow analysis Ion suppression	Cells Biofluids Tissues Organoids	±2–10 μL	High	Low	High	Covington et al. (2017)
Capillary electrophoresis-mass spectrometry (CE-MS)	Rapid analysis with low sample size	Affected by salt Low stability compared to GM- and LC-MS Poor reproducibility and sensitivity	Cells Biofluids Tissues Organoids	±1–2 nL (nanoliter)	Low	Low	High	Pang et al. (2023)
Nuclear magnetic resonance (NMR)	Provide structural information Non-destructive	Expensive instrument Some chemical classes are not detected	Cells Biofluids Tissues Organoids	±100–200 μL	Low	High	Low	Segers et al. (2019)
Matrix-assisted laser desorption ionization-mass spectrometry (MALDI-MS)	Fast analysis with low sample volume Identify high MW metabolites	Complex matrix	Cells Biofluids Tissues Organoids	±5–10 μL	High	Low	High	Bauermeister et al. (2022)

(GC-Q-MS) and triple quadruple instruments (GC-QqQ-MS) are the mass analyzers used in GC-MS (Pasikanti et al. 2008; Zhang et al. 2012). Moreover, the identification of the metabolites is corroborated with the known standards run under similar conditions as for the samples. Multivariate statistical analyses (MVSA) are used to analyze the metabolomics data. Partial least squares (PLS) and principal component analyses (PCA) are two major methods by which data is analyzed and compiled. Omics data can be used as a platform for the integration of the changes in the metabolic pathways as well as identification of a variety of metabolites. The comprehensive information of the raw data can be affirmed by various powerful bioinformatic tools referred as multi-omics software/algorithm (Aggarwal et al. 2023; Manzoni et al. 2018). Accordingly, this analytical tool can be applied for the phenotypic fingerprinting of microbial metabolites. Many authors have established that diverse array of transport proteins is categorized as antimicrobial compounds under multidrug resistant (MDR) conditions. The tolerance ability of *Pseudomonas putida* toward antibiotics was related to the efflux pumps of the bacteria. GC-MS metabolic profiling data of *Pseudomonas putida* substantiated ornithine was the metabolites secreted by *P. putida* cells under the propranolol toxic conditions (Sayqal et al. 2016). In another report, GC-MS of urine sample of gout patients was analyzed to study the metabolic pathway and its changes under the diseased condition. It was affirmed that various metabolic pathways (TCA cycle, lipid and amino acid) in gout patients are carried out by 30 different metabolites as compared to that in healthy individual. The identified metabolites can be scientifically incorporated for the diagnosis of the disease (Li et al. 2018).

In light to enhance versatility in metabolomic studies, mass spectroscopy (MS) was coupled with effective separation process of liquid chromatography (LC) for the identification of diverse complex high molecular weight and non-volatile metabolites. In high-performance liquid chromatography (HPLC), separation of complex molecules is achieved by isocratic or gradient approach on the basis of polarities of the metabolite with the wide variety of organic solvents (formic acid, acetonitrile, methanol, and water). Separated chemical moieties are further analyzed under MS through detectors (MALDI, chemical ionization, atmospheric pressure chemical ionization (APCI)) for the identification of metabolites (Clasquin et al. 2012; Zhang and Watson 2015). The coupling of LC-MS is a great application in understanding the behavior of all the complex metabolites due to increased signal response less noise and thereby high peak resolution. In a further advancement of LC-MS techniques, reverse phase liquid chromatography (RPLC) has shown to elute the complex phytochemicals categorized as semi-polar metabolites (Aretz and Meierhofer 2016). The highlight of LC-MS analytical tool is its ability to bioanalyze small sample size with complex composition with high selectivity and sensitivity in a short time duration. Various sugars and amino acids are assorted by hydrophilic interaction liquid chromatography (HILIC), while saturated and unsaturated lipids are looked into by normal phase liquid chromatography (NPLC) coupled with APCI-MS detectors (Du and Huang 2019). In one of the reported data, metabolic profiling of diverse array of complex metabolites in prokaryotic and eukaryotic microorganisms was proposed by using highly sensitive ion-free pentafluoro phenyl propyl (PFPP) column in LC-MS technique. Around 113 primary and intermediate metabolites produced by microorganisms were investigated by PFPP (Takenaka et al. 2021). The reproducibility of the large number of metabolite quantification with high rate of recovery was attained by LC-MS. In another approach, HPLC was employed to probe the presence of secondary metabolites by *Salinispora arenicola* strains grown under 1% and 3% NaCl condition for 43 days for rifamycin production. Various secondary metabolites, viz., derivatives of rifamycin W, rifamycin B, and rifamycin S during the rifamycin production, were detected by HPLC-DAD (Bose et al. 2015).

Nuclear magnetic resonance (NMR) is an exceptionally promising analytical tool for metabolite analysis due to its reproducibility and quantitative accuracy. In NMR, the generated spectrum facilitates the identification and absolute quantification of unknown metabolites through a single reference (Nagana Gowda and Raftery 2021). Furthermore, NMR allows comprehensive metabolic profiling of the biological samples as it enables the direct processing of samples without any need for separation tissues, cells, and biological fluids. This ensures that full metabolic profiling is achieved of the biological samples which can be left unidentified during preparation and separation of samples (Barding et al. 2012; Tulpan et al. 2011).

In addition, use of stable isotope-labelled precursors in NMR alleviates the in-depth analysis metabolic pathways and metabolic flux of biological samples. Wide range of atomic nuclei, including 1H, 13C, 31P, or 15 N, are tagged to detect and analyze complex metabolites, and ^{1}H NMR is considered as the best candidate of choice due its presence in nearly all metabolites and high NMR sensitivity as compared to other known nuclei (Huang et al. 2015).

In NMR-based metabolomics, high-throughput analysis is predominantly carried out using one-dimensional (1D) NMR. Nuclear overhauser enhancement spectroscopy (1D NOESY) and Carr-Purcell-Meiboom-Gill (CPMG) are experimentally favored for the metabolic analysis of low macromolecular content in urine, cells, and tissue extracts and macromolecule content in blood serum or plasma, respectively. NMR methods are gaining traction in metabolomics applications for 2D NMR as they greatly enhance spectral resolution and mitigate peak overlap issues enabling identification of unknown metabolites and enhanced quantitation of low-abundance metabolites (Vignoli et al. 2019). Correlation spectroscopy (COSY) and total correlation spectroscopy (TOCSY) are two variants of 2D NMR-based spectroscopy. Numerous strategies have been implemented for the data acquisition in NMR. It has been quoted that 2D HSQC, SOFAST (band-selective optimized flip angle short transient), and SOFAST HMQC (heteronuclear multiple quantum correlation) based data acquisition was attained in just 15 s and enables real-time metabolic studies in biological live cells experimental set-up (Mahrous and Farag 2015).

A global threat of antibiotic-resistant bacteria is escalating at an alarming rate where substantial proportion of bacteria are getting resistant to at least one class of antimicrobial agents. The identification of microbial metabolites by advance metabolomic techniques has led to the emergence of novel drug candidates (Fig. 2). In context to this, metabolomic study of wild-type *Bacillus cereus* (B. cereus) and mutated *B. cereus* in the presence of a multivalent antimicrobial compound

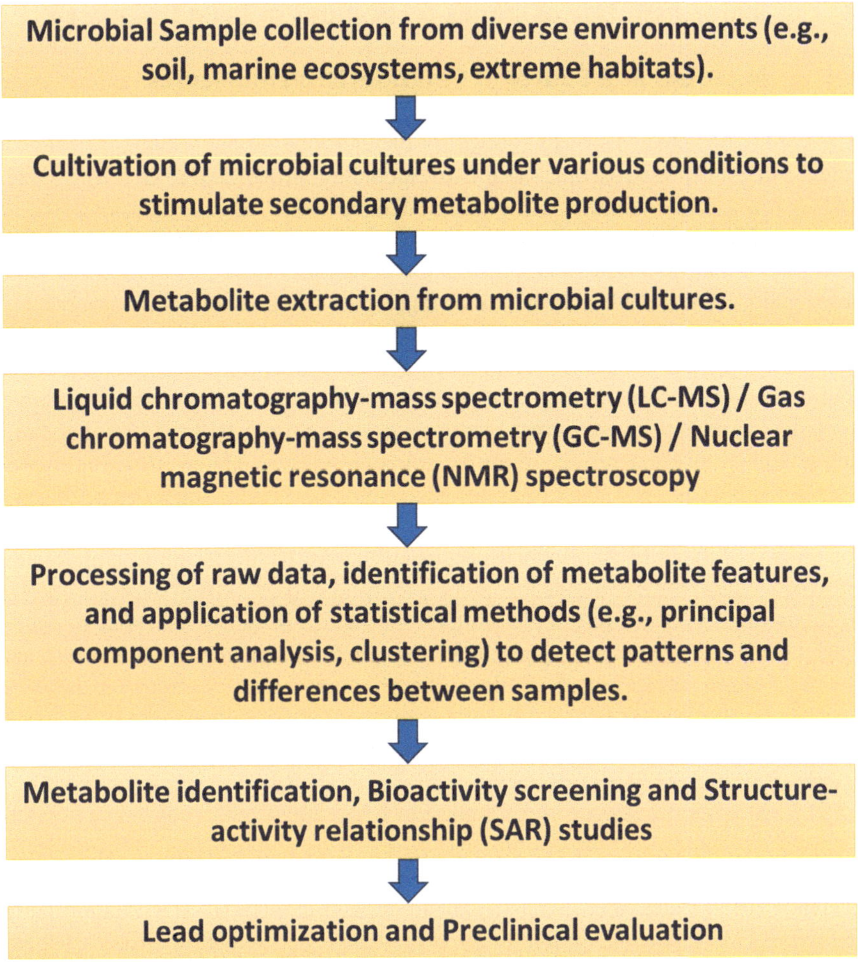

Fig. 2 Workflow for the systematic exploration of microbial metabolomes for novel drug candidates

C16-DABCO (1,4-diazabicyclo2,2,2-octane) and mannose functionalized poly(amidoamine) dendrimer (DABCOMD) was reported. The alterations in metabolomic pathways were examined by an untargeted proton (1H) NMR hydrophilic metabolomics investigation following the exposure of *Bacillus cereus* to the multivalent antimicrobial agent DABCOMD and C16-DABCO (Aries and Cloninger 2021). Many hydrophilic metabolites associated with energy metabolism were identified and quantified by 1H NMR which are correlated with key resistance mechanisms in Gram-positive bacteria. Recently, NMR metabolomics used for isolation of intracellular (IC) and extracellular (EC) metabolites from a range of pathogenic Gram-positive and Gram-negative bacteria has been reported (Pinu et al. 2017).

4 Synergistic Use of Omics Techniques in Drug Discovery

4.1 Microbes as a Source of Potential Drug Molecules

The earth's biosphere is a vast repository of microbial biomass that constitutes a broad category of microscopic organisms, including viruses, bacteria, fungi, algae, and protozoa. Due to their extensive metabolic repertoire, microorganisms produce a vast array of distinctive compounds. These specialized metabolites play a significant role in microbial fitness and profoundly affect microbial physiology, metabolism, and stress responses. In addition, they are a valuable source for innovative biotechnology that substantially impacts almost all facets of plant, animal, and human life. Although natural products can be derived from any living organism, microorganisms encompassing both prokaryotes and eukaryotes show remarkable versatility in producing bioactive compounds. Approximately 23,000 biologically active compounds are produced by the microorganisms. Among the prokaryotic unicellular bacteria, the *Bacillus* and *Pseudomonas* species, filamentous actinomycetes, myxobacteria, and cyanobacteria are the most frequent producers. Notably, the species of the order *Actinomycetales* produce over 10,000 bioactive compounds of which about 7600 are derived from *Streptomyces* and 2500 from the rarer actinomycetes species collectively accounting for 45% of the bioactive microbial metabolites. On the other hand, eukaryotic fungi, especially endophytic fungal species, are the most significant producers, while yeasts, phycomycetes, and slime molds rarely produce bioactive metabolites. The total number of bioactive metabolites produced by fungal species is estimated at 8600, representing 38% of all microbial products (Abdel-Razek et al. 2020).

The serendipitous discovery of penicillin by Alexander Fleming in 1928 from *Penicillium* mold was a turning point in drug discovery and human health and established the importance of microorganisms as sources of therapeutic compounds (Gaynes 2017). Since then, microorganism-derived compounds have found many applications in medicine, agriculture, food industry, and scientific research (Pham et al. 2019). Early investigations into antibiotics yielded many discoveries including cephalosporin C from *Cephalosporium acremonium* (Newton and Abraham 1955); erythromycin from *Saccharopolyspora erythraea* (Lü et al. 2020); vancomycin from *Amycolatopsis orientalis* (Geraci et al. 1956); and streptomycin from *Streptomyces griseus*, chloramphenicol from *Streptomyces venezuelae*, and chlortetracycline from *Streptomyces aureofaciens* (Quinn et al. 2020). In addition to antibiotics, microorganisms over the years have yielded a plethora of other essential drugs, including statins derived from *Aspergillus terreus*, immunosuppressants like cyclosporine from *Tolypocladium inflatum*, and anticancer agents including doxorubicin and bleomycin from bacteria of the genus *Streptomyces*. The three antidiabetic drugs of microbial origin include acarbose, miglitol, and voglibose. Acarbose is one of the most employed digestive enzyme inhibitors for treating type 2 diabetes mellitus (T2DM). It is derived from the *Actinoplanes* species and involves the inhibition of α-glucosidase, α-amylase, sucrase, and maltase, without exhibiting any insulinotropic effects. Miglitol, a structural analog of glucose, is a second-generation α-glucosidase inhibitor. It was initially isolated from various *Bacillus* and *Streptomyces* strains. Its current production involves the oxidation of 1-amino-D-sorbitol to 6-amino-L-sorbose through fermentation with *Gluconobacter oxydans*. Voglibose is another α-glucosidase inhibitor synthesized from valiolamine obtained from the fermentation broth of *Streptomyces hygroscopicus subsp. limoneus*. It acts as a competitive and reversible inhibitor of glucoamylase, sucrase, and isomaltase, while it does not affect α-amylases and is demonstrated to reduce plasma glucose and insulin levels in a dose-dependent manner (Ríos et al. 2015).

The metabolites produced by microorganism act as a vital reservoir for a multitude of compounds with therapeutic applications, thus representing a rich domain for drug discovery endeavors. The unprecedented progress in experimental technologies including genomics, transcriptomics, proteomics, and metabolomics and the emergence of the idea of multi-omics analysis now offer the chance to reveal the untapped pharmaceutical potential embedded within microbial diversity. As researchers and scientists explore more and more metabolic pathways, we can see the emergence of novel therapeutic compounds capable of addressing a multitude of human diseases in the future.

4.2 Use of Metabolomics and Multi-omics Techniques for the Discovery of Potential Drug Molecules from Microbial Sources

Metabolomic analyses provide a wealth of information regarding the activity and responses of microorganisms. The in-depth study of small molecules within a microbial system allows researchers to identify and quantify the vast array of low-molecular-weight (<1500 Da) metabolites and signalling molecules produced by microorganisms in the entire life cycle or during specific physiological phases (Covington et al. 2017). In the previous section, we elaborated nuclear magnetic resonance (NMR)- and mass spectrometry (MS)-based metabolomic analyses, two analytical platforms to facilitate metabolome profiling for natural product discovery. Through the comparative study of the metabolic profiles, it is possible to prioritize organisms, identify compounds that are activated following exposure to stimuli, rank features through bioactivity spectrums or molecular class, and even dereplicate prioritized secondary metabolites. Microbial secondary metabolites significantly impact human health through isolated pharmaceuticals and compounds produced in situ from within human microbiomes. Metabolomics may also provide valuable insights into the ways these microorganisms adapt their secondary metabolic processes in response to dietary influences, medicinal treatments, and endogenous host conditions (Clarke et al. 2010; Covington et al. 2017; Donia et al. 2014; Hooper et al. 2012; Newman 2016).

Over the last decade, microbial metabolomics has seen a lot of advancement and has been applied extensively in numerous biomedical contexts including the investigation of microbial antibiotic resistance, disease mechanisms, and identification of new bacterial species, functional genes, and natural product (Pang and Hu 2023; Ye et al. 2022). The first study in the field of microbial metabolomics was reported by Tweeddale et al. in the year 1998 to investigate the shift in the global metabolic profile of *Escherichia coli* under different growth conditions using two-dimensional thin-layer chromatography (Tweeddale et al. 1998). While the analysis did not yield a comprehensive perspective of the metabolome, it successfully identified metabolic trends that demonstrate the stringent physiological regulation of metabolic pools. The results of this study suggest for the first time that examining the total metabolite pool, referred to as the "metabolome," can reveal previously unrecognized facets of cellular metabolism and its global regulatory frameworks. The new methodologies for conducting targeted metabolomics experiments were developed. The potential of NMR in targeted metabolomics was demonstrated by using ^{19}F NMR for the elucidation of microbial degradation pathways of fluorophenols. The ^{19}F NMR technique demonstrated considerable potential for conducting studies on metabolomics in newly isolated microorganisms and the identification of relatively unstable pathway intermediates like fluoromuconolactones and fluoromaleylacetates (Boersma et al. 2001). The critical role of sulfated molecules in cell-cell communication inspired Mougous et al. to apply Fourier transform-ion cyclotron resonance mass spectrometry (FT-ICRMS) for the identification of sulfated metabolites in microorganisms, particularly focusing on pathogenic mycobacteria. The initial detection of sulfur-containing molecules was performed via mass spectral analysis of cell extracts from bacteria that were metabolically labeled with a stable sulfur isotope ($^{34}SO_4^{2-}$). This technique enabled the identification of several previously unknown sulfated molecules in *Mycobacterium tuberculosis* and *Mycobacterium smegmatis* (Mougous et al. 2002). The application of the metabolic labeling/MS approach provided a context for discovering secondary metabolites with relatively little information about their chemical structure. As the technology continued to develop, many different approaches were evaluated including 1H NMR spectroscopy, solid phase microextraction (SPME) techniques coupled to GC-MS analysis, electrospray mass spectrometry (ES-MS), and so on (Castrillo et al. 2003; Defernez and Colquhoun 2003; Verdonk et al. 2003). The ability to measure the entire complement of metabolites within microorganisms started to pave the way for assessing their biochemical diversity. The metabolite-profiling analyses were subsequently reported for the *Corynebacterium glutamicum*, *Bacillus subtilis*, *Pseudomonas aeruginosa*, and *Pseudomonas putida S12* (Choi et al. 2007; Soga et al. 2003; Strelkov et al. 2004; Van Der Werf et al. 2008). The technological advancements in omics techniques opened opportunities to tap into the biotechnological potential of the microbes and use these yet untapped reservoirs of metabolic diversity for future natural product drug discovery.

With the increasing metabolomic studies, the microbial metabolite repertoire is expanding, and metabolomics is emerging as a powerful tool in identifying natural products that are key to drug development. However, metabolite identification remains the primary bottleneck in metabolomics with only a small fraction of metabolites in the metabolomics datasets being usually identified, and many remain to be discovered (Reverter et al. 2020). The annotation of known metabolites can be effectively accomplished through the utilization of open-source databases such as the Human Metabolome Database (HMDB), the Global Natural Products Social (GNPS), the Dictionary of Natural Products (DNP), the Dictionary of Marine Natural Products, and the Collection of Open Natural Products (COCONUT) (Chassagne et al. 2019; Sorokina et al. 2021). The annotation allows for the early-stage compound dereplication curtailing the rediscovery of already known compounds. The integration of bioinformatics tools, such as Metabolic in Silico Network Expansion (MINE), into the workflow allows for the annotation of anticipated metabolites. This process involves comparing metabolic characteristics with an extensive database derived from computational predictions of molecules that are probable to exist, informed by biochemical reactions and metabolites classified in the KEGG (Kyoto Encyclopedia of Genes and Genomes) database (Jeffryes et al. 2015). However, the complexity and diversity in the structure and chemical properties of these metabolites properties make their analysis particularly challenging, and metabolite annotation offers only initial insights into the potential identities of molecules (Rochfort 2005). As a next step, targeted fragmentation of metabolites through mass spectrometry (MS2) allows for a more detailed structural characterization, but the MS2 data interpretation is incredibly difficult. Various bioinformatics platforms such as UNIFI, GNPS, and Qemistree are available which can generate valid suggestions for candidate structures, and in silico approaches based on the computation of MS2 fragmentation trees, machine learning (ML), and Bayesian models can help generate limited structural hypotheses (Palermo 2023) (Table 2).

A metabolomic approach in drug discovery offers a distinctive opportunity to connect metabolite profiles with the biology of organisms. However, a significant challenge in the field of metabolomics is that many natural product biosynthetic gene clusters (BGCs) remain inactive under typical laboratory conditions. This results in their expression being minimal, which in turn means that the associated products do not reach detectable levels. Consequently, the natural products identified to date are derived primarily from a limited subset of biosynthetically active genes, and the metabolites produced by the silent BGCs remain a largely untapped reservoir of microbial compounds (Yoshimura et al. 2020). In addition, the production of metabolites is multi-dimensional and is influenced by the genomic interactions of the organism with the environment. These interactions regulate the production of metabolites in an integrated manner. Therefore, it is imperative to understand the ecological and biological influences on the production of metabolites to enhance yields, optimize extraction methods, or select bioactive compounds. This complexity calls for the integration of multiple technological frameworks and integrated metabolomic techniques to ensure a detailed investigation into the discovery of microbial natural products.

Due to its inherent limitations, metabolomics alone cannot provide a complete picture of the biological system under study. The ultimate potential of metabolomics can only be unlocked by combining it with other “omics” techniques. In a biological system, the metabolome is the downstream product of genomic and epigenomic processes, as well as transcriptomic and proteomic activities, and can be considered as the chemical phenotype (metabotype) of an organism. The integration of the data generated from multiple omics approaches including genomics,

Table 2 List of metabolomics-based data analysis tools

Microbial metabolite prediction tools		Metabolomics data analysis tools		
Tool	Features	Tool	Features	References
CO-ED	Identifying untargeted multifunctional enzymes	Mzmine3	User-friendly graphical	Sorokina et al. (2021)
MetDNA 2	Metabolite annotation from knowns to unknowns through knowledge-guided multi-layer metabolic networking	MetaboAnalyst 5.0	Comprehensive web-based tool for comprehensive metabolomics data analysis, interpretation, and integration with other omics data	Ren et al. (2018)
SIRIUS 4	Metabolite identification by decomposing isotope patterns	MetFlow	Interactive and integrated web server for metabolomics data cleaning and differential metabolite discovery	Pinu et al. (2019)
Metabo-lights	Generates metabolite structures and reference spectra	LipidSig	Lipidomic data analysis	Medema and Fischbach (2015)
MassBank	High-quality mass spectral database	MS-DIAL 4.0	Lipidome atlas in MS-DIAL 4.0	Manzoni et al. (2018)
BiG-FAM	Assigning BGCs to gene cluster families based on the output from anti SMASH	El-MAVEN	Fast, robust, and user-friendly mass spectrometry data. Processing engine for metabolomics	Jeffryes et al. (2015)
Anti-SMASH 6.0	Predicting secondary metabolites such as antibiotics, with an additional prediction of RiPP cleavage enzyme	LION	Web-based ontology enrichment tool for lipidomic data analysis	Fortuin and Soares (2022)
MIBiG 3.0	Database of annotate experimentally validated BGCs	Omicsnet	Web-based tool for creation and visual analysis of biological networks in 3D space	Durmus et al. (2015)
IMG-ABC 5.0	Integrated data management and comparative analysis system for microbial genomes and microbiomes	METLIN	Technology platform for the identification of known and unknown metabolites and other chemical entities	Wu et al. (2024)

transcriptomics, proteomics, and metagenomics can overcome the limitations of metabolomics and provide a more holistic understanding of the studied biological systems. The proficient integration of multi-omics technologies, resulting in the formulation of novel panomics strategies, commonly known as integrative omics or multi-omics can thus significantly enhance the successful discoveries of novel natural products (Pinu et al. 2019; Resurreccion and Fong 2022; Zhang et al. 2021).

4.3 Integrative Multi-omics Approaches in Natural Product Discovery from Microorganisms

4.3.1 Genomics- and Metabolomics-Guided Discovery of Microbial Natural Products

Integrating metabolomics with genomics provides a more effective approach to identifying novel metabolites and developing new leads for drug discovery. Advances in sequencing technologies have generated a wealth of genomic information. Genomics is the study of all of a person's genes (the genome), including interactions of those genes with each other and with the person's environment (definition adapted from https://www.genome.gov/about-genomics/fact-sheets/A-Brief-Guide-to-Genomics). In the drug discovery process, the genomic methodologies not only provide insights into protein targets, cellular pathways, and disease resistance mechanisms but also help uncover a significant diversity in microbial metabolic processes. The integration of metabolomic and genomic information can provide a link between the genome sequence and the chemical phenome of an organism. This metabologenomic approach therefore correlates the genetic diversity with the metabolites allowing for the metabolic characterization of genes of interest, confirming known pathways, and permitting the identification of new ones (Avalon et al. 2022; Drenos 2017; Van Der Hooft et al. 2020).

Several studies have reported the use of paired metabolomics and genomics datasets for the discovery of microbial natural products and their biosynthetic pathways (Fig. 3). An estimated up to 90% of gene clusters responsible for secondary metabolite production remain inactive under typical growth conditions (Walsh and Fischbach 2010). A metabolomic study conducted under specific conditions can only identify metabolites that are actively expressed; thus the success of such studies relies on enhancing the gene expression of the organism(s). In this context, the genome mining approach computationally detects and analyzes the BGCs of the chemical compounds and combined with metabolomics connects the genes to molecules (Medema and Fischbach 2015). Over the years several approaches including pattern-based, weighted pattern-based (correlation-based), and feature-based methods have been developed to correlate genetic and chemical data to extract meaningful information from integrated datasets, and many compounds have been discovered using these approaches (Caesar et al. 2021). The integration of genomics and metabolomics data facilitated the identification of three novel antibiotic natural products, namely, columbamides A, B, and C (Kleigrewe et al. 2015), as well as a new variant of thiomarinol (Maansson et al. 2016) derived from marine bacterial

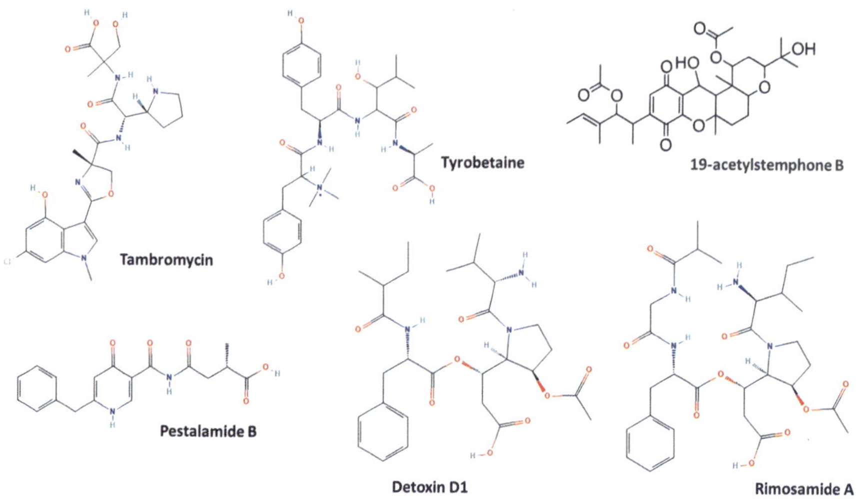

Fig. 3 Microbial bioactive compounds discovered using integrative metabolomics and genomics approaches

sources. Goering et al. performed a large-scale correlation of gene clusters from 178 actinomycete genomes with metabolites. This metabologenomics approach led to the isolation and characterization of an antiproliferative agent tambromycin, a new chlorinated natural product, composed of several nonstandard amino acid monomeric units, including a unique pyrrolidine-containing amino acid named tambroline (Goering et al. 2016). McClure et al. utilizing a metabolite/gene cluster correlation platform, discovered the biosynthetic origins of a new family of natural products—rimosamides in *Streptomyces rimosus* and detoxin in *Streptomyces mobaraensis* exhibiting anti-antibiotic activity (McClure et al. 2016). Combining the genomic and metabolomic data also resulted in the discovery of the tyrobetaine (nonribosomal peptides with an unusual trimethylammonium tyrosine residue) natural products from *Streptomyces* sp. NRRL WC-3773 (Parkinson et al. 2018). In multiple studies of fungal metabologenomics, the researchers correlated genomics and metabolomics datasets from 110 fungi and reported the isolation of pestalamides from *Aspergillus brasiliensis* and three novel stemphone analogs, 19-acetylstemphones G, B, and E with antiproliferative activity from *Aspergillus biplanus* NRRL 5071 (Ayon et al. 2024; Caesar et al. 2023). Thus, the combined approach has been instrumental in enhancing the discovery of bioactive compounds from microbial sources.

4.3.2 Transcriptomics- and Metabolomics-Guided Discovery of Microbial Natural Products

The rapid advances in RNA-sequencing technology have facilitated microbial expression profiling by compiling the types and levels of genes expressed in cells. The core objective of any transcriptomic study is to generate a comprehensive registry of all the transcript species, including mRNAs, non-coding RNAs, and small

RNAs, and delve into the transcriptional structure of genes by highlighting various sequence elements including the transcription start sites, the 5′ and 3′ ends, splicing variations, and post-transcriptional modifications. Both hybridization (microarrays) and sequence-based approaches (RNA-seq) can be used to analyze the expression level variations and deliver accurate, "digital" quantifications of gene expression for each transcript across various developmental stages and physiological conditions (Wang et al. 2009). To adapt to the physical, chemical, and biological conditions found in their environments, microbes develop new adaptive strategies and synthesize new metabolites which are essential for the interaction of organisms with their environment. The production of these secondary metabolites which is not vital for the growth of microbial cells is regulated at the transcriptional level. To adapt to environmental changes and stressors, microbes dynamically produce and activate proteins/enzymes from the otherwise transcriptionally silent secondary metabolite biosynthetic gene clusters. These expression variations in microbial ecosystems can be comprehensively evaluated by transcriptome analysis (Mózsik et al. 2022; Sato 2024). In recent years there has been an increasing interest in combining the data generated from transcriptomics and metabolomics studies to identify a range of bioactive metabolites from different systems along with the key genes and enzymes that facilitate their biosynthesis. Although a direct association cannot be derived between the metabolite and transcript by integrating the transcriptomics and metabolomic data, the synergistic analysis of the two datasets is still powerful in providing phenotypic measurements to anchor the global measurements of the transcriptome (Cavill et al. 2016). This information can also be combined with recombinant DNA tools for engineering metabolic networks to produce desired metabolites and increase pharmaceutical production.

4.3.3 Proteomics- and Metabolomics-Guided Discovery of Microbial Natural Products

Advances in mass spectrometry-based proteomics and metabolomics platforms have contributed significantly to our understanding of microbial physiology, biochemistry, and environmental interactions. Mass spectrometry-based proteomic analysis of in vitro cultured bacterial isolates enables quantitative comparison of proteomes under various controlled conditions. Moreover, the development of the label-free quantitative LC-MS/MS allows the analysis of complex proteomic samples and directly identifies thousands of proteins quantitatively (Fortuin and Soares 2022). Metabolites can be considered as the products of the active proteome of an organism, and global metabolomics is used to generate a metabolic fingerprint using the overall metabolites identified in a single sample. While each technique is powerful on its own, integrated proteomic and metabolomic analysis to profile the biosynthetic enzymes provides scientists with insights into the full biosynthetic potential of microorganisms under specific conditions (Du and van Wezel 2018). In an integrative proteometabolomic study, proteomics investigates the global protein landscape, while metabolomics characterizes the small-molecule metabolites collectively offering a comprehensive view of the molecular interactions that govern biological systems. The combination of quantitative proteomics and metabolomics can define correlations between the abundance of natural products and changes in the microbial protein poll, allowing the detection of biosynthetic enzyme clusters of the producing strains (Paillié-Jiménez et al. 2020). The synergistic analysis of proteomics and metabolomics datasets can provide a comprehensive understanding of microbial metabolism, biosynthetic pathways, and protein-metabolite interactions. This approach can thus enable researchers to unlock the full potential of microbial secondary metabolomes, leading to the discovery of novel natural products with promising biological activities and applications in various industries revolutionizing the field of microbial product discovery.

5 Bacteria as Targets for Treating Infectious Diseases

The alarming increase in the rate antibiotic resistance of bacteria against all infectious diseases is contributing to both morbidity and mortality on a global scale. As per the annual data released by European Centre for Disease Prevention and Control (ECDC), in Europe multidrug-resistant (MDR) bacteria has caused around 25,000 deaths in case of infectious diseases. Global rise in the instance of MDR bacteria has been accounted for vancomycin-resistant enterococci (VRE), New Delhi metallo-beta-lactamase-1 (NDM-1), and broad-spectrum beta-lactamase (ESBL)-producing Enterobacteriaceae (Bright Esegbuyota et al. 2022; van Duin and Paterson 2016). The major cause of MDR arises when pathogen starts evading the standard treatment of infection caused by the bacteria by modifying the target sites, altering the drug transport mechanism and lastly inactivating the drug. Alteration in the target sites occurs due the high ability of the pathogens to get mutated during their growth cycle. These mutations mainly influence the target sites of antibiotics by diminishing the antibiotic's affinity to the bacterial ribosomes, furthermore decreasing its efficacy and consequently resulting in resistance (Molton et al. 2013). It is noteworthy that the activity of efflux pump for specific protein can be enhanced by the mutations which will ensue in the outflow of the antibiotics from bacterial cells (Oberhardt et al. 2009). For instance, rifampicin inhibits bacterial RNA synthesis by binding to bacterial RNA polymerase and is extensively used to control the infectious disease "tuberculosis" caused by *Mycobacterium tuberculosis*. MDR has been extensively reported in *M. tuberculosis* known to mutate as very high frequency. *M. tuberculosis* has shown resistance against rifampicin due to mutations in the β-subunit of the RNA polymerase gene, which in turn causes the structural alteration in the binding site, leading to a decreased affinity of rifampicin for its target and hindering its ability to effectively inhibit RNA synthesis (Yelamanchi et al. 2022).

The demand of innovative therapeutic approaches for the treatment of infectious diseases has become a necessity due to inadequacies of the existing treatment options along with rapidly evolving ability of microbes. The new era of post-antibiotic phase has started due to slow development of new antibiotics toward the highly mutated resistant strains of bacteria causing deadly infectious diseases. Therapeutic solutions for bacterial infections can be dealt with the various metabolomic analytical techniques designed to collect the data of bacterial intercellular and extracellular metabolites in comprehension of bacterial pathogenesis. Recently, there has been a resurgence in various scientific strategies that mainly focus on understanding the host-pathogen interactions and involvement of cellular complex metabolites to combat the infectious diseases (Muteeb et al. 2023; Yang et al. 2021). Currently, the rapid shift toward the translation of metabolites through metabolomics has led to the therapeutic development and enhanced the efficacy of existing treatments for highly infectious diseases. Metabolic profiling of the metabolome in the pathogenic bacteria can be attained by using a combination of high-resolution techniques like NMR, GC, MS, and LC-MS. Metabolomics has provided valuable insights into the complex metabolites involved in the fundamental mechanisms of pathogenesis for bacterial, viral, and parasitic infections (Baidoo and Teixeira Benites 2019).

5.1 Metabolic Profiling of Pathogenic Bacteria

Metabolomics is considered as the innovative and comprehensive scientific domain which enables in detail understanding of how metabolome of the pathogens interacts with host during the infection scenarios. Protein, nucleotides, complex metabolites, ligands, infection-specific gene regulation can be conceptualized as an operative factor during pathogen-host interactions (PHIs) (Durmus et al. 2015). Advancement in multi-omics techniques with special reference to mass spectrometry (MS) has enabled to understand the molecular adaptation processes of the pathogenic bacteria under challenging environments. Microorganisms can survive under different ecological niches by metabolically adapting to the environment (Shah et al. 2015). Moreover, pathogens alter their genomic composition and metabolic pathways and later on generate variety of cellular metabolites to circumvent the host's immune system. Metabolomics studies have revealed that virulence trait is due to various post-translation modifications undergoing in the pathogenic bacteria. The integration of multi-omics techniques has underscored the identification of novel targets, surface biomolecules and metabolites in pathogen-specific pathways, thereby retracing alternative approaches for the development of next-generation therapeutics for MDR. Identification of critical metabolic pathways necessary for pathogen survival may open new possibilities in therapy development (Wang and de Carvalho 2023) (Fig. 4).

Accordingly, NMR spectroscopy-based metabolic profiling of cystic fibrosis patient infected with *Pseudomonas aeruginosa* strains for a particular length of infection was investigated. About 29 metabolites were identified, and data validation leavened that these metabolites lead to alterations in metabolic pathways by enhanced uptake of amino acid uptake. NMR-based profiling approach has given a wider comparative analysis picture of inter-patient differences and the progression of changes throughout the cystic fibrosis infection during prolonged infection (Depke et al. 2020). In a similar report, GC/MS-based profiling of intracellular and extracellular metabolomes was carried for two strains of *Pseudomonas aeruginosa*, PAO1 and the clinical isolate TBCF10839 cultured under different physiological

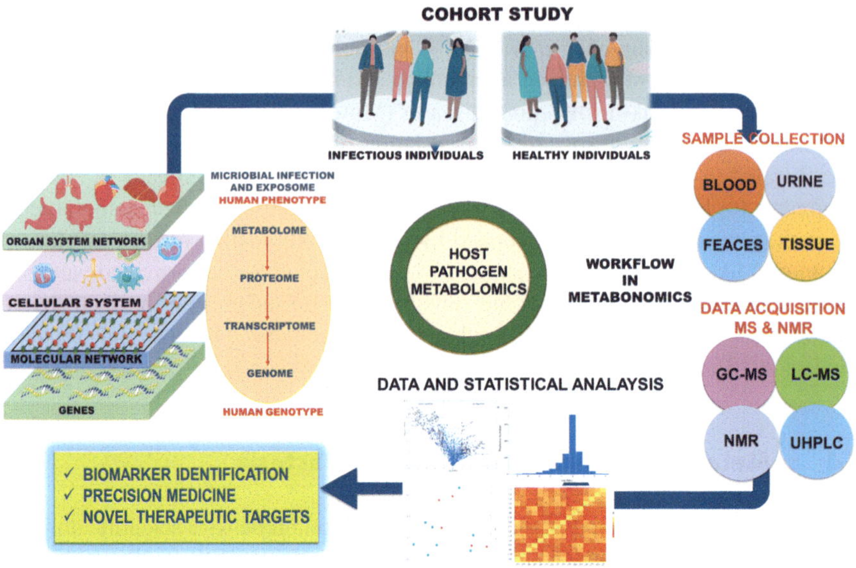

Fig. 4 Schematic representation of metabolic profiling of human samples by metabolomics

conditions. In at least 243 distinct compounds, 65 compounds were found to be subsisting under the typical growth condition in a specific medium. The rest of the metabolites were matched with pre-existing metabolite libraries. In evidence to the clinical challenges faced due to antibiotic resistance and MDR, LC-MS profiling of polymyxin-susceptible PAK and the resistant PAKpmrB6 strains of *Pseudomonas aeruginosa* was employed to evaluate the data associated with polymyxin resistance (Han et al. 2019). It was revealed alterations in the metabolic profiles of both strains were observed on exposure to polymyxin B for 1 h. Furthermore, a total of 427 hydrophobic and 871 hydrophilic metabolite data was recorded and analyzed. It was concluded that there might be the development of a new class of polypeptide antibiotics due to reduction in the synthesis of lipopolysaccharides and peptidoglycan in PAK strain and elevated levels of trehalose-6-phosphate in both the strains under osmotic stress (Lybbert et al. 2020; Xiao et al. 2022). Accordingly, rapid and thorough metabolite screening of *Campylobacter jejuni* strains bearing ermB resistance gene was employed through ultra-high-performance liquid chromatography quadrupole time-of-flight mass spectrometer (UHPLC-QTOF/MS). The 36 distinct identified metabolites were predominantly associated with the biological processes of *Campylobacter jejuni*. They were further categorized into groups including sugars, alkaloids, cyclic compounds, fatty acid amides, peptides, and benzoic acids (Fu et al. 2018).

Escherichia coli is a highly diverse and versatile bacterial species exhibiting both beneficial and pathogenic characteristics depending upon the strains. Various strains of *E. coli*, viz., enterohemorrhagic *E. coli* (EHEC) and uropathogenic *E. coli* (UPEC), are considered pathotype, while strain, namely, *E. coli* strain Nissle 1917, is familiar for its probiotic potential. Phylogenetically all the strains *E. coli* strain Nissle are closely related to pathogenic strains of *E. coli*. High-resolution LC-MS and LC-MS/MS metabolomic tool was utilized to understand the metabolic profiling of phylogenetically linked *E. coli* strains (AlRabiah et al. 2014; Xu et al. 2019). The finding indicates that despite their phylogenetic similarities, these strains exhibited functional distinctions attributed to the presence of metabolites that may confer benefits to the host giving an insight to host-pathogen interaction.

5.2 Identifying Virulence Factors for Effective Therapeutics

In particular bacterial infections remain a critical public health concern, and understanding the underlying mechanisms of bacterial virulence is significant for developing effective therapeutic strategies. In clinical microbiology, the identification and characterization of pathogenic bacteria were carried on by the conventional time-consuming biochemical assays. The rapid upgradation of metabolomic technologies has revolutionized the field of bacterial pathogenesis research, providing underlying metabolite profiling for the virulence factors associated with the bacterial infection. Researches can get insight into the relationship between the bacterial metabolism and expression of virulence factors under various mutative stages which can be investigated by combining metabolomic analyses with other omics approaches. In a similar context, metabolomics and omics data analysis was carried for the genetic screening of 29 ExPEC-associated virulence factors (VF) furnished by GC-MS and LC-MS analytical techniques. It was corroborated that the gene expression of these virulence factors was associated with characteristic features of bacterial pathogens which include adhesins, toxins, and siderophores. The metabolic profiling of flagellar formation of enzyme Jhp0106 in wild-type *H. pylori* and Djhp0106 in mutated *H. pylori* was investigated. Exometabolomic profile data analysis revealed that Djhp0106 was directly involved in the uptake and secretion of metabolites leading to alterations in the metabolic pathways during pathogenesis. Significant difference in the metabolic profile of the wild-type Jhp0106 and mutated strain Djhp0106 was identified my exometabolomics (Kao et al. 2021). In another approach, metabolomic profiling was used to identify and quantitate the metabolites linked with gut microbiota to correlate the symbiotic association of the superbugs. Analytical data of metabolites in human serum, urine, and feces in comparison to the *E. coli* (gut microbe) was constructed by GC/time-of-flight mass spectrometry (GC/TOFMS). Structurally variant 145 compounds were identified which were further categorized as derivatives of alkyl chloroformate derivatives (methyl- and ethyl-). The identification of 92, 103, 118, and 52 microbial metabolites in human serum, urine, and fecal samples, respectively, has firmly established the phenomenon of co-metabolism between the host and gut microbiota. This information can further be utilized for the formulation of targeted therapy for the treatment of inflammatory bowel disease (IBD). Furthermore, metabolomics has been extensively explored to underpin the gene expression of gut microbiome during bowel diseases (Bjerrum et al. 2021; Puig-Castellví et al. 2023). Recently researchers have deciphered the critical role of gene expression for adaptations in bacterial pathogenicity of IBD. They analyzed metagenomic and metabolomic cohorts of bowel disease patients from different countries and successfully identified 31 bacterial features, 25 KO genes, and 13 metabolites. A new theranostics model against bowel diseases can be produced by compiling the comprehensive data of gut microbial metabolites and KO genes (Ning et al. 2023).

Another study was reported based on the high-throughput metabolic fingerprinting of uropathogenic *Escherichia coli* (UPEC) strains by FT-IR spectroscopy, GC-MS, and LC-MS. Genotypic and phenotypic aspects of pathogenic strain of *E. coli were* measured on the basis of metabolites and the virulence factor data collected during metabolic profiling by all the three tools. It was further substantiated that the use of multiple metagenomic techniques (GC-MS for polar molecules, LC-MS for lipophilic molecules, and FTIR for virulence factor) can improve the application of metabolomics in theranostics (Torres-Puig et al. 2022).

In an innovative approach, to validate the involvement of virulence factor genes in bacterial pathogenesis, a virulence factor gene database (VFDB 2.0) was instituted. The VFDB 2.0 is constituted of approximate 62,332 species related alleles with high nucleotide identity. In an advancement to the proposed VFDB 2.0, a MetaVF toolkit was introduced to identify the pathogen (*C. perfringens*, *E. coli*, *F. tularensis*, *K. pneumoniae*, *S. dysenteriae*, *S. flexneri*, *S. mutans*, *S. pneumoniae*, *S. sonnei*, *F. nucleatum*, *C. difficile*, and *H. influenza*)-specific VFGs (Dong et al. 2024). An insight into the role of gastrointestinal diseases and toxins in triggering the colorectal cancer (CRC) has been reported by various researchers. A genotoxin "Colibactin" secreted by *E. coli* and adhesin FadA carried by *Fusobacterium nucleatum* acted as an activator for CRC tumor growth (Arthur et al. 2012; Rubinstein et al. 2013).

Similarly, the formation of Lectin-Glycan Interaction Network (LGI) of *E. coli* was affiliated with the interaction of host mucin and proteins secreted by pathogen. LGI was known to be the major factor for the antibiotic resistance and prolonged pathogenesis (Cho et al. 2020).

Antibiotic resistance of *Salmonella typhimurium* (*S. typhimurium*) has become a global threat, and lack for new therapeutics had added to this worse situation. Metabolomic analysis and omics have become a new way of understanding adaptation mechanism and the virulence factors involved during *S. typhimurium* infection. Metabolic analysis and identification of prominent genes expressed in *S. Typhimurium* toward resistance to H_2O_2 and to by-pass the phagocytosis defense mechanism of host were reported by Karash et al. (2017). Metabolomic investigation unveiled the participation of *Salmonella enterica* specific 52 proteins, which can further be formulated as a drug therapy against the infection caused by *S. enterica* (Hossain et al. 2017).

Therapeutics of many pathogenic bacteria has become challenging due to their MDR characteristics. Novel and effective drug therapy can be fabricated with the structured information of all the pathogenic bacteria, viz., *Klebsiella pneumoniae*,

S. Typhimurium, *P. aeruginosa*, *Acinetobacter baumannii*, and many more. The detailed metabolic information can be attained by the multi-omics analysis which includes proteomics, metabolomics, and transcriptomics. On the basis of various metabolite data gathered by metabolomic profiling, around 29 proteins and various gene clusters have been identified in the case of *K. pneumoniae* for the development of new therapeutics (Cesur et al. 2020; Ramos et al. 2018). The role of mutation in the metabolic pathways of lipid, peptidoglycan biosynthesis in the case of *P. aeruginosa* resistance against polymyxins was well established by metabolomics under various environmental stresses (Han et al. 2019). A new potential targeted therapy can be explicated once the detail framework of metabolite fingerprints is generated by multi-omics. Similarly, clinical prognosis of *Acinetobacter baumannii* can be emphasized by discovering metabolic correlation of proteins and genes expressed in resistance to triclosan (Fernando et al. 2017).

Various types of mathematical simulations and databases have been preceded with various metabolites encoding the genome of various organisms. Genome-scale metabolic network reconstructions (GENREs) are one of mathematical formalizations to study the metabolic reactions and growth patterns of the pathogenic bacteria. These databases can be explored for extensive analysis of MDR in various clinical isolates for the development of novel target specific therapeutics. Data analysis of various virulence factors associated with *C. difficile* metabolism was during its infection and was generated by GENREs (Jenior et al. 2021).

The discovery and development of new therapeutics are critical challenges to address the growing threat of MDR in various bacterial infections. A new source of drug candidate is required for effective treatment of infectious diseases. Bacterial metabolites are considered as potential candidates for theranostics as they exhibit potent biological activities. Probiotics have recently been bioengineered to deliver various therapeutic agents for the treatment of metabolic diseases. *Lactococcus lactis*, *Escherichia coli* Nissle 1917 (EcN), and *Lactobacillus casei* BL23-based bioengineered probiotics are few of such examples. These bioengineered probiotics have the ability to release various antimicrobial peptides which inhibit the adhesion and colonization of pathogenic bacteria. Along with this they were known to regulate expression of various anti-inflammatory factors. Metabolomic analysis of these probiotics has helped to improve their therapeutic properties. In context to this, live biotherapeutic products (LBPs) can be formulated from the pre-existing bioengineered probiotics as an effective therapy against various bacterial infections. Likewise, Ye et al. 2021, group has developed a bio-engineered LBP referred as SBT001/BioPersist and SBT002/BioColoniz, to prolong the impact of Ecn and L. reuteri's based probiotics in IBD infection (Ye et al. 2021). The rapid advancement of multi-omics tools and techniques has strengthened the understanding of host-pathogen interaction for the discovery and development of targeted-based therapy.

6 Conclusion

The emergence and advancement of metabolomics and omics have revolutionized the personalized medicine from a one-size-fits-approach to a more targeted and individualized strategies of disease prevention and treatment. By providing detailed characterization of metabolic phenotypes, metabolomics can enable precision medicine at multiple levels, including the identification of metabolic derangements that contribute to disease and the discovery of new therapeutic targets. Microbial enzymes are known to be associated with gene clusters for the synthesis of secondary metabolites by the microorganisms. Majority of the secondary metabolite affiliated gene clusters have not been identified and linked with the natural product synthesized by microorganism. These natural products of microorganism could be utilized in the development of human medicine. Recently, theoretical and experimental studies have revealed that metabolomics can unriddle the metabolome of complex microbial-based metabolites. The metabolomic studies of gut microbes have substantiated that uncharacterized gene of these microbes directly participates in the metabolism of proteins and carbohydrates. Metabolomics, the comprehensive analysis of metabolites in a biological specimen, offers a unique perspective on the intricate biochemical processes that underlie health and disease.

References

Abdelghani Z, Hourani N, Zaidan Z, Dbaibo G, Mrad M, Hage-Sleiman R (2021) Therapeutic applications and biological activities of bacterial bioactive extracts. Arch Microbiol 203(8). https://doi.org/10.1007/s00203-021-02505-1

Abdel-Razek AS, El-Naggar ME, Allam A, Morsy OM, Othman SI (2020) Microbial natural products in drug discovery. Processes 8(4). https://doi.org/10.3390/PR8040470

Aggarwal N, Kitano S, Puah GRY, Kittelmann S, Hwang IY, Chang MW (2023) Microbiome and human health: current understanding, engineering, and enabling technologies. Chem Rev 123(1). https://doi.org/10.1021/acs.chemrev.2c00431

AlRabiah H, Xu Y, Rattray NJW, Vaughan AA, Gibreel T, Sayqal A, Upton M, Allwood JW, Goodacre R (2014) Multiple metabolomics of uropathogenic E. Coli reveal different information content in terms of metabolic potential compared to virulence factors. Analyst 139(17). https://doi.org/10.1039/c4an00176a

Aminov R (2022) Metabolomics in antimicrobial drug discovery. Expert Opin Drug Discov 17(9). https://doi.org/10.1080/17460441.2022.2113774

Aretz I, Meierhofer D (2016) Advantages and pitfalls of mass spectrometry based metabolome profiling in systems biology. Int J Mol Sci 17(5). https://doi.org/10.3390/ijms17050632

Aries ML, Cloninger MJ (2021) NMR hydrophilic metabolomic analysis of bacterial resistance pathways using multivalent antimicrobials with challenged and unchallenged wild type and mutated gram-positive bacteria. Int J Mol Sci 22(24). https://doi.org/10.3390/ijms222413606

Arthur JC, Perez-Chanona E, Mühlbauer M, Tomkovich S, Uronis JM, Fan TJ, Campbell BJ, Abujamel T, Dogan B, Rogers AB, Rhodes JM, Stintzi A, Simpson KW, Hansen JJ, Keku TO, Fodor AA, Jobin C (2012) Intestinal inflammation targets cancer-inducing activity of the microbiota. Science 338(6103). https://doi.org/10.1126/science.1224820

Avalon NE, Murray AE, Baker BJ (2022) Integrated Metabolomic-genomic workflows accelerate microbial natural product discovery. Anal Chem 94(35). https://doi.org/10.1021/acs.analchem.2c02245

Ayon NJ, Earp CE, Gupta R, Butun FA, Clements AE, Lee AG, Dainko D, Robey MT, Khin M, Mardiana L, Longcake A, Rangel-Grimaldo M, Hall MJ, Probert MR, Burdette JE, Keller NP, Raja HA, Oberlies NH, Kelleher NL, Caesar LK (2024) Bioactivity-driven fungal metabologenomics identifies antiproliferative stemphone analogs and their biosynthetic gene cluster. Metabolomics 20(5):90. https://doi.org/10.1007/s11306-024-02153-8

Azhar F, Busharat M, Chaudhary SRA, Waheed Z, Jamil MN (2023) Metabolomics in drug discovery: restoring antibiotic pipeline. Asian Pac J Trop Biomed 13(9). https://doi.org/10.4103/2221-1691.385568

Baidoo EEK, Teixeira Benites V (2019) Mass spectrometry-based microbial metabolomics: techniques, analysis, and applications. Methods Mol Biol 1859. https://doi.org/10.1007/978-1-4939-8757-3_2

Barding GA, Salditos R, Larive CK (2012) Quantitative NMR for bioanalysis and metabolomics. Anal Bioanal Chem 404(4). https://doi.org/10.1007/s00216-012-6188-z

Bauermeister A, Mannochio-Russo H, Costa-Lotufo LV, Jarmusch AK, Dorrestein PC (2022) Mass spectrometry-based metabolomics in microbiome investigations. Nat Rev Microbiol 20(3). https://doi.org/10.1038/s41579-021-00621-9

Bjerrum JT, Wang YL, Seidelin JB, Nielsen OH (2021) IBD metabonomics predicts phenotype, disease course, and treatment response. EBioMedicine 71. https://doi.org/10.1016/j.ebiom.2021.103551

Boersma MG, Solyanikova IP, Van Berkel WJH, Vervoort J, Golovleva L, Rietjens IMCM (2001) 19F NMR metabolomics for the elucidation of microbial degradation pathways of fluorophenols. J Ind Microbiol Biotechnol 26(1–2). https://doi.org/10.1038/sj.jim.7000027

Bose U, Hewavitharana AK, Ng YK, Shaw PN, Fuerst JA, Hodson MP (2015) LC-MS-based metabolomics study of marine bacterial secondary metabolite and antibiotic production in Salinispora arenicola. Mar Drugs 13(1). https://doi.org/10.3390/md13010249

Bright Esegbuyota I, Hope O, Oyama G (2022) Occurrence of New Delhi Metallo-Beta-Lactamase 1 producing enterococcus species in Oghara water nexus: an emerging environmental implications of resistance dynamics. Microbiol Insights 15. https://doi.org/10.1177/11786361221133731

Caesar LK, Montaser R, Keller NP, Kelleher NL (2021) Metabolomics and genomics in natural products research: complementary tools for targeting new chemical entities. Nat Prod Rep 38(11). https://doi.org/10.1039/d1np00036e

Caesar LK, Butun FA, Robey MT, Ayon NJ, Gupta R, Dainko D, Bok JW, Nickles G, Stankey RJ, Johnson D, Mead D, Cank KB, Earp CE, Raja HA, Oberlies NH, Keller NP, Kelleher NL (2023) Correlative metabologenomics of 110 fungi reveals metabolite–gene cluster pairs. Nat Chem Biol 19(7). https://doi.org/10.1038/s41589-023-01276-8

Castrillo JI, Hayes A, Mohammed S, Gaskell SJ, Oliver SG (2003) An optimized protocol for metabolome analysis in yeast using direct infusion electrospray mass spectrometry. Phytochemistry 62(6). https://doi.org/10.1016/S0031-9422(02)00713-6

Cavill R, Jennen D, Kleinjans J, Briedé JJ (2016) Transcriptomic and metabolomic data integration. Brief Bioinform 17(5):891–901. https://doi.org/10.1093/bib/bbv090

Cesur MF, Siraj B, Uddin R, Durmuş S, Çakır T (2020) Network-based metabolism-centered screening of potential drug targets in Klebsiella pneumoniae at genome scale. Front Cell Infect Microbiol 9. https://doi.org/10.3389/fcimb.2019.00447

Chassagne F, Cabanac G, Hubert G, David B, Marti G (2019) The landscape of natural product diversity and their pharmacological relevance from a focus on the Dictionary of Natural Products®. Phytochem Rev 18(3). https://doi.org/10.1007/s11101-019-09606-2

Cho SH, Lee KM, Kim CH, Kim SS (2020) Construction of a lectin–glycan interaction network from enterohemorrhagic Escherichia coli strains by multi-omics analysis. Int J Mol Sci 21(8). https://doi.org/10.3390/ijms21082681

Choi C, Münch R, Leupold S, Klein J, Siegel I, Thielen B, Benkert B, Kucklick M, Schobert M, Barthelmes J, Ebeling C, Haddad I, Scheer M, Grote A, Hiller K, Bunk B, Schreiber K, Retter I, Schomburg D, Jahn D (2007) SYSTOMONAS – an integrated database for systems biology analysis of Pseudomonas. Nucleic Acids Res 35(SUPPL. 1). https://doi.org/10.1093/nar/gkl823

Clarke TB, Davis KM, Lysenko ES, Zhou AY, Yu Y, Weiser JN (2010) Recognition of peptidoglycan from the microbiota by Nod1 enhances systemic innate immunity. Nat Med 16(2). https://doi.org/10.1038/nm.2087

Clasquin MF, Melamud E, Rabinowitz JD (2012) LC-MS data processing with MAVEN: a metabolomic analysis and visualization engine. Curr Protoc Bioinformatics. https://doi.org/10.1002/0471250953.bi1411s37

Covington BC, McLean JA, Bachmann BO (2017) Comparative mass spectrometry-based metabolomics strategies for the investigation of microbial secondary metabolites. Nat Prod Rep 34(1). https://doi.org/10.1039/c6np00048g

Defernez M, Colquhoun IJ (2003) Factors affecting the robustness of metabolite fingerprinting using 1H NMR spectra. Phytochemistry 62(6). https://doi.org/10.1016/S0031-9422(02)00704-5

Depke T, Thöming JG, Kordes A, Häussler S, Brönstrup M (2020) Untargeted LC-MS metabolomics differentiates between virulent and avirulent clinical strains of Pseudomonas aeruginosa. Biomol Ther 10(7). https://doi.org/10.3390/biom10071041

Dong W, Fan X, Guo Y, Wang S, Jia S, Lv N, Yuan T, Pan Y, Xue Y, Chen X, Xiong Q, Yang R, Zhao W, Zhu B (2024) An expanded database and analytical toolkit for identifying bacterial virulence factors and their associations with chronic diseases. Nat Commun 15(1):8084. https://doi.org/10.1038/s41467-024-51864-y

Donia MS, Cimermancic P, Schulze CJ, Wieland Brown LC, Martin J, Mitreva M, Clardy J, Linington RG, Fischbach MA (2014) A systematic analysis of biosynthetic gene clusters in the human microbiome reveals a common family of antibiotics. Cell 158(6). https://doi.org/10.1016/j.cell.2014.08.032

Drenos F (2017) Mechanistic insights from combining genomics with metabolomics. Curr Opin Lipidol 28(2). https://doi.org/10.1097/MOL.0000000000000393

Du CX, Huang Z (2019) Analysis of amino acids in human tears by hydrophilic interaction liquid chromatography and quadrupole orbitrap mass spectrometry. RSC Adv 9(63). https://doi.org/10.1039/c9ra05956c

Du C, van Wezel GP (2018) Mining for Microbial Gems: integrating proteomics in the Postgenomic natural product discovery pipeline. Proteomics 18(18). https://doi.org/10.1002/pmic.201700332

Durmus S, Çakir T, Özgür A, Guthke R (2015) A review on computational systems biology of pathogen-host interactions. Front Microbiol 6(APR). https://doi.org/10.3389/fmicb.2015.00235

Fernando DM, Chong P, Singh M, Spicer V, Unger M, Loewen PC, Westmacott G, Kumar A (2017) Multi-omics approach to study global changes in a triclosan-resistant mutant strain of Acinetobacter baumannii ATCC 17978. Int J Antimicrob Agents 49(1). https://doi.org/10.1016/j.ijantimicag.2016.10.014

Fortuin S, Soares NC (2022) The integration of proteomics and metabolomics data paving the way for a better understanding of the mechanisms underlying microbial acquired drug resistance. Front Med 9. https://doi.org/10.3389/fmed.2022.849838

Fu Q, Liu D, Wang Y, Li X, Wang L, Yu F, Shen J, Xia X (2018) Metabolomic profiling of campylobacter jejuni with resistance gene ermB by ultra-high performance liquid chromatography-quadrupole time-of-flight mass spectrometry and tandem quadrupole mass spectrometry. J Chromatogr B Analyt Technol Biomed Life Sci 1079:62–68. https://doi.org/10.1016/j.jchromb.2018.02.009

Gaynes R (2017) The discovery of penicillin—new insights after more than 75 years of clinical use. Emerg Infect Dis 23(5). https://doi.org/10.3201/eid2305.161556

Geraci JE, Heilman FR, Nichols DR, Ross GT, Wellman WE (1956) Some laboratory and clinical experiences with a new antibiotic, vancomycin. Proc Staff Meet Mayo Clin 31(21):564

Goering AW, McClure RA, Doroghazi JR, Albright JC, Haverland NA, Zhang Y, Ju KS, Thomson RJ, Metcalf WW, Kelleher NL (2016) Metabologenomics: correlation of microbial gene clusters with metabolites drives discovery of a nonribosomal peptide with an unusual amino acid monomer. ACS Cent Sci 2(2). https://doi.org/10.1021/acscentsci.5b00331

Gomes PWP, de Tralia Medeiros TC, Maimone NM, Leão TF, de Moraes LAB, Bauermeister A (2023) Microbial metabolites annotation by mass spectrometry-based metabolomics. Adv Exp Med Biol 1439. https://doi.org/10.1007/978-3-031-41741-2_9

Han M-L, Zhu Y, Creek DJ, Lin Y-W, Gutu AD, Hertzog P, Purcell T, Shen H-H, Moskowitz SM, Velkov T, Li J (2019) Comparative metabolomics and Transcriptomics reveal multiple pathways associated with polymyxin killing in Pseudomonas aeruginosa. MSystems 4(1). https://doi.org/10.1128/msystems.00149-18

Hooper LV, Littman DR, Macpherson AJ (2012) Interactions between the microbiota and the immune system. Science 336(6086). https://doi.org/10.1126/science.1223490

Hossain T, Kamruzzaman M, Choudhury TZ, Mahmood HN, Nabi AHMN, Hosen MI (2017) Application of the subtractive genomics and molecular docking analysis for the identification of novel putative drug targets against salmonella enterica subsp. enterica serovar Poona. Biomed Res Int 2017. https://doi.org/10.1155/2017/3783714

Huang Y, Zhang Z, Chen H, Feng J, Cai S, Chen Z (2015) A high-resolution 2D J-resolved NMR detection technique for metabolite analyses of biological samples. Sci Rep 5. https://doi.org/10.1038/srep08390

Jeffryes JG, Colastani RL, Elbadawi-Sidhu M, Kind T, Niehaus TD, Broadbelt LJ, Hanson AD, Fichn O, Tyo KEJ, Henry CS (2015) MINEs: open access databases of computationally predicted enzyme promiscuity products for untargeted metabolomics. J Chem 7(1). https://doi.org/10.1186/s13321-015-0087-1

Jenior ML, Leslie JL, Powers DA, Garrett EM, Walker KA, Dickenson ME, Petri WA, Tamayo R, Papin JA (2021) Novel drivers of virulence in clostridioides difficile identified via context-specific metabolic network analysis. MSystems 6(5). https://doi.org/10.1128/msystems.00919-21

Kao CY, Kuo PY, Liao HW (2021) Untargeted microbial exometabolomics and metabolomics analysis of helicobacter pylori j99 and jhp0106 mutant. Metabolites 11(12). https://doi.org/10.3390/metabo11120808

Karash S, Liyanage R, Qassab A, Lay JO, Kwon YM (2017) A comprehensive assessment of the genetic determinants in salmonella Typhimurium for resistance to hydrogen peroxide using Proteogenomics. Sci Rep 7(1). https://doi.org/10.1038/s41598-017-17149-9

Kathera C (2018) Microbial proteomics and their importance in medical microbiology. In: Recent developments in applied microbiology and biochemistry. https://doi.org/10.1016/B978-0-12-816328-3.00003-9

Kleigrewe K, Almaliti J, Tian IY, Kinnel RB, Korobeynikov A, Monroe EA, Duggan BM, Di Marzo V, Sherman DH, Dorrestein PC, Gerwick L, Gerwick WH (2015) Combining mass spectrometric metabolic profiling with genomic analysis: a powerful approach for discovering natural products from cyanobacteria. J Nat Prod 78(7). https://doi.org/10.1021/acs.jnatprod.5b00301

Kobras CM, Fenton AK, Sheppard SK (2021) Next-generation microbiology: from comparative genomics to gene function. Genome Biol 22(1). https://doi.org/10.1186/s13059-021-02344-9

Krug D, Müller R (2014) Secondary metabolomics: the impact of mass spectrometry-based approaches on the discovery and characterization of microbial natural products. Nat Prod Rep 31(6). https://doi.org/10.1039/c3np70127a

Li Q, Wei S, Wu D, Wen C, Zhou J (2018) Urinary metabolomics study of patients with gout using gas chromatography-mass spectrometry. Biomed Res Int 2018. https://doi.org/10.1155/2018/3461572

Liu R, Bao ZX, Zhao PJ, Li GH (2021) Advances in the study of metabolomics and metabolites in some species interactions. Molecules 26(11). https://doi.org/10.3390/MOLECULES26113311

Locey KJ, Lennon JT (2016) Scaling laws predict global microbial diversity. Proc Natl Acad Sci USA 113(21). https://doi.org/10.1073/pnas.1521291113

Lü J, Long Q, Zhao Z, Chen L, He W, Hong J, Liu K, Wang Y, Pang X, Deng Z, Tao M (2020) Engineering the erythromycin-producing strain Saccharopolyspora erythraea HOE107 for the heterologous production of Polyketide antibiotics. Front Microbiol 11. https://doi.org/10.3389/fmicb.2020.593217

Lybbert AC, Williams JL, Raghuvanshi R, Daniel Jones A, Quinn RA (2020) Mining public mass spectrometry data to characterize the diversity and ubiquity of P. Aeruginosa specialized metabolites. Metabolites 10(11). https://doi.org/10.3390/metabo10110445

Maansson M, Vynne NG, Klitgaard A, Nybo JL, Melchiorsen J, Nguyen DD, Sanchez LM, Ziemert N, Dorrestein PC, Andersen MR, Gram L (2016) An integrated metabolomic and genomic mining workflow to uncover the biosynthetic potential of bacteria. MSystems 1(3). https://doi.org/10.1128/msystems.00028-15

Mahrous EA, Farag MA (2015) Two dimensional NMR spectroscopic approaches for exploring plant metabolome: a review. J Adv Res 6(1). https://doi.org/10.1016/j.jare.2014.10.003

Manzoni C, Kia DA, Vandrovcova J, Hardy J, Wood NW, Lewis PA, Ferrari R (2018) Genome, transcriptome and proteome: the rise of omics data and their integration in biomedical sciences. Brief Bioinform 19(2). https://doi.org/10.1093/BIB/BBW114

Mashego MR, Rumbold K, De Mey M, Vandamme E, Soetaert W, Heijnen JJ (2007) Microbial metabolomics: past, present and future methodologies. Biotechnol Lett 29(1). https://doi.org/10.1007/s10529-006-9218-0

McClure RA, Goering AW, Ju KS, Baccile JA, Schroeder FC, Metcalf WW, Thomson RJ, Kelleher NL (2016) Elucidating the rimosamide-detoxin natural product families and their biosynthesis using metabolite/gene cluster correlations. ACS Chem Biol 11(12). https://doi.org/10.1021/acschembio.6b00779

Medema MH, Fischbach MA (2015) Computational approaches to natural product discovery. Nat Chem Biol 11(9). https://doi.org/10.1038/nchembio.1884

Molton JS, Tambyah PA, Ang BSP, Ling ML, Fisher DA (2013) The global spread of healthcare-associated multidrug-resistant bacteria: a perspective from Asia. Clin Infect Dis 56(9). https://doi.org/10.1093/cid/cit020

Mougous JD, Leavell MD, Senaratne RH, Leigh CD, Williams SJ, Riley LW, Leary JA, Bertozzi CR (2002) Discovery of sulfated metabolites in mycobacteria with a genetic and mass spectrometric approach. Proc Natl Acad Sci USA 99(26). https://doi.org/10.1073/pnas.252514899

Mózsik L, Iacovelli R, Bovenberg RAL, Driessen AJM (2022) Transcriptional activation of biosynthetic gene clusters in filamentous fungi. Front Bioeng Biotechnol 10. https://doi.org/10.3389/fbioe.2022.901037

Muteeb G, Rehman MT, Shahwan M, Aatif M (2023) Origin of antibiotics and antibiotic resistance, and their impacts on drug development: a narrative review. Pharmaceuticals 16(11). https://doi.org/10.3390/ph16111615

Nagana Gowda GA, Raftery D (2021) NMR-based metabolomics. Adv Exp Med Biol 1280:19–37. https://doi.org/10.1007/978-3-030-51652-9_2

Newman DJ (2016) Predominately uncultured microbes as sources of bioactive agents. Front Microbiol 7(NOV). https://doi.org/10.3389/fmicb.2016.01832

Newton GGF, Abraham EP (1955) Cephalosporin C, a new antibiotic containing sulphur and D-α-aminoadipic acid [1]. Nature 175(4456). https://doi.org/10.1038/175548a0

Ning L, Zhou YL, Sun H, Zhang Y, Shen C, Wang Z, Xuan B, Zhao Y, Ma Y, Yan Y, Tong T, Huang X, Hu M, Zhu X, Ding J, Zhang Y, Cui Z, Fang JY, Chen H, Hong J (2023) Microbiome and metabolome features in inflammatory bowel disease via multi-omics integration analyses across cohorts. Nat Commun 14(1). https://doi.org/10.1038/s41467-023-42788-0

Oberhardt MA, Chavali AK, Papin JA (2009) Flux balance analysis: interrogating genome-scale metabolic networks. Methods Mol Biol 500(1). https://doi.org/10.1007/978-1-59745-525-1_3

Paillié-Jiménez ME, Stincone P, Brandelli A (2020) Natural pigments of microbial origin. Front Sustain Food Syst 4. https://doi.org/10.3389/fsufs.2020.590439

Palermo A (2023) Metabolomics- and systems-biology-guided discovery of metabolite lead compounds and druggable targets. Drug Discov Today 28(2). https://doi.org/10.1016/j.drudis.2022.103460

Pang H, Hu Z (2023) Metabolomics in drug research and development: the recent advances in technologies and applications. Acta Pharm Sin B 13(8). https://doi.org/10.1016/j.apsb.2023.05.021

Parkinson EI, Tryon JH, Goering AW, Ju KS, McClure RA, Kemball JD, Zhukovsky S, Labeda DP, Thomson RJ, Kelleher NL, Metcalf WW (2018) Discovery of the Tyrobetaine natural products

and their biosynthetic gene cluster via Metabologenomics. ACS Chem Biol 13(4). https://doi.org/10.1021/acschembio.7b01089
Pasikanti KK, Ho PC, Chan ECY (2008) Gas chromatography/mass spectrometry in metabolic profiling of biological fluids. J Chromatogr B Analyt Technol Biomed Life Sci 871(2). https://doi.org/10.1016/j.jchromb.2008.04.033
Pham JV, Yilma MA, Feliz A, Majid MT, Maffetone N, Walker JR, Kim E, Cho HJ, Reynolds JM, Song MC, Park SR, Yoon YJ (2019) A review of the microbial production of bioactive natural products and biologics. Front Microbiol 10(JUN). https://doi.org/10.3389/fmicb.2019.01404
Pinu FR, Villas-Boas SG, Aggio R (2017) Analysis of intracellular metabolites from microorganisms: quenching and extraction protocols. Metabolites 7(4). https://doi.org/10.3390/metabo7040053
Pinu FR, Beale DJ, Paten AM, Kouremenos K, Swarup S, Schirra HJ, Wishart D (2019) Systems biology and multi-omics integration: viewpoints from the metabolomics research community. Meta 9(4). https://doi.org/10.3390/metabo9040076
Puig-Castellví F, Pacheco-Tapia R, Deslande M, Jia M, Andrikopoulos P, Chechi K, Bonnefond A, Froguel P, Dumas ME (2023) Advances in the integration of metabolomics and metagenomics for human gut microbiome and their clinical applications. TrAC Trends Anal Chem 167. https://doi.org/10.1016/j.trac.2023.117248
Quinn GA, Banat AM, Abdelhameed AM, Banat IM (2020) Streptomyces from traditional medicine: sources of new innovations in antibiotic discovery. J Med Microbiol 69(8). https://doi.org/10.1099/jmm.0.001232
Ramos PIP, Fernández Do Porto D, Lanzarotti E, Sosa EJ, Burguener G, Pardo AM, Klein CC, Sagot MF, De Vasconcelos ATR, Gales AC, Marti M, Turjanski AG, Nicolás MF (2018) An integrative, multi-omics approach towards the prioritization of Klebsiella pneumoniae drug targets. Sci Rep 8(1). https://doi.org/10.1038/s41598-018-28916-7
Ren JL, Zhang AH, Kong L, Wang XJ (2018) Advances in mass spectrometry-based metabolomics for investigation of metabolites. RSC Adv 8(40). https://doi.org/10.1039/c8ra01574k
Resurreccion EP, Fong KW (2022) The integration of metabolomics with other Omics: insights into understanding prostate cancer. Metabolites 12(6). https://doi.org/10.3390/metabo12060488
Reverter M, Rohde S, Parchemin C, Tapissier-Bontemps N, Schupp PJ (2020) Metabolomics and marine biotechnology: coupling metabolite profiling and organism biology for the discovery of new compounds. Front Mar Sci 7. https://doi.org/10.3389/fmars.2020.613471
Ríos JL, Francini F, Schinella GR (2015) Natural products for the treatment of type 2 diabetes mellitus. Planta Med 81(12–13). https://doi.org/10.1055/s-0035-1546131
Roberts LD, Souza AL, Gerszten RE, Clish CB (2012) Targeted metabolomics. Curr Protoc Mol Biol 1(SUPPL.98). https://doi.org/10.1002/0471142727.mb3002s98
Rochfort S (2005) Metabolomics reviewed: a new “omics” platform technology for systems biology and implications for natural products research. J Nat Prod 68(12). https://doi.org/10.1021/np050255w
Rubinstein MR, Wang X, Liu W, Hao Y, Cai G, Han YW (2013) Fusobacterium nucleatum promotes colorectal carcinogenesis by modulating E-cadherin/β-catenin signaling via its FadA Adhesin. Cell Host Microbe 14(2). https://doi.org/10.1016/j.chom.2013.07.012
Sato Y (2024) Transcriptome analysis: a powerful tool to understand individual microbial behaviors and interactions in ecosystems. Biosci Biotechnol Biochem 88(8):850–856. https://doi.org/10.1093/bbb/zbae064
Sayqal A, Xu Y, Trivedi DK, Almasoud N, Ellis DI, Rattray NJW, Goodacre R (2016) Metabolomics analysis reveals the participation of efflux pumps and ornithine in the response of pseudomonas putida DOT-T1E cells to challenge with propranolol. PLoS One 11(6). https://doi.org/10.1371/journal.pone.0156509
Segers K, Declerck S, Mangelings D, Heyden YV, Van Eeckhaut A (2019) Analytical techniques for metabolomic studies: a review. Bioanalysis 11(24). https://doi.org/10.4155/bio-2019-0014
Shaffer JP, Nothias LF, Thompson LR, Sanders JG, Salido RA, Couvillion SP, Brejnrod AD, Lejzerowicz F, Haiminen N, Huang S, Lutz HL, Zhu Q, Martino C, Morton JT, Karthikeyan S, Nothias-Esposito M, Dührkop K, Böcker S, Kim HW et al (2022) Standardized multi-omics of

Earth's microbiomes reveals microbial and metabolite diversity. Nat Microbiol 7(12). https://doi.org/10.1038/s41564-022-01266-x

Shah NJ, Sureshkumar S, Shewade DG (2015) Metabolomics: a tool ahead for understanding molecular mechanisms of drugs and diseases. Indian J Clin Biochem 30(3). https://doi.org/10.1007/s12291-014-0455-z

Shih CJ, Chen PY, Liaw CC, Lai YM, Yang YL (2014) Bringing microbial interactions to light using imaging mass spectrometry. Nat Prod Rep 31(6). https://doi.org/10.1039/c3np70091g

Soga T, Ohashi Y, Ueno Y, Naraoka H, Tomita M, Nishioka T (2003) Quantitative Metabolome analysis using capillary electrophoresis mass spectrometry. J Proteome Res 2(5). https://doi.org/10.1021/pr034020m

Sorokina M, Merseburger P, Rajan K, Yirik MA, Steinbeck C (2021) COCONUT online: collection of open natural products database. *Journal of.* Cheminformatics 13(1). https://doi.org/10.1186/s13321-020-00478-9

Strelkov S, von Elstermann M, Schomburg D (2004) Comprehensive analysis of metabolites in Corynebacterium glutamicum by gas chromatography/mass spectrometry. Biol Chem 385(9). https://doi.org/10.1515/BC.2004.111

Takenaka M, Yoshida T, Hori Y, Bamba T, Mochizuki M, Vavricka CJ, Hattori T, Hayakawa Y, Hasunuma T, Kondo A (2021) An ion-pair free LC-MS/MS method for quantitative metabolite profiling of microbial bioproduction systems. Talanta 222. https://doi.org/10.1016/j.talanta.2020.121625

Tanaka K, Bamba T, Kondo A, Hasunuma T (2024) Metabolomics-based development of bioproduction processes toward industrial-scale production. Curr Opin Biotechnol 85. https://doi.org/10.1016/j.copbio.2023.103057

Torres-Puig S, García V, Stærk K, Andersen TE, Møller-Jensen J, Olsen JE, Herrero-Fresno A (2022) "Omics" technologies—what have they told us about Uropathogenic Escherichia coli fitness and virulence during urinary tract infection? Front Cell Infect Microbiol 12. https://doi.org/10.3389/fcimb.2022.824039

Tulpan D, Léger S, Belliveau L, Culf A, Čuperlović-Culf M (2011) MetaboHunter: an automatic approach for identification of metabolites from 1H-NMR spectra of complex mixtures. BMC Bioinformatics 12. https://doi.org/10.1186/1471-2105-12-400

Tweeddale H, Notley-Mcrobb L, Ferenci T (1998) Effect of slow growth on metabolism of Escherichia coli, as revealed by global metabolite pool ('metabolome') analysis. J Bacteriol 180(19). https://doi.org/10.1128/jb.180.19.5109-5116.1998

Van Der Hooft JJJ, Mohimani H, Bauermeister A, Dorrestein PC, Duncan KR, Medema MH (2020) Linking genomics and metabolomics to chart specialized metabolic diversity. Chem Soc Rev 49(11). https://doi.org/10.1039/d0cs00162g

Van Der Werf MJ, Overkamp KM, Muilwijk B, Koek MM, Der Werff-Van V, Der Vat BJC, Jellema RH, Coulier L, Hankemeier T (2008) Comprehensive analysis of the metabolome of Pseudomonas putida S12 grown on different carbon sources. Mol BioSyst 4(4). https://doi.org/10.1039/b717340g

van Duin D, Paterson DL (2016) Multidrug-resistant bacteria in the community: trends and lessons learned. Infect Dis Clin N Am 30(2). https://doi.org/10.1016/j.idc.2016.02.004

Verdonk JC, De Vos CHR, Verhoeven HA, Haring MA, Van Tunen AJ, Schuurink RC (2003) Regulation of floral scent production in petunia revealed by targeted metabolomics. Phytochemistry 62(6). https://doi.org/10.1016/S0031-9422(02)00707-0

Vignoli A, Ghini V, Meoni G, Licari C, Takis PG, Tenori L, Turano P, Luchinat C (2019) High-throughput metabolomics by 1D NMR. Angew Chem Int Ed 58(4). https://doi.org/10.1002/anie.201804736

Walsh CT, Fischbach MA (2010) Natural products version 2.0: connecting genes to molecules. J Am Chem Soc 132(8). https://doi.org/10.1021/ja909118a

Wang H, de Carvalho LPS (2023) Metabolomic profiling reveals bacterial metabolic adaptation strategies and new metabolites. Curr Opin Chem Biol 74. https://doi.org/10.1016/j.cbpa.2023.102287

Wang Z, Gerstein M, Snyder M (2009) RNA-Seq: a revolutionary tool for transcriptomics. Nat Rev Genet 10(1). https://doi.org/10.1038/nrg2484
Wu S, Zhou H, Chen D, Lu Y, Li Y, Qiao J (2024) Multi-omic analysis tools for microbial metabolites prediction. Brief Bioinform 25(4). https://doi.org/10.1093/bib/bbae264
Xiao C, Zhu Y, Yang Z, Shi D, Ni Y, Hua L, Li J (2022) Prevalence and molecular characteristics of Polymyxin-resistant Pseudomonas aeruginosa in a Chinese Tertiary Teaching Hospital. Antibiotics 11(6). https://doi.org/10.3390/antibiotics11060799
Xu G, Guo H, Lv H (2019) Metabolomics assay identified a novel virulence-associated Siderophore encoded by the high-Pathogenicity Island in Uropathogenic Escherichia coli. J Proteome Res 18(5). https://doi.org/10.1021/acs.jproteome.9b00190
Yang B, Fang D, Lv Q, Wang Z, Liu Y (2021) Targeted therapeutic strategies in the Battle against pathogenic bacteria. Front Pharmacol 12. https://doi.org/10.3389/fphar.2021.673239
Ye J, Erland LAE, Gill SK, Bishop SL, Verdugo-Meza A, Murch SJ, Gibson DL (2021) Metabolomics-guided hypothesis generation for mechanisms of intestinal protection by live biotherapeutic products. Biomol Ther 11(5). https://doi.org/10.3390/biom11050738
Ye D, Li X, Shen J, Xia X (2022) Microbial metabolomics: from novel technologies to diversified applications. TrAC Trends Anal Chem 148. https://doi.org/10.1016/j.trac.2022.116540
Yelamanchi SD, Mishra A, Behra SK, Karthikkeyan G, Prasad TSK, Surolia A (2022) Rifampicin-mediated metabolic changes in mycobacterium tuberculosis. Metabolites 12(6). https://doi.org/10.3390/metabo12060493
Yoshimura A, Covington BC, Gallant É, Zhang C, Li A, Seyedsayamdost MR (2020) Unlocking cryptic metabolites with mass spectrometry-guided transposon mutant selection. ACS Chem Biol 15(10). https://doi.org/10.1021/acschembio.0c00558
Zhang T, Watson DG (2015) A short review of applications of liquid chromatography mass spectrometry based metabolomics techniques to the analysis of human urine. Analyst 140(9). https://doi.org/10.1039/c4an02294g
Zhang A, Sun H, Wang P, Han Y, Wang X (2012) Modern analytical techniques in metabolomics analysis. Analyst 137(2). https://doi.org/10.1039/c1an15605e
Zhang HW, Lv C, Zhang LJ, Guo X, Shen YW, Nagle DG, Zhou YD, Liu SH, Zhang WD, Luan X (2021) Application of omics- and multi-omics-based techniques for natural product target discovery. Biomed Pharmacother 141. https://doi.org/10.1016/j.biopha.2021.111833

Metabolomics in Food Fermentation and Designing Functional Foods

Beatriz Adriana Andrade-Espinoza, Jenny Arratia-Quijada, Karla Nuño, Víctor Ruíz-Santoyo, and Lucía Flores Contreras

Abstract

Fermentation is a technique applied in the food industry for preservation, processing, and novel food design. During fermentation, microorganisms generate substrate from the ingredient or source raw materials, such as enzymes, amino acids, vitamins, bioactive peptides, exopolysaccharides, organic acids, short-chain fatty acids, and phenolic compounds, among others, which have been associated with health-promoting properties. Therefore, fermented foods can be considered functional foods because they contain bioactive compounds that may improveconsumer health. To understand the development process of fermented food, metabolomics has been used. Metabolomics is an *omic* science currently applied in the food industry, such as food safety, food quality, and food science, in order to determine the presence of metabolite molecules smaller than 1500 Da in biological systems. In this chapter, we examine the most recent applications and advantages of metabolomics during the fermentation process and the design of fermented functional foods.

B. A. Andrade-Espinoza (✉) · J. Arratia-Quijada · L. F. Contreras
Departamento de Ciencias Biomédicas, Centro Universitario de Tonalá,
Tonalá, Jalisco, Mexico
e-mail: beatriz.andrade@academicos.udg.mx; jenny.arratia@academicos.udg.mx;
lucia.fcontreras@academicos.udg.mx

K. Nuño
Departamento de Psicología, Educación y Salud, Nutrición, ITESO, Universidad Jesuita de Guadalajara, Guadalajara, Mexico
e-mail: karlanuno@iteso.mx

V. Ruíz-Santoyo
Laboratorio de Investigación Interdisciplinaria, Área de Nanoestructuras y Biomateriales, Escuela Nacional de Estudios Superiores, Unidad León, UNAM, León Guanajuato, CP, Mexico
e-mail: vruizs@enes.unam.mx

S. Kaur et al. (eds.), *Microbial Metabolomics*,
https://doi.org/10.1007/978-981-96-4824-5_11

Keywords

Metabolomics · Metabolites · Fermentation · Analytical techniques · Metabolite analysis · Food industry

1 Introduction

In recent years, the food industry has witnessed a significant transformation, fueled by progress in scientific research and technological innovations. Among these developments, the rise of "omics" technologies, particularly metabolomics, has revolutionized our understanding of food production, safety, and quality (Chi et al. 2024). Metabolomics, which involves the comprehensive analysis of small molecules within biological systems, provides unparalleled insights into the metabolic processes that underpin the characteristics and functionalities of food (Adamyan et al. 2024). The present chapter explores the pivotal role of metabolomics in the food industry, focusing on its applications in food resources, production, and processing of fermented foods. By examining the intricate relationships between metabolites, microorganisms, and environmental factors, metabolomics has opened new avenues for enhancing the nutritional and functional properties of food. The chapter delves into the emerging discipline of foodomics, which integrates various omics technologies to study food and nutrition, and discusses how these approaches are reshaping our approach to food safety, classification, and traceability. Furthermore, the chapter highlights the importance of metabolomics in the development of fermented functional foods, where the interplay between microbial communities and fermentation conditions leads to the production of bioactive compounds with significant health benefits. Through advanced analytical techniques such as nuclear magnetic resonance (NMR) and gas chromatography-mass spectrometry (GC-MS), researchers can now monitor and optimize the fermentation process, ensuring the quality, safety, and sensory attributes of the final products (Ovbude et al. 2024). As we move forward, the continuous advancement of metabolomics methodologies and the expansion of multi-omics databases will play a crucial role in standardizing and enhancing food production at an industrial scale. This chapter provides a comprehensive overview of the current state and future prospects of metabolomics in the food industry, offering valuable insights for both researchers and practitioners in this dynamic field.

2 Applications of Metabolomics in the Food Industry

Since the inception of the Human Genome Project, the development and financing of a large number of sequencing projects have led to the development, improvement, and progress of technologies to obtain information in the bioinformatic and biology fields (Mozzi et al. 2013). The advances in these areas have profound implications capable of causing a major paradigm shift in how we approach all biological aspects including health, disease, agriculture, nutrition, and food production (Weimer 2013). This change is evident in the emergence of the "Omics age." The term omics derives from the Latin voice *-omne* which means whole, totality, everything. Nowadays this word is used as a suffix for referring to those disciplines such as genomics, transcriptomics, proteomics, and metabolomics. These disciplines study a certain set of molecules globally (genes, transcripts, proteins, and metabolites, respectively) or even seek to define the relationships among them (Mozzi et al. 2013; Rizo et al. 2018).

Metabonomics, also known as metabolome analysis, or metabolomics constitutes one of the four "omics" technologies. Metabolomics was first proposed in 1999 (Wu et al. 2022a) and is a term coined from "metabolite +omics." Metabolomics is the in-depth study and analysis of the complete set of small molecules. Metabolites are low-molecular-mass compounds (<1500 Da) found within a biological system. These metabolites are the products of metabolism, which means that are not genetically encoded and are produced and modified by the metabolism of living organisms, encompassing a wide range of endogenous and exogenous compounds synthesized by organisms such as amino acids, nucleic acids, peptides, carbohydrates, polyphenols, organic acids, vitamins, alkaloids, and minerals (Mozzi et al. 2013; Oikawa 2022).

Metabolomics focuses on understanding the metabolic processes and interactions within cells, organisms, and communities of organisms (Selamat 2021). Metabolomics provides valuable insights into how cells and organisms function within their environments by examining the unique chemical patterns that specific cellular processes generate. This holistic approach offers a deeper understanding of the complex relationships between metabolism and overall biological function (Mozzi et al. 2013; Selamat 2021; Wu et al. 2022a). Originally, metabolomics was used to identify biomarkers of pathologic processes in the pharmaceutical field. Additionally, it was employed to elucidate and understand genes whose roles were unknown. Nowadays, the scope of metabolomics includes not only living organisms but also water, rivers, soil, and foods (Oikawa 2022). Due to this, metabolomics is applied in various industries, such as pharmaceuticals, agriculture, and food.

In the realm of food, metabolomics is used in food resources, production, and processing, as well as in human diets (Selamat 2021). Current research studies concerning nutrition and food science mainly involve the use of various robust technologies and routine analytical techniques to determine food identity markers (Mozzi et al. 2013; Selamat 2021). Finding metabolic markers of food is greatly influenced by these instruments. By using advanced food metabolomics techniques, changes in food composition in terms of compositional analysis, food safety, food classification, and traceability can be clearly shown, providing a new paradigm for answering questions in this field (Mozzi et al. 2013; Wu et al. 2022a). In this context, has recently emerged the foodomics; this new discipline studies nutrition and food through the implementation of omics technologies (Mozzi et al. 2013).

On the other hand, metabolomics food research focuses not only on ensuring the composition, quality, and safety of food but also on supplying adequate nutrients to effectively combat diseases. This has given rise to the emergence of functional foods and nutraceuticals (FFNs) (Kumar et al. 2018). Functional foods include foods with at least some benefits to human functions and can reduce or prevent diseases or also improve health. Functional foods are ideally manufactured and distributed using animals, plants, and microorganisms. Functional foods include a wide variety of foods including fermented foods (Fu et al. 2022).

Fermented foods are those products that undergo biochemical changes as a result of the action of microorganisms or their respective metabolites (Fu et al. 2022; Wan-Mohtar et al. 2022). Fermented foods offer many advantages for health, like better digestion, prevention of diarrhea, and reinforced immunity. The mechanism of action generally involves the decomposition of proteins in food through the action of protease enzymes produced by microorganisms, resulting in the formation of different metabolites and essential nutrients (Wan-Mohtar et al. 2022).

In this context, metabolomics has a significant impact on the food industry (Fig. 1) as it provides critical insights into the physiology of organisms, helping prevent and diagnose diseases and determining the likelihood of developing certain pathologies in the future. Additionally, it enhances the quality, safety, and development of new food products (Fung et al. 2018).

However, while metabolite analysis can be effective through the metabolomics approach, there are specific challenges that must be addressed when conducting a metabolomics analysis, to give an example the sensitivity of metabolites to environmental conditions, the organic response of the metabolome to various genetic and

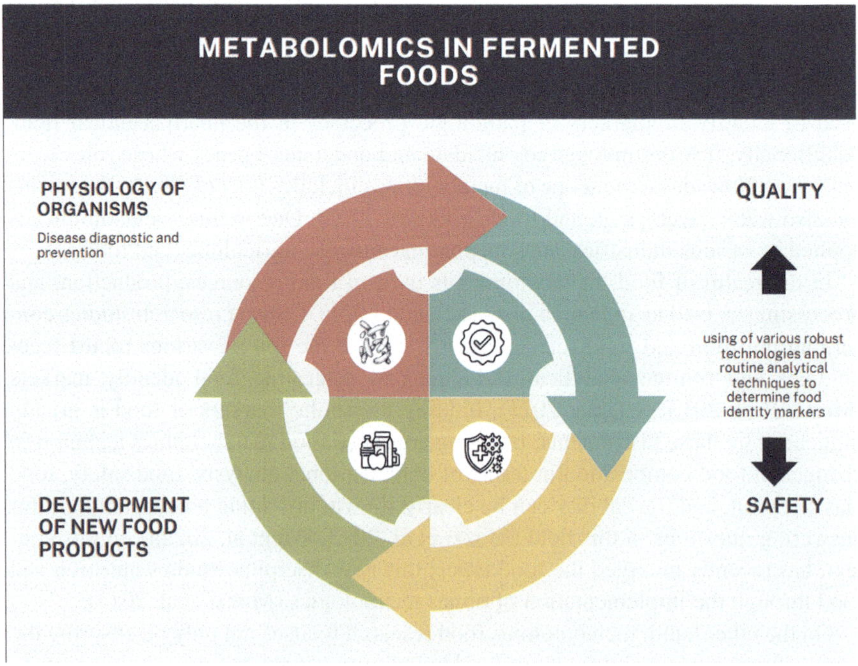

Fig. 1 Impact of metabolomics in fermented foods

environmental stimuli, and the stability of metabolites, which display different rates of replacement within the cell. Some may be volatile in the presence of oxygen, light, or temperature. Even the choice of technique for conducting the analysis and the cost of the analytical instrument can pose challenges in identifying metabolic compounds (González-Peña and Brennan 2019).

For several years, the development of analytical technologies has seen significant advancements in the measurement and interpretation of complex metabolic profiles. The most suitable analytical technologies for comprehensive metabolite analysis must combine high sensitivity, robustness, and high-performance ability, essential for efficiently processing a large number of samples. It is important to note that no single analytical technique can isolate and detect all different molecules simultaneously (Johanningsmeier et al. 2016).

Metabolomics analyses can be classified as targeted or untargeted. Untargeted metabolomics is more comprehensive and unbiased, detecting all possible metabolite groups to identify patterns or fingerprints without needing to identify or quantify specific compounds. It provides high performance, facilitating the identification of novel compounds; however, the interpretation of the results is more complex. Targeted metabolomics is easier to interpret and can even be linked to specific pathways. This type of analysis evaluates the behavior of specific groups of compounds in samples but is limited to a certain number of compounds and cannot assess untargeted compounds (Selamat 2021).

Among the techniques most frequently employed in metabolomics research are nuclear magnetic resonance (NMR) spectroscopy and mass spectrometry (MS). These techniques enable simultaneous high-throughput analysis of a wide variety of metabolites, providing highly accurate spectroscopic/structural information (Larive et al. 2014).

Nuclear Magnetic Resonance (NMR)

Nuclear magnetic resonance (NMR) spectroscopy is a powerful and versatile technique used for the identification of metabolites in a wide range of food products, including fermented foods such as yogurt, bread, cheese, and coffee, as well as meat and honey, among others. This technique enables rapid evaluation of combination at the molecular level lacking the need for complex separation or purification processes for sample preparation. Some of its features such as the simplicity of sample preparation, the ability to quantify metabolite levels, its inherently non-destructive nature, and the high experimental reproducibility of NMR have established it as the preferred methodology in metabolomics analyses. This technique uses sample volumes of 300–500 μL, and in a single experimental evaluation, a wide variety of different chemical entities can be identified, and most importantly, nearly all food constituents can be detected through a quantitative signal. Additionally, NMR is a highly sensitive and high-throughput technique. One of its disadvantages is that compound analysis requires a significant amount of time, and it is an expensive technique (Fraga-Corral et al. 2020; Gao et al. 2021; Selamat 2021).

GC-MS and LC-MS

Mass spectrometry (MS) stands out as one of the most common techniques in metabolomics due to its versatility, high sensitivity, and ability to analyze a wide variety of metabolites in plants and animal tissues. It allows for the separation of thermolabile compounds with higher molecular weight and greater polarity without the need for a pre-preparation stage, using volumes of less than 100 μL. This technique allows for the identification of unknown compounds in a food sample, while known compounds can be accurately quantified. Similarly, it enables the determination of the molecular structure and chemical properties of the identified compounds by measuring their molecular weight through mass spectrometry.

Mass spectrometry (MS) is used in combination with chromatographic techniques, which can be either gas-based (GC-MS) or liquid-based (LC-MS). In this process, the components of a mixture are separated using either GC or LC, after which MS detects and characterizes each one. The key difference between GC-MS and LC-MS is that GC-MS uses inert gases, such as helium, while LC-MS employs solvents that correspond to the mobile phase in liquid chromatography. MS can create numerous ionizations with different ion sources and can calculate chemical compounds with different ion sources (Fraga-Corral et al. 2020; Gao et al. 2021; Selamat 2021).

The analytical technique GC-MS has been extensively used for a long time due to its high reproducibility and robustness. It is primarily employed for analyzing volatile organic mixtures and their primary derivatives. In the food industry, GC-MS plays a crucial role in detecting adulterated foods, toxic substances, and additives, as well as in assessing nutritional quality, enhancing food quality, and implementing various food varieties. Conversely, LC-MS is a chemical method that facilitates the separation of complex mixtures and the molecular identification of various chemical components with high sensitivity and specificity. Due to greater separation and sensitivity, metabolomics based on these GC-MS/LC-MS techniques has been extensively applied to study various foods, such as fermented foods (Lisanti et al. 2021).

As previously mentioned, there are many applications that these metabolomics tools can have in the food industry, specifically in the field of fermented products and functional foods. With the continuous advancement of these techniques and improvements in databases and data processing software, the application of metabolomics in food research is increasingly being accepted by researchers as a method to empirically delineate the properties of fermented foods and probe functional metabolomes (Gao et al. 2021).

3 Designing Fermented Functional Foods Using Metabolomics

The fermentation process depends on different variables, such as the technique used, pH, temperature, enzyme used, food ingredients, and starter cultures, to obtain alcohols, carbon dioxide, and some organic acids in fermented foods. However, the starter microorganisms could be one of the most important variables because of the impact on odor, flavor, bioactive components, and even the safety and shelf-life of the final product.

Production on an industrial scale is well-characterized; meanwhile, in other situations such as in small care production in local industries, when the ingredients or equipment are not sterilized, when fermentation occurs with native microorganisms (non-starter organisms) of the raw food matrix, or when consortia is the started culture, results can be different at quality control and sensory patterns because of the dynamic of the microorganisms. Thus, the complexity of distinct microbial communities can result in a new product (Leeuwendaal et al. 2022; Srinivas et al. 2022).

Fermentation pathways of organisms are described in the literature, in which the main microorganisms enlisted are *Lactococcus*, *Lactobacillus*, yeast, and some fungi genera. However, metabolites obtained or degraded depend on the response of microorganisms or interaction between organisms to biological or environmental stimuli (Mozzi et al. 2013). Therefore, the use of omics can be an approach to how the organisms work in the different phases of fermentation with certain conditions and improve the culture system. Metabolomics of the species can be conducted to identify or quantify intracellular metabolites and what the cell excretes under defined conditions. Microbial metabolomics, as the study of metabolic changes in organisms, can give a metabolic pattern with metabolite identification, including novel molecules, and pathway mapping during fermentation under microenvironment changes (Bertini et al. 2013; Halouska et al. 2023).

Metabolomics techniques such as nuclear magnetic resonance (NMR) and gas chromatography-mass spectrometry (GC-MS) commonly use solvents to identify and quantify metabolites from cells, tissues, biofluids, or organisms.NMR can be useful to identify and measure metabolites in both biological samples and microorganisms, even in microorganisms. It allows the identification and measurement of metabolites in biological samples as microorganisms from their compartmentalization and understanding of the exchange dynamics between metabolites and macromolecules or regulatory events. The process consists of obtaining bacterial biomass, extracting their metabolites, results bioanalysis, and then the identification and quantification. Also, data obtained can be analyzed using multivariate statistical analysis to detect the differentiation of metabolites (Bertini et al. 2013; Halouska et al. 2013; Gowda and Raftery 2021; Moco 2022; Tatulli et al. 2023).

For example, NMR-based metabolomics was conducted to identify metabolite profiles of *Lactobacillus* strains that are associated with sensorial properties in fermented vegetable juice. *Lactobacillus* spp. and *Lactobacillus brevis* species were used considering the differences in lactic fermentative characteristics. The analysis showed variations among metabolites in the *Lactobacillus* metabolic pathways in fermented food production, such as fermentation from carbohydrates, decarboxylation, and transamination of amino acids. *L. brevis* strains showed the ability to utilize sucrose and citric acid and convert glutamic acid and tyrosine into gamma-aminobutyric acid (GABA) and tyramine, respectively (Tomita et al. 2017).

On the other hand, *Baijiu* and vinegar contain several metabolites from different source materials, such as sorghum and *Daqu*. It can even be produced by the autochthonous microorganisms' consortia from the *Daqu*. Such diversity of microorganisms can have different contributions to the final product. Therefore, selected bacteria (*Bacillus licheniformis*, *Pediococcus pentosaceus*, *Lactobacillus plantarum*, *Pichia kudriavzevii*, *Wickerhamomyces anomalus*, *Saccharomyces cerevisiae*, and *Saccharomycopsis fibuligera*) were blended in different combination during sorghum-based laboratory-scale alcoholic fermentation. After, H-NMR was used to know how different conditions resulted in variations of the metabolites produced. Among the findings, it was reported that *S. cerevisiae* was superior to ethanol production, meanwhile, lactic acid bacteria had high amylolytic and proteolytic activities, increasing saccharification and protein degradation (Li et al. 2018).

In another study, NMR metabolomics approach was applied to know changes in cheese ripening with added autochthonous microorganisms and with a different kind of milk. Recently, NMR combined with chemometrics was used to study metabolite changes of *Phaseolus vulgaris* flour fermented by LAB with homo- and heterofermentative metabolism and some yeast. It was observed that beneficial metabolites such as gamma-aminobutyric acid (GABA) and free amino acids were obtained in the fermentation process, especially *Lactiplantibacillus plantarum* species, which was found to be the most effective in bean flour fermentation (Tatulli et al. 2023).

Meanwhile, gas chromatography/time-of-flight-mass spectrometry (GC-MS) is a technique that is sensitive toward volatiles and semi-volatile compounds in biological samples that can be combined with multivariate statistical analyses to obtain functional metabolite profiles. For example, a study evaluated with seven different batch fermentations the change of metabolites using *Streptococcus thermophilus*, and results were the identification of 64 metabolites (Khakimov et al. 2017; Parlindungan and Jones 2023).

In fermented vegetables, it was identified 78 metabolites with GS/TOF-MS, among which were organic acids, sugars, amino acids, fatty acids, and miscellaneous metabolites derived from the degradation of high molecular polymers from vegetables during fermentation with *Lactobacillus plantarum* (Kim et al. 2019). Metabolomics analysis of the microbial consortium with *Ketogulonicigenium vulgare*, *Gluconobacter oxydans*, and *Bacillus endophyticus* was conducted using GS/TOF-MS, to identify and quantify in the early stage of fermentation intracellular substrates such as glucose, phosphoric acid, tetradecanoic acid, hexadecenoic acid, and octadecanoic acid. Meanwhile, lactic acid, trehalose, inositol, and fumaric acid were present in the middle stage of fermentation. In addition, it was found that there were differences between two versus three-species consortium, mostly in amino acids, which showed intracellular abundance in the three-species consortium. Finally, it was observed that *B. endophyticus* and *G. oxydans* together promoted the growth of *K. vulgare* and then the secretion of the 2-keto-L-gulonic acid (2-KGA), the precursor of vitamin C (Ma et al. 2019).

The yeast *Saccharomyces cerevisiae* has been used in wine production because of its fermentation capacity; nevertheless, there are chemical compounds in low concentrations that have not been characterized. Thus, GC-MS along with transcriptomic techniques of four *S. cerevisiae* strains from different beverages was conducted. Results showed metabolic differences between the origin of the strains, particularly cachaça with higher production of ethyl esters, ethyl decanoate, and ethyl octanoate (Mendes et al. 2017).

Products fermented with probiotic bacteria have organoleptic characteristics that are very different from those that are not produced by fermentation. The fermentation process allows extending the shelf life of food because it avoids spoilage and thus food waste, in addition to providing health benefits due to its nutritional and functional properties attributed to the secondary metabolites generated by the interaction between raw materials and probiotics (Gao et al. 2021; Sionek et al. 2023).

Metabolites produced during the fermentation process are organic compounds of low molecular weight (less than 1000 Da), which can provide nutritional value, modify sensory properties, or serve as energy sources for survival. The analysis of the metabolomic profile of fermented products using analytical tools such as NMR and various chromatography-mass spectrometry (MS) techniques allows the monitoring of metabolites responsible for qualitative properties during fermentation in order to predict the nutritional, functional, taste, and safety characteristics of the final products (Gao et al. 2021; Park and Kim 2021).

Among the metabolites that modify the flavor of fermented foods are volatile compounds, which are identified by GC-MS. Some of these are aromatic compounds (ethyl cinnamate and ethyl 3-phenylpropionate), pyrazines, esters, and alcohols; their production also gives characteristics of maturity to the product, and this depends entirely on the microbial metabolism during fermentation (Gao et al. 2021). In products fermented by *B. subtilis*, there is a characteristic flavor associated with the production of compounds such as ammonium, 2,5-dimethylpyrazine, isovalerate, isobutyrate, and 2-methylbutyrate, which confer an undesirable flavor and modify the nutritional and functional composition of the product (Gao et al. 2020). In cheeses and dairy products, the identification of lactose, lactate and citrate, amino acids, and fatty acids during the fermentation process determines not only the flavor but also the maturity of the product (Khattab et al. 2019; Gao et al. 2021).

Through high-throughput sequencing (HTS), researchers have identified the microbial community structure and the

impact of ingredients during the fermentation of traditional Korean foods such as kimchi, fermented soy products, and kombucha. These HTS studies revealed that added ingredients can positively modulate the composition of microbiota and chemicals. In particular, the increased growth of Leuconostoc rapi improves flavor properties due to the production of mannitol, an antioxidant that gives the food a fresh taste and allows for a longer shelf life, while a decrease in acid metabolites associated with spoilage was observed (Mannaa et al. 2021). Through an omics analysis, it has been found in particular that there are about 130 overregulated metabolites and 160 downregulated proteins associated with the soybean fermentation process that specifically impact the flavor of the product (Wu et al. 2022b).

Concerning aroma, it is important to consider the interaction between microbial communities and environmental factors during the fermentation process. In particular, in the case of wine fermentation by *Saccharomyces cerevisiae* and *Oenococcus oeni*, the use of appropriate probiotics is considered to promote the formation of specific aromas according to the industrial production (Sirén et al. 2019). Among the metabolites responsible for aroma are volatile compounds corresponding to esters, aldehydes, alcohols, ketones, aromatic compounds, furans, acids, phenols, terpenes, and compounds containing nitrogen and sulfur (Diez-Simon et al. 2019). The presence of these compounds is found in low concentrations and contributes to both aroma and flavor. In particular, a study by Zhang et al. (2024) found that the flavor of different types of milk (goat, sheep, cow, and buffalo) is due to the differential presence of the metabolites 3-hydroxyphenyl-3-hydroxypropano ic acid, inosine 5-triphosphate, methylcysteine, N-cinnamylglycine, and small peptide (Tyr-Asn).

Another important aspect to take care of during the elaboration of fermented foods is their safety since the fermentation could be observed as an increase of toxic compounds depending on the substrates and the inoculated microorganisms used before and during the process. Product analysis by NMR and LC-MS allows the identification of these toxic compounds.

In particular, the cleavage of glycosides from soybean, rice, or wheat to free compounds such as saponins and flavonoids is characteristic during fermentation. A decrease in alkaloid production is feasible when the fermentation process is carried out by yeasts (Seo et al. 2018; Gao et al. 2021). Also, metabolites representing toxicity such as biogenic amines, ethyl carbamate, and the presence of pathogenic microorganisms can be determined early in the process, as these could be modulated by probiotic action (Oyedeji et al. 2021). Metabolomics has also allowed the identification of preservative chemicals in fermented products such as benzoic acid, sorbic acid, ethyl paraben, and propionic acid (Tungkijanansin et al. 2020).

The quality of fermented food is influenced by metabolites from protein and fat degradation (bioactive peptides, amino acids, and fats), which vary according to the fermentation time; in this sense a prolonged fermentation time correlates with a higher production of functional actives such as gamma-aminobutyric acid and isoflavones and saponins, which bring health benefits as antioxidants and also contribute to the digestibility (Gao et al. 2021).

The role of microorganisms during the development of fermented food has an impact on the quality, taste, functionality, safety, and health of the products. Therefore, the analysis of microbial communities will allow knowing the relationship between metabolites and microorganisms in relation to flavor, qualities, and function, in order to achieve a better standardized production at the industrial level. In this regard, a multi-omics approach is required, where it is crucial that the multi-omics database of microorganisms in fermented foods (MDMFF) continues to grow for application in the fermented food industry (Shi et al. 2022).

4 Conclusions and Future Prospects

Metabolomics is a tool in food research for analysis, safety, classification, design, and food traceability. The use of metabolomic techniques will help to understand how variation in parameters during fermentation affects the metabolites originating from one microorganism, two-species consortium, or three-species consortia. The above would allow the fermentation process to be optimized, reduce costs and food waste, achieve scale production, generate new functional products, and assure quality in the final product.

The fermentation of foods using probiotic bacteria significantly alters their organoleptic properties, extending shelf life and enhancing health benefits through the production of beneficial metabolites. These metabolites, ranging from volatile compounds to organic acids, play an important role in defining the flavor, aroma, and safety of the final product. Advanced analytical techniques have been instrumental in understanding the complex interactions between microbial communities, raw materials, and fermentation conditions. These tools help monitor the metabolites responsible for the qualitative characteristics of fermented products and ensure their nutritional and functional benefits. The continuous development of multi-omics databases and methodologies will further enhance our ability to standardize and optimize the production of high-quality fermented foods, ultimately benefiting both consumers and producers.

Finally, it is necessary to conduct more studies on the metabolomics of fermented foods in other countries, as most of them are conducted in countries such as China or Japan.

References

Adamyan L et al (2024) Metabolomic biomarkers of endometriosis: a systematic review. J Endometriosis Uterine Disord 7(April):100077. https://doi.org/10.1016/j.jeud.2024.100077

Bertini I, Hu X, Luchinat C (2013) Global metabolomics characterization of bacteria: pre-analytical treatments and profiling. Metabolomics 10(2):241–249. https://doi.org/10.1007/s11306-013-0571-4

Chi J et al (2024) Artificial intelligence in metabolomics: a current review. TrAC – Trends Anal Chem 178:117852. https://doi.org/10.1016/j.trac.2024.117852

Diez-Simon C, Mumm R, Hall RD (2019) Mass spectrometry-based metabolomics of volatiles as a new tool for understanding aroma and flavour chemistry in processed food products. Metabolomics 15(3):1–20. https://doi.org/10.1007/s11306-019-1493-6

Fraga-Corral M et al (2020) Analytical metabolomics and applications in health, environmental and food science. Crit Rev Anal Chem 52(4):712–734. https://doi.org/10.1080/10408347.2020.1823811

Fu J et al (2022) Application of metabolomics for revealing the interventional effects of functional foods on metabolic diseases. Food Chem 367:130697. https://doi.org/10.1016/j.foodchem.2021.130697

Fung F, Wang HS, Menon S (2018) Food safety in the 21st century. Biom J 41(2):88–95. https://doi.org/10.1016/j.bj.2018.03.003

Gao YX et al (2020) 1H NMR-based chemometric metabolomics characterization of soymilk fermented by Bacillus subtilis BSNK-5. Food Res Int 138(PA):109686. https://doi.org/10.1016/j.foodres.2020.109686

Gao Y et al (2021) Metabolomics approaches for the comprehensive evaluation of fermented foods: a review. Food Secur 10(10):1–18. https://doi.org/10.3390/foods10102294

González-Peña D, Brennan L (2019) Recent Advances in the Application of Metabolomics for Nutrition and Health. Annu Rev Food Sci Technol 10:479–519. https://doi.org/10.1146/annurev-food-032818-121715

Gowda GAN, Raftery D (2021) NMR based metabolomics – cancer metabolomics. Adv Exp Med Biol 1280:19–37. https://doi.org/10.1007/978-3-030-51652-9

Halouska S, Bo Z, Rosmarie G, Shulei L, Emily S, Fenton RJ, Barletta RG, Somerville GA, Powers R (2013) Revisiting protocols for the NMR analysis of bacterial metabolomes. J Integr OMICS 3(2):120–137. https://doi.org/10.5584/jiomics.v3i2.139.Revisiting

Halouska S, Bo Z, Rosmarie G, Shulei L, Emily S, Fenton RJ, Barletta RG, Somerville GA, Powers R (2023) Metabolomic profiling reveals bacterial metabolic adaptation strategies and new metabolites. Curr Opin Chem Biol 74:1–7. https://doi.org/10.1016/j.cbpa.2023.102287

Johanningsmeier SD, Harris GK, Klevorn CM (2016) Metabolomic technologies for improving the quality of food: practice and promise. Annu Rev Food Sci Technol 7:413–438. https://doi.org/10.1146/annurev-food-022814-015721

Khakimov B et al (2017) Untargeted GC-MS metabolomics reveals changes in the metabolite dynamics of industrial scale batch fermentations of streptoccoccus thermophilus broth. Biotechnol J 12(10):1–9. https://doi.org/10.1002/biot.201700400

Khattab AR et al (2019) Cheese ripening: a review on modern technologies towards flavor enhancement, process acceleration and improved quality assessment. Trends Food Sci Technol 88(March):343–360. https://doi.org/10.1016/j.tifs.2019.03.009

Kim J et al (2019) Metabolite profile changes and increased antioxidative and antiinflammatory activities of mixed vegetables after fermentation by Lactobacillus plantarum. PLoS One 14(5):e0217180. https://doi.org/10.1371/journal.pone.0217180

Kumar A et al (2018) Metabolomics-assisted biotechnological interventions for developing plant-based functional foods and nutraceuticals. Crit Rev Food Sci Nutr 58(11):1791. https://doi.org/10.1080/10408398.2017.1285752

Larive CK, Barding GA, Dinges MM (2014) NMR spectroscopy for metabolomics and metabolic profiling. Anal Chem 87(1):133–146. https://doi.org/10.1021/ac504075g

Leeuwendaal NK et al (2022) Fermented foods, health and the gut microbiome. Nutrients 14(7):1–26. https://doi.org/10.3390/nu14071527

Li RY et al (2018) Characterization of bacteria and yeasts isolated from traditional fermentation starter (Fen-Daqu) through a 1H NMR-based metabolomics approach. Food Microbiol 76:11–20. https://doi.org/10.1016/j.fm.2018.03.015

Lisanti MT et al (2021) Minty aroma compounds in red wine: development of a novel automated HS-SPME-arrow and gas chromatography-tandem mass spectrometry quantification method. Food Chem 361. https://doi.org/10.1016/j.foodchem.2021.130029

Ma Q et al (2019) Integrated proteomic and metabolomic analysis of a reconstructed three-species microbial consortium for one-step fermentation of 2-keto-l-gulonic acid, the precursor of vitamin C. J Ind Microbiol Biotechnol 46(1):21–31. https://doi.org/10.1007/s10295-018-2096-3

Mannaa M et al (2021) Evolution of food fermentation processes and the use of multi-omics in deciphering the roles of the microbiota. Food Secur 10(11). https://doi.org/10.3390/foods10112861

Mendes I et al (2017) Integrating transcriptomics and metabolomics for the analysis of the aroma profiles of Saccharomyces cerevisiae strains from diverse origins. BMC Genomics 18(1):1–13. https://doi.org/10.1186/s12864-017-3816-1

Moco S (2022) Studying metabolism by NMR-based metabolomics. Front Mol Biosci 9(April):1–12. https://doi.org/10.3389/fmolb.2022.882487

Mozzi F et al (2013) Metabolomics as a tool for the comprehensive understanding of fermented and functional foods with lactic acid bacteria. Food Res Int 54:1152. https://doi.org/10.1016/j.foodres.2012.11.010

Oikawa A (2022) Food Metabolomics. J Nutr Sci Vitaminol 68:S128–S130. https://doi.org/10.3177/jnsv.68.S128
Ovbude ST et al (2024) Applications of chromatographic methods in metabolomics: a review. J Chromatogr B Analyt Technol Biomed Life Sci 1239:124124. https://doi.org/10.1016/j.jchromb.2024.124124
Oyedeji AB et al (2021) Metabolomic approaches for the determination of metabolites from pathogenic microorganisms: a review. Food Res Int 140:110042. https://doi.org/10.1016/j.foodres.2020.110042
Park MK, Kim YS (2021) Mass spectrometry based metabolomics approach on the elucidation of volatile metabolites formation in fermented foods: a mini review. Food Sci Biotechnol 30(7):881–890. https://doi.org/10.1007/s10068-021-00917-9
Parlindungan E, Jones OAH (2023) Using metabolomics to understand stress responses in Lactic Acid Bacteria and their applications in the food industry. Metabolomics 19(12):1–17. https://doi.org/10.1007/s11306-023-02062-2
Rizo J et al (2018) Omics in traditional vegetable fermented foods and beverages. Crit Rev Food Sci Nutr 60(5):791–809. https://doi.org/10.1080/10408398.2018.1551189
Selamat J (2021) Application of the metabolomics approach in food authentication. Molecules 26(24):7565
Seo HS et al (2018) Untargeted metabolite profiling for koji-fermentative bioprocess unravels the effects of varying substrate types and microbial inocula. Food Chem 266(February):161–169. https://doi.org/10.1016/j.foodchem.2018.05.048
Shi H et al (2022) Advances in fermented foods revealed by multi-omics: a new direction toward precisely clarifying the roles of microorganisms. Front Microbiol 13(December):1–9. https://doi.org/10.3389/fmicb.2022.1044820
Sionek B et al (2023) Traditional and new microorganisms in lactic acid fermentation of food. Fermentation 9(12):1–21. https://doi.org/10.3390/fermentation9121019
Sirén K et al (2019) Multi-omics and potential applications in wine production. Curr Opin Biotechnol 56:172–178. https://doi.org/10.1016/j.copbio.2018.11.014
Srinivas M et al (2022) The application of metagenomics to study microbial communities and develop desirable traits in fermented foods. Food Secur 11(20):1–28. https://doi.org/10.3390/foods11203297
Tatulli G et al (2023) NMR-based metabolomic study on phaseolus vulgaris flour fermented by lactic acid bacteria and yeasts. Molecules 28(12). https://doi.org/10.3390/molecules28124864
Tomita S et al (2017) Rapid discrimination of strain-dependent fermentation characteristics among Lactobacillus strains by NMR-based metabolomics of fermented vegetable juice. PLoS One 12(7):e0182229. https://doi.org/10.1371/journal.pone.0182229
Tungkijanansin N et al (2020) Simultaneous determination of benzoic acid, sorbic acid, and propionic acid in fermented food by headspace solid-phase microextraction followed by GC-FID. Food Chem 329:127161. https://doi.org/10.1016/j.foodchem.2020.127161
Wan-Mohtar WAAQI et al (2022) Fermented foods as alternative functional foods during post-pandemic in Asia. Front Food Sci Technol 2. https://doi.org/10.3389/frfst.2022.1047970
Weimer BC (2013) Food science, technology and nutrition. In: Metabolomics in food and nutrition. https://doi.org/10.1016/b978-1-84569-512-5.50014-8
Wu W et al (2022a) Emerging applications of metabolomics in food science and future trends. Food Chem: X 16(August):100500. https://doi.org/10.1016/j.fochx.2022.100500
Wu Y et al (2022b) Multi-omics analyses of the mechanism for the formation of soy sauce-like and soybean flavor in Bacillus subtilis BJ3-2. BMC Microbiol 22(1):1–13. https://doi.org/10.1186/s12866-022-02555-5
Zhang F et al (2024) Widely targeted metabolomic analysis revealed the diversity in milk from goats, sheep, cows, and buffaloes and its association with flavor profiles. Food Secur 13(9). https://doi.org/10.3390/foods13091365

Metabolomics for Plant Growth-Promoting Microbe Profiling and Designing Biofertilizer

Siddhartha Das and Sweety Maity

Abstract

Metabolomics, the global study of metabolites in biological systems, plays a very crucial role in profiling and designing plant growth-promoting microbes. The comprehensive study of metabolic pathways and interactions between plants and related microbial populations is made possible by this cutting-edge omics technology. Certain metabolites, such as phytohormones, siderophores, and antibiotics, that promote plant growth can be found by studying the metabolic profiles of microorganisms and plants. These insights allow for the characterization of beneficial microbes and their symbiotic relationships with plants, leading to the identification of key microbial traits that enhance nutrient acquisition, stress resistance, and overall plant health. Moreover, metabolomics-driven approaches enable the engineering of microbial consortia with optimized metabolic capabilities, tailored to support specific plant species or cultivars under various environmental conditions. This targeted design of PGPMs not only improves agricultural productivity but also promotes sustainable farming practices by reducing the reliance on chemical fertilizers and pesticides. As a powerful tool in plant-microbe interaction studies, metabolomics facilitates a deeper understanding of the complex biochemical networks involved, paving the way for innovative strategies in crop improvement and sustainable agriculture. We draw attention to current research that has enhanced models of crop microorganism structure and function through the use of multi-omics-based methodologies. Lastly, we suggest using comprehensive methods to combine host and microbial omics infor-

S. Das (✉)
Department of Plant Pathology, M.S. Swaminathan School of Agriculture, Centurion University of Technology and Management, Paralakhemundi, Odisha, India

S. Maity
Department of Molecular Biology and Biotechnology, Odisha University of Agriculture and Technology, Bhubaneswar, Odisha, India

S. Kaur et al. (eds.), *Microbial Metabolomics*,
https://doi.org/10.1007/978-981-96-4824-5_12

mation in order to attain accuracy and efficiency in data processing. This is important for controlling both abiotic and biotic stresses as well as comprehending how the microbiota influences plant fitness.

Keywords

Metabolomics · Plant growth-promoting microbe · Nutrient · Stress · Fertilizer

1 Introduction

Metabolomics is therefore extensive research of metabolites; the tiny molecular substrates or the intermediates appear as a metabolic product within a biological system. Cellular metabolites, which might include amino acids, nucleotides, carbohydrates, lipids, and other macromolecules, are identified and quantified in this process. The ultimate aim of metabolomics is to provide an imprint of the metabolic state of cells, tissues, organs, or organisms under specific conditions, offering insights into physiological and pathological states. With the rapid increase in human population, it is predicted that by 2050, the global food consumption would have increased by 70% (FAO 2019). This implies that there will be a substantial increase in food production required over the next 30 years. More food productivity has been achieved for many years due to enhanced farming practices including planting better seeds that come from breeding process (Yu and Li 2022; Zhang et al. 2023). For example, the present worldwide growth rates per year of 1.1% to 1.8% for wheat, soybeans, rice, and maize are predicted to decrease in the upcoming years. The continuous declination in crop yields caused by both biotic and abiotic stresses, along with geopolitical conflicts across the globe, had a major impact (FAO 2017; Arif et al. 2020; Nasir et al. 2022). There are certain examples like Ug99, a novel strain of wheat stem rust, which is rapidly proliferating worldwide due to its exceptional resistance to the majority of wheat resistance genes. This poses a significant threat to the world's food security (Li et al. 2019). Similarly, the extremely virulent *Fusarium oxysporum* (TR4), which is proliferating throughout the main banana-growing regions of the world, is endangering the global output of Cavendish bananas. Additionally, the disease known as maize fatal necrosis has severely harmed the output of maize in a significant portion of sub-Saharan Africa (Wamaitha et al. 2018). High-throughput molecular techniques have been extensively used to study the array of microbes of different portions of plants (Abdelfattah et al. 2016; Yuan et al. 2023; Regalado et al. 2020; Matsumoto et al. 2021). The shape and exudation of roots, in particular, significantly influence the rhizosphere's chemical and physical characteristics (Lundberg and Teixeira 2018; Yuan et al. 2023; Chai and Schachtman 2022). In particular, various plant species have been associated with distinct microbial communities when grown in the same soil, which defines their rhizosphere microbiomes. Furthermore, distinct microbial populations inhabit distinct crop genotypes (Zhang et al. 2020; Cordovez et al. 2021). Unfortunately, a lot of these researches only used one omics technology, which led to a narrow knowledge of the potential and complexity found in the plant microbiome. Thus, it is imperative to go beyond one-dimensional functional and taxonomic investigations and toward integrative studies. Adoption of multi-omics is probably not going to be enough in and of itself to boost food production to feed the world's population, but well-executed integrative omics research can help clarify biological roles of the plant microbiome at a deeper level.

2 An Overview of the Research Gaps in Agricultural Microbiomes

There is quite less information available on the efficacy of plant-associated microbiomes on the productivity of food crops. Realizing the full potential of crop microorganisms in agricultural productivity requires a comprehensive comprehension of the structural and functional components of connections among plants, microorganisms, soil, and agricultural management methods. Depending on the environment, plant-associated microbial species can live as either pathogens or mutualists when they coexist with or invade their host plants. Thus, new high-throughput approaches will need to be developed in order to characterize plant microbial members beyond recognition, to functionally characterize particular species (Hacquard et al. 2016; Hiruma et al. 2016; Tian et al. 2022). Using further microbiome engineering could boost agricultural output and enhance nutrient efficiency as well as biotic and abiotic stress tolerance (Arif et al. 2020; Albright et al. 2022). This growth has been driven in part by technological innovation, which further empowers research into plant health and plant-microbe interactions through gene editing, high-throughput sequencing (HTS: metagenomics, metatranscriptomics, bioinformatics) (Bose et al. 2024), etc. However, research on food crop microbiomes (rice, wheat, maize, potatoes, cassava, beans, etc.) has not kept pace with this rapid pace of growth (Busby et al. 2017). Every major crop category underwent a literature search utilizing the FAO's Evidence Crop Classification, Version 1.1 (Da Silva 2012; FAO 2019). Here, terminologies like "plant name," "microbial communities," and "plant name microbiome" were used in searching in the Proteome Exchange, Massive systems, Integrated Microbial Genomes, and NCBI BioProjects databases. According to a recent assessment by Brunel et al. (2020), the number of rhizosphere metagenome datasets published annually from every continent has significantly increased.

3 Plant Microbiomes and Abiotic/Biotic Stresses: An Intersection of Integrated Omics

The molecular processes underlying crop microbiomes' responses to biotic and abiotic stressors remain largely unknown, despite breakthroughs in HTS omics. The number of studies using integrated omics techniques should rise as a solution to close this gap. Our recommendation is to connect multiomics datasets and the plant stress response by means of multiomics techniques along with stress hormone profiling, plant immune output, machine learning, and network analysis. By doing this, simulations of the biological mechanisms underlying plant-microbe relationships under various environmental stress scenarios will be developed. This is because, in addition to altering the relationships between plants, microbiomes, and environments, plant immunity and hormones play a major role in recruiting and assembling the plant-associated microbiomes (Carrión et al. 2019). Furthermore, as microbial responses have been shown to guard against desiccation and death, future research should concentrate on examining the roles that microbes play in plants to understand their responses to biotic and abiotic stresses. The attributes of bacteria used to tolerate stress vary depending on the type of stress treatment; for example, in high-salinity environments, germs use distinct strategies to maintain osmotic equilibrium (Malik et al. 2020). This has been noted in the coevolution of crops, crop root endophytes, and actinomycetes and fungi. Using omics techniques, Gonin et al. (2023), for example, demonstrated how the microbiota of the roots of *Arabidopsis* affected auxin-independent root branching. Trait-based omics offer a superior substitute for conventional species-based research since they assign greater weight to functional characteristics in microbes and plants than to microbial species IDs. As an example, Hou et al. (2021) discovered a connection between light perception in *Arabidopsis* and microbial root commensals. In order to demonstrate how low light focused on leaves causes shoot-root-microbiota modification of the root system by bacterial communities, but not by fungi or oomycetes, this work employed a synthetic microbial community. Microbial commensals were also discovered to enhance plant growth in low-light environments. Understanding the dynamics of crop microbiome communities in relation to agroecosystem activities can also be aided through the combination of omics datasets across species (Defrenne et al. 2021). This is because interactions between diseases and crops take place within a broader pathosystem. Thus, host crops and pathogen interactions can be analyzed with the aid of integrated omics, such as co-transcriptomic datasets (Campos et al. 2021; Gorshkov and Tsers 2022). A useful predictor for modeling techniques could be the alterations in the plant rhizosphere microbiome in response to biotic and abiotic stimuli. For instance, Trivedi et al. (2017) used the results in identifying bacteria belonging to the *Acidobacteria*, *Firmicutes*, and *Actinobacteria* taxa, using a semi-quantitative integration approach as key microbial predictors of agroecosystem soil suppressiveness toward *Fusarium oxysporum*. The present investigation underscores the observation that the microbial communities correlated with plants exhibit lower levels of diversity in contrast to their surroundings (Fitzpatrick et al. 2018; Carrión et al. 2019; Gao et al. 2021; Shalev et al. 2022). This is caused by the more plant-selective recruitment of certain microbiota and the adaptation of certain bacteria to colonize plant tissues more than others (Waller et al. 2005; Brader et al. 2017; Sarkar et al. 2019; Liu et al. 2021). Notably, current research is showing how plant roots allow beneficial and commensal microorganisms to colonize while initiating effective defense responses against harmful microbes (Sarkar et al. 2019; Mahdi et al. 2022). Microbe-associated molecular patterns (MAMPs) trigger immunity in a synthetic bacterial community (SynCom), ultimately controlling efficient root mobilization by commensal microbes (Teixeira et al. 2021). Additionally, they discovered that, in contrast to MAMP-triggered immunity-inducing strains, massive strains that suppressed immunity were effective root colonizers. According to other research, root exudates directly control the rhizosphere microbiomes of several crops in sick soils (Du et al. 2021). Plant microbiomes are balanced by a complex chemical network that is formed by rhizosphere microbiomes and the environment. Using machine learning methods, AI platforms may be utilized to isolate a large number of microbial taxa that are retrieved via microbiome investigations. For example, in human microbiome studies, Huang et al. (2023) introduced the unique idea of culturomics by automated microbiota imaging and isolation (CAMII).

4 Plant Microbiome Investigations Utilizing the Latest Developments in Individual Omics

Plant microbial communities are complicated; however datasets that have been generated over the past 10 years through the application of metagenomics and metatranscriptomics procedures in genomics technologies have contributed to our understanding of this. When needed, genomes generated by metagenomics can be utilized as reference sequences in metatranscriptomics, and it also provides new insights into possible profiles of microbial communities. Measuring gene expression levels also depends heavily on metatranscriptomics. Indeed, due to the regulation of the gene expression level, microbes such as bacteria can respond quickly to changes in their environment, even though the control of the protein synthesis could have occurred in post-transcriptional and post-translational gene expression. Accordingly, real-time regulating plant-microbe interactions alongside additional environmental response alterations may be possible through the use of metatranscriptomics. Metagenomics, metatranscriptomics, metaproteomics, and metagenomics are a few examples of analytical techniques that can be used to directly extract genomic DNA, RNA, or metabolites from soil samples. The basic processes of interactions between microbes and hosts, also the metabolic pathways involve in the interactions between different species, are comprehensively illustrated by metabolomics (Fig. 1). Among the substances released during beneficial relationships between plants and their associated bacteria are those.

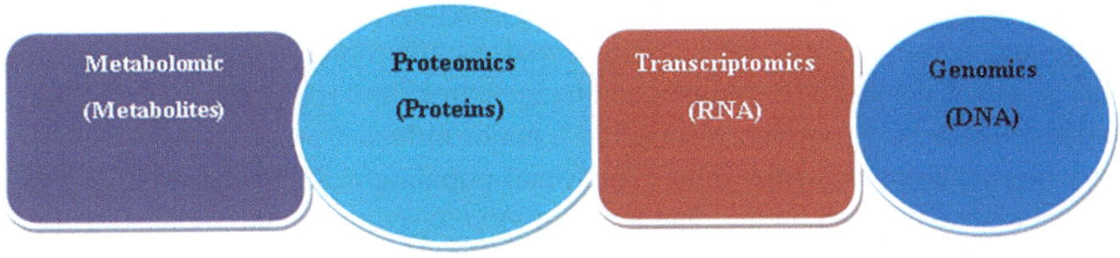

Fig. 1 Components of metabolic study

4.1 Metagenomics

The most popular method in the realm of omics is metagenomics. Our knowledge and collection of microorganisms linked with plants have significantly increased since the development of HTS technology (Lundberg et al. 2012; Ofek-Lalzar et al. 2014). The molecular bases of plant-microbe interactions, including plant genotype and related environmental variables that drive microbiome composition, structure, and evolution, are now more easily resolved with the availability of large-scale plant-associated microorganism genetic information. More deeper understanding into the structure and operation of core plant microbiomes includes the existence of both helpful and disease-causing microbes, which have been revealed using HTS datasets generated through targeted or shotgun sequencing (Hu et al. 2020). Similarly, to assess the relationship among exogenous enzyme activity and microbiological gene functioning in soil microbial communities, Trivedi et al. (2016) used qPCR, GeoChip 4.0 analysis, and 16S rRNA sequencing methodologies. An additional potent technique for determining gene function in plant-microbe relationships is transposon sequencing (Tn-Seq). The identification of host plant quantitative loci for traits related along with microbiome richness, characteristics, and community organization in the phyllosphere and roots is made possible by genome-wide association studies (GWAS), which are an extremely effective method. The structure and function of the root zone can be altered by applying GWAS and metagenome investigations to the identification of particular microbial species and prospective plant genes.

4.2 Metatranscriptomics

Metatranscriptomics is the crucial technique that captures gene expression up to community level, whereas transcriptomics helps to analyze host transcriptome. With sophisticated techniques, we may use transcriptomics to analyze the patterns of expression of many transcripts in the host organism under various environmental conditions. Transcriptomics has been utilized to interpret interactions between plants and microbes in order to comprehend the functions of root exudates in influencing the microbiome of plants and to provide insight into the temporal dynamics of transcription involved in the modulation of the host immune system, particularly under conditions of biotic stress. This finding revealed that certain properties that are able to be transmitted to next generations, such as promotion of growth, specificity, and heterogeneity, are communicated by microbial communities in soil. Specifically, Li et al. (2019) employed metagenomics and root metatranscriptomics to demonstrate how previous sowing has a major influence on the composition and activities of the peanut rhizosphere microbial population. Another recent example of such a breakthrough used global transcriptome responses in the root and leaf tissues of sorghum, elicited in response to drought throughout the growth season, and showed signs of a disruption of the plant-arbuscular mycorrhizal symbiosis (Varoquaux et al. 2019). It was found that, besides the loss of the fungal mass in roots, drought circumstances significantly depleted essential genes associated with arbuscular mycorrhizal symbiosis. The microbiome response—in the case of arbuscular mycorrhizal fungus—correlates with the genomic responses in the plant host, indicating that drought limits the symbiotic interaction between sorghum and these fungi (Varoquaux et al. 2019). In another study, Maier et al. (2021) used a combination of metabolomics and transcriptomics to identify a unique molecular response known as the general non-self-response (GNSR). GNSR and particular metabolite properties are being investigated as a result of their discovery of a substantial association between transcription and metabolic responses as well as a possible relationship between GNSR and secondary metabolism. Abiotic and biotic stress involvement by GNSR demonstrates the unrealized potential of plant microbes. *Medicago truncatula's* arbuscular mycorrhizal symbiosis is governed by regulatory circuits and transcriptome signatures that have been uncovered by Mohammadi-Dehcheshmeh et al. (2018) using an integrative framework of promoter analysis, machine learning, and meta-analysis in another extended investigation. A comprehensive method was used to identify important transcription factor genes, such as MTR_6g029180, MTR_1g098300, MTR_5g031160, MTR_3g079620, and MTR_3g045440, that are associated with arbuscular mycorrhizal colonization. Furthermore, the combination of metagenomics and metatranscriptomics with machine learning techniques presents chances to create new microbial biomarkers for ecological monitoring tasks, such as detecting heavy metal buildup in soil. Starr et al. (2019) discovered using metatranscriptomics techniques that RNA viruses mostly invade fungus in grassland environments. This is especially helpful because it has been discovered that some RNA viruses can infect pathogenic fungus and induce them to convert to a nonvirulent form (Koskella and Taylor 2018; Pratama et al. 2020; Zhang et al. 2020; Koskella et al. 2022).

4.3 Metabolomics

The study of cellular metabolism's intermediates and tiny molecular products is known as metabolomics. Twenty percent of the carbon that plants synthesize is used by them to produce molecules taken from their roots. This process is termed as the "rhizosphere effect," and it encourages the production of unique root microbiota from the soil (Stringlis et al. 2018). The transfer of metabolites between the microbiota and those of the plant hosts is a crucial mechanism that regulates the involvement, communication, and change of these symbiotic partners (Geier et al. 2020; Aharoni et al. 2023). It has been demonstrated that the rhizosphere microbiome is selectively modulated by metabolites emitted by roots. These metabolites are made up of a diverse array of low-molecular-weight chemicals, which can be classified as secondary (terpenoids, phenols, coumarins, etc.) or primary (organic acids, carboxylates, amino acids, sugars, etc.) (Chai and Schachtman 2022). These substances function as sources of carbon and nitrogen needed for microbial development. The types of bioactive metabolites and their content range greatly throughout plant species. The formation of specific communities of micro-rhizosphere in various plant species is mostly caused by root exudates. The root exudates are essential for controlling the microbiota in the rhizosphere, promoting interactions with helpful microorganisms, and inhibiting pathogens. For example, 2, 4-dihydroxy-7-methoxy-2H-1,4-benzoxazin-3(4H)-one, a significant secondary metabolite found in maize root exudates, is produced when benzoxazinoid is broken down. It functions as a potent allelochemical against nearby neighboring plants and soil bacteria in addition to acting as a chemical lure for beneficial plant growth-promoting microbes (Neal et al. 2012). Additionally, it has been discovered that 6-methoxybenzoxazolin-2-one, a different breakdown product of benzoxazinoid compounds, provides resistance against a dangerous pest that feeds on maize leaves (*Spodoptera frugiperda*). A basic concern in plant microbiome studies is how the communities and activities of the rhizosphere microbiome react to the expulsion of chemical nutrients and discharge from host plants. Currently, metabolomics education would be very crucial for studying the use or application of plant microbiomes and generally host-microbiome interactions (Hu et al. 2020; Geier et al. 2020). Huang et al.'s extensive investigation from 2019 showed that specific triterpenes produced by *Arabidopsis* plants regulate and assemble the microbiota in the roots of the plants. The scientists discovered distinct metabolic networks, composed of various triterpene biosynthesis gene clusters, expressed only in the roots of *Arabidopsis* by the use of metabolomics, metagenomics, and mutant lines. For example, an antibiotic coumarin was found to specifically modify the root zone microbial population in *Arabidopsis* through the use of metabolite fingerprinting in conjunction with shotgun metagenome sequencing (Stringlis et al. 2018). Scopoletin, a coumarin, has the ability to specifically inhibit *Fusarium oxysporum* and *Verticillium dahlia*, two plant fungal diseases. However, *Pseudomonas capeferrum* WCS358 and *Pseudomonas simiae* WCS417, the strains that produce systemic resistance and promote development, demonstrated a significant tolerance to the antimicrobial activities of scopoletin. As an illustration, consider the recently proposed "panomics" techniques, which combine system modeling, post-translational modifications, proteomics, metabolomics, and epigenomics in investigations of plant-microbe interactions.

5 Linking the Metabolome and Genome Directly

As noted above, defining the true size of the metabolome is a daunting task. This was to be borne out by efforts to do the same for *E. coli*, which produced a complement of some 750 metabolites, although the metabolome of that species was probably at least more than two times in size. In sharp contrast to proteomics, this case study clearly demonstrates that at least for the most part, the metabolome is genome-independent. Both horizontal and allosteric controls further complicate this link; the former suggests a relationship between transcripts and metabolites, while the latter is regulated by effector molecules that can be difficult to evaluate. In fact, metabolomics in application within a multiomics approach in plants has been instrumental

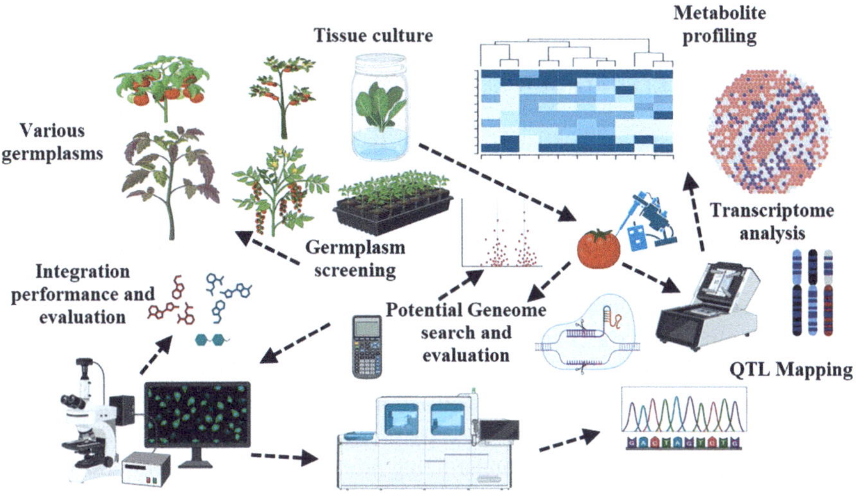

Fig. 2 Typical representation of a plant genome-wide association study. The investigated population consists of natural variants that could hold sufficient genetic variation to allow metabolomics per se or in combination with transcriptomics. These studies do not only give information about gene function but also the identification of key gene-trait correlations that are predictive of a plant's capacity to bear biotic or abiotic stresses, yield of crops, or susceptibility to disease

in identifying important genes for future breeding purposes, besides aiding in the comprehension of the effects of domestication on the metabolome (Fig. 2). On the other hand, because it frequently identifies the gene that causes accumulation of a certain metabolite, it can be very helpful in the identification of undiscovered genes. It should be noted that in the biological function context, the vast majority of more than a million metabolites remain unknown. Two other approaches, however, warrant mention. In one, armed with knowledge of enzyme abundance, investigators paired quantitative proteomics in yeast kinase knockouts with machine learning to predict the metabolome of kinase-deficient cells. It had also shown, other than the very specific signature of the enzyme abundance of the different kinases, that it could model levels of amino and organic acids, adenosine triphosphate, adenosine diphosphate, AMP, adenosine monophosphate, and glycolytic and pentose phosphate intermediates. A second means is the incorporation of genome-scale metabolic models, which could further narrow down the search space for unknown metabolites or metabolite features in MS that are currently not yet annotated. One obtains this information through computational methods, literature-based research, or jamborees. Since more thorough genome-scale modelling of extracellular metabolism in *Streptomyces* species has been produced, this technique has already been used to support systems metabolic engineering efforts. Further, the method was applied to the predictive evolution of metabolic pathway exploration for lineages having elevated secretion of metabolites.

6 Designing Aspects of Metabolomics for PGPM Profiling

6.1 Sample Preparation

Proper sample preparation is essential for developing metabolomics strategies aimed at profiling plant growth-promoting microorganisms (PGPM) (Miller et al. 2019). This process starts with careful collection and preservation of samples to avoid any metabolic changes, often utilizing rapid freezing or stabilization methods. Choosing the right extraction techniques is critical to effectively capture a wide array of microbial metabolites; common solvent systems such as methanol, ethanol, or their mixtures with water are usually optimized for optimal metabolite recovery (Tian et al. 2022). Techniques for homogenization must ensure complete disruption of microbial cells while protecting sensitive compounds from degradation. Isolating PGPM from plant tissues may be necessary to reduce background interference and improve metabolite specificity. Standardizing protocols is crucial to minimize variability and ensure consistency across samples. The addition of internal standards and quality controls significantly enhances the reliability of the metabolomics data. Ultimately, customized sample preparation protocols support precise and thorough metabolite profiling, allowing for a deeper understanding of PGPM roles and interactions within the plant microbiome (Miller et al. 2019).

6.2 Analytical Techniques

Analytical techniques are essential for thoroughly profiling the metabolome, enabling a detailed understanding of the biochemical dynamics within biological systems. Prominent methods include mass spectrometry (MS) and nuclear magnetic resonance (NMR) spectroscopy. Mass spectrometry, frequently combined with chromatography methods like gas chromatography (GC-MS) and liquid chromatography (LC-MS), delivers exceptional sensitivity and resolution for the detection and quantification of metabolites in a variety of biological samples (Patel et al. 2021). Although NMR spectroscopy is generally less sensitive, it provides unmatched reproducibility and the capability to analyze samples without destruction. Advanced techniques, such as tandem mass spectrometry (MS/MS), facilitate the structural elucidation of metabolites, thereby enhancing pathway analyses. Emerging technologies, including high-resolution mass spectrometry (HRMS) and ion mobility spectrometry (IMS), offer further improvements in metabolite characterization by distinguishing isobaric compounds and refining separation processes (Camunas-Alberca et al. 2023). Within metabolomics workflows, untargeted strategies aim for extensive coverage of metabolites, whereas targeted approaches concentrate on measuring specific metabolites of interest. Data acquired from these techniques undergo processing through bioinformatics tools for tasks such as peak detection, alignment, and identification, integrating results into biological pathways. Ultimately, these analytical platforms are crucial for elucidating metabolic fluxes, identifying biomarkers, and uncovering mechanisms of disease in systems biology.

6.3 Data Analysis and Interpretation

The analysis of plant growth-promoting microorganisms (PGPM) entails a methodical examination of data to reveal the biochemical interactions that affect plant growth and well-being. This analysis starts with sophisticated techniques such as mass spectrometry (MS) or nuclear magnetic resonance (NMR) to identify and measure the metabolites that PGPM produce (Singh et al. 2023). The resulting high-dimensional datasets undergo thorough normalization and analysis using statistical methods like principal component analysis (PCA) or partial least squares-discriminant analysis (PLS-DA) to uncover meaningful metabolic patterns and biomarkers (Martínez-Sena et al. 2023). Subsequently, bioinformatics platforms are utilized to connect these metabolites with specific biochemical pathways, enriching our understanding of PGPM's functional roles in nutrient cycling, stress resilience, and pathogen defense. Moreover, combining this data with genomic and transcriptomic information provides a systems biology viewpoint, associating metabolite profiles with gene expression and regulatory processes. Ultimately, this analysis enhances our grasp of the intricate molecular interactions between PGPM and host plants, offering essential insights for refining agricultural practices, boosting crop resilience, and promoting sustainable farming methods.

6.4 Integration with Other Omics

Integrating various omics disciplines, including genomics, transcriptomics, and proteomics, enriches the profiling of plant growth-promoting microorganisms (PGPM) in metabolomics by providing a comprehensive view of microbial roles and plant-microbe interactions (Sharma et al. 2020). This unified approach aids in the identification of critical metabolic pathways, biomarkers, and regulatory networks, shedding light on the contributions of PGPMs to nutrient acquisition, stress resilience, and overall plant vitality. Such an in-depth analysis promotes advancements in sustainable agriculture and precision microbiome management practices.

6.5 Metabolite Engineering

Metabolite engineering refers to the intentional alteration of metabolic pathways in plants to improve the synthesis of both primary and secondary metabolites vital for promoting growth and enhancing stress resistance (Hartline et al. 2021). Within the framework of profiling plant growth-promoting microbes (PGPM), metabolomics emerges as an effective method for detecting and quantifying the microbial metabolites that impact plant health. Through investigating the metabolic changes triggered by PGPM, researchers can uncover essential biochemical pathways and pinpoint bioactive compounds that drive plant-microbe interactions (Hartline et al. 2021). This method supports the enhancement of microbial strains or the development of metabolic engineering techniques aimed at increasing agricultural productivity and sustainability while also minimizing reliance on chemical fertilizers and pesticides.

6.6 Field Application and Validation

Field data through metabolomics approaches or data-driven methods to examine biochemical pathways, metabolic fluxes, and biomarkers are extremely useful. By combining PGPM with metabolomics data, forecast on phenotypic characteristics, evaluate the influence of environmental conditions, and enhance metabolic engineering techniques (Miller et al. 2019). The validation process requires thorough testing with experimental datasets to guarantee accuracy and dependability. The uses of these models extend across agriculture, healthcare, and biotechnology, facilitating improvements in crop yield, disease detection, and drug development (Al-Turki et al. 2023). Data-driven models serve as a link between intricate metabolic networks and practical applications, underscoring their significant impact on systems biology and precision medicine.

7 Importance in Understanding Plant-Microbe Interactions

Understanding plant-microbe interactions is crucial for several reasons, ranging from enhancing agricultural productivity to maintaining ecosystem health. Here are key points highlighting the importance of this field of study:

7.1 Enhancing Agricultural Productivity

7.1.1 Improved Crop Yields

Interactions between plants and microbes are essential for sustainably increasing crop yields. Beneficial microorganisms, including rhizobacteria and mycorrhizal fungi, enhance nutrient availability, support plant growth, and strengthen resistance against pests and environmental stresses (Poudel et al. 2021). Nitrogen-fixing bacteria enhance soil fertility by transforming atmospheric nitrogen into absorbable forms for plants. Mycorrhizal fungi extend root systems, improving the absorption of water and nutrients. Additionally, certain microbes produce phytohormones and antimicrobial substances that enhance plant health and decrease dependence on chemical fertilizers. By leveraging these natural interactions, farmers can achieve greater yields while fostering environmentally friendly agricultural methods. It allows for the development of biofertilizers and biostimulants, decreasing the need for chemical fertilizers (Kaur and Kaur 2023).

7.1.2 Biological Pest and Disease Control

Interactions between plants and microbes are essential for the natural management of pests and diseases. Beneficial microbes, including rhizobacteria and fungi, generate bioactive substances that either inhibit pathogens or outcompete harmful organisms. Certain microbes trigger systemic resistance in plants, boosting their capacity to fend off pests and diseases. For instance, *Trichoderma* fungi safeguard plants by producing antifungal compounds (Pattanayak et al. 2021), while *Bacillus* species secrete enzymes that break down pathogen cell walls (Iqbal et al. 2023). These interactions decrease the dependency on chemical pesticides, fostering sustainable agriculture and reducing environmental harm. Utilizing plant-microbe symbiosis presents an eco-friendly approach to improving crop health and productivity.

7.2 Sustainable Agriculture

7.2.1 Reduced Chemical Inputs

Minimizing chemical inputs in sustainable agriculture means decreasing reliance on synthetic fertilizers, pesticides, and herbicides to safeguard both the environment and human health (Getahun et al. 2024). This method fosters natural soil fertility, promotes biodiversity, and maintains ecosystem balance. Techniques such as crop rotation, integrated pest management, and the application of organic amendments like compost aid in lessening the dependence on chemicals while still ensuring productivity. By implementing biological pest control and precision farming practices, farmers can enhance resource efficiency, cut costs, and lower pollution levels (Zhou et al. 2024). In the end, reducing chemical inputs strengthens resilient farming systems, contributing to long-term sustainability and healthier food production.

7.2.2 Soil Health and Fertility

Microbes are essential for the cycling of nutrients and the simplification of organic matter, which enhances soil fertility and structure. Comprehending these mechanisms facilitates the sustainable management of soil health, sustaining agricultural yield over a period of time.

7.3 Plant Stress Tolerance

7.3.1 Abiotic Stress Resistance

Plants exhibit complex metabolic adjustments to withstand abiotic stressors such as drought, salinity, and extreme temperatures (Verslues et al. 2023). Metabolomics provides valuable insights into how plants respond to these challenges by identifying changes in metabolites triggered by stress, including osmolytes like proline and glycine betaine, antioxidants such as flavonoids and ascorbate, and stress hormones like abscisic acid (Poudel et al. 2021). The increased production of secondary metabolites, including phenolics and alkaloids, helps to reduce oxidative damage and maintain cellular stability. Through metabolomics profiling, researchers can identify biomarkers associated with stress tolerance and uncover the strategies that plants employ for survival.

7.3.2 Biotic Stress Resistance

The resistance of plants to biotic stress is intricately connected to their metabolomics changes, which encompass fluctuations in both primary and secondary metabolites (Al-Khayri et al. 2023). These metabolites serve essential functions in defending against pathogens, pests, and herbivores. For example, compounds such as phenolics, alkaloids, and terpenoids exhibit antimicrobial properties or function as deterrents. Moreover, phytohormones, including jasmonic acid and salicylic acid, play a crucial role in regulating defense signaling pathways (Luo et al. 2019). Metabolic alterations induced by stress also boost the production of antioxidants, which help to reduce oxidative damage. Recent advancements in plant metabolomics have shed light on these biochemical responses, facilitating the identification of traits linked to stress resistance and contributing to crop enhancement for sustainable agricultural practices.

7.4 Ecological Balance and Biodiversity

7.4.1 Ecosystem Stability

The stability of ecosystems is vital for preserving ecological balance and biodiversity. It encompasses an ecosystem's capacity to withstand, bounce back from, or adjust to disturbances like natural disasters, human interventions, or shifts in climate. Healthy ecosystems sustain a wide variety of species, facilitating ongoing energy flow, nutrient cycling, and ecological relationships (Freschet et al. 2021). Biodiversity is crucial for stability, as a diverse array of species strengthens the ecosystem's ability to adapt to changes. Safeguarding ecosystem stability is imperative for maintaining natural equilibrium and ensuring that resources remain sustainable for generations to come (Prescott et al. 2021).

7.4.2 Restoration Ecology

Restoration ecology is dedicated to rehabilitating and revitalizing ecosystems that have been degraded, with the goal of restoring ecological balance and enhancing biodiversity (McGeoch et al. 2024). This field seeks to undo the harm inflicted by human activities, such as deforestation, pollution, and habitat destruction. By reinstating native species, reviving natural processes, and improving habitat conditions, restoration ecology strives to return ecosystems to their original states. This area of study is crucial for preserving biodiversity, boosting ecosystem services, and alleviating the effects of climate change. Initiatives can vary from reforestation and wetland restoration to the removal of invasive species. Restoration ecology prioritizes long-term sustainability, promoting resilience within ecosystems to benefit both wildlife and human populations (Alexander et al. 2016).

7.5 Advancements in Biotechnology

7.5.1 Genetic Engineering

Genetic engineering represents a significant breakthrough in biotechnology, allowing for the precise alteration of an organism's DNA to change its traits (Gill et al. 2014). This innovation has transformed various sectors, including agriculture, medicine, and environmental science. By altering or inserting specific genes, researchers can develop genetically modified crops that offer greater yields, enhanced pest resistance, or better nutritional qualities. Genetic engineering within the field of metabolomics entails the alteration of genes to investigate and modify metabolic pathways (Liu et al. 2023). This technique allows scientists to change enzyme activities, enhance the production of metabolites, and comprehend cellular reactions. Its applications span drug development, agriculture, and biotechnology, facilitating the creation of superior crops, biofuels, and focused treatments for metabolic conditions. Additionally, genetic engineering has the potential to tackle environmental issues by engineering organisms that can decompose pollutants or produce sustainable biofuels.

7.5.2 Synthetic Biology

Synthetic biology is a cutting-edge field within biotechnology that combines biology, engineering, and computer science to design and construct new biological parts, devices, and systems or to re-engineer existing ones for useful purposes. By leveraging tools like CRISPR, DNA synthesis, and computational modeling, synthetic biology enables advancements such as biofuel production, personalized medicine, synthetic vaccines, and sustainable agriculture (Usman 2024). It holds potential for addressing global challenges, including climate change, food security, and healthcare innovation. This interdisciplinary approach is transforming how we understand and manipulate life at a molecular level, driving progress in science and industry.

7.6 Economic and Social Impact

7.6.1 Farmer Livelihoods

Agriculture serves as the foundation of rural economies and has a direct influence on the livelihoods of farmers. Variations in agricultural policies, market accessibility, and climatic conditions have a considerable impact on income stability and food security (Zheng et al. 2022). Smallholder farmers frequently encounter difficulties, including variable crop prices, restricted access to credit, and increasing input expenses. To enhance productivity and resilience, it is vital to invest in sustainable farming methods, fair market systems, and rural infrastructure. Supporting the livelihoods of farmers is essential for alleviating poverty and achieving wider economic and social development objectives.

7.6.2 Food Security

Food security is essential for maintaining economic and social stability (Wahbeh et al. 2022). It guarantees that everyone has access to an adequate supply of safe and nutritious food, which promotes health and productivity. A deficiency in food security can worsen issues like poverty, malnutrition, and social inequality, ultimately obstructing economic growth and community cohesion. Global food systems face considerable challenges from climate change, population growth, and disruptions in supply chains (Farooq et al. 2022). Tackling food security requires the implementation of sustainable agricultural methods, fair distribution practices, and innovative technologies to enhance resilience.

8 Profiling Plant Growth-Promoting Microbes (PGPM)

8.1 Identifying Key Metabolites Produced by PGPM

Identifying key metabolites produced by plant growth-promoting microbes (PGPM) is essential for understanding and harnessing their beneficial effects on plants. PGPMs secrete a variety of metabolites that play important role in enhancing plant growth, improving nutrient uptake, and protecting against pathogens and environmental stress. Among the most significant metabolites are phytohormones such as indole-3-acetic acid (IAA), which triggers the growth and elongation of the root, and gibberellins, which stimulate stem elongation and seed germination (Duca et al. 2014). Siderophores are another vital group of metabolites; these compounds chelate iron from the soil, making it more accessible to plants, thus aiding in their growth and development. Additionally, PGPMs produce antibiotics and antifungal compounds that inhibit pathogenic microbes, thereby reducing plant diseases and promoting healthier growth. Volatile organic compounds (VOCs) emitted by PGPMs are also critical as they can improve plant growth by inducing systemic resistance or directly stimulating plant metabolism (Russo et al. 2022). For instance, compounds like acetoin and 2,3-butanediol have been shown to trigger systemic resistance in plants against a range of pathogens. Exopolysaccharides (EPS) are another class of important metabolites that improve soil structure by promoting soil aggregation, enhancing water retention, and facilitating better root colonization (Cania et al. 2019). Identification of these key metabolites requires state-of-the-art analytical techniques like mass spectrometry, nuclear magnetic resonance spectroscopy, and chromatography. Ultimately, the identification and application of key metabolites produced by PGPMs hold great promise for sustainable agriculture, offering eco-friendly solutions to improve crop yield, resilience, and soil health.

8.2 Techniques for Metabolite Extraction and Analysis

Metabolite extraction and analysis techniques are fundamental in metabolomics, enabling the detailed study of metabolic profiles in various biological systems.

Aforesaid process begins with sample preparation, which involves selecting appropriate tissues or fluids and employing specific methods to quench metabolic activity, such as rapid cooling or the addition of solvents. Common extraction methods include solvent extraction, solid-phase extraction (SPE), and liquid-liquid extraction (LLE). Solvent extraction typically uses solvents like methanol, acetonitrile, or chloroform to dissolve metabolites from biological matrices (Aly and Górecki 2020). SPE involves passing a sample through a solid-phase sorbent to selectively retain and elute metabolites, while LLE separates metabolites based on their differential solubility in two immiscible solvents. Once extracted, metabolites are analyzed using various advanced analytical techniques. Chromatography, particularly gas chromatography (GC) and liquid chromatography (LC), is widely used to separate complex mixtures of metabolites based on their chemical properties (Abdulhussain et al. 2021). GC is appropriate for both volatile and thermally stable substances, whereas LC is ideal for a broader range of metabolites, including nonvolatile and thermolabile compounds. These chromatographic techniques are often coupled with mass spectrometry (MS), which provides high sensitivity and specificity for metabolite identification and quantification. MS detects and measures the mass-to-charge ratios of ionized metabolites, offering detailed structural information. Another powerful technique is nuclear magnetic resonance spectroscopy, giving all the information on molecular structure and dynamics of metabolites. NMR is highly reproducible and non-destructive, making it suitable for comprehensive metabolomics studies (Letertre et al. 2021).

8.3 Role of Metabolomics in Characterizing Beneficial Microbes

Metabolomics plays a pivotal role in characterizing beneficial microbes, offering deep insights into their metabolic functions and interactions with host organisms (Winder et al. 2022). This would provide an overall understanding of the biochemical activities underlying the beneficial effects exerted by microbes through the analysis of the complete set of produced metabolites. This holistic approach enables researchers to identify specific metabolites responsible for promoting plant growth, enhancing nutrient uptake, and providing resistance against pathogens. For instance, through metabolomics profiling, key metabolites such as indole-3-acetic acid (IAA), gibberellins, and siderophores have been identified, which are crucial for root development, seed germination, and iron acquisition, respectively (Lenin and Jayanthi 2012). To accurately and sensitively identify and measure these metabolites, sophisticated analytical methods such as nuclear magnetic resonance (NMR) spectroscopy and mass spectrometry (MS) are used.

9 Metabolomics Techniques

Mass Spectrometry (MS) in Metabolite Detection

Mass spectrometry is one of the power-packed analytical techniques applied in detecting, identifying, and quantifying metabolites in complex biological samples (Yan et al. 2022). The process involves the ionization of chemical compounds into charged molecules or fragments of molecules and their mass-to-charge ratio measurement. Here's a detailed overview of the technique:

9.1 Sample Preparation

Proper sample preparation is critical for accurate MS analysis (Tamara et al. 2022). The steps typically include the following:

- *Extraction*: Metabolites are extracted from biological matrices (e.g., tissues, cells, biofluids) using solvents such as methanol, acetonitrile, or water. The choice of solvent depends on the polarity and solubility of target metabolites.
- *Quenching*: Rapidly halting metabolic activity to preserve the current state of metabolites, often achieved by adding cold solvents or using liquid nitrogen.
- *Filtration/centrifugation*: Removing particulates to prevent interference with the MS analysis.
- *Concentration*: Reducing the volume of the sample to increase metabolite concentration, if necessary, using techniques like evaporation or lyophilization.

9.2 Ionization Techniques

Ionization is the process of converting metabolites into ions (Müller et al. 2022). Common ionization methods include the following:

- *Electrospray ionization (ESI)*: Soft ionization technique suitable for large, polar, and non-volatile molecules. By creating an aerosol out of a liquid and applying a high voltage to it, ions are produced.
- *Matrix-assisted laser desorption/ionization (MALDI)*: Suitable for analyzing large biomolecules like proteins and polymers. It uses a laser energy-absorbing matrix to create ions (Zhu et al. 2022).
- *Gas chromatography-mass spectrometry (GC-MS) ionization*: Typically involves electron impact (EI) ionization, where high-energy electrons ionize gas-phase molecules. Suitable for volatile and thermally stable compounds (Marcillo et al. 2023).

9.3 Mass Analyzers

The mass analyzer separates ions based on their mass-to-charge ratios (Yan et al. 2022). Different types of mass analyzers include the following:

- *Quadrupole*: In brief, uses oscillating electric fields and sorts the ions according to their mass-to-charge ratio; it is a strong, easy technique and hence finds wide application.
- *Time of flight (TOF)*: It measures the time of flight of ions over a fixed distance, related to their mass-to-charge ratio. It provides high resolution and mass accuracy.
- *Orbitrap*: Traps ions in an electrostatic field and measures their oscillations, offering high resolution and mass accuracy.
- *Ion trap*: Captures ions in a three-dimensional electrostatic field, allowing for multiple stages of mass spectrometry (MS/MS).

9.4 Data Acquisition and Analysis

The generated data includes a spectrum of ion signals that represent the metabolites in the sample (Misra 2021). Key steps in data handling are the following:

- *Peak detection*: Identifying peaks in the mass spectrum corresponding to different metabolites.
- *Deconvolution*: Resolving overlapping peaks to correctly identify individual metabolites.
- *Quantification*: The height or area for each peak is determined to provide the relative concentration of metabolites.
- *Database searching*: Comparing the mass spectra to known databases (e.g., METLIN, HMDB) to metabolites can be distinguished using their mass-to-charge ratios and fragmentation patterns.
- *Statistical analysis*: Using bioinformatics tools to interpret complex data, identify significant changes in levels of metabolite, and correlate them with biological conditions or treatments.

9.5 Applications in Metabolomics

MS is extensively used in various fields of metabolomics, including the following:

- Biomarker discovery: The process of locating metabolites that may be used as markers of physiological or pathological states (Katam et al. 2022).
- Pathway analysis: Knowing the modifications made to metabolic pathways in various situations.

- Analyzing how diet affects metabolic profiles is the focus of nutritional studies (Tamara et al. 2022).
- Environmental studies: Determining how the environment affects metabolism.
- Agricultural research: Examining how microbes interact with plants and how their metabolism changes in response to different treatments.

One effective and flexible tool in metabolomics is mass spectrometry, providing detailed insights into the metabolic composition of biological samples, which is essential for advancing our understanding of biochemical processes in health, disease, and the environment.

9.5.1 Nuclear Magnetic Resonance (NMR) Spectroscopy Applications

In metabolomics, NMR spectroscopy is among the most applied non-destructive analytic techniques used in identifying and quantifying metabolites within intricate biological materials. NMR spectroscopy detects the magnetic properties of some atomic nuclei to give information on molecular dynamics, environment, and structure at the atomic level (Stryeck et al. 2017). This paper gives an in-depth review of the application of NMR spectroscopy in metabolomics.

9.5.1.1 Principles of NMR Spectroscopy

NMR spectroscopy is based on the magnetic properties of nuclei, particularly those of hydrogen (^{1}H) and carbon (^{13}C), which are commonly used in metabolomics (Peng et al. 2024). When placed in a strong magnetic field, these nuclei resonate at characteristic frequencies when exposed to a radiofrequency pulse. The resulting NMR signals provide information about the chemical environment of the nuclei, which can be interpreted to deduce molecular structures and concentrations.

9.5.1.2 Sample Preparation

Proper sample preparation is crucial for obtaining high-quality NMR spectra:

- *Sample type*: Common samples include biofluids (e.g., blood serum, urine), tissue extracts, and cell lysates (Emwas et al. 2019).
- *Solvents*: Deuterated solvents (e.g., D_2O, deuterated chloroform) are used to avoid interference from solvent protons in the NMR spectra.
- *pH adjustment*: The pH of the sample may need to be adjusted and buffered to ensure reproducibility and stability of the metabolites.
- *Concentration*: Samples are often concentrated to achieve sufficient signal-to-noise ratio. The typical volume for NMR analysis is 0.5–1 mL.

9.5.1.3 NMR Data Acquisition

Key steps and parameters in NMR data acquisition include the following:

- *Magnetic field strength*: Higher magnetic field strengths (e.g., 600 MHz, 800 MHz) provide better resolution and sensitivity.

- *Pulse sequences*: Different pulse sequences are used to obtain various types of information. Common sequences in metabolomics include ^{1}H NMR, ^{13}C NMR, and ^{2}D NMR (e.g., COSY, HSQC, TOCSY).
- *Number of scans*: Increasing the number of scans improves the signal-to-noise ratio but also increases the acquisition time.
- *Temperature control*: Samples are typically maintained at a fixed temperature (e.g., 25 °C) to ensure stability and reproducibility.

9.5.1.4 Data Processing and Analysis

NMR data processing involves several steps:

- *Fourier transform*: Converts the time-domain NMR data into frequency-domain spectra.
- *Phase and baseline correction*: Adjusts the spectra for accurate peak identification.
- *Peak assignment*: Identifies peaks corresponding to specific metabolites using chemical shift databases and reference spectra.
- *Quantification*: Measures peak areas to quantify metabolite concentrations. Internal standards or known reference compounds are often used for accurate quantification.

9.6 Advantages of NMR in Metabolomics

- *Non-destructive*: NMR is non-destructive, allowing for repeated analysis of the same sample.
- *Quantitative*: NMR provides absolute quantification of metabolites without the need for external calibration.
- *Reproducibility*: High reproducibility and reliability in metabolite identification and quantification.
- *Structural information*: Provides detailed structural information, aiding in the identification of unknown metabolites.

9.7 Challenges and Limitations

- *Sensitivity*: NMR is less sensitive compared to mass spectrometry, requiring higher sample concentrations (Peng et al. 2024).
- *Complexity of spectra*: Overlapping peaks in complex mixtures can complicate spectral interpretation.
- *Cost and accessibility*: High-field NMR instruments are expensive and require specialized maintenance and operation.

NMR spectroscopy is a vital tool in metabolomics, offering unique advantages in the comprehensive analysis of metabolites.

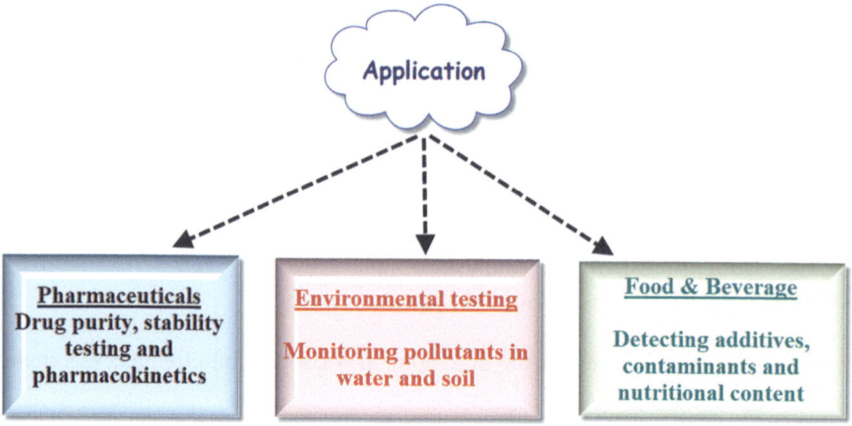

Fig. 3 Applications in diverse sectors

9.8 High-Performance Liquid Chromatography (HPLC) for Compound Separation

High-performance liquid chromatography is an instrumental method of analysis that allows separating, identification, and quantitative determination of different mixture components (Hirose et al. 2019). This technique features the exploitation of the following characteristics in a huge number of applications: pharmaceuticals, environmental tests, food and beverage testing, and clinical diagnosis to its high resolution, speed, and sensitivity. Application in different sectors is depicted under Fig. 3.

10 Plant-Microbe Interaction Dynamics

10.1 Metabolomics Pathways Involved in Plant Growth Promotion

Primary metabolism includes fundamental processes such as photosynthesis, respiration, and the synthesis of essential biomolecules like amino acids, nucleotides, and lipids. Photosynthesis is a critical pathway where light energy is converted into chemical energy, producing glucose and oxygen. The Calvin cycle, a part of photosynthesis, generates key intermediates like glyceraldehyde-3-phosphate, which serve as precursors for various biosynthetic pathways (Balcke et al. 2017). The tricarboxylic acid cycle (TCA cycle), glycolysis, and oxidative phosphorylation break down glucose during respiration to produce ATP, the cell's energy currency.

10.2 Secondary Metabolism

Secondary metabolites, though not directly involved in growth and development, play crucial roles in plant defense, signaling, and interactions with the environment. Alkaloids, flavonoids, terpenoids, and phenolic chemicals are a few of them. Flavonoids, for instance, are involved in UV protection and pigmentation, contributing to the plant's ability to attract pollinators (Chen et al. 2023) . Terpenoids, which include essential oils, serve as defense compounds against herbivores and pathogens. Moreover, phenolic compounds like lignin strengthen the plant cell wall, providing structural integrity and resistance to environmental stresses.

10.3 Hormone Biosynthesis

Plant hormones, or phytohormones, are central to regulating growth, development, and responses to environmental stimuli. Auxins, gibberellins, cytokinins, abscisic acid, and ethylene are important phytohormones. For root development, apical dominance, and cell elongation, auxins like indole-3-acetic acid (IAA) are essential. They modulate the expression of genes intrinsically involved in cell division and differentiation (Jahn et al. 2021). Gibberellins (GAs) promote stem elongation, seed germination, and flowering (Alferez et al. 2021). The biosynthesis of GAs involves complex pathways, including the conversion of geranylgeranyl diphosphate (GGPP) to bioactive GAs through multiple enzymatic steps. Cytokinins promote cell division and shoot formation, often working in conjunction with auxins to balance root and shoot growth (Zhao et al. 2024). The isopentenyl adenine pathway is key in cytokinin biosynthesis. Abscisic acid (ABA) is vital for stress responses, particularly in regulating stomatal closure to prevent water loss during drought conditions (Gupta et al. 2022a, 2022b). Ethylene, a gaseous hormone, regulates fruit ripening, leaf abscission, and response to mechanical stress.

10.4 Symbiotic and Microbial Interactions

Plant growth promotion is also significantly influenced by interactions with symbiotic microorganisms, such as mycorrhizal fungi and rhizobacteria. These microorganisms enhance nutrient uptake, particularly nitrogen and phosphorus, and induce systemic resistance against pathogens (Adomako et al. 2022). For instance, rhizobacteria like *Bacillus* and *Pseudomonas* species produce siderophores, organic compounds that chelate iron from the soil, making it more available to plants. They also synthesize phytohormones, such as IAA, and volatile organic compounds (VOCs) that modulate plant growth and stress responses.

10.5 Signaling Pathways

Plant signaling pathways integrate external and internal signals to modulate growth and development. Exogenous NO and ROS act as signaling molecules against biotic and abiotic stresses. Defense-related gene expression and the production of secondary metabolites are just two of the many downstream processes that are triggered by activated mitogen-activated protein kinases (Giri et al. 2019) and calcium-dependent protein kinases (Atif et al. 2019).

10.6 Metabolomics Analysis Techniques

Advanced analytical methods such as nuclear magnetic resonance (NMR) spectroscopy and mass spectrometry (MS) are used in the investigation of metabolomics pathways (Peng et al. 2024). These techniques enable the identification and quantification of metabolites, providing a comprehensive overview of metabolic changes under various conditions. Metabolomic pathways are integral to plant growth promotion, encompassing primary and secondary metabolism, hormone biosynthesis, and intricate signaling networks.

10.7 Impact of PGPM on Plant Metabolic Profiles

Plant growth-promoting microorganisms (PGPM) profoundly influence plant metabolic profiles through various mechanisms that enhance growth, stress tolerance, and overall plant health (Kaya et al. 2023). PGPMs, including bacteria, fungi, and actinomycetes, interact with plants in the rhizosphere, leading to significant changes in primary and secondary metabolism.

10.8 Primary Metabolism Enhancement

PGPMs promote primary metabolism by enhancing nutrient uptake. For instance, rhizobacteria like *Bacillus* (Santander et al. 2024) and *Pseudomonas* species fix atmospheric nitrogen and solubilize phosphates, increasing the availability of essential nutrients (Mehmood et al. 2023). This leads to increased photosynthetic activity, higher chlorophyll content, and improved carbohydrate synthesis. Enhanced nutrient acquisition also boosts amino acid and protein synthesis, which turn to constructive framework for plant growth and development.

10.9 Hormone Modulation

PGPMs influence plant hormonal balance by synthesizing phytohormones such as indole-3-acetic acid (IAA), gibberellins, cytokinins, and ethylene (Pandey et al.

2023). For example, IAA produced by rhizobacteria stimulates root elongation and branching, enhancing root surface area and nutrient uptake. Cytokinins promote cell division and shoot formation, while gibberellins facilitate stem elongation and seed germination (Zhao et al. 2024). Additionally, some PGPMs can modulate ethylene levels, reducing plant stress responses and promoting growth.

10.10 Secondary Metabolism and Stress Responses

PGPMs induce the production of secondary metabolites, which play crucial roles in plant defense and stress tolerance. These include phenolics, flavonoids, terpenoids, and alkaloids, which have antimicrobial properties and protect against pathogens and herbivores (Liu et al. 2023). For instance, mycorrhizal fungi can enhance the production of phenolic compounds, strengthening the plant cell wall and increasing resistance to pathogens.

10.11 Systemic Resistance

PGPMs can trigger induced systemic resistance (ISR) in plants, leading to an enhanced defensive state. This involves the accumulation of pathogenesis-related proteins and secondary metabolites that fortify the plant against future pathogen attacks. Signaling molecules like jasmonic acid and salicylic acid play key roles in this process, integrating external microbial signals with internal metabolic responses (Luo et al. 2019).

10.12 Metabolomic Shifts

Advanced metabolomic techniques, like MS and NMR spectroscopy, showed a considerable shift in metabolic profiling of the plant because of PGPM (Badillo-Sanchez et al. 2023). These shifts include alterations in the levels of primary metabolites like sugars and amino acid components, as well as secondary metabolites involved in defense and signaling. Such changes underscore the broad impact of PGPMs on plant metabolism, highlighting their potential in sustainable agriculture for improving crop growth, yield, and resilience to environmental stresses.

10.13 Case Studies of Successful PGPM Profiling

Profiling plant growth-promoting microorganisms (PGPM) has yielded several successful case studies demonstrating the benefits of PGPM in enhancing plant growth, nutrient uptake, and stress resistance. These case studies provide valuable insights into the practical applications of PGPM in agriculture.

10.13.1 Bacillus spp. in Tomato Cultivation

In a notable study, *Bacillus subtilis* and *Bacillus amyloliquefaciens* were applied to tomato plants (Zalila-Kolsi et al. 2023). The profiling revealed that these bacteria significantly enhanced root and shoot growth by producing indole-3-acetic acid (IAA), a crucial phytohormone. Moreover, they increased the availability of phosphorus and nitrogen in the soil. Metabolomics analysis showed elevated levels of essential amino acids and improved photosynthetic pigments, leading to higher fruit yield and quality. The study highlighted the potential of *Bacillus* spp. in sustainable tomato farming by eliminating the need for chemical fertilizers and enhancing crop resilience (Awan et al. 2023).

10.13.2 Mycorrhizal Fungi in Maize

According to the case study conducted on maize, inoculated with arbuscular mycorrhizal fungi, it indicates that it has improved significantly plant growth and nutrient uptake. Mycorrhizal inoculation improved phosphorus and other micronutrient absorption. Profiling the metabolites showed an increase in primary metabolites such as sugars and amino acids, as well as secondary metabolites like flavonoids and phenolics, which play a role in stress resistance (Dhaliwal et al. 2022). The symbiotic relationship also induced systemic resistance, enhancing the plant's ability to withstand pathogens and environmental stresses.

10.13.3 *Pseudomonas fluorescens* in Wheat

In wheat cultivation, the application of *Pseudomonas fluorescens* was profiled for its growth-promoting effects (Gruet et al. 2022). This bacterium was found to produce siderophores, which chelate iron from the soil and make it more accessible to the plant. The metabolomics profiling showed increased chlorophyll content and, therefore, increased the rate of photosynthesis. In addition, the presence of proline and betaines, stress-related metabolites that are known to be involved in drought tolerance, was also indicated. The use of *Pseudomonas fluorescens* in wheat cultivation highlighted its potential in improving crop yield and resilience under abiotic stress conditions (Wu et al. 2022).

10.13.4 *Azospirillum brasilense* in Rice

The application of *Azospirillum brasilense* in rice cultivation demonstrated notable improvements in plant growth and productivity (Guimarã et al. 2020). Profiling revealed that this bacterium fixed atmospheric nitrogen, enriching the soil and promoting plant growth. Metabolomics analysis showed higher levels of key metabolites such as amino acids and organic acids, which are vital for plant development. Additionally, *Azospirillum* enhanced root architecture, leading to better nutrient and water uptake. These case studies underscore the significant impact of PGPM on plant health and productivity.

11 Designing PGPM Using Metabolomics

11.1 Strategies for Engineering Microbes with Enhanced Growth-Promoting Traits

Engineering microbes with enhanced growth-promoting traits is a cutting-edge approach in agricultural biotechnology aimed at improving crop productivity and resilience. Several strategies are employed to optimize these microbes for better performance. One simple approach is genetic modification—introduction or overexpression of specific genes responsible for promotion of desired plant growth traits like nitrogen fixation, phosphate solubilization, and phytohormone production in microbial genomes. For instance, genes encoding for the production of indole-3-acetic acid (IAA), a crucial phytohormone, can be introduced into bacteria like *Bacillus* and *Pseudomonas* to enhance their ability to stimulate root growth (Fanai et al. 2024). Another strategy involves the use of synthetic biology tools to design and construct microbial consortia with complementary functions. These consortia can be engineered to perform multiple growth-promoting activities simultaneously, such as nitrogen fixation by *Azospirillum*, phosphate solubilization by *Bacillus*, and siderophore production by *Pseudomonas* (Timofeeva et al. 2023). This approach ensures a holistic enhancement of plant growth by addressing multiple nutrient requirements and stress factors. Metabolic engineering is also employed to optimize metabolic pathways within microbes to increase the production of beneficial compounds. CRISPR-Cas9 technology has revolutionized microbial engineering by allowing precise genome editing (Usman 2024).

11.2 Synthetic Biology Approaches to Metabolite Production

Synthetic biology has revolutionized metabolite production by enabling precise manipulation of biological systems to produce valuable compounds efficiently. By redesigning microorganisms such as bacteria, yeast, and algae, scientists can create biosynthetic pathways that convert simple substrates into complex metabolites. This approach leverages tools like CRISPR-Cas9 for genome editing, synthetic gene circuits for regulatory control, and metabolic engineering to optimize pathway fluxes (Usman 2024). For instance, the production of pharmaceuticals like artemisinin, originally derived from plants, has been significantly enhanced through engineered yeast strains. Similarly, synthetic biology facilitates the sustainable production of biofuels, biodegradable plastics, and specialty chemicals by optimizing the metabolic networks of host organisms (Han et al. 2022). Integrating computational models and machine learning further refines these processes, predicting the most efficient genetic modifications and environmental conditions. Despite challenges like metabolic burden and pathway bottlenecks, ongoing advancements in synthetic biology continue to push the boundaries of metabolite production, promising a future where custom-designed biological systems can meet diverse industrial and medical needs (Dhaliwal et al. 2022).

11.3 Examples of Metabolomics-Driven PGPM Design

Metabolomics-driven plant growth-promoting microbe (PGPM) design leverages comprehensive metabolic profiling to enhance agricultural productivity and sustainability. By thoroughly checking the complete set of metabolites in both plants and their associated microbes, researchers can identify key biochemical interactions that promote plant health and growth. For instance, *Pseudomonas fluorescens* has been optimized using metabolomics to produce specific siderophores that enhance iron uptake in plants, boosting their growth in iron-deficient soils (Gruet et al. 2022). Another example is the design of *Bacillus subtilis* strains that produce tailored phytohormones like auxins and cytokinins, which are crucial for root development and stress resilience (Zalila-Kolsi et al. 2023). In a notable case, a strain of *Trichoderma harzianum* was engineered to produce secondary metabolites that induce systemic resistance in plants against pathogens, effectively functioning as a biocontrol agent (Xiao et al. 2023). Furthermore, combining metabolomics data with genomic and proteomic insights allows for the creation of multi-functional PGPMs that address various plant needs simultaneously, from nutrient acquisition to stress tolerance and disease resistance. These advances in metabolomics-driven PGPM design are paving the way for more efficient, sustainable, and eco-friendly agricultural practices.

12 Applications in Agriculture

12.1 Enhancing Crop Yield and Stress Resistance

Enhancing crop yield and stress resistance through metabolomics studies represents a cutting-edge approach in modern agriculture. Metabolomics, the comprehensive analysis of metabolites within an organism, provides critical insights into the biochemical pathways that underpin plant growth, development, and responses to environmental stressors. By identifying specific metabolites associated with high yield and stress tolerance, researchers can pinpoint key metabolic pathways and regulatory networks that can be targeted for crop improvement. One of the foremost applications of metabolomics in crop enhancement is the identification of biomarkers for stress resistance (Aina et al. 2024). For example, metabolomics profiling has revealed that certain amino acids, such as proline, accumulate in plants under drought conditions, contributing to osmotic balance and stress protection. By understanding these metabolic adjustments, scientists can breed or engineer crops with enhanced capacity to produce these protective compounds, thereby improving their resilience to water scarcity. Enhanced production of these compounds through metabolic engineering or selective breeding can lead to crops with better innate resistance, reducing the need for chemical pesticides (Yin et al. 2024). Integration of metabolomics data with genomic and proteomic information allows for a holistic understanding of plant physiology. This systems biology approach accelerates the identification of genetic loci and regulatory mechanisms that can be manipulated to enhance both yield and stress resistance. Consequently, metabolomics studies are crucial for developing next-generation crops that are more productive, resilient, and environmentally sustainable, addressing the challenges of food security at global level and climate change (Farooq et al. 2022).

12.2 Reducing Dependence on Chemical Fertilizers and Pesticides

Reducing dependence on chemical fertilizers and pesticides is crucial for fostering sustainable agriculture and mitigating environmental damage. Chemical fertilizers, while effective at boosting crop yields, contribute to soil fertility degradation, water pollution, and greenhouse gas emissions (Kaur and Kaur 2023). Pesticides, similarly, can harm non-target species, including beneficial insects, and lead to the development of resistant pest populations. For these reasons, integrated pest management and organically produced strategies are increasingly being put into action. IPM combines biological, cultural, physical, and chemical tools to lessen the environmental and economic toxicity of pests (Khokhar et al. 2024). By utilizing natural predators, crop rotation, and resistant crop varieties, IPM reduces the need for chemical interventions. For example, introducing ladybugs to control aphid populations and employing pheromone traps to disrupt pest mating cycles are effective IPM strategies that minimize pesticide use. Bt crops, which produce a bacterial toxin harmful to specific pests, significantly lower the need for insecticides. Similarly, biofertilizers and biopesticides derived from natural organisms or substances offer eco-friendly alternatives (Koch et al. 2015). Rhizobia bacteria, for example, enhance nitrogen fixation in legumes, reducing the need for synthetic nitrogen fertilizers (Babcock-Jackson et al. 2023). Moreover, precision agriculture technologies, such as soil sensors and satellite imaging, allow for the targeted application of fertilizers and pesticides only where needed, optimizing usage and minimizing environmental impact. Cover cropping and green manures also improve soil health and fertility naturally, reducing dependency on chemical inputs.

13 Case Studies of Metabolomics in Agricultural Practices

Metabolomics, the comprehensive study of metabolites within a biological system, has increasingly been applied to agricultural practices, offering insights that enhance crop production, disease resistance, and overall plant health (Singh et al. 2022). Several case studies illustrate its transformative impact. One prominent example involves the use of metabolomics to understand and combat citrus greening disease (Huanglongbing), a devastating ailment affecting citrus crops globally. In another case, metabolomics was employed to optimize the growth conditions of tomatoes. By profiling the metabolites under various environmental conditions, scientists identified the optimal combination of soil nutrients and irrigation practices that maximized the concentration of beneficial compounds like lycopene and flavonoids in the fruit (Wang et al. 2023). This data-driven approach not only improved yield but also enhanced the nutritional value of the tomatoes. Furthermore, metabolomics has been instrumental in the developing of biocontrol agents. For instance, the metabolic profiling of *Trichoderma* spp., a genus of fungi used as biocontrol agents, revealed the production of specific metabolites that suppress plant pathogens. In rice agriculture, metabolomics studies have identified stress-responsive metabolites that help plants withstand drought and salinity (Gupta et al. 2022a, 2022b). By understanding these metabolic adjustments, researchers have been able to breed or engineer rice varieties with enhanced resilience to environmental stressors, ensuring stable food production in the face of global warming and climate changes. These case studies underscore the power of metabolomics in revolutionizing agricultural practices, offering targeted solutions for crop improvement, disease management, and sustainable farming (Gupta et al. 2022a, 2022b). The combination of metabolomics with other omics technologies will probably lead to further deeper understandings and uses in agriculture as technology develops.

14 Challenges and Future Directions

14.1 Technical and Analytical Challenges in Metabolomics

Comprehensive research on metabolites in biological systems is called metabolomics and faces several technical and analytical challenges despite its transformative potential in fields such as systems biology, medicine, and agriculture (Gupta et al. 2022a, 2022b). One primary technical challenge is the inherent chemical diversity and wide concentration range of metabolites, which complicates their detection and quantification. Metabolites can vary greatly in polarity, molecular weight, and abundance, requiring sophisticated and often multiple analytical platforms, such as mass spectrometry (MS) and nuclear magnetic resonance (NMR) spectroscopy, to capture the full metabolic profile (Peng et al. 2024).

Sample preparation is another critical hurdle. Metabolites are highly sensitive to environmental changes, and their concentrations can be rapidly altered by factors such as temperature, pH, and enzymatic activity (Poudel et al. 2021). Ensuring consistent and accurate sample collection, storage, and processing is crucial to prevent artificial alterations of the metabolome. Additionally, the extraction methods must be tailored to preserve the integrity of diverse metabolite classes, necessitating a balance between comprehensiveness and specificity. Data analysis in metabolomics is another complex challenge. Statistical analysis and interpretation of metabolomics data also pose significant challenges. Due to the high dimensionality of the data, advanced statistical methods, such as multivariate analysis and machine learning, are often employed to discern meaningful biological patterns (Ren et al. 2015). However, these methods require careful validation to avoid overfitting and ensure reproducibility. The challenge of integration—another key point—is the metabolomics data integration with other omics datasets in order to draw a holistic picture of how activities are interrelated in biological systems.

While metabolomics holds great potential for further advancing our understanding of biological processes and disease mechanisms, addressing these technical and analytical challenges is essential for realizing its full potential. Continuous advancements in analytical technologies, computational tools, and integrative approaches are key to overcoming these hurdles and enhancing the reliability and applicability of metabolomics studies.

14.2 Integrating Multi-omics Approaches for Comprehensive Profiling

Integrating multi-omics approaches—genomics, transcriptomics, proteomics, metabolomics, and more—has emerged as a powerful strategy for comprehensive biological profiling (Yan et al. 2022). This holistic approach provides a detailed, multi-layered understanding of biological systems, capturing the complexity of molecular interactions and their impact on cellular functions. Genomics provides the blueprint of an organism by detailing its complete DNA sequence and gives insights into the functions that may be encoded by the genome. Transcriptomics builds on this by examining RNA transcripts, revealing which genes are actively revealed under certain specified conditions Liu et al. (2020). Proteomics takes the next step by analyzing the proteome, the entire set of proteins expressed, which directly perform cellular functions and interact with various biomolecules. This makes it complementary to profiling small molecules and metabolites, which are, in fact, very end products of cellular processes and can provide a snapshot of metabolic activities. When integrated, these omics layers reveal the flow of biological information from DNA to RNA to protein to metabolite, allowing researchers to understand how genetic variations manifest as phenotypic traits through complex biochemical networks. For example, in cancer research, multi-omics integration can identify mutations (genomics), their effects on gene expression (transcriptomics), subsequent changes in protein abundance (proteomics), and alterations in metabolic pathways (metabolomics) (Winder et al. 2022). This comprehensive profiling can uncover biomarkers for early detection, therapeutic targets, and insights into drug resistance mechanisms.

In agriculture, multi-omics approaches can enhance crop improvement by linking genotype with phenotype through intermediate biological processes. By understanding how genetic diversity influences traits like yield, stress tolerance, and nutritional content, scientists can design better breeding programs and develop crops suited to changing environmental conditions (Poudel et al. 2021). Technological advancements, such as high-throughput sequencing and mass spectrometry, coupled with sophisticated bioinformatics tools, have made multi-omics integration feasible. However, challenges remain, including the need for standardized protocols, effective data integration methods, and the interpretation of vast datasets. Despite these hurdles, multi-omics approaches hold tremendous potential for advancing personalized medicine, biotechnology, and environmental science by providing a comprehensive understanding of life's molecular underpinnings (Pretzsch and Ecker 2023).

14.3 Future Prospects for Metabolomics in Sustainable Agriculture

Metabolomics, the comprehensive study of metabolites within biological systems, holds immense promise for advancing sustainable agriculture. By offering a detailed snapshot of the biochemical activities within plants, soil, and associated microbes, metabolomics provides valuable insights that can drive more efficient and environmentally friendly farming practices. One key area where metabolomics is all set to make a significant impact is in the optimization of plant growth-promoting microbes (PGPM) (Krishna et al. 2024). By understanding the metabolic pathways that confer drought tolerance, heat resistance, or improved photosynthetic efficiency, researchers can breed or engineer plants that thrive under adverse environmental conditions. In addition, metabolomics can help reduce agricultural waste and increase resource use efficiency. By analyzing the metabolic processes involved in plant growth and development, farmers can fine-tune irrigation schedules, optimize fertilization regimes, and implement precision farming techniques that minimize inputs while maximizing outputs (Getahun et al. 2024). This not only conserves water and reduces the runoff of nutrients into waterways but also enhances overall farm productivity and sustainability. Another promising application of metabolomics is in the realm of biocontrol. By identifying the metabolites involved in plant-pathogen interactions, researchers can develop natural biopesticides and disease-resistant crop varieties, reducing dependence on synthetic chemicals that are hazardous to environment and human health (Thakur et al. 2023). Overall, the future prospects for metabolomics in sustainable agriculture are bright.

15 Conclusion

Plant growth-promoting microorganism (PGPM) research benefits greatly from the thorough study of metabolites within biological systems, known as metabolomics, which has led to critical discoveries and developments in sustainable agriculture. This field focuses on the whole set of small-molecule chemicals found in plants and microbes, providing a clear understanding of their interactions and functions. The importance of metabolomics in PGPM research is multifaceted, enhancing our ability to improve crop yield, health, and resilience. One of the major inputs of metabolomics has been to decipher the complicated biochemical pathways engendered in plant-microbe interactions. This will enable profiling of the metabolites produced by the plants and their associated microbes, trying to identify key compounds that facilitate beneficial interactions. For example, specific metabolites like siderophores, phytohormones, and antibiotics can be linked to enhanced nutrient acquisition, growth regulation, and disease suppression in plants. Understanding these interactions at a molecular level allows for the targeted design of PGPMs that can optimize these beneficial traits. Furthermore, metabolomics enables the screening and identification of novel PGPM strains with superior efficacy. By comparing the metabolic profiles of different microbial strains, scientists can pinpoint those that produce the most effective metabolites for plant growth promotion. This approach not only accelerates the discovery of potent PGPMs but also ensures their compatibility with specific crops and environmental conditions. Another significant aspect of metabolomics in PGPM research is its role in improving the consistency and reliability of bioinoculants. Metabolomics helps overcome these issues by providing an extensive understanding of the metabolic stability and activity of PGPMs under different conditions, leading to more robust and predictable outcomes. Metabolomics is also one of integral approaches to systems biology when combined with other omics technologies like genomics and proteomics. In summary, metabolomics is indispensable in PGPM research, offering a detailed molecular insight that drives the design and optimization of effective, sustainable agricultural solutions. Its integration into PGPM development promises to revolutionize crop management and enhance food security.

15.1 Potential Impact on Future Agricultural Productivity and Sustainability

The potential impact of advanced agricultural technologies on future agricultural productivity and sustainability is profound, promising to address the growing global demand for food while mitigating environmental challenges. Innovations such as precision farming, genetically modified organisms (GMOs), and synthetic biology are revolutionizing how crops are grown and managed. This approach optimizes resource use, which in turn reduces the need for water, fertilizers, and pesticides, while maximizing yield and minimizing environmental impact. For example, variable rate technology (VRT) enables precise application of inputs, ensuring that crops receive exactly what they need, where they need it, and in the right amounts. This not only boosts productivity but also reduces runoff and soil degradation, promoting long-term sustainability. Genetically modified organisms (GMOs) hold significant promise for enhancing crop resilience and yield. By introducing traits such as drought tolerance, pest resistance, and enhanced nutritional content, GMOs can improve food security and reduce the reliance on chemical inputs. For instance, Bt crops, engineered to express a natural insecticide from the bacterium *Bacillus thuringiensis*, have significantly reduced the need for synthetic insecticides, lowering production costs and environmental impact. Synthetic biology further extends these capabilities by enabling the design of custom microbial consortia and engineered plants that can perform specific functions, such as nitrogen fixation or phosphate solubilization, enhancing soil fertility and reducing dependence on synthetic fertilizers. One such potential example may be metabolomics-driven design for plant growth-promoting microbes, and this may help in providing bespoke solutions for improving plant health and greater yield in a sustainable way. Moreover, the integration of such technologies into agricultural practices that include crop rotation, reduced or no tillage, and agroforestry, though, can eventually result in robust farming systems that adapt to climate change and reduce GHG emissions.

References

Abdelfattah A, Wisniewski M, Nicosia MGLD, Cacciola SO, Schena L (2016) Metagenomic analysis of fungal diversity on strawberry plants and the effect of management practices on the fungal community structure of aerial organs. PLoS One 11:e0160470. https://doi.org/10.1371/journal.pone.0160470

Abdulhussain N, Nawada S, Schoenmakers P (2021) Latest trends on the future of three-dimensional separations in chromatography. Chem Rev 121(19):12016–12034. https://doi.org/10.1021/acs.chemrev.0c01244

Adomako MO, Roiloa S, Yu FH (2022) Potential roles of soil microorganisms in regulating the effect of soil nutrient heterogeneity on plant performance. Microorganisms 10(12):2399. https://doi.org/10.3390/microorganisms10122399

Aharoni A, Goodacre R, Fernie AR (2023) Plant and microbial sciences as key drivers in the development of metabolomics research. Proc Natl Acad Sci USA 120:e2217383120. https://doi.org/10.1073/pnas.2217383120

Aina O, Bakare OO, Fadaka AO, Keyster M, Klein A (2024) Plant biomarkers as early detection tools in stress management in food crops: a review. Planta 259(3):60. https://doi.org/10.1007/s00425-024-04333-1

Albright MBN, Louca S, Winkler DE, Feeser KL, Haig S-J, Whiteson KL, Emerson JB, Dunbar J (2022) Solutions in microbiome engineering: prioritizing barriers to organism establishment. ISME J 16:331–338. https://doi.org/10.1038/s41396-021-01088-5

Alexander S, Aronson J, Whaley O, Lamb D (2016) The relationship between ecological restoration and the ecosystem services concept. Ecol Soc 21(1). https://doi.org/10.5751/ES-08288-210134

Alferez F, de Carvalho DU, Boakye D (2021) Interplay between Abscisic acid and gibberellins, as related to ethylene and sugars, in regulating maturation of non-climacteric fruit. Int J Mol Sci 22(2):669. https://doi.org/10.3390/ijms22020669

Al-Khayri JM, Rashmi R, Toppo V, Chole PB, Banadka A, Sudheer WN, Nagella P, Shehata WF, Al-Mssallem MQ, Alessa FM, Almaghasla MI, Rezk AA (2023) Plant secondary metabolites: the weapons for biotic stress management. Metabolites 13(6):716. https://doi.org/10.3390/metabo13060716

Al-Turki A, Murali M, Omar AF, Rehan M, Sayyed RZ (2023) Recent advances in PGPR-mediated resilience toward interactive effects of drought and salt stress in plants. Front Microbiol 14:1214845. https://doi.org/10.3389/fmicb.2023.1214845

Aly AA, Górecki T (2020) Green approaches to sample preparation based on extraction techniques. Molecules (Basel, Switzerland) 25(7):1719. https://doi.org/10.3390/molecules25071719

Arif I, Batool M, Schenk PM (2020) Plant microbiome engineering: expected benefits for improved crop growth and resilience. Trends Biotechnol 38:1385–1396. https://doi.org/10.1016/j.tibtech.2020.04.015

Atif RM, Shahid L, Waqas M, Ali B, Rashid MAR, Azeem F, Nawaz MA, Wani SH, Chung G (2019) Insights on calcium-dependent protein kinases (CPKs) signaling for abiotic stress tolerance in plants. Int J Mol Sci 20(21):5298. https://doi.org/10.3390/ijms20215298

Babcock-Jackson L, Konovalova T, Krogman JP, Bird R, Díaz LL (2023) Sustainable fertilizers: publication landscape on wastes as nutrient sources, wastewater treatment processes for nutrient recovery, biorefineries, and Green ammonia synthesis. J Agric Food Chem 71(22):8265–8296. https://doi.org/10.1021/acs.jafc.3c00454

Badillo-Sanchez D, Serrano Ruber M, Davies-Barrett A, Jones DJL, Hansen M, Inskip S (2023) Metabolomics in archaeological science: a review of their advances and present requirements. Sci Adv 9(32):eadh0485. https://doi.org/10.1126/sciadv.adh0485
Balcke GU, Bennewitz S, Bergau N, Athmer B, Henning A, Majovsky P, Jiménez-Gómez JM, Hoehenwarter W, Tissier A (2017) Multi-omics of tomato glandular Trichomes reveals distinct features of central carbon metabolism supporting high productivity of specialized metabolites. Plant Cell 29(5):960–983. https://doi.org/10.1105/tpc.17.00060
Bose KJC, Kaur S, Sharma S, Sarwan J, Uddin N (2024) Methods from the Field of synthetic biology that aim to improve plant growth and resistance to stress through the use of genetic engineering. In: Kaur S, Dwibedi V, Sahu PK (eds) Metabolomics, proteomics and gene editing approaches in biofertilizer industry. Springer, Singapore, pp 99–121. https://doi.org/10.1007/978-981-97-2910-4_6
Brader G, Compant S, Vescio K, Mitter B, Trognitz F, Ma L-J, Sessitsch A (2017) Ecology and genomic insights into plant-pathogenic and plant-nonpathogenic endophytes. Annu Rev Phytopathol 55:61–83. https://doi.org/10.1146/annurev-phyto-080516-035641
Brunel C, Pouteau R, Dawson W, Pester M, Ramirez KS, van Kleunen M (2020) Towards unraveling macroecological patterns in rhizosphere microbiomes. Trends Plant Sci 25:1017–1029. https://doi.org/10.1016/j.tplants.2020.04.015
Busby PE, Soman C, Wagner MR, Friesen ML, Kremer J, Bennett A, Morsy M, Eisen JA, Leach JE, Dangl JL (2017) Research priorities for harnessing plant microbiomes in sustainable agriculture. PLoS Biol 15:e2001793. https://doi.org/10.1371/journal.pbio.2001793
Campos MD, do Rosario Félix M, Patanita M, Materatski P, Varanda C (2021) High throughput sequencing unravels tomato-pathogen interactions towards a sustainable plant breeding. Horticult Res 8:171. https://doi.org/10.1038/s41438-021-00607-x
Camunas-Alberca SM, Moran-Garrido M, Sáiz J, Gil-de-la-Fuente A, Barbas C, Gradillas A (2023) Integrating the potential of ion mobility spectrometry-mass spectrometry in the separation and structural characterisation of lipid isomers. Front Mol Biosci 10:1112521. https://doi.org/10.3389/fmolb.2023.1112521
Cania B, Vestergaard G, Krauss M, Fliessbach A, Schloter M, Schulz S (2019) A long-term field experiment demonstrates the influence of tillage on the bacterial potential to produce soil structure-stabilizing agents such as exopolysaccharides and lipopolysaccharides. Environ Microbiome 14(1):1. https://doi.org/10.1186/s40793-019-0341-7
Carrión VJ, Perez-Jaramillo J, Cordovez V, Tracanna V, Hllander MD, Buck DR, Mendes LW, Vanijcken WFJ, Exposito RG, Elsayed SS, Mohanraju P, Arifah A, Vanderoost J, Paulson JN, Mendes R, Vanwezel GP, Medema MH, Raaijmakers JM (2019) Pathogen induced activation of disease-suppressive functions in the endophytic root microbiome. Science 366:606–612. https://doi.org/10.1126/science.aaw9285
Chai YN, Schachtman DP (2022) Root exudates impact plant performance under abiotic stress. Trends Plant Sci 27:80–91. https://doi.org/10.1016/j.tplants.2021.08.003
Chen S, Wang X, Cheng Y, Gao H, Chen X (2023) A review of classification, biosynthesis, biological activities and potential applications of flavonoids. Molecules (Basel, Switzerland) 28(13):4982. https://doi.org/10.3390/molecules28134982
Cordovez V, Rotoni C, Dini-Andreote F, Oyserman B, Carrión VJ, Raaijmakers JM (2021) Successive plant growth amplifies genotype specific assembly of the tomato rhizosphere microbiome. Sci Total Environ 772:144825. https://doi.org/10.1016/j.scitotenv.2020.144825
Da Silva JG (2012) Feeding the world sustainably. UN chronicle XLIX. United Nations Department of Global Communications, New York. https://www.un.org/en/chronicle/article/feeding-world-sustainably, https://doi.org/10.18356/89c2a819-en
Defrenne CE, Abs E, Cordeiro AL, Dietterich L, Hough M, Jones JM, Kivlin SN, Chen W, Cusack D, Franco ALC, Khasanova A, Stover D, Olivares ALR (2021) The ecology underground coalition: building a collaborative future of belowground ecology and ecologists. New Phytol 229:3058–3064. https://doi.org/10.1111/nph.17163

Du J, Li Y, Ur-Rehman S, Mukhtar I, Yin Z, Dong H, Wang H, Zhang X, Gao Z, Zhao X, Xin X, Ding XB (2021) Synergistically promoting plant health by harnessing synthetic microbial communities and prebiotics. IScience 24:102918. https://doi.org/10.1016/j.isci.2021.102918

Duca D, Lorv J, Patten CL, Rose D, Glick BR (2014) Indole-3-acetic acid in plant–microbe interactions. Antonie Van Leeuwenhoek 106:85–125

Emwas AH, Roy R, McKay RT, Tenori L, Saccenti E, Gowda GAN, Raftery D, Alahmari F, Jaremko L, Jaremko M, Wishart DS (2019) NMR spectroscopy for metabolomics research. Metabolites 9(7):123. https://doi.org/10.3390/metabo9070123

Fanai A, Bohia B, Lalremruati F, Lalhriatpuii N, Lalrokimi, Lalmuanpuii R, Singh PK, Zothanpuia (2024) Plant growth promoting bacteria (PGPB)-induced plant adaptations to stresses: an updated review. PeerJ 12:e17882. https://doi.org/10.7717/peerj.17882

FAO (2017) The future of food and agriculture: trends and challenges. Food and Agriculture Organization of the United Nations, Rome

FAO (2019) Microbiome: the missing link?: Science and innovation for health, climate and sustainable food systems. Food and Agriculture Organization of the United Nations, Rome. http://www.fao.org/publications/card/ en/c/CA6767EN/

Farooq MS, Uzair M, Raza A, Habib M, Xu Y, Yousuf M, Yang SH, Ramzan Khan M (2022) Uncovering the research gaps to alleviate the negative impacts of climate change on food security: a review. Front Plant Sci 13:927535. https://doi.org/10.3389/fpls.2022.927535

Fitzpatrick CR, Copeland J, Wang PW, Guttman DS, Kotanen PM, Johnson MTJ (2018) Assembly and ecological function of the root micro biome across angiosperm plant species. Proc Natl Acad Sci USA 115:E1157–E1165. https://doi.org/10.1073/pnas.1717617115

Freschet GT, Pagès L, Iversen CM, Comas LH, Rewald B, Roumet C, Klimešová J, Zadworny M, Poorter H, Postma JA, Adams TS, Bagniewska-Zadworna A, Bengough AG, Blancaflor EB, Brunner I, Cornelissen JHC, Garnier E, Gessler A, Hobbie SE, Meier IC et al (2021) A starting guide to root ecology: strengthening ecological concepts and standardising root classification, sampling, processing and trait measurements. New Phytol 232(3):973–1122. https://doi.org/10.1111/nph.17572

Gao M, Xiong C, Gao C, Tsui CKM, Wang M-M, Zhou X, Zhang A-M, Cai L (2021) Disease-induced changes in plant microbiome assembly and functional adaptation. Microbiome 9:187. https://doi.org/10.1186/s40168-021-01138-2

Geier B, Sogin EM, Michellod D, Janda M, Kompauer M, Spengler B, Dubilier N, Liebeke M (2020) Spatial metabolomics of in situ host–microbe interactions at the micrometre scale. Nat Microbiol 5:498–510. https://doi.org/10.1038/s41564-019-0664-6

Getahun S, Kefale H, Gelaye Y (2024) Application of precision agriculture technologies for sustainable crop production and environmental sustainability: a systematic review. Sci World J 2024:2126734. https://doi.org/10.1155/2024/2126734

Gill SS, Gill R, Tuteja R, Tuteja N (2014) Genetic engineering of crops: a ray of hope for enhanced food security. Plant Signal Behav 9(3):e28545. https://doi.org/10.4161/psb.28545

Giri H, Cai X, Panicker SR, Biswas I, Rezaie AR (2019) Thrombomodulin regulation of mitogen-activated protein kinases. Int J Mol Sci 20(8):1851. https://doi.org/10.3390/ijms20081851

Gonin M, Salas-González I, Gopaulchan D, Frene JP, Roden S, Van de Poel B, Salt DE, Castrillo G (2023) Plant microbiota controls an alternative root branching regulatory mechanism in plants. Proc Natl Acad Sci USA 120:e2301054120. https://doi.org/10.1073/pnas.2301054120

Gorshkov V, Tsers I (2022) Plant susceptible responses: the underestimated side of plant–pathogen interactions. Biol Rev Camb Philos Soc 97:45–66. https://doi.org/10.1111/brv.12789

Gruet C, Muller D, Moënne-Loccoz Y (2022) Significance of the diversification of wheat species for the assembly and functioning of the root-associated microbiome. Front Microbiol 12:782135. https://doi.org/10.3389/fmicb.2021.782135

Gupta A, Mishra R, Rai S, Bano A, Pathak N, Fujita M, Kumar M, Hasanuzzaman M (2022a) Mechanistic insights of plant growth promoting bacteria mediated drought and Salt stress tolerance in plants for sustainable agriculture. Int J Mol Sci 23(7):3741. https://doi.org/10.3390/ijms23073741

Gupta K, Wani SH, Razzaq A, Skalicky M, Samantara K, Gupta S, Pandita D, Goel S, Grewal S, Hejnak V, Shiv A, El-Sabrout AM, Elansary HO, Alaklabi A, Brestic M (2022b) Abscisic acid: role in fruit development and ripening. Front Plant Sci 13:817500. https://doi.org/10.3389/fpls.2022.817500

Hacquard S, Kracher B, Hiruma K, Munch PC, Oter RG, Thon MR, Weimann A, Damm U, Dallery JF, Hainaut M, Henrissat B, Lespinet O, Sacristan S, Thematt EVLV, Kemen E, McHardy AC, Lefert PS, OConnell RJ. (2016) Survival trade-offs in plant roots during colonization by closely related beneficial and pathogenic fungi. Nat Commun 7:11362. https://doi.org/10.1038/ncomms11362

Han X, Liu J, Tian S, Tao F, Xu P (2022) Microbial cell factories for bio-based biodegradable plastics production. iScience 25(11):105462. https://doi.org/10.1016/j.isci.2022.105462

Hartline CJ, Schmitz AC, Han Y, Zhang F (2021) Dynamic control in metabolic engineering: theories, tools, and applications. Metab Eng 63:126–140. https://doi.org/10.1016/j.ymben.2020.08.015

Hirose T, Keck D, Izumi Y, Bamba T (2019) Comparison of retention behavior between supercritical fluid chromatography and Normal-phase high-performance liquid chromatography with various stationary phases. Molecules (Basel, Switzerland) 24(13):2425. https://doi.org/10.3390/molecules24132425

Hiruma K, Gerlach N, Sacristán S (2016) Root endophyte *Colletotrichum tofieldiae* confers plant fitness benefits that are phosphate status dependent. Cell 165:464–474. https://doi.org/10.1016/j.cell.2016.02.028

Hou S, Thiergart T, Vannier N, Mesny F, Ziegler J, Pickel B, Hacquard S (2021) A microbiota–root–shoot circuit favours *Arabidopsis* growth over defence under suboptimal light. Nat Plants 7:1078–1092. https://doi.org/10.1038/s41477-021-00956-4

Hu J, Wei Z, Kowalchuk GA, Xu Y, Shen Q, Jousset A (2020) Rhizosphere microbiome functional diversity and pathogen invasion resistance build up during plant development. Environ Microbiol 22:5005–5018. https://doi.org/10.1111/1462-2920.15097

Huang Y, Sheth RU, Zhao S, Cohen LA, Dabaghi K, Moody T, Wang HH (2023) High-throughput microbial culturomics using automation and machine learning. Nat Biotechnol 41:1424–1433. https://doi.org/10.1038/s41587-023-01674-2

Iqbal S, Begum F, Rabaan AA, Aljeldah M, Al Shammari BR, Alawfi A, Alshengeti A, Sulaiman T, Khan A (2023) Classification and multifaceted potential of secondary metabolites produced by Bacillus subtilis group: a comprehensive review. Molecules (Basel, Switzerland) 28(3):927. https://doi.org/10.3390/molecules28030927

Jahn L, Hofmann U, Ludwig-Müller J (2021) Indole-3-acetic acid is synthesized by the endophyte Cyanodermella asteris via a tryptophan-dependent and -independent way and mediates the interaction with a non-host plant. Int J Mol Sci 22(5):2651. https://doi.org/10.3390/ijms22052651

Katam R, Lin C, Grant K, Katam CS, Chen S (2022) Advances in plant metabolomics and its applications in stress and single-cell biology. Int J Mol Sci 23(13):6985. https://doi.org/10.3390/ijms23136985

Kaur R, Kaur S (2023) Carrier based biofertilizers. In: Kaur S, Dwibedi V, Sahu PK, Kocher GS (eds) Metabolomics, proteomes and gene editing approaches in biofertilizer industry. Springer, pp 57–75. https://doi.org/10.1007/978-981-99-3561-1

Kaya C, Uğurlar F, Adamakis IS (2023) Epigenetic and hormonal modulation in plant-plant growth-promoting microorganism symbiosis for drought-resilient agriculture. Int J Mol Sci 24(22):16064. https://doi.org/10.3390/ijms242216064

Khokhar MK, Kumar R, Kumar A, Sehgal M, Singh SP, Meena PN, Singh N, Acharya LK, Birah A, Singh K, Bana RS, Gurjar MS, Chander S, Choudhary M (2024) Impact of IPM practices on microbial population and disease development in transplanted and direct-seeded rice. Front Microbiol 15:1388754. https://doi.org/10.3389/fmicb.2024.1388754

Koch MS, Ward JM, Levine SL, Baum JA, Vicini JL, Hammond BG (2015) The food and environmental safety of Bt crops. Front Plant Sci 6:283. https://doi.org/10.3389/fpls.2015.00283

Koskella B, Taylor TB (2018) Multifaceted impacts of bacteriophages in the plant microbiome. Annu Rev Phytopathol 56:361–380. https://doi.org/10.1146/annurev-phyto-080417-045858

Koskella B, Hernandez CA, Wheatley RM (2022) Understanding the impacts of bacteriophage viruses: from laboratory evolution to natural eco systems. Ann Rev Virol 9:57–78. https://doi.org/10.1146/annurev-virology-091919-075914

Lenin G, Jayanthi M (2012) Indole acetic acid, gibberellic acid and siderophore production by PGPR isolates from rhizospheric soils of Catharanthus roseus. Int J Pharm Biol Arch 3(4):933–938

Letertre MPM, Giraudeau P, de Tullio P (2021) Nuclear magnetic resonance spectroscopy in clinical metabolomics and personalized medicine: current challenges and perspectives. Front Mol Biosci 8:698337. https://doi.org/10.3389/fmolb.2021.698337

Li X, Jousset A, de Boer W, Carrión VJ, Zhang T, Wang X, Kuramae EE (2019) Legacy of land use history determines reprogramming of plant physiology by soil microbiome. ISME J 13:738–751. https://doi.org/10.1038/s41396-018-0300-0

Liu H-J, Wang X, Xiao Y, Luo J, Quio F, Yang W, Zhang R, Meng Y, Sun J, Yan S, Peng Y, Niu L, Jian L, Song W, Yan J, Li C, Zhao Y, Liu Y, Warburton ML, Zhao J, Yan J (2020) CUBIC: an atlas of genetic architecture promises directed maize improvement. Genome Biol 21:20. https://doi.org/10.1186/s13059-020-1930-x

Liu H, Li J, Carvalhais LC, Percy CD, Verma JP, Schenk PM, Singh BK (2021) Evidence for the plant recruitment of beneficial microbes to sup press soil-borne pathogens. New Phytol 229:2873–2885. https://doi.org/10.1111/nph.17057

Liu S, Zenda T, Tian Z, Huang Z (2023) Metabolic pathways engineering for drought or/and heat tolerance in cereals. Front Plant Sci 14:1111875. https://doi.org/10.3389/fpls.2023.1111875

Lundberg DS, Teixeira PJPL (2018) Root-exuded coumarin shapes the root microbiome. Proc Natl Acad Sci USA 115:5629–5631. https://doi.org/10.1073/pnas.1805944115

Lundberg DS, Lebeis SL, Paredes SH, Yourstone S, Gehring J, Malfatti S, Tremblay J, Engelbrektson A, Kunin V, Del Rio TG, Edgar RC, Eickhorst T, Ley RE, Hugenholtz P, Tringe SG, Dangl JL (2012) Defining the core Arabidopsis thaliana root microbiome. Nature 488:86–90. https://doi.org/10.1038/nature11237

Luo J, Xia W, Cao P, Xiao Z, Zhang Y, Liu M, Zhan C, Wang N (2019) Integrated transcriptome analysis reveals plant hormones Jasmonic acid and salicylic acid coordinate growth and defense responses upon fungal infection in poplar. Biomol Ther 9(1):12. https://doi.org/10.3390/biom9010012

Mahdi LK, Miyauchi S, Uhlmann C, Oter RG, Langen G, Wawra S, Niu Y, Guan R, Albertyn SR, Bulgarelli D, Parker JE, Zuccaro A (2022) The fungal root endophyte *Serendipita vermifera* displays inter-kingdom synergistic beneficial effects with the microbiota in Arabidopsis thaliana and barley. ISME J 16:876–889. https://doi.org/10.1038/s41396-021-01138-y

Maier BA, Kiefer P, Field CM, Hemmerle L, Bortfeldmiller M, Emmenegger B, Schafer M, Pfeilmeier S, Sunagawa S, Vogel CM, Vorholt JA (2021) A general non-self-response as part of plant immunity. Nat Plants 7:696–705. https://doi.org/10.1038/s41477-021-00913-1

Malik AA, Martiny JBH, Brodie EL, Martiny AC, Treseder KK, Allison SD (2020) Defining trait-based microbial strategies with consequences for soil carbon cycling under climate change. ISME J 14:1–9. https://doi.org/10.1038/s41396-019-0510-0

Marcillo A, Baca Cabrera JC, Widdig A, Birkemeyer C (2023) A comparison between mobile and stationary gas chromatography-mass spectrometry devices for analysis of complex volatile profiles. Anal Bioanal Chem 415(1):137–155. https://doi.org/10.1007/s00216-022-04391-y

Martínez-Sena T, Moro E, Moreno-Torres M, Quintás G, Hengstler J, Castell JV (2023) Metabolomics-based strategy to assess drug hepatotoxicity and uncover the mechanisms of hepatotoxicity involved. Arch Toxicol 97(6):1723–1738. https://doi.org/10.1007/s00204-023-03474-8

Matsumoto H, Fan X, Wang Y, Kusstatscher P, Duan J, Wu S, Chen S, Qiao K, Wang Y, Ma B, Zhu G, Hashidoko Y, Berg, Cernava, Wang M (2021) Bacterial seed endophyte shapes disease resistance in rice. Nat Plants 7:60–72. https://doi.org/10.1038/s41477-020-00826-5

McGeoch MA, Clarke DA, Mungi NA, Ordonez A (2024) A nature-positive future with biological invasions: theory, decision support and research needs. Philos Trans R Soc Lond Ser B Biol Sci 379(1902):20230014. https://doi.org/10.1098/rstb.2023.0014

Mehmood N, Saeed M, Zafarullah S, Hyder S, Rizvi ZF, Gondal AS, Jamil N, Iqbal R, Ali B, Ercisli S, Kupe M (2023) Multifaceted impacts of plant-beneficial pseudomonas spp. in managing various plant diseases and crop yield improvement. ACS Omega 8(25):22296–22315. https://doi.org/10.1021/acsomega.3c00870

Miller SB, Heuberger AL, Broeckling CD, Jahn CE (2019) Non-targeted metabolomics reveals sorghum rhizosphere-associated exudates are influenced by the belowground interaction of substrate and sorghum genotype. Int J Mol Sci 20(2):431. https://doi.org/10.3390/ijms20020431

Misra BB (2021) New software tools, databases, and resources in metabolomics: updates from 2020. Metabolomics 17(5):49. https://doi.org/10.1007/s11306-021-01796-1

Mohammadi-Dehcheshmeh M, Niazi A, Ebrahimi M, Tahsili M, Nurollah Z, Ebrahimi Khaksefid R, Ebrahimi M, Ebrahimie E (2018) Unified transcriptomic signature of arbuscular mycorrhiza colonization in roots of Medicago truncatula by integration of machine learning, promoter analysis, and direct merging meta-analysis. Front Plant Sci 9:1550. https://doi.org/10.3389/fpls.2018.01550

Müller WH, Verdin A, De Pauw E, Malherbe C, Eppe G (2022) Surface-assisted laser desorption/ionization mass spectrometry imaging: A review. Mass Spectrom Rev 41(3):373–420. https://doi.org/10.1002/mas.21670

Nasir MA, Nugroho AD, Lakner Z (2022) Impact of the Russian-Ukrainian conflict on global food crops. Food Secur 11:2979. https://doi.org/10.3390/foods11192979

Neal AL, Ahmad S, Gordon-Weeks R, Ton J (2012) Benzoxazinoids in root exudates of maize attract Pseudomonas putida to the rhizosphere. PLoS One 7:e35498. https://doi.org/10.1371/journal.pone.0035498

Ofek-Lalzar M, Sela N, Goldman-Voronov M, Green SJ, Hadar Y, Minz D (2014) Niche and host-associated functional signatures of the root surface microbiome. Nat Commun 5:4950. https://doi.org/10.1038/ncomms5950

Pandey P, Tripathi A, Dwivedi S, Lal K, Jhang T (2023) Deciphering the mechanisms, hormonal signaling, and potential applications of endophytic microbes to mediate stress tolerance in medicinal plants. Front Plant Sci 14:1250020. https://doi.org/10.3389/fpls.2023.1250020

Patel MK, Pandey S, Kumar M, Haque MI, Pal S, Yadav NS (2021) Plants metabolome study: emerging tools and techniques. Plants (Basel, Switzerland) 10(11):2409. https://doi.org/10.3390/plants10112409

Pattanayak S, Das S, Banerjee P (2021) Crop protection through fungal biopesticides: an endeavor of modern organic agriculture. In: Biopesticides in organic farming. CRC Press, pp 87–90

Peng Y, Zhang Z, He L, Li C, Liu M (2024) NMR spectroscopy for metabolomics in the living system: recent progress and future challenges. Anal Bioanal Chem 416(9):2319–2334. https://doi.org/10.1007/s00216-024-05137-8

Poudel M, Mendes R, Costa LAS, Bueno CG, Meng Y, Folimonova SY, Garrett KA, Martins SJ (2021) The role of plant-associated bacteria, fungi, and viruses in drought stress mitigation. Front Microbiol 12:743512. https://doi.org/10.3389/fmicb.2021.743512

Pratama AA, Terpstra J, de Oliveria ALM, Salles JF (2020) The role of rhizosphere bacteriophages in plant health. Trends Microbiol 28:709–718. https://doi.org/10.1016/j.tim.2020.04.005

Prescott SL, Wegienka G, Kort R, Nelson DH, Gabrysch S, Hancock T, Kozyrskyj A, Lowry CA, Redvers N, Poland B, Robinson J, Moubarac JC, Warber S, Jansson J, Sinkkonen A, Penders J, Erdman S, Nanan R, van den Bosch M, Schneider K et al (2021) Project earthrise: proceedings of the ninth annual conference of inVIVO planetary health. Int J Environ Res Public Health 18(20):10654. https://doi.org/10.3390/ijerph182010654

Pretzsch CM, Ecker C (2023) Structural neuroimaging phenotypes and associated molecular and genomic underpinnings in autism: a review. Front Neurosci 17:1172779. https://doi.org/10.3389/fnins.2023.1172779

Regalado J, Lundberg DS, Deusch O, Kersten S, Karasov T, Poersch K, Shirsekar G, Weigel D (2020) Combining whole-genome shotgun sequencing and rRNA gene amplicon analyses to improve detection of microbe–microbe interaction networks in plant leaves. ISME J 14:8. https://doi.org/10.1038/s41396-020-0665-8

Ren S, Hinzman AA, Kang EL, Szczesniak RD, Lu LJ (2015) Computational and statistical analysis of metabolomics data. Metabolomics 11:1492–1513

Russo A, Pollastri S, Ruocco M, Monti MM, Loreto F (2022) Volatile organic compounds in the interaction between plants and beneficial microorganisms. J Plant Interact 17(1):840–852

Santander C, González F, Pérez U, Ruiz A, Aroca R, Santos C, Cornejo P, Vidal G (2024) Enhancing water status and nutrient uptake in drought-stressed lettuce plants (Lactuca sativa L.) via inoculation with different bacillus spp. Isolated from the Atacama Desert. Plants (Basel, Switzerland) 13(2):158. https://doi.org/10.3390/plants13020158

Sarkar D, Rovenich H, Jeena G, Nizam S, Tissier A, Balcke GU, Mahdi LK, Bonkowski M, Langen G, Zuccaro A (2019) The inconspicuous gatekeeper: endophytic Serendipita vermifera acts as ex tended plant protection barrier in the rhizosphere. New Phytol 224:886–901. https://doi.org/10.1111/nph.15904

Shalev O, Ashkenazy H, Neumann M, Weigel D (2022) Commensal Pseudomonas protect Arabidopsis thaliana from a coexisting pathogen via multiple lineage-dependent mechanisms. ISME J 16:1235–1244. https://doi.org/10.1038/s41396-021-01168-6

Sharma M, Sudheer S, Usmani Z, Rani R, Gupta P (2020) Deciphering the omics of plant-microbe interaction: perspectives and new insights. Curr Genomics 21(5):343–362. https://doi.org/10.2174/1389202921999200515140420

Singh DP, Bisen MS, Shukla R, Prabha R, Maurya S, Reddy YS, Singh PM, Rai N, Chaubey T, Chaturvedi KK, Srivastava S, Farooqi MS, Gupta VK, Sarma BK, Rai A, Behera TK (2022) Metabolomics-driven mining of metabolite resources: applications and prospects for improving vegetable crops. Int J Mol Sci 23(20):12062. https://doi.org/10.3390/ijms232012062

Singh A, Mazahar S, Chapadgaonkar SS, Giri P, Shourie A (2023) Phyto-microbiome to mitigate abiotic stress in crop plants. Front Microbiol 14:1210890. https://doi.org/10.3389/fmicb.2023.1210890

Starr EP, Nuccio EE, Pett-Ridge J, Banfield JF, Firestone MK (2019) Metatranscriptomic reconstruction reveals RNA viruses with the potential to shape carbon cycling in soil. Proc Natl Acad Sci USA 116:25900–25908. https://doi.org/10.1073/pnas.1908291116

Stringlis IA, Yu K, Feussner K, de Jonge R, Bentum SV, Verk MCV, Berendsen RL, Bakker PAHM, Feussner I, Pieterse CMJ (2018) MYB72-dependent coumarin exudation shapes root microbiome assembly to promote plant health. Proc Natl Acad Sci USA 115:E5213–E5222. https://doi.org/10.1073/pnas.1908291116

Stryeck S, Birner-Gruenberger R, Madl T (2017) Integrative metabolomics as emerging tool to study autophagy regulation. Microbial Cell (Graz, Austria) 4(8):240–258. https://doi.org/10.15698/mic2017.08.584

Tamara S, den Boer MA, Heck AJR (2022) High-resolution native mass spectrometry. Chem Rev 122(8):7269–7326. https://doi.org/10.1021/acs.chemrev.1c00212

Teixeira PJPL, Colaianni NR, Law TF, Conway JM, Gilbert S, Li H, Gonzalez IS, Panda D, Risco NMD, Finkel OM, Castrillo G, Mieczkowski P, Jones CD, Dangl JL (2021) Specific modulation of the root immune system by a community of commensal bacteria. Proc Natl Acad Sci USA 118:e2100678118. https://doi.org/10.1073/pnas.2100678118

Thakur N, Nigam M, Mann NA, Gupta S, Hussain CM, Shukla SK, Shah AA, Casini R, Elansary HO, Khan SA (2023) Host-mediated gene engineering and microbiome-based technology optimization for sustainable agriculture and environment. Funct Integr Genomics 23(1):57. https://doi.org/10.1007/s10142-023-00982-9

Tian Y, Liu Y, Yue L, Uwaremwe C, Zhao X, Zhou Q, Wang Y, Wang R (2022) Bacterial inoculant and sucrose amendments improve the growth of Rheum palmatum L. by reprograming its metabolite composition and altering its soil microbial community. Int J Mol Sci 23(3):1694. https://doi.org/10.3390/ijms23031694

Timofeeva AM, Galyamova MR, Sedykh SE (2023) Plant growth-promoting soil bacteria: nitrogen fixation, phosphate Solubilization, Siderophore production, and other biological activities. Plants (Basel, Switzerland) 12(24):4074. https://doi.org/10.3390/plants12244074

Trivedi P, Delgado-Baquerizo M, Trivedi C, Hu H, Anderson IC, Jeffries TC, Zhou J, Singh BK (2016) Microbial regulation of the soil carbon cycle: evidence from gene–enzyme relationships. ISME J 10:2593–2604. https://doi.org/10.1038/ismej.2016.65

Trivedi P, Delgado-Baquerizo M, Trivedi C, Hamonts K, Anderson IC, Singh BK (2017) Keystone microbial taxa regulate the invasion of a fungal pathogen in agro-ecosystems. Soil Biol Biochem 111:10–14. https://doi.org/10.1016/j.soilbio.2017.03.013

Usman M (2024) The impact of synthetic biology on genetic research and biotechnology. Front Biotechnol Genet 1(02):107–125

Varoquaux N, Cole B, Gao C et al (2019) Transcriptomic analysis of field droughted sorghum from seedling to maturity reveals biotic and metabolic responses. Proc Natl Acad Sci USA 116:27124–27132. https://doi.org/10.1073/pnas.1907500116

Verslues PE, Bailey-Serres J, Brodersen C, Buckley TN, Conti L, Christmann A, Dinneny JR, Grill E, Hayes S, Heckman RW, Hsu PK, Juenger TE, Mas P, Munnik T, Nelissen H, Sack L, Schroeder JI, Testerink C, Tyerman SD, Umezawa T et al (2023) Burning questions for a warming and changing world: 15 unknowns in plant abiotic stress. Plant Cell 35(1):67–108. https://doi.org/10.1093/plcell/koac263

Wahbeh S, Anastasiadis F, Sundarakani B, Manikas I (2022) Exploration of food security challenges towards more sustainable food production: a systematic literature review of the major drivers and policies. Foods (Basel, Switzerland) 11(23):3804. https://doi.org/10.3390/foods11233804

Waller F, Achatz B, Baltruschat H, Fodor J, Becker K, Fischer M, Heier T, Huckewhoven R, Neumann C, Wettstein DV, Franken P, Kogel KH (2005) The endophytic fungus *Piriformospora indica* reprograms barley to salt-stress tolerance, disease resistance, and higher yield. Proc Natl Acad Sci USA 102:13386–13391. https://doi.org/10.1073/pnas.0504423102

Wamaitha MJ, Nigam D, Maina S, Stomeo F, Wangai A, Njuguna JN, Garcia-Ruiz H (2018) Metagenomic analysis of viruses associated with maize lethal necrosis in Kenya. Virol J 15:90. https://doi.org/10.1186/s12985-018-0999-2

Wang YH, Zhang RR, Yin Y, Tan GF, Wang GL, Liu H, Zhuang J, Zhang J, Zhuang FY, Xiong AS (2023) Advances in engineering the production of the natural red pigment lycopene: a systematic review from a biotechnology perspective. J Adv Res 46:31–47. https://doi.org/10.1016/j.jare.2022.06.010

Winder CL, Witting M, Tugizimana F, Dunn WB, Reinke SN, Metabolomics Society Education and Training Committee (2022) Providing metabolomics education and training: pedagogy and considerations. Metabolomics 18(12):106. https://doi.org/10.1007/s11306-022-01957-w

Wu Y, Xiao S, Qi J, Gong Y, Li K (2022) Pseudomonas fluorescens BsEB-1: an endophytic bacterium isolated from the root of Bletilla striata that can promote its growth. Plant Signal Behav 17(1):2100626. https://doi.org/10.1080/15592324.2022.2100626

Xiao Z, Zhao Q, Li W, Gao L, Liu G (2023) Strain improvement of Trichoderma harzianum for enhanced biocontrol capacity: strategies and prospects. Front Microbiol 14:1146210. https://doi.org/10.3389/fmicb.2023.1146210

Yan S, Bhawal R, Yin Z, Thannhauser TW, Zhang S (2022) Recent advances in proteomics and metabolomics in plants. Mol Hortic 2(1):17. https://doi.org/10.1186/s43897-022-00038-9

Yu H, Li J (2022) Breeding future crops to feed the world through de novo domestication. Nat Commun 13:1171. https://doi.org/10.1038/s41467-022-28732-8

Yuan H, Wang X, Gao L, Wang T, Liu B, Fang D, Gao Y (2023) Progress towards the Sustainable Development Goals has been slowed by indirect effects of the COVID-19 pandemic. Commun Earth Environ 4:184. https://doi.org/10.1038/s43247-023-00846-x

Zhang L, Zhang J, Wei Y, Hu W, Liu G, Zeng H, Shi H (2020) Microbiome- wide association studies reveal correlations between the structure and metabolism of the rhizosphere microbiome and disease resistance in cassava. Plant Biotechnol J 19:689–701. https://doi.org/10.1111/pbi.13495

Zhang S, Huang G, Zhang Y, Lv X, Wan K, Liang J, Feng Y, Dao J, Wu S, Zhang L, Yang X, Lian X, Huang L, Shao L, Zhang J, Qin S, Tao D, Crews TE, Sacks EJ, Lyu J, Wade LJ, Hu F (2023)

Sustained productivity and agronomic potential of perennial rice. Nat Sustainability 6:81–88. https://doi.org/10.1038/s41893-022-00997-3
Zhao J, Wang J, Liu J, Zhang P, Kudoyarova G, Liu CJ, Zhang K (2024) Spatially distributed cytokinins: metabolism, signaling, and transport. Plant Commun 5(7):100936. https://doi.org/10.1016/j.xplc.2024.100936
Zheng S, Yin K, Yu L (2022) Factors influencing the farmer's chemical fertilizer reduction behavior from the perspective of farmer differentiation. Heliyon 8(12):e11918. https://doi.org/10.1016/j.heliyon.2022.e11918
Zhou W, Arcot Y, Medina RF, Bernal J, Cisneros-Zevallos L, Akbulut MES (2024) Integrated Pest management: an update on the sustainability approach to crop protection. ACS Omega 9(40):41130–41147. https://doi.org/10.1021/acsomega.4c06628
Zhu X, Xu T, Peng C, Wu S (2022) Advances in MALDI mass spectrometry imaging single cell and tissues. Front Chem 9:782432. https://doi.org/10.3389/fchem.2021.782432

Microbial Metabolomics for Pest Management: Leads and Flaws

C. V. Babu, Balenahalli Narasingappa Ramesh, B. N. Ganavi, and Gatta Vishnu Vyas

Abstract

Microbial metabolomics presents a promising frontier in the realm of pest management, offering novel solutions for combating agricultural pests and vectors of disease. This chapter provides an overview of microbial metabolomics and its applications in pest control. It explores the role of microbial metabolites as potential biocontrol agents, elucidating their mechanisms of action and showcasing successful case studies. The chapter also examines the use of microbial metabolomics in identifying new bioactive compounds and screening microbial communities for pest control purposes. However, this burgeoning field faces challenges such as the variability in microbial metabolite production and the complexity of microbial interactions in natural environments. Strategies to overcome these challenges, including advancements in analytical techniques and interdisciplinary collaboration, are discussed. Looking ahead, the chapter highlights future perspectives for microbial metabolomics in precision pest management and emphasizes the importance of continued research and innovation in this area. Overall, microbial metabolomics holds great promise as a sustainable and effective approach to pest management, but further exploration and collaboration are needed to realize its full potential.

C. V. Babu · B. N. Ganavi · G. V. Vyas
Department of Plant Biotechnology, University of Agricultural Sciences, GKVK, Bengaluru, Karnataka, India

B. N. Ramesh (✉)
ICAR- Post-Harvest Engineering and Technology, University of Agricultural Sciences, GKVK, Bengaluru, Karnataka, India
e-mail: rameshbn20@uasbangalore.edu.in

S. Kaur et al. (eds.), *Microbial Metabolomics*,
https://doi.org/10.1007/978-981-96-4824-5_13

Keywords

Microbial metabolomics · Biocontrol · Biopesticide

1 Introduction

Microbial diversity represents an extensive and largely untapped reservoir of bioactive compounds, with significant potential across various sectors, including industrial, pharmaceutical, and agricultural applications. Within agriculture, microbial biocontrol agents have gained prominence for their role in managing insect pests and plant diseases. Although the potential of microbial biocontrol was recognized as early as the mid-1990s, its widespread adoption has been overshadowed by the dominance of chemical pesticides, initially favored for their rapid and broad-spectrum efficacy (Arthurs and Dara 2019; Harish et al. 2021). However, the environmental and health risks associated with chemical pesticides, as famously highlighted in Rachel Carson's *Silent Spring*, have led to a paradigm shift toward sustainable pest management practices that prioritize ecological balance and long-term agricultural sustainability (Kumar et al. 2021a, b).

The current challenges in modern agriculture, such as the development of pesticide-resistant pest populations and the global spread of invasive species, underscore the urgent need for alternative, eco-friendly crop protection strategies. Microbial pesticides, particularly those derived from microbial pathogens and their metabolites, offer significant advantages over chemical pesticides, including high host specificity, biodegradability, and minimal environmental impact (Carlini and Grossi-de-Sá 2002; Olson 2015; Mnif and Ghribi 2015). Despite these benefits, the adoption of microbial pesticides has been limited due to inconsistent field performance, short shelf life, limited commercial availability, and a general lack of awareness among farmers (Kumar et al. 2021a, b).

The global biopesticide market, reflecting a growing interest in sustainable alternatives, was valued at approximately $3 billion in 2013 and is projected to exceed $4.5 billion by 2030, with microbial products representing the fastest-growing segment (Olson 2015; Harish et al. 2021). Among the most prominent microbial biopesticides are strains of *Bacillus thuringiensis* and various entomopathogenic fungi, such as *Beauveria bassiana* and *Metarhizium anisopliae*, which target a broad spectrum of arthropod pests (Dunham 2015; Pan et al. 2023). The efficacy of microbial biocontrol agents is often mediated by their secondary metabolites, which have garnered attention for their unique structures and modes of action. Notably, compounds such as avermectins and spinosyns, produced by *Streptomyces* species, have demonstrated significant pesticidal activity and environmental safety, underscoring the potential of microbial metabolites as viable alternatives to synthetic pesticides (Tanaka and Omura 1993; Idris et al. 2022; Vachon et al. 2012).

While synthetic pesticides have long been effective in controlling pests, their adverse effects on soil biodiversity, aquatic ecosystems, and human health are well-documented (Syromyatnikov et al. 2020). The overuse of these chemicals can lead to soil degradation, reduced fertility, and the disruption of beneficial soil microorganisms, ultimately impacting crop productivity and ecosystem health (Kumar et al. 2021a, b). Moreover, pesticide contamination in water bodies through runoff and leaching poses significant risks to aquatic life and human health, with the potential for bioaccumulation and biomagnification leading to chronic diseases, such as cancer and diabetes (Jayaraj et al. 2017; Sabarwal et al. 2018).

In contrast, biopesticides, which include microbial pesticides, phytopesticides, and nanobiopesticides, present a more sustainable and environmentally friendly alternative to synthetic chemicals. These biopesticides are characterized by their specificity in targeting pests, reduced environmental persistence, and lower risk of residues in food products (Harish et al. 2021; Hummadi et al. 2021). However, the specificity that makes biopesticides desirable can also be a limitation, as it may necessitate the use of multiple products to control different pests simultaneously. Additionally, their short shelf life, while advantageous in terms of environmental impact, may hinder their effectiveness in long-term pest management strategies (Idris et al. 2022).

Given these complexities, a critical assessment of microbial metabolomics in pest management is essential. This chapter explores the potential of microbial metabolites as sustainable pest management tools, examining both the promising leads and the inherent challenges. By analyzing the current state of microbial pesticides and discussing strategies to overcome existing limitations, this chapter aims to contribute to developing more effective and sustainable pest management practices that align with the goals of modern agriculture.

2 Importance of Understanding Microbial Metabolomics

Microbial metabolomics is a rapidly advancing field that provides comprehensive insights into the metabolites produced by microorganisms, collectively known as the metabolome. These metabolites, which are the end products of cellular metabolism, offer crucial information about the organism's physiological state and its interactions with the environment. In pesticide production, understanding these metabolites is essential for identifying and optimizing bioactive compounds that can be developed into effective biopesticides.

Key techniques in microbial metabolomics include mass spectrometry (MS), which is widely used due to its high sensitivity and specificity. MS is often coupled with gas chromatography (GC) or liquid chromatography (LC) to enhance the separation of complex mixtures, allowing for more accurate identification of metabolites. Nuclear magnetic resonance (NMR) spectroscopy also plays a critical role by providing detailed structural information through a non-destructive method, while chromatography techniques like LC and GC are essential for separating metabolites before detection (Chen et al. 2023). Bioinformatics tools are increasingly vital in this field, enabling the interpretation of vast datasets and linking metabolomic data with other omics data to offer a holistic view of microbial metabolic functions (Noecker et al. 2019).

The application of microbial metabolomics in pesticide production has been demonstrated through several recent studies. For example, research on *Bacillus thuringiensis* has shown that specific metabolites, such as delta-endotoxins, possess strong insecticidal properties, making them suitable for biopesticide formulations (Caballero et al. 2020). Similarly, studies on *Trichoderma* species have revealed that secondary metabolites like trichodermin not only enhance plant growth but also increase resistance to diseases, thereby optimizing biopesticide formulations (Hamrouni et al. 2024). Additionally, investigations into *Pseudomonas fluorescens* have identified metabolites such as 2,4-diacetylphloroglucinol, which inhibit pathogenic fungi and protect crops from infections, further illustrating the potential of microbial metabolites in biopesticide development (Pandiyan et al. 2024).

The adaptability of microorganisms in response to environmental stressors has also been highlighted in recent research. For instance, Wang et al. (2022) identified that certain bacteria produce novel metabolites like 3-hydroxybutyric acid, which aid in their adaptation to environmental challenges. This adaptability is crucial for understanding how microbial metabolism can be leveraged for biotechnological applications, including the development of biopesticides and environmental remediation.

Recent advancements in microbial metabolomics have also been facilitated by the integration of machine learning (ML) techniques, which have significantly improved the identification and annotation of microbial metabolites with pesticidal properties. ML approaches address key challenges such as data preprocessing and metabolite identification in complex microbial communities, which are essential for the successful development of biopesticides. For instance, Shen et al. (2023) developed a random forest model to predict the pesticidal potential of metabolites derived from various microbial species, including *Bacillus* and *Pseudomonas*. This model utilized features such as molecular structure, physicochemical properties, and functional groups to predict bioactivity, leading to the identification of several novel metabolites with broad-spectrum pesticidal properties, which were experimentally validated through bioassays.

Moreover, the importance of standardized protocols for sample preparation in microbial metabolomics has been increasingly recognized. Shankar and Sharma (2023) emphasize the need for optimized extraction methods tailored to specific bacterial strains to ensure reliable metabolite analysis. These standardized protocols are crucial for the reproducibility and comparability of metabolomic studies across different research settings, which is vital for the consistent production and application of microbial pesticides. Patel et al. (2022) further highlight the role of standardized data processing pipelines in enhancing the accuracy of metabolomic studies, thereby supporting the robust identification of metabolites with pesticidal properties.

Ongoing research in microbial metabolomics continues to drive innovations in sustainable agricultural practices by identifying and optimizing microbial metabolites with pesticidal properties. For example, Zhang et al. (2024) demonstrated the application of a support vector machine model to discover new microbial metabolites with pesticidal activity from soil-derived bacteria. The identified metabolites showed significant efficacy against a range of agricultural pests, underscoring the potential of ML techniques in facilitating the development of biopesticides as alternatives to chemical pesticides.

3 Role of Microbial Metabolites in Pest Control

Microbial metabolites have emerged as promising alternatives to chemical pesticides in agricultural pest management due to their high target specificity and ecological safety (Subbanna et al. 2020). These metabolites, produced by various bacteria and fungi, exhibit insecticidal, antifungal, antibacterial, and nematicidal activities (Verma et al. 2020). Some commercially successful examples include avermectins and spinosad (Dara 2021). Biopesticides based on fermentation solids and solubles, without live microbes, are also available in the market (Dara 2021). Plant secondary metabolites have gained interest as potential biopesticides, offering novel chemical frameworks and modes of action (Pino et al. 2013). However, challenges in commercialization include sustainable use of raw materials, standardization of complex extracts, and regulatory requirements. Despite these challenges, microbial metabolites play a critical role in integrated pest management and maintaining crop productivity (Dara 2021).

Microbial metabolites play a crucial role in pest control, offering a natural and eco-friendly alternative to chemical pesticides. These metabolites are secondary products produced by microorganisms such as bacteria, fungi, and actinomycetes, and they can have various biological activities that are effective against pests. Recent studies have discovered novel secondary metabolites from various microbial agents, primarily using conventional activity monitoring and basic molecular tools. A variety of microbial metabolites with pesticidal activity have seen both insecticides and fungicides.

4 Microbial Metabolites with Insecticidal Activity

Bacteria produce various secondary metabolites, many effective against insect pests and plant pathogens. Entomopathogenic bacteria like *Bacillus*, *Xenorhabdus*, and *Photorhabdus*, as well as soil bacteria like *Pseudomonas*, are extensively researched for their ability to produce these metabolites. For instance, *Bacillus* species produce various secondary metabolites such as Cry toxins, Cytolysins, Vegetative Insecticidal Proteins (VIPs), Thurungiensin, and Phospholipase C (Table 1), which disrupt insect midgut cells, synergize other insecticidal proteins, or interfere with RNA polymerase activity, leading to insect death. *Xenorhabdus* and *Photorhabdus* bacteria, which are symbiotically associated with nematodes, produce secondary metabolites that suppress the immune system of insect hosts after invading their hemolymph. Compounds like benzylideneacetone (BZA) and p-hydroxyphenylpropionic acid (PHPP) inhibit key enzymes.

Insects are the largest and most diverse animal group on Earth, with over 700 fungal species from 100 genera adapting to the entomopathogenic fungi (EPF)

Table 1 Microbial metabolites obtained from different microorganisms (bacteria, fungi, actinomycetes) showing insecticidal activity

Secondary metabolite	Source	Mode of action	Target order	References
Bacteria				
Cry toxin	*Bacillus thuringiensis, B. papillae, Clostridium bifermentans*	Disruption of midgut cell and causing osmotic imbalance	Lepidoptera, diptera, and coleoptera	Bravo et al. (2007)
Cytolysins	*Bacillus thuringiensis*	Causes colloidal osmotic lysis	Lepidoptera and coleoptera	Knowles and Ellar (1987)
Vegetative insecticidal proteins (VIPs)	*Bacillus thuringiensis, B. cereus*	Prevent microfilament formation (VIP 2). apoptotic cell death of midgut epithelial cells (VIP 3)	Lepidoptera and coleoptera	Warren (1997)
Thuringiensin/β- exotoxin	*Bacillus thuringiensis*	Interfering the RNA polymerase activity by competing with ATP binding sites	Diptera, Lepidoptera, Coleoptera, Orthoptera, Hymenoptera, and Isoptera and some nematode pests.	Liu et al. (2014)
Phospholipase C	*Bacillus cereus*	Hydrolysis of glycerophospholipids which directly influence membrane dynamics and cellular signaling	Coleoptera	Binnington and Baule (1993)
Benzylideneacetone (BZA), p-hydroxyphenylpropionic acid (PHPP	*Photorhabdus luminescens. Xenorhabdus nematophila*	Inhibit phospholipase A2 (PLA2) and shut down eicosanoid biosynthesis	Lepidoptera	Maharana et al. (2022)
Fungi				
Cordycepins	*Cordyceps*	Inhibitory action on chitin synthesis	Lepidoptera	Charnley (2003)
Dustanin, Hypocrellins, 3-hopane- triterpenes	*Hypocrella Aschersonia*	Cytotoxicity against insect cells	Aleyrodidae, Coccidae families of Hemiptera, and nematodes	Butt et al. (2001)
Beauvericin, Bassianin, Oosporein, and bassianolide	*Beauveria*	Malfunctioning of different enzymes by disorienting tertiary structures through SH group redox reaction of amino acids	Lepidoptera, Coleoptera, Hemiptera, Homoptera, and hymenoptera	Steinrauf (1985)

Swainsonine and Destruxins	*Metarhizium*	Weakening the host immune defense and damaging muscular and digestive systems	Coleoptera, Hemiptera, Isoptera, Homoptera, Heteroptera, Diptera (mosquitoes). hymenoptera, Siphonaptera and Lepidoptera	Jeffs and Khachatourians (1997)
Beauvericin, Beauverolides, and dipicolinic acid (DPA)	*Paecilomyces (Isaria)*	Melanized spots and structural deformities on tracheal system	Hemiptera	Steinrauf (1985)
Hydroxycarboxylic acid, cyclosporine, and dipicolinic acid, Bassianolide	*Verticillium*	Solubilizing specific cuticular proteins and can synergize proteases and chitinases	Hemiptera and Thysanoptera (thrips)	Butt et al. (2001)
Efrapeptins, Tolypin, Diketopiperazines	*Tolypocladium*	Inhibitors of intracellular protein transport system and ATPase of mitochondria	Diptera (mosquitoes), Ephemeroptera (mayflies)	Charnley (2003)
Hirsutellin A and B	*Hirsutella*	Inhibit protein synthesis by specific cleavage of rRNA and ribotoxins	Mites (citrus rust mites-Phyllocoptruta oleivora), Lepidotera	Omoto and McCoy (1998)
Actinomycetes				
Spinosyns	*Saccharopolyspora spinosa*	Involuntary Muscle contractions, tremors, and paralysis	Lepidoptera, Diptera, Thysanoptera, and some species of Coleoptera and Orthoptera	Sparks et al. (2012)
Avermectins	*Streptomyces avermitilis*	Affect neural and neuromuscular transmissions in Central nervous system	Mite species	Wang et al. (2011)
Polyoxins and Nikkomycins		Insect growth regulators, Inhibiting chitin formation	Lepidoptera and Coleoptera	Binnington and Baule (1993)
Chitinase	*Streptomyces griseus*	Degradation of chitin	Lepidoptera, Coleoptera, Hemiptera, and hymenoptera	Gohel et al. (2006)

lifestyle. Most insect-pathogenic fungi belong to the order *Hypocreales*, particularly in the families *Cordycipitaceae*, *Clavicipitaceae*, and *Ophiocordycipitaceae*. Key genera like *Beauveria*, *Cordyceps*, and *Isaria* produce entomotoxins, including non-ribosomal peptides and polyketides. *Beauveria* and *Metarhizium* (*Clavicipitaceae*) are the most commercially important bio-insecticides, infecting over 200 insect species, many of which are significant agricultural pests or disease vectors.

5 Microbial Metabolites with Antifungal and Antibacterial Activity

Microbial metabolites can have various biological activities, including antibiotic properties that are effective against plant pathogens. By suppressing the growth of these harmful microorganisms, microbial metabolites can protect crops from diseases, which in turn reduces their vulnerability to pest damage. These show antibiotic properties that help in the disruption of cell walls or membranes, inhibition of enzyme activity, and suppression of plant pathogens by biocontrol agents (Table 2). Using microbial metabolites for disease and pest management is an environmentally friendly approach. Unlike chemical pesticides, which can have adverse effects on non-target organisms and the environment, microbial metabolites are generally more specific and biodegradable.

6 Applications of Microbial Metabolomics in Pest Management

Microbial metabolomics is emerging as a powerful tool for identifying bioactive compounds and screening potential biocontrol agents. This approach allows for global profiling of low-molecular-weight metabolites, facilitating the discovery of novel antimicrobial natural products (Wu et al. 2016). Metabolomics, combined with genomics and transcriptomics, can enhance the bioprospecting and characterization of microbial biocontrol agents, particularly for sustainable agriculture. Modern screening methods have evolved to include high-content screens and phenotypic assays, which can reveal cellular interactions and potential off-target effects of bioactive compounds (Kellogg and Kang 2020). The integration of metabolomics with genome sequencing and bioinformatics has led to the discovery of novel bioactive substances, including antimicrobial peptides (AMPs) that show promise as biopesticides (Caesar et al. 2021). These advancements in omics technologies are driving innovation in the identification and development of microbial biocontrol agents for agricultural applications.

This approach involves the use of sophisticated analytical techniques such as mass spectrometry (MS), nuclear magnetic resonance (NMR) spectroscopy, and chromatography (liquid or gas) coupled with mass spectrometry (LC-MS or GC-MS). These techniques enable the detection, quantification, and identification

Table 2 Microbial metabolites obtained from different microorganisms (bacteria, fungi, actinomycetes) showing antifungal and antibacterial activity

Secondary metabolite	Origin microorganism	Target pest	References
Macrolactin A	*Bacillus* sp. *sunhua*	Potato scab pathogen (Streptomyces scabies)	Han et al. (2005)
Syringomycin E	*Pseudomonas syringae*	Citrus green mold (Penicillium digitatum)	Bull et al. (1998)
Blasticidin S	*Streptomyces griseochromogenes*	Rice blast caused by Pyricularia oryzae	Fu et al. (2011)
Kasugamycin	*Streptomyces kasugaensis*	Rice blast (Pyricularia oryzae), leaf spot in sugar beet and celery (Cercospora sp.), and scab in pears and apples (Venturia sp.)	Umezawa et al. (1965)
Cryptocin	*Cryptosporiopsis quercina*	Rice blast (Pyricularia oryzae)	Strobel et al. (1999)
Cytochalsins	*Phomopsis* sp.	Sclerotinia sclerotiorum, Fusarium oxysporum, and Botrytis cinerea	Fu et al. (2011)
Colletotric acid	*Colletotrichum gloesporioides*	Brown spot of rice (Helminthosporium sativum)	Zou et al. (2000)
Rufuslactone	*Lactarius rufus*	Alternaria brassicae, Botrytis cinerea	Luo et al. (2005)
Oxytetracycline	*Streptomyces rimosus*	Fire blight of apple (Erwinia amylovora)	Finlay et al. (1950)
Streptomycin	*Streptomyces griseus*	Xanthomonas oryzae, X. citri, Pseudomonas tabaci	Saxena (2014)
Gluconic acid	*Pseudomonas* strain ANS	*Gaeumannomyces graminis var. tritici*	Barret et al. (2009)
Volatile organic compound	*Pseudomonas* spp. *Serratia* spp. *Stenotrophomonas* spp., etc	Many soil-borne fungal pathogens	Naz et al. (2022)
Siderophore	*Pseudomonas* sp. *Rhizobium meliloti*	*Verticillium, Pythium, Fusarium, Macrophomina phaseolina*	Boruah and Hage (2024)
Bacillomycin, fengycin	*Bacillus amyloliquefaciens* FZB42	*Fusarium oxysporum*	Chen et al. (2009)
Bacillomycin D	*Bacillus vallismortis* ZZ185	*Fusarium graminearum, Alternaria alternata. Rhizoctonia solani, Cryphonectria parasitica,* and *Phytophthara capsici Aspergillus flavus*	Zhao et al. (2010)
Iturin, fengycin, and surfactin A	*Bacillus subtilis*	*Colletotrichum glocosporioides, B. cinerea* and *R solane*	Kim et al. (2010)
Mycosubtilin	*B subtitis* BBG100	*Pythium aphanidermatum*	Leclère et al. (2005)

(continued)

Table 2 (continued)

Secondary metabolite	Origin microorganism	Target pest	References
Zwittermicin A Kanasomine	*Bacillus cereus* UW85	*Phytophthora medicaginis and P. aphanidermatum*	Stabb et al. (1994)
Wuyiencin	*S. hygroscopias var. wwyiensis*	*Botrytis cinerea*	Shi et al. (2020)
Xanthobaccin A	*Lysobacter* sp. strain SB- K88	*Aphanomyces cochlioides*	Tofazzal Islam (2010)

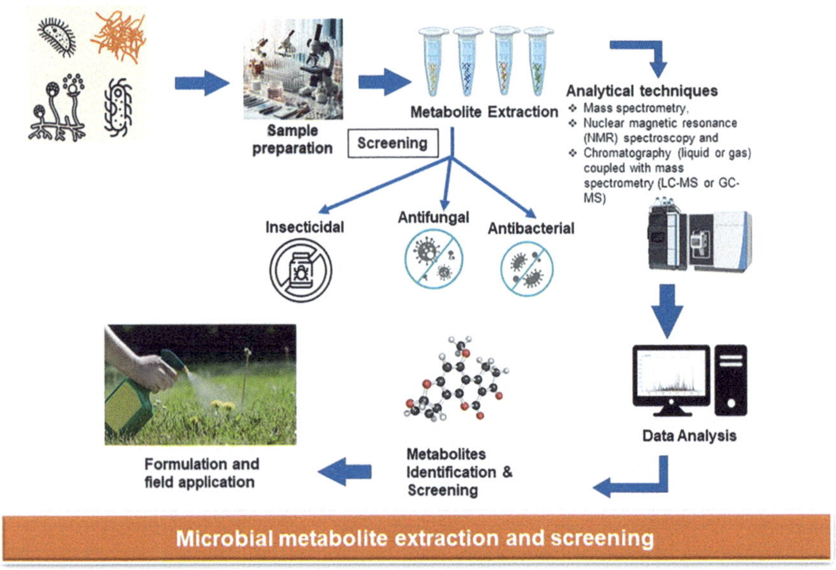

Fig. 1 Pictorial representation of the development of microbial metabolite-based biopesticide

of a wide range of metabolites produced by microorganisms depicted in Fig. 1 (Leiss et al. 2011).

7 Screening for Biocontrol Agents

Biocontrol agents are microorganisms or their metabolites that can suppress the growth of plant pathogens, pests, or weeds, offering a sustainable alternative to chemical pesticides. The use of biocontrol agents is becoming increasingly important in agriculture as part of integrated pest management (IPM) strategies aimed at reducing the reliance on chemical inputs and promoting environmental sustainability.

7.1 Microbial Antagonists

Microbial metabolomics can identify metabolites produced by microbial antagonists that inhibit the growth of plant pathogens. For example, *Bacillus* species are known for producing lipopeptides such as surfactin, iturin, and fengycin, which have strong antifungal properties. Metabolomic profiling of *Bacillus* strains can reveal the presence of these lipopeptides and other metabolites that contribute to their biocontrol efficacy. *Bacillus*-based nano-bioformulations can be used for effective and eco-friendly management of plant diseases and insect pests (Kumar et al. 2021a, b).

7.2 Induced Systemic Resistance (ISR)

Certain metabolites produced by beneficial microorganisms can trigger ISR in plants, enhancing their resistance to pathogens. Microbial metabolomics can help identify these ISR-inducing metabolites and the conditions under which they are produced. For instance, metabolites from *Pseudomonas fluorescens* have been shown to induce ISR in plants, leading to increased resistance to bacterial and fungal pathogens (Naz et al. 2022).

7.3 Suppression of Soil-Borne Pathogens

Soil-borne pathogens pose a significant challenge in agriculture, and biocontrol agents can play a key role in their management. Microbial metabolomics can identify secondary metabolites that suppress soil-borne pathogens, such as nematicidal compounds produced by *Streptomyces* species. By understanding the metabolite profiles of these microorganisms, researchers can select the most effective biocontrol agents for specific agricultural applications (Subbanna et al. 2020).

8 Plant Growth-Promoting Rhizobacteria (PGPR)

PGPR are beneficial bacteria that enhance plant growth by various mechanisms, including the production of phytohormones, solubilization of nutrients, and suppression of pathogens. Microbial metabolomics can be used to analyze the metabolites produced by PGPR and identify those that contribute to plant growth promotion and disease suppression. This information can be used to formulate effective biocontrol products that improve crop yields and reduce the need for chemical fertilizers and pesticides (Dara 2021).

9 Challenges and Limitations

9.1 Variability in Microbial Metabolite Production

The variability in microbial metabolite production is a significant barrier to the consistent and effective use of microbial metabolites in pest management. This variability arises from several factors, including genetic diversity within microbial populations, environmental conditions, and the availability of nutrients (Boness et al. 2023). Microorganisms are highly responsive to their environment, and even small changes in temperature, pH, or nutrient availability can lead to substantial differences in the types and quantities of metabolites produced (Rusu et al. 2023; Fiorini et al. 2022; Zhou and Zhao 2021).

This variability is particularly challenging when attempting to standardize the production of microbial metabolites for commercial biocontrol products. In laboratory settings, controlled conditions can be maintained to optimize metabolite production. However, when these microorganisms are introduced into field conditions, the environmental variability can result in unpredictable changes in metabolite profiles (Thukral et al. 2023). For example, a microbial strain that produces a high yield of pesticidal compounds in the lab may show significantly reduced production under field conditions, leading to inconsistent pest control outcomes. To address this issue, research is increasingly focused on identifying the key regulatory pathways that control metabolite production and developing strategies to stabilize these pathways under varying conditions. This could involve genetic engineering to create more robust microbial strains that are less sensitive to environmental fluctuations or the development of formulation techniques that protect and stabilize the active metabolites. Additionally, there is growing interest in the use of synthetic biology to construct synthetic microbial communities that can maintain stable metabolite production in diverse environments (Rusu et al. 2023; Fiorini et al. 2022; Zhou and Zhao 2021; Santamaria et al. 2022).

Another approach being explored is the use of elicitors or inducers of chemical compounds that can trigger or enhance metabolite production. By applying these elicitors in conjunction with microbial biocontrol agents, it may be possible to achieve more consistent metabolite production in the field (Benchlih et al. 2023). However, this approach requires a deep understanding of the specific metabolic pathways involved and the conditions under which they are activated. The development of reliable methods to monitor and quantify metabolite production in real time is also crucial (Zehra et al. 2021). Advances in biosensor technology and in-field diagnostic tools could provide the necessary means to track metabolite levels and adjust application strategies accordingly. These tools could enable more precise and adaptive pest management approaches, where the application of microbial metabolites is fine-tuned based on real-time feedback from the field (Whipps 2001; Abdul Malik et al. 2020; Pršić and Ongena 2020).

9.2 Complexity of Microbial Interactions in Natural Environments

The natural environment is a complex and dynamic ecosystem where microorganisms interact with each other, with plants, and with pests in multifaceted ways. These interactions can have profound effects on the production and efficacy of microbial metabolites used in pest management (Weiland-Bräuer 2021). In natural environments, microorganisms are rarely found in isolation; they exist within communities where they compete for resources, communicate through chemical signals, and sometimes cooperate in ways that can either enhance or inhibit metabolite production (Weiland-Bräuer 2021).

For instance, in a soil environment, a biocontrol microorganism may interact with other soil microbes in ways that affect its ability to produce pesticidal metabolites (Geesink et al. 2024). These interactions can be competitive, with other microbes outcompeting the biocontrol agent for nutrients or inhibiting its growth through the production of antimicrobial compounds. Conversely, interactions can be synergistic, where the presence of other microbes stimulates the production of certain metabolites through mechanisms such as quorum sensing or metabolic cross-feeding (Weiland-Bräuer 2021). The complexity of these interactions is further compounded by the influence of environmental factors such as soil type, moisture levels, and the presence of organic matter. These factors can affect the composition and activity of the microbial community, leading to variability in metabolite production and pest control outcomes. Additionally, the interaction between the biocontrol agent and the target pest is not static; it can change over time as the pest adapts to the presence of the microbial metabolite or as the pest population shifts in response to environmental pressures. To effectively harness microbial metabolites for pest management, it is essential to understand and manage these complex interactions (Tang et al. 2023; Gallardo-Navarro et al. 2024; Arıkan and Muth 2023). This requires a multidisciplinary approach that integrates microbiology, ecology, and agronomy. Techniques such as metagenomics, transcriptomics, and metabolomics are increasingly being used to study the composition and function of microbial communities in natural environments. These omics technologies provide insights into the metabolic networks that drive metabolite production and reveal how these networks are influenced by interactions within the microbial community and with the environment. Moreover, mathematical modeling and systems biology approaches are being employed to predict the outcomes of microbial interactions and to design microbial consortia that can achieve desired outcomes in pest management. These models can help identify key factors that need to be controlled to ensure consistent metabolite production and can guide the development of strategies to manipulate microbial communities in favor of biocontrol agents (Abbas et al. 2024; Deshayes et al. 2017; Zehra et al. 2021).

9.3 Technological and Methodological Challenges

The technological and methodological challenges associated with microbial metabolomics are significant, particularly in the context of its application to pest management (Gomes et al. 2023). Advanced analytical technologies such as mass spectrometry (MS) and nuclear magnetic resonance (NMR) spectroscopy are the cornerstones of metabolomics analysis, but they come with inherent challenges that must be addressed to fully realize the potential of microbial metabolites in agricultural applications. Mass spectrometry, for example, is a highly sensitive technique capable of detecting and quantifying a wide range of metabolites (Gomes et al. 2023; Noecker et al. 2019). However, the complexity of the data generated by MS can be overwhelming (Guo et al. 2022) Metabolomics profiles typically consist of hundreds to thousands of peaks, each corresponding to different metabolites, and distinguishing between these metabolites particularly isobaric compounds that share the same mass but have different structures requires high-resolution instruments and advanced data analysis techniques (Parker et al. 2023). Additionally, the identification of unknown metabolites remains a significant challenge, as many microbial metabolites are not yet represented in existing databases (Li et al. 2023). NMR spectroscopy, on the other hand, provides detailed information about the structure of metabolites but is less sensitive than MS and requires larger sample sizes. NMR is particularly useful for the elucidation of complex structures and the study of dynamic interactions between metabolites and other molecules. However, the high cost of NMR equipment and the need for specialized technical expertise limit its accessibility, particularly in resource-limited settings (Chen et al. 2022). The integration of MS and NMR data with other omics data (e.g., genomics, proteomics) is another challenge. While this integration can provide a more comprehensive understanding of microbial metabolism, it requires sophisticated bioinformatics tools and a multidisciplinary approach to data analysis. The development of standardized protocols for sample preparation, data acquisition, and analysis is also needed to ensure reproducibility and comparability of results across different studies (Guo et al. 2022; Parker et al. 2023; Li et al. 2023; Chen et al. 2022).

Furthermore, the scalability of these technologies for field applications is limited. While laboratory-based metabolomics studies can provide detailed insights into microbial metabolism, translating these findings into practical pest management solutions requires the development of portable and cost-effective analytical tools that can be used in the field (Parker et al. 2023). Recent advances in microfluidics and portable MS technologies are promising in this regard, but further development and validation are needed to bring these tools to maturity. In addition to technological challenges, there are methodological challenges related to the experimental design and interpretation of metabolomics data (Li et al. 2023). The complexity of microbial metabolism and the influence of environmental factors make it difficult to establish clear cause-and-effect relationships between specific metabolites and pest control outcomes. Experimental designs that account for this complexity, such as factorial experiments and time-series analyses, are needed to unravel the intricate relationships between microbial metabolites, environmental conditions, and pest dynamics (Weiland-Bräuer 2021; Wang et al. 2022).

9.4 Case Studies Highlighting These Challenges

Numerous case studies highlight the practical challenges associated with the use of microbial metabolites in pest management. For instance, the use of lipopeptides produced by *Bacillus subtilis* as biocontrol agents for root-knot nematodes (*Meloidogyne* spp.) in tomato plants has shown promise in laboratory studies. However, field trials have revealed significant variability in the efficacy of these lipopeptides, which has been attributed to differences in soil composition, moisture levels, and the presence of competing microorganisms. These factors affect both the production of lipopeptides by *Bacillus subtilis* and their stability in the soil, leading to inconsistent pest control outcomes (Leconte et al. 2022; Zhang et al. 2022).

Similarly, the application of microbial metabolites for the control of insect pests in maize has faced challenges related to the complexity of the soil microbiome. In some cases, the introduction of biocontrol agents has led to unintended shifts in the microbial community, resulting in the suppression of beneficial microbes or the emergence of opportunistic pathogens (Köhl et al. 2019). These case studies underscore the importance of considering the broader ecological context when developing and deploying microbial metabolites for pest management. In another example, the use of fungal metabolites for controlling plant pathogens has been complicated by the environmental stability of these compounds. Many fungal metabolites are sensitive to environmental conditions such as UV radiation, temperature fluctuations, and moisture levels. As a result, their efficacy can be significantly reduced in field conditions compared to controlled laboratory environments. Strategies to enhance the stability of these metabolites, such as encapsulation or the development of more stable analogs, are being explored to overcome these challenges (Huynh et al. 2022; Coolen et al. 2022; Tang 2011).

These case studies highlight the need for a more holistic approach to the development of microbial metabolites for pest management, one that considers not only the microbial strain and metabolite of interest but also the complex interactions with the environment and other organisms. This approach requires a combination of laboratory research, field trials, and ecological studies to ensure that biocontrol agents are effective, stable, and safe in real-world conditions.

10 Overcoming Challenges

10.1 Advancements in Analytical Techniques

Recent advancements in analytical techniques have significantly enhanced the ability to detect, identify, and quantify microbial metabolites, addressing some of the challenges associated with their variability and complexity (Tang 2011; Canciu et al. 2021). High-resolution mass spectrometry (HRMS), for instance, has improved the sensitivity and specificity of metabolite detection, enabling the identification of even minor components in complex mixtures. HRMS coupled with tandem mass spectrometry (MS/MS) has become a powerful tool for the structural elucidation of metabolites, allowing researchers to determine the precise molecular structure of novel compounds. One of the key benefits of HRMS is its ability to provide accurate mass measurements, which are crucial for distinguishing between isobaric compounds. This capability, combined with advanced fragmentation techniques, allows for the comprehensive characterization of metabolites and the identification of previously unknown compounds. The development of HRMS-based metabolomics workflows has also facilitated the discovery of novel bioactive metabolites with potential applications in pest management. Nuclear magnetic resonance (NMR) spectroscopy has also seen significant advancements, particularly in the development of cryogenically cooled probes that enhance sensitivity and reduce sample volume requirements (Vervoort et al. 2021). This has made NMR more accessible for the study of microbial metabolites, particularly those present in low concentrations. Additionally, the integration of NMR with other analytical techniques, such as liquid chromatography (LC) and MS, has enabled more comprehensive metabolic profiling and structural elucidation. In parallel, advances in data analysis and bioinformatics tools have improved the ability to interpret complex metabolomics data (Bharti and Grimm 2021). Machine learning algorithms and network-based approaches are increasingly being used to analyze metabolomics datasets, identify patterns, and predict the biological activity of metabolites. These tools can help to uncover relationships between metabolites and their effects on pests, providing new insights into the mechanisms of action of microbial biocontrol agents (Canciu et al. 2021; Vervoort et al. 2021; Bharti and Grimm 2021).

Moreover, the development of portable analytical devices, such as miniaturized mass spectrometers and lab-on-a-chip technologies, holds great promise for field applications. These devices can provide real-time, on-site analysis of microbial metabolites, enabling more precise and adaptive pest management strategies. For example, portable MS devices can be used to monitor the levels of key metabolites in the field, allowing for immediate adjustments to application rates or the timing of biocontrol agent deployment (Wang et al. 2022; Mulligan and Vircks 2012).

These advancements in analytical techniques are not only improving the detection and characterization of microbial metabolites but are also paving the way for more effective and targeted pest management strategies. By providing a deeper understanding of the metabolic processes involved in biocontrol, these techniques are helping to overcome some of the key challenges associated with the variability and complexity of microbial metabolomics.

10.2 Innovative Approaches to Enhance Metabolite Stability and Production

Enhancing the stability and production of microbial metabolites is critical for their successful application in pest management. Innovative approaches are being developed to address the challenges associated with metabolite degradation and inconsistent production under field conditions (Tamano 2014). One approach involves the use of genetic engineering to modify microbial strains, making them more robust and capable of producing higher yields of metabolites (Alper et al. 2010). By targeting key regulatory genes and pathways involved in metabolite biosynthesis, researchers can create strains that are less sensitive to environmental fluctuations and more consistent in their metabolite production (Rusu et al. 2023). For example, the overexpression of genes involved in the production of secondary metabolites, such as no ribosomal peptide synthetases or polyketide synthases, can lead to increased yields of bioactive compounds. Another strategy focuses on the stabilization of metabolites through formulation techniques (Naseri and Koffas 2020). Encapsulation of microbial metabolites in protective matrices, such as nanoparticles or polymer coatings, can enhance their stability by shielding them from environmental factors like UV radiation, temperature changes, and microbial degradation. These formulations can also provide a controlled release of the active metabolites, ensuring a sustained and effective concentration in the field over time. This approach is particularly useful for metabolites that are prone to rapid degradation under field conditions, such as certain lipopeptides or volatile organic compounds (Kim et al. 2021; Singh et al. 2019; Sekurova et al. 2019).

In addition to genetic and formulation strategies, the use of synthetic biology offers new possibilities for optimizing metabolite production. Synthetic biology allows for the design of custom microbial strains with tailored metabolic pathways that can produce desired metabolites at high yields (Rojo et al. 2023). This can involve the introduction of synthetic gene circuits that regulate metabolite production in response to specific environmental cues or the construction of synthetic microbial consortia that work together to enhance metabolite production. For example, researchers have developed engineered microbial consortia where one strain produces a precursor metabolite that is then converted into a bioactive compound by another strain (Chubukov et al. 2016; Alper et al. 2010). This division of labor can lead to more efficient and stable metabolite production, as each strain can be optimized for its specific function. Additionally, synthetic biology techniques can be used to create synthetic pathways that bypass regulatory bottlenecks or reduce the production of unwanted byproducts, further improving the efficiency of metabolite production (Alper et al. 2010). The development of novel cultivation techniques also holds promise for enhancing metabolite production. For example, the use of bioreactors with controlled environmental conditions can optimize the growth of microbial strains and the production of metabolites. These bioreactors can be designed to mimic specific environmental conditions, such as soil pH or moisture levels, allowing for the study of metabolite production under conditions that closely resemble field environments. This approach can help identify the optimal conditions for metabolite production and improve the scalability of microbial metabolite production for commercial applications (Naseri and Koffas 2020; Sekurova et al. 2019; Chubukov et al. 2016; Rojo et al. 2023).

These innovative approaches are helping to overcome some of the key challenges associated with the use of microbial metabolites in pest management. By improving the stability and production of these metabolites, researchers are paving the way for more consistent and effective biocontrol strategies that can be applied in diverse agricultural settings.

10.3 Future Directions in Addressing Challenges

As the field of microbial metabolomics continues to evolve, several future directions are emerging that have the potential to address the current challenges and limitations in the application of microbial metabolites for pest management. One promising direction is the integration of advanced machine learning and artificial intelligence (AI) techniques with metabolomics (Petrick and Shomron 2022). Machine learning algorithms can be used to analyze complex metabolomics datasets, identify patterns, and make predictions about metabolite production and activity. For example, AI models can be trained to predict the effects of specific environmental conditions on metabolite production or to identify the key metabolites that contribute to biocontrol efficacy. These models can then be used to guide the design of microbial strains or the optimization of application strategies. Another emerging direction is the exploration of new microbial sources for metabolite discovery. While many studies have focused on well-characterized microbial species, there is a vast diversity of microorganisms in nature that remain unexplored (Galal et al. 2022). Advances in metagenomics and single-cell sequencing are making it possible to discover new microbial species and their associated metabolites directly from environmental samples, without the need for culturing. These techniques are opening up new avenues for the discovery of novel bioactive compounds with potential applications in pest management (Li et al. 2022; Galal et al. 2022; Petrick and Shomron 2022).

The development of more sophisticated synthetic biology tools is also expected to play a significant role in addressing the challenges associated with metabolite production and stability. For example, CRISPR-based genome editing techniques can be used to precisely modify microbial genomes, enabling the creation of strains with enhanced metabolite production or the ability to produce novel metabolites (Siddiqui et al. 2012; Chubukov et al. 2016). Additionally, the use of synthetic biology to engineer microbial communities with specific functions such as the production of a cocktail of metabolites with synergistic effects offers new possibilities for improving the efficacy and consistency of biocontrol agents. Collaborative research initiatives that bring together experts from different fields will continue to be important for addressing the multidisciplinary challenges of microbial metabolomics. These collaborations can help bridge the gap between laboratory research and field application, ensuring that biocontrol agents are both effective and practical for use in diverse agricultural settings. Public-private partnerships, in particular, have the potential to accelerate the translation of research findings into commercial products that can be deployed at scale (Rojo et al. 2023; Siddiqui et al. 2012; Chubukov et al. 2016).

Finally, the development of regulatory frameworks that support the use of microbial metabolites in agriculture will be crucial for the widespread adoption of these biocontrol agents. Clear guidelines for the evaluation and approval of microbial metabolites, along with incentives for their use, can help to overcome barriers to adoption and encourage the transition from chemical to biological pest control methods (Sundh et al. 2021). Policymakers and industry stakeholders will need to work together to create an enabling environment that supports innovation in microbial metabolomics and promotes the use of sustainable pest management strategies. These future directions highlight the potential for microbial metabolomics to continue advancing as a powerful tool for pest management. By addressing the current challenges and exploring new opportunities, researchers can develop more effective and sustainable biocontrol strategies that contribute to the long-term health of agricultural ecosystems (Tang 2011; Sundh et al. 2021).

11 Conclusion

Microbial metabolomics represents a cutting-edge approach to pest management, offering the potential to discover and utilize bioactive compounds from microbial sources as precise and effective biocontrol agents. These metabolites, due to their specificity, can target agricultural pests with minimal off-target effects on non-pest species and reduced environmental impact. However, the implementation of microbial metabolomics in pest management is challenged by several factors, including the variability in metabolite production and the complex interactions within microbial communities in natural ecosystems. Overcoming these challenges necessitates the advancement of high-resolution analytical techniques and robust experimental frameworks that can reliably identify and characterize these bioactive compounds. Furthermore, interdisciplinary collaboration among microbiologists, chemists, and agronomists is critical to fully harness the potential of microbial metabolites for pest control. The integration of microbial metabolomics with current pest management strategies, such as integrated pest management (IPM), could lead to enhanced precision and sustainability in pest control practices, thereby reducing reliance on synthetic chemical pesticides and promoting environmentally sustainable agricultural systems. Continued research and innovation in this field are essential for overcoming current limitations and fully realizing the potential of microbial metabolomics in agricultural pest management.

References

Abbas HS, Abou Baker DH, Elshimy R, Abou Elazm FI, Khan J (2024) Bacterial secondary metabolites: recent advances and agricultural applications. In: Bacterial secondary metabolites, pp 399–414

Abdul Malik NA, Kumar IS, Nadarajah K (2020) Elicitor and receptor molecules: orchestrators of plant defense and immunity. Int J Mol Sci 21(3):963

Alper H, Cirino P, Nevoigt E, Sriram G (2010) Applications of synthetic biology in microbial biotechnology. J Biomed Biotechnol 2010:1

Arıkan M, Muth T (2023) Integrated multi-omics analyses of microbial communities: a review of the current state and future directions, vol 19. Mol Omics, p 607

Arthurs S, Dara SK (2019) Microbial biopesticides for invertebrate pests and their markets in the United States. J Invertebr Pathol 165:13–21

Barret M, Frey-Klett P, Guillerm-Erckelboudt AY, Boutin M, Guernec G, Sarniguet A (2009) Effect of wheat roots infected with the pathogenic fungus Gaeumannomyces graminis var.

tritici on gene expression of the biocontrol bacterium Pseudomonas fluorescens Pf29Arp. Mol Plant-Microbe Interact 22(12):1611–1623
Benchlih S, Esmaeel Q, Aberkani K, Tahiri A, Belabess Z, Lahlali R, Barka EA (2023) Modes of action of biocontrol agents and elicitors for sustainable protection against bacterial canker of tomato. Microorganisms 11(3):726
Bharti R, Grimm DG (2021) Current challenges and best-practice protocols for microbiome analysis. Brief Bioinform 22(1):178–193
Binnington KC, Baule VJ (1993) Naturally occurring insecticidal molecules as candidates for genetic engineering. In: Molecular approaches to fundamental and applied entomology. New York, Springer, pp 38–89
Boness HV, de Sá HC, Dos Santos EK, Canuto GA (2023) Sample preparation in microbial metabolomics: advances and challenges. In: Microbial natural products chemistry: a metabolomics approach, pp 149–183
Boruah RR, Hage A (2024) Role of rhizobial metabolites in control of soil-borne phytopathogenic fungi. In: Bioactive microbial metabolites. Academic, pp 51–65
Bravo A, Gill SS, Soberón M (2007) Mode of action of Bacillus thuringiensis Cry and Cyt toxins and their potential for insect control. Toxicon 49(4):423–435
Bull CT, Wadsworth ML, Sorensen KN, Takemoto JY, Austin RK, Smilanick JL (1998) Syringomycin E produced by biological control agents controls green mold on lemons. Biol Control 12(2):89–95
Butt TM, Jackson C, Magan N (eds) (2001) Fungi as biocontrol agents: progress, problems and potential. CABI Publishing
Caballero J, Jiménez-Moreno N, Orera I, Williams T, Fernández AB, Villanueva M, Ferré J, Caballero P, Ancín-Azpilicueta C (2020) Unraveling the composition of insecticidal crystal proteins in Bacillus thuringiensis: a proteomics approach. Appl Environ Microbiol 86(12):e00476–e00420
Caesar LK, Montaser R, Keller NP, Kelleher NL (2021) Metabolomics and genomics in natural products research: complementary tools for targeting new chemical entities. Nat Prod Rep 38(11):2041–2065
Canciu A, Tertis M, Hosu O, Cernat A, Cristea C, Graur F (2021) Modern analytical techniques for detection of bacteria in surface and wastewaters. Sustain For 13(13):7229
Carlini CR, Grossi-de-Sá MF (2002) Plant toxic proteins with insecticidal properties. A review on their potentialities as bioinsecticides. Toxicon 40(11):1515–1539
Charnley AK (2003) Fungal pathogens of insects: cuticle degrading enzymes and toxins. Adv Bot Res 40:241–321
Chen XH, Koumoutsi A, Scholz R, Schneider K, Vater J, Süssmuth R, Piel J, Borriss R (2009) Genome analysis of Bacillus amyloliquefaciens FZB42 reveals its potential for biocontrol of plant pathogens. J Biotechnol 140(1–2):27–37
Chen Y, Li EM, Xu LY (2022) Guide to metabolomics analysis: a bioinformatics workflow. Metabolites 12(4):357
Chen N, Zhang X, Xi J, Yang Y, Yuan Q (2023) Recent advances of microbial metabolism analysis: from metabolic molecules to environments. SCIENCE CHINA Chem 66(11):2941–2950
Chubukov V, Mukhopadhyay A, Petzold CJ, Keasling JD, Martín HG (2016) Synthetic and systems biology for microbial production of commodity chemicals. NPJ Syst Biol Appl 2(1):1–11
Coolen S, Rogowska-van der Molen M, Welte CU (2022) The secret life of insect-associated microbes and how they shape insect–plant interactions. FEMS Microbiol Ecol 98(9):fiac083
Dara SK (2021) Microbial metabolites as pesticides. In: Microbes for sustainable Insect pest management: hydrolytic enzyme & secondary metabolite–volume 2, pp 75–88
Deshayes C, Siegwart M, Pauron D, Froger JA, Lapied B, Apaire-Marchais V (2017) Microbial pest control agents: are they a specific and safe tool for insect pest management? Curr Med Chem 24(27):2959–2973
Dunham B (2015) Microbial pesticides: a key role in the multinational portfolio. New Ag Int:32–36
Finlay AC, Hobby GL, P'an SY, Regna PP, Routien JB, Seeley DB, Shull GM, Sobin BA, Solomons IA, Vinson JW, Kane JH (1950) Terramycin, a new antibiotic. Science 111(2874):85–85

Fiorini F, Bajerski F, Jeske O, Lepleux C, Overmann J, Brönstrup M (2022) A metabolomics-based toolbox to assess and compare the metabolic potential of unexplored, Difficult-to-Grow Bacteria. Mar Drugs 20(11):713

Fu J, Zhou Y, Li HF, Ye YH, Guo JH (2011) Antifungal metabolites from Phomopsis sp. By254, an endophytic fungus in Gossypium hirsutum. Afr J Microbiol Res 5(10):1231–1236

Galal A, Talal M, Moustafa A (2022) Applications of machine learning in metabolomics: disease modeling and classification. Front Genet 13:1017340

Gallardo-Navarro O, Aguilar-Salinas B, Rocha J, Olmedo-Álvarez G (2024) Higher-order interactions and emergent properties of microbial communities: the power of synthetic ecology. Heliyon 10(14):e33896

Geesink P, Ter Horst J, Ettema TJ (2024) More than the sum of its parts: uncovering emerging effects of microbial interactions in complex communities. FEMS Microbiol Ecol 100(4):fiae029

Gohel V, Singh A, Vimal M, Ashwini P, Chhatpar HS (2006) Bioprospecting and antifungal potential of chitinolytic microorganisms. Afr J Biotechnol 5(2):54–72

Guo J, Yu H, Xing S, Huan T (2022) Addressing big data challenges in mass spectrometry-based metabolomics. Chem Commun 58(72):9979–9990

Hamrouni R, Regus F, Farnet Da Silva AM, Orsiere T, Boudenne JL, Laffont-Schwob I, Christen P, Dupuy N (2024) Current status and future trends of microbial and nematode-based biopesticides for biocontrol of crop pathogens. Crit Rev Biotechnol 45:1–20

Han JS, Cheng JH, Yoon TM, Song J, Rajkarnikar A, Kim WG, Yoo ID, Yang YY, Suh JW (2005) Biological control agent of common scab disease by antagonistic strain Bacillus sp. sunhua. J Appl Microbiol 99(1):213–221

Harish S, Murugan M, Kannan M, Parthasarathy S, Prabhukarthikeyan S, Elango K (2021) Entomopathogenic viruses. In: Omkar (ed) Microbial approaches for insect pest management. Springer, New York, pp 1–57

Hummadi EH, Dearden A, Generalovic T, Clunie B, Harrott A, Cetin Y, Demirbek M, Khoja S, Eastwood D, Dudley E, Hazir S (2021) Volatile organic compounds of Metarhizium brunneum influence the efficacy of entomopathogenic nematodes in insect control. Biol Control 155:104527

Huynh MP, Hibbard BE, Ho KV, Shelby KS (2022) Toxicometabolomic profiling of resistant and susceptible western corn rootworm larvae feeding on Bt maize seedlings. Sci Rep 12(1):11639

Idris H, Suryani E, Gustia H, Ramadhan AI (2022) The effect of various essential oil and solvent additives on the botanical pesticide of Piper Aduncum essential oil on formulation antifungal activity. Results Eng 16:100644

Jayaraj R, Megha P, Sreedev P (2017) Organochlorine pesticides their toxic effects on living organisms and their fate in the environment. Interdiscip Toxicol 9:90–100. https://doi.org/10.1515/intox-2016-0012

Jeffs LB, Khachatourians GG (1997) Toxic properties of Beauveria pigments on erythrocyte membranes. Toxicon 35(8):1351–1356

Kellogg J, Kang S (2020) Metabolomics, an essential tool in exploring and harnessing microbial chemical ecology. Phytobiomes J 4(3):195–210

Kim PI, Ryu JW, Kim YH, Chi YT (2010) Production of biosurfactant lipopeptides iturin A, fengycin, and surfactin A from Bacillus subtilis CMB32 for control of Colletotrichum gloeosporioides. J Microbiol Biotechnol 20(1):138–145

Kim JH, Lee N, Hwang S, Kim W, Lee Y, Cho S, Palsson BO, Cho BK (2021) Discovery of novel secondary metabolites encoded in actinomycete genomes through coculture. J Ind Microbiol Biotechnol 48(3–4):kuaa001

Knowles BH, Ellar DJ (1987) Colloid-osmotic lysis is a general feature of the mechanism of action of bacillus thuringiensis δ-endotoxins with different insect specificity. Biochim Biophys Acta (BBA)-General Subjects 924(3):509–518

Köhl J, Kolnaar R, Ravensberg WJ (2019) Mode of action of microbial biological control agents against plant diseases: relevance beyond efficacy. Front Plant Sci 10:845

Kumar P, Pandhi S, Mahato DK, Kamle M, Mishra A (2021a) Bacillus-based nano-bioformulations for phytopathogens and insect–pest management. Egypt J Biol Pest Control 31:1–11

Kumar J, Ramlal A, Mallick D, Mishra V (2021b) An overview of some biopesticides and their importance in plant protection for commercial acceptance. Plan Theory 10(6):1185

Leclère V, Béchet M, Adam A, Guez JS, Wathelet B, Ongena M, Thonart P, Gancel F, Chollet-Imbert M, Jacques P (2005) Mycosubtilin overproduction by Bacillus subtilis BBG100 enhances the organism's antagonistic and biocontrol activities. Appl Environ Microbiol 71(8):4577–4584

Leconte A, Tournant L, Muchembled J, Paucellier J, Héquet A, Deracinois B, Deweer C, Krier F, Deleu M, Oste S, Jacques P (2022) Assessment of lipopeptide mixtures produced by Bacillus subtilis as biocontrol products against apple scab (Venturia inaequalis). Microorganisms 10(9):1810

Leiss KA, Choi YH, Verpoorte R, Klinkhamer PG (2011) An overview of NMR-based metabolomics to identify secondary plant compounds involved in host plant resistance. Phytochem Rev 10:205–216

Li P, Luo H, Ji B, Nielsen J (2022) Machine learning for data integration in human gut microbiome. Microb Cell Factories 21(1):241

Li S, Siddiqa A, Thapa M, Chi Y, Zheng S (2023) Trackable and scalable LC-MS metabolomics data processing using asari. Nat Commun 14(1):4113

Liu X, Ruan L, Peng D, Li L, Sun M, Yu Z (2014) Thuringiensin: a thermostable secondary metabolite from Bacillus thuringiensis with insecticidal activity against a wide range of insects. Toxins 6(8):2229–2238

Luo DQ, Wang F, Bian XY, Liu JK (2005) Rufuslactone, a new antifungal sesquiterpene from the fruiting bodies of the basidiomycete Lactarius rufus. J Antibiot 58(7):456–459

Maharana C, Padala VK, Hubballi AB, Nikhil Raj M, Paschapur A, Bhat C, Singh AK, Subbanna ARNS (2022) Secondary metabolites of microbials as potential pesticides. In: Sustainable management of potato pests and diseases. Springer Singapore, Singapore, pp 111–142

Mnif I, Ghribi D (2015) Potential of bacterial derived biopesticides in pest management. Crop Prot 77:52–64

Mulligan CC, Vircks KE (2012) Advances in field-portable mass spectrometers for on-site analytics. Inform 23(9):613–617

Naseri G, Koffas MA (2020) Application of combinatorial optimization strategies in synthetic biology. Nat Commun 11(1):2446

Naz R, Khushhal S, Asif T, Mubeen S, Saranraj P, Sayyed RZ (2022) Inhibition of bacterial and fungal phytopathogens through volatile organic compounds produced by pseudomonas sp. In: Secondary metabolites and volatiles of PGPR in plant-growth promotion. Springer International Publishing, Cham, pp 95–118

Noecker C, Chiu HC, McNally CP, Borenstein E (2019) Defining and evaluating microbial contributions to metabolite variation in microbiome-metabolome association studies. mSystems 4(6):10–1128

Olson S (2015) An analysis of the biopesticide market now and where it is going. Outlooks Pest Manag 26(5):203–206

Omoto C, McCoy CW (1998) Toxicity of purified fungal toxin Hirsutellin a to the citrus rust mite Phyllocoptruta oleivora (Ash.). J Invertebr Pathol 72(3):319–322

Pan X, Guo X, Zhai T, Zhang D, Rao W, Cao F, Guan X (2023) Nanobiopesticides in sustainable agriculture: developments, challenges, and perspectives. Environ Sci Nano 10(1):41–61

Pandiyan A, Sarsan S, Durga GGS, Ravikumar H (2024) Biofertilizers and biopesticides as microbial inoculants in integrated pest management for sustainable agriculture. In: Microbial essentialism. Academic, pp 485–518

Parker EJ, Billane KC, Austen N, Cotton A, George RM, Hopkins D, Lake JA, Pitman JK, Prout JN, Walker HJ, Williams A (2023) Untangling the complexities of processing and analysis for untargeted LC-MS data using open-source tools. Metabolites 13(4):463

Patel S, Kumar H, Desai N (2022) Optimizing data processing pipelines in microbial metabolomics for biopesticide production. Appl Microb Technol 18(2):289–302

Petrick LM, Shomron N (2022) AI/ML-driven advances in untargeted metabolomics and exposomics for biomedical applications. Cell Rep Phys Sci 3(7):100978

Pino O, Sánchez Y, Rojas MM (2013) Plant secondary metabolites as an alternative in pest management. I: background, research approaches and trends. Revista de Protección Vegetal 28(2):81–81

Pršić J, Ongena M (2020) Elicitors of plant immunity triggered by beneficial bacteria. Front Plant Sci 11:594530

Rojo FP, Vuong P, Pillow JJ, Kaur P (2023) Microbial synthetic biology for plant metabolite production: a strategy to reconcile human health with the realization of the UN Sustainable Development Goals. Biofuels Bioprod Biorefin 17(6):1485–1495

Rusu AV, Trif M, Rocha JM (2023) Microbial secondary metabolites via fermentation approaches for dietary supplementation formulations. Molecules 28(16):6020

Sabarwal A, Kumar K, Singh RP (2018) Hazardous effects of chemical pesticides on human health–cancer and other associated disorders. Environ Toxicol Pharmacol 63:103–114. https://doi.org/10.1016/j.etap.2018.08.018

Santamaria G, Liao C, Lindberg C, Chen Y, Wang Z, Rhee K, Pinto FR, Yan J, Xavier JB (2022) Evolution and regulation of microbial secondary metabolism. Elife 11:e76119

Saxena S (2014) Microbial metabolites for development of ecofriendly agrochemicals. Allelopath J 33(1):1–24

Sekurova ON, Schneider O, Zotchev SB (2019) Novel bioactive natural products from bacteria via bioprospecting, genome mining and metabolic engineering. Microb Biotechnol 12(5):828–844

Shankar V, Sharma P (2023) Standardized protocols for metabolite extraction in microbial metabolomics. Microb Biotechnol Rev 12(3):320–335

Shen W, Li X, Zhao Y (2023) Predicting the bioactivity of microbial metabolites using machine learning models. J Microb Metab 18(3):567–580

Shi L, Liu B, Wei Q, Ge B, Zhang K (2020) Genome-wide transcriptomic analysis of the response of Botrytis cinerea to wuyiencin. PLoS One 15(4):e0224643

Siddiqui MS, Thodey K, Trenchard I, Smolke CD (2012) Advancing secondary metabolite biosynthesis in yeast with synthetic biology tools. FEMS Yeast Res 12(2):144–170

Singh BP, Rateb ME, Rodriguez-Couto S, Polizeli MDLTDM, Li WJ (2019) Microbial secondary metabolites: recent developments and technological challenges. Front Microbiol 10:914

Sparks TC, Dripps JE, Watson GB, Paroonagian D (2012) Resistance and cross-resistance to the spinosyns–a review and analysis. Pestic Biochem Physiol 102(1):1–10

Stabb EV, Jacobson LM, Handelsman JO (1994) Zwittermicin A-producing strains of Bacillus cereus from diverse soils. Appl Environ Microbiol 60(12):4404–4412

Steinrauf LK (1985) Beauvericin and the other enniatins. Met Ions Biol Syst 19:139–171

Strobel GA, Miller RV, Martinez-Miller C, Condron MM, Teplow DB, Hess WM (1999) Cryptocandin, a potent antimycotic from the endophytic fungus Cryptosporiopsis cf. quercina. Microbiology 145(8):1919–1926

Subbanna ARNS, Stanley J, Rajasekhara H, Mishra KK, Pattanayak A, Bhowmick R (2020) Perspectives of microbial metabolites as pesticides in agricultural pest management. In: Co-evolution of secondary metabolites, pp 925–952

Sundh I, Del Giudice T, Cembalo L (2021) Reaping the benefits of microorganisms in cropping systems: is the regulatory policy adequate? Microorganisms 9(7):1437

Syromyatnikov MY, Isuwa MM, Savinkova OV, Derevshchikova MI, Popov VN (2020) The effect of pesticides on the microbiome of animals. Agriculture 10(3):79

Tamano K (2014) Enhancing microbial metabolite and enzyme production: current strategies and challenges. Front Microbiol 5:718

Tanaka Y, Omura S (1993) Agroactive compounds of microbial origin. Ann Rev Microbiol 47(1):57–87

Tang J (2011) Microbial metabolomics. Curr Genomics 12(6):391–403

Tang L, Zhao X, Chen X, Jiang Y, Gudda F, Wang Y, Ling W (2023) Distribution of bound-PAH residues and their correlations with the bacterial community at different depths of soil from an abandoned chemical plant site. J Hazard Mater 453:131328

Thukral M, Allen AE, Petras D (2023) Progress and challenges in exploring aquatic microbial communities using non-targeted metabolomics. ISME J 17(12):2147–2159

Tofazzal Islam M (2010) Mode of antagonism of a biocontrol bacterium Lysobacter sp. SB-K88 toward a damping-off pathogen Aphanomyces cochlioides. World J Microbiol Biotechnol 26:629–637

Umezawa H, Okami Y, Hashimoto T, Suhara Y, Hamada M, Takeuchi T (1965) A new antibiotic, kasugamycin. J Antibiot Ser A 18(2):101–103

Vachon V, Laprade R, Schwartz JL (2012) Current models of the mode of action of bacillus thuringiensis insecticidal crystal proteins: a critical review. J Invertebr Pathol 111(1):1–12

Verma C, Jandaik S, Gupta BK, Kashyap N, Suryaprakash VS, Kashyap S, Kerketta A (2020) Microbial metabolites in plant disease management: review on biological approach. Int J Chem Stud 8:2570–2581

Vervoort N, Goossens K, Baeten M, Chen Q (2021) Recent advances in analytical techniques for high throughput experimentation. Anal Sci Adv 2(3–4):109–127

Wang JB, Pan HX, Tang GL (2011) Production of doramectin by rational engineering of the avermectin biosynthetic pathway. Bioorg Med Chem Lett 21(11):3320–3323

Wang J, Pursell ME, DeVor A, Awoyemi O, Valentine SJ, Li P (2022) Portable mass spectrometry system: instrumentation, applications, and path to 'omics analysis'. Proteomics 22(23–24):2200112

Warren GW (1997) Vegetative insecticidal proteins: novel proteins for control of corn pests. In: Advances in insect control: the role of transgenic plants, pp 109–121

Weiland-Bräuer N (2021) Friends or foes—microbial interactions in nature. Biology 10(6):496

Whipps JM (2001) Microbial interactions and biocontrol in the rhizosphere. J Exp Bot 52(suppl_1):487–511

Wu C, Choi YH, van Wezel GP (2016) Metabolic profiling as a tool for prioritizing antimicrobial compounds. J Ind Microbiol Biotechnol 43(2–3):299–312

Zehra A, Raytekar NA, Meena M, Swapnil P (2021) Efficiency of microbial bio-agents as elicitors in plant defense mechanism under biotic stress: a review. Curr Res Microb Sci 2:100054

Zhang D, Qiang R, Zhou Z, Pan Y, Yu S, Yuan W, Cheng J, Wang J, Zhao D, Zhu J, Yang Z (2022) Biocontrol and action mechanism of Bacillus subtilis lipopeptides' fengycins against Alternaria solani in potato as assessed by a transcriptome analysis. Front Microbiol 13:861113

Zhang T, Liu Q, Huang W (2024) Discovery and optimization of pesticidal metabolites from soil bacteria using machine learning. Appl Microb Technol 22(1):29–41

Zhao Z, Wang Q, Wang K, Brian K, Liu C, Gu Y (2010) Study of the antifungal activity of Bacillus vallismortis ZZ185 in vitro and identification of its antifungal components. Bioresour Technol 101(1):292–297

Zhou M, Zhao J (2021) A review on the health effects of pesticides based on host gut microbiome and metabolomics. Front Mol Biosci 8:632955

Zou WX, Meng JC, Lu H, Chen GX, Shi GX, Zhang TY, Tan RX (2000) Metabolites of Colletotrichum gloeosporioides, an endophytic fungus in Artemisia mongolica. J Nat Prod 63(11):1529–1530

Omics for Thriving Plant Microbiomes: Growing Food Security for a Sustainable Future

Debdut Manna, Rosalin Laishram, S. R. Arpitha, Monika Rai, and Anil Dahuja

Abstract

Food security is under unprecedented threat due to factors such as rapid population growth, climate change, and significant postharvest food losses. The plant microbiome, a complex community of microorganisms associated with plants, holds significant potential for enhancing agricultural sustainability and ensuring food security. This chapter provides a comprehensive review of omics technologies—genomics, transcriptomics, proteomics, metabolomics, and metagenomics—as tools for understanding plant–microbiome interactions and their role in improving plant resilience to abiotic stresses, including drought, salinity, and heat. By integrating omics data, we can decode the intricate relationships between plants and their associated microbes, revealing mechanisms by which beneficial microbes enhance plant stress tolerance. We discuss nondestructive methods like phenomics and imaging, which offer valuable insights into plant responses under various stress conditions without compromising plant integrity. Additionally, the chapter highlights the transformative potential of web-based omics platforms and artificial intelligence in advancing microbiome research, facilitating the development of predictive models for stress tolerance, and guiding crop improvement strategies. The application of these advanced technologies promises to drive innovative solutions for sustainable agriculture. However, significant challenges remain, including data integration, scalability, and translating omics insights into practical agricultural practices.

Keywords

Plant microbiome · Omics · Food security · Plant–microbe interaction

D. Manna · R. Laishram (✉) · S. R. Arpitha · M. Rai (✉) · A. Dahuja
Division of Biochemistry, ICAR-Indian Agricultural Research Institute, New Delhi, India

S. Kaur et al. (eds.), *Microbial Metabolomics*,
https://doi.org/10.1007/978-981-96-4824-5_14

1 Introduction

The agricultural sector confronts a multitude of complex challenges including unsustainable production practices, biodiversity loss, the impacts of climate change, issues of malnutrition, environmental pollution, shifting precipitation patterns, rising global temperatures, and the occurrence of severe weather events, all of which hinder progress toward achieving food security and eliminating hunger and malnutrition. As climate change causes abiotic stressors and plant diseases, protecting crop plants is becoming more important (Singh et al. 2023a, b). Therefore, it is imperative to increase food production in the upcoming year as agriculture supports the economy, lays the framework for other businesses, and is a part of the problem with the world's food supply being depleted due to increasing population (Kimotho and Maina 2024). Excessive use of agrochemicals to achieve high productivity has led to contamination of the environment, including air, water, and soil pollution (Gomiero 2016). The production of crops has been severely limited as a result of all these difficulties. Thus, there will be severe long-term effects on environmental sustainability, food security, and climate conditions if these issues are not resolved. The sustainability of these issues is essential for maintaining environmental stability, climate resilience, and global food systems.

Our biosphere depends on the relationships between plants and microbes because they support sustainable agriculture. Many different types of microbes, from mutualists to pathogens, are linked to plants. Relationships with beneficial microbes that are mutualistic and symbiotic show positive interaction, whereas interactions with pathogenic microbes show negative interaction (Rodriguez et al. 2019). Many microbes live alongside plants in both in their internal and external tissues; this coexistence is referred to as holobionts. The metabolism of the plants and the microbes that are linked to it are closely related because they have coevolved. They assist one another to keep the holobionts operating normally (Berg et al. 2016). Their genetic flexibility and ability to produce metabolites enable them to flourish in a wider range of ecologicalniches while interacting with plants to offer additional nutrients and aid in stress tolerance (Diwan et al. 2022). These microorganisms are frequently referred to as the "second genome" of plants and are typically found in the rhizosphere, endosphere, rhizoplane, spermosphere, and phytosphere. In addition, manipulating the microbial sect in the rhizosphere has been extensively studied as a method to enhance plant growth, increase yield, and boost resistance to different types of stresses. These findings will provide valuable insights into the complete plant–microbe interaction process and help identify potential microorganisms that could enhance crop productivity in the near future (Sandrini et al. 2022). Growing evidence indicates that the soil microbiome, despite being imperceptible to the unaided eye, is essential to the interactions among people, healthy soils, and the climate. Therefore, it is essential that current and accurate information in this area be made available in order to take various initiatives that address the complex issues outlined in the introduction. The UN Decade on Ecosystem Restoration, the Global Soil Partnership, and policies that support healthy soils' role in attaining the 2030 Agenda for Sustainable Development's Sustainable Development Goals

(SDGs) are examples (Summit 2020). Many intergovernmental groups that address soils and biological variety for food security recognize the relevance of microbial variation for healthy and functional soils. But given the knowledge gaps in this emerging field of science, providing details about the complexity of the soil microbiome has proven to be a challenging task thus far. Scientists are particularly focused on plant–microbe interactions and biostimulants derived from microbes as a means of promoting sustainable farming practices because of the enormous benefits these discoveries hold for modern agriculture (Berg et al. 2017).

Although the unique tenets of plant–microbe association are well understood, little is known about the variables influencing these associations and how they interact with one another. Similar to this, a deeper comprehension of the helpful interactions and communication between them would expand our views on how they might be used to achieve agricultural sustainability (Trivedi et al. 2020; Thoms et al. 2021). Advances in technology have made it possible to study microbial variation and the composition of a given plant microbiome. In contrast to low-throughput prior technologies such as different molecular approaches and one-dimensional omics methods, which had limitations in providing a comprehensive scenario of the microbiome, we can now obtain more sophisticated methods to obtain higher-quality images of the microbiome. The application of omics techniques, such as transcriptomics, metabolomics, and genomics, has shown to be extremely beneficial in investigating the physiological, biochemical, and plant–microbe molecular interactions under diverse circumstances. By combining the abovementioned omics techniques, one can better analyze, visualize, and interpret data, leading to a more profound comprehension of biological dynamics and their interaction (Chao et al. 2024). Successful integrative omics research will advance our understanding of diverse functions of the plant microbiome.

Every omics strategy has its own set of restrictions that could compromise the method's sensitivity or specificity. Nevertheless, a few of these drawbacks can be solved by combining various strategies. In order to enhance understanding of the interrelationships between microbes and plants, this book chapter will explore the transformative implications of omics technology. This chapter focuses on how diverse omics techniques might improve and achieve sustainable agriculture practices by allowing the unwinding of numerous mechanisms such as microbe–host interactions, nitrogen fixation, systemic resistance induction, and mycorrhizal association.

2 Adaptation of Plants to Various Stressors and Other Unfavorable Conditions

Environmental obstacles, including biotic (such as pests and viruses) and abiotic ones, continuously impact the growth and yield of plants. Numerous factors, including salinity, drought, heavy metals, and temperature, can act as abiotic stresses and reduce crop productivity while modifying a range of biochemical, molecular, and physiological mechanisms (Umar et al. 2021). Out of various factors, salinity and drought are the two main environmental factors limiting crop productivity. The importance of drought is growing due to rising temperatures and accelerating processes of desertification, mostly in arid regions. On the other hand, agricultural soils are increasingly at risk from soil salinity, which is caused by salt deposition following either natural or man-made processes (Sandrini et al. 2022).

Reduced crop yields due to water scarcity and other agricultural risks brought on by climate change pose a threat to food security. The extreme heat and physical dryness of recent years have contributed to a notable decrease in global production (Lesk et al. 2016). Salt stress, heavy metal, and other stressors have an impact on crop growth and productivity much like drought does. At the moment, saline land is growing quickly due to glacier melting and heat stress-induced salt buildup in soils (Trenberth et al. 2014). The overuse of chemical fertilizers damages soil's physiological and biochemical processes, alters the interactions between plants and soil microbiota, and increases pollution in the environment. The properties of soil, including temperature, pH, moisture, and organic matter content, are also affected (Singh et al. 2020). Agrochemicals can rapidly and clearly increase biomass and pest resistance in a range of crops when applied as easily accessible chemical fertilizers and pesticides, but they can also seriously jeopardize the wellbeing of living organisms and the ecosystem in which they operate. Nitrogen-fixing bacteria unique to plants and uptake of nutrients were found to be negatively impacted by certain agrochemical treatments used in agriculture, accounting roughly half of soil microorganisms (Li et al. 2013).

3 Sustainable Agricultural Practices

Plants exhibit a phenomenon called "rhizosphere effect" where they use 20% of their synthesized carbon to attract specific microbiota to roots from soil (Kimotho and Maina 2024). Incorporating signalling molecules to the soil that attract microorganisms may provide plants with an additional biological defense mechanism against salt stress (Liu et al. 2019). When combined with social and technological advancements in agriculture, the capacity to sustainably improve plant nutrition and pollination efficiency may be crucial in helping to reduce production gaps and prevent present and future global food shortages (Orrell and Bennett 2013). Together with improvements in plant breeding, the addition of intercropping to agricultural methods and the inoculation of soil with beneficial bacteria may lead to higher crop production yields.

The soil microbiome is influenced and maintained by many factors, primarily the physiochemical aspects of soil and environmental factors, including soil moisture, pH, and nitrogen content. High organic carbon levels influence higher microbial diversity. While a large amount of fertilizer application affects fungal communities and influence the growth of denitrifiers, tillage practices can diminish the efficiency of the native microbiome in degrading agrochemicals. Therefore, strategic tillage represents a balanced approach (Batool et al. 2024). Beneficial bacteria like *Bacillus* and *Pseudomonas* species can work in concert with soil compost to provide synergistic effects that improve soil fertility and reduce the spread of soil-borne diseases (Liu et al. 2022). Integrating these microbial interactions and management strategies into sustainable agricultural practices can enhance soil health, increase crop resilience, and improve overall productivity.

4 Mitigating Food Loss Through Plant Microbiome Management

The rapid growth of the world's population has raised significant concern regarding the ability to ensure food security for everyone. To meet this demand, food production must increase substantially from the available arable land. However, the overuse of chemical fertilizers, technological advancement, intensive agricultural practices, and soil erosion have significantly contributed to the decline in soil fertility (Astapati and Nath 2023). Additionally, disease outbreaks and pest infestation severely affect crop productivity, resulting in extensive economic losses (Huang McBeath et al. 2010). In response to these challenges, microorganisms emerged as promising biocontrol agents and biofertilizer with the potential to effectively replace chemical-based agricultural fertilizers.

Following this, many microbial products and carrier have been developed along with application strategies. Despite these development and effective microbial technology advancement, it is often challenging to translate lab-based trial into field applications due to varying climatic and edaphic conditions. It is crucial to select specific microbial strain according to the plant requirements, type, and the composition of the native microbiome while also refining the type of application techniques to ensure the effective propagation of these microorganisms under native field conditions (Trivedi et al. 2016).

Varieties of organisms present in plant rhizosphere and soil play a major role in controlling the availability, movement, and uptake of nutrients by plants as well as recycling organic matter. The application of herbicides and pesticides and salinization along with the depletion of water resources create major environmental problems; thus, increasing plant productivity in a manner focusing of environmental sustainability is of utmost important (Morrissey et al. 2004). However, it is essential to identify and isolate key species that enhance plant growth and to design synthetic plant communities that improve plant performance including disease resistance and nutrient uptake (Gopal and Gupta 2016). By utilizing a variety of cutting-edge technologies, such as metagenome profiling and advanced sequencing, these studies demonstrate that bacteria, viruses, fungi, oomycetes, nematodes, and algae make up the majority of the microbiomes found in plants, and out of that, bacterial communities like Proteobacteria, Actinobacteria, and Bacteroidetes continue to be the most prevalent communities (Nadarajah 2021). The fact that certain microbial taxa are conserved in all plant–soil environmental cross talk was also brought out by these investigations. More research is necessary to fully understand the potential effects of microbe-to-microbe interaction and microbe-to-plant interactions on plant growth and development.

4.1 Plant Symbiotic Associations

Plants alter soil pH and secrete a wide range of carbon sources like citrate, malate, and oxalate which attract beneficial microbial communities while suppressing pathogens through antimicrobial and chelating compounds. Selecting a microbe-friendly plant variety involves considering several factors like crop rotation recommendation, synergistic crop rotation, and suitability to the climate or habitat. Laboratory-scale experiment alone may not fully capture the belowground interactions occurring in actual agricultural land, highlighting the need to expand experiments to field and farm levels (Batool et al. 2024).

4.1.1 Arbuscular Mycorrhizal (AM) Fungi

Among various symbiotic associations reported in plants, Arbuscular mycorrhizal (AM) fungi facilitates improved nutrient uptake particularly phosphorus, which is often limited in soil by forming symbiotic association with plant roots. AM fungi also enhance the absorption of other essential nutrients such as nitrogen, potassium, and certain trace minerals. Plants gain indirect benefit from AM fungi which reduce herbivore visitation rates and lower the severity of herbivore damage. AM fungi can encourage indirect defenses, such as emitting volatile compounds that attract herbivore adversaries impacting yield (Orrell and Bennett 2013; Bardgett and Gibson 2017). AM fungi alter the amount, quality, and segregation of jasmonic acid, ethylene, and phenolic compounds which alter plant defense mechanisms effectively similar to agrochemicals (Van der Ent et al. 2009). Once they invade host plants, AM fungi and rhizobia set off the salicylic acid pathway's signal cascade, preparing plants for protection against potential foes (Bardgett and Gibson 2017). Although AM fungi generally reduce harm from fungal infections, outcome interactions of AM fungi and pathogen depend on the pathogen class. The development of AM fungal seed coatings allows for customized applications tailored to specific crop varieties and anticipated pressures.

4.1.2 Rhizobium

Rhizobium species colonize leguminous plant root and form root nodules that establish a symbiotic relationship with the host through reducing atmospheric nitrogen into ammonium. The capacity of rhizobia to fix nitrogen significantly reduces the need for chemical fertilizers in agriculture (Verma et al. 2020). The rhizosphere is the site where the initial interaction between the host plant and rhizobia occurs, particularly when nitrogen levels are low.

Flavonoids which are frequently released into the soil by legume roots draw rhizobia into the rhizosphere and cause them to secrete lipo-chitooligosaccharides (LCOs, also known as Nod factors (NFs)). Two separate but related processes are activated concurrently when NFs are detected in the susceptible area of the root: epidermal bacterial infection and cortex cell division to form nodules that contain rhizobial symbionts (Poole et al. 2018). In the low oxygen environment inside mature nodules, the bacteria in organelle-like symbiosomes develop into bacteroids after being liberated from infection threads (ITs) and convert nitrogen gas into ammonia. To enable nutrition exchange between the plant and rhizobia, two or more peripheral vascular bundles that surround the nodule fixation zone converge toward the nodule apex (Yang et al. 2022).

4.1.3 Plant Growth-Promoting Endophytes

Bacterial endophytes reside inside plants for at least a portion of their life cycle. Plant endophyte interaction can display a variety of symbiotic association patterns from neutral (commensal), beneficial (mutualism), and pathogenic (Hardoim et al. 2008). Endophytes encompass a variety of microorganisms comprising bacteria, fungi, actinomycetes, archaea, and protists (Mishra et al. 2021). Among these, plant growth-promoting rhizobacteria (PGPR) are endophytic, free-living rhizosphere bacteria that can stimulate plant growth directly and indirectly by reducing stress. They enhance nutrient uptake, improve resistance to pathogens, and help plants tolerate abiotic stresses such as heavy metal contamination, salinity, and drought. Through induced systemic resistance (ISR), PGPR can enhance plant tolerance to herbivores and pathogens (Van der Ent et al. 2009). Seed coatings of rhizobia are used to raise yield of crop, yet PGPR seed coatings could boost plant tolerance or defense. However, transferring greenhouse results to actual field experiments remains challenging (Orrell and Bennett 2013).

Rhizobacteria interact with plants to have additional health promoting impact beyond the direct effect on plant growth. Induced systemic resistance (ISR) occurs when plants interact with nonpathogenic microorganisms to improve their disease resistance. 1-Aminocyclopropane-1-carboxylate (ACC) deaminase produced by rhizosphere microbes aids in converting ACC into ammonia and α-ketobutyrate contributing to plant nourishment and lowering ethylene content, thus lessening negative effects of ethylene. PGPR that produce ACC deaminase greatly increase plant resistance to various abiotic stresses (Kumari and Kumawat 2024). Additionally, rhizosphere microbes influence plant defense against herbivory and pathogens and impact nutrient availability and transformation (Bardgett and Gibson 2017).

Bacteria belonging to the genus *Azospirillum* in response to IAA (indole-3-acetic acid) encourage larger roots and more effective uptake of nitrogen from the soil. Groundwater and soil nitrate contamination through leachates is a major concern, thus utilizing these organisms could provide a better alternative to the synthetic application of nitrogenous and phosphorous fertilizers. While commercial fertilizer products made using *Azospirillum* are currently accessible, their limited host range and inconsistent performance prevent wider use. Gaining more knowledge about the symbiotic relationship between plants and microbes may aid in solving these issues (Cruz-Hernández et al. 2022).

Pseudomonas aeruginosa secretes lipopolysaccharides recognized by plant pattern recognition receptors, triggering mechanisms like pattern-triggered immunity, which contributes to plant defense mechanisms (Zhou et al. 2023). The exopolysaccharides produced encourage interaction with host plant and other microbes, improving nutrient availability, shielding against environmental stressors, and facilitating microbial adhesion (Niazi et al. 2023). Similarly, *Pseudomonas syringae* through type III secretion system secretes effector proteins, which promotes pathogen virulence or symbiotic interactions. It also secretes pectinolytic enzymes, facilitating interactions with host plants encouraging growth of microbes and its metabolism, and aiding in pectin-rich plant tissue breakdown.

Agrobacterium tumefaciens employs type IV secretion system to transfer effector proteins into the host cell cytoplasm and on the other hand utilizes adhesins to establish either symbiotic or pathogenic relationships ensuring attachment to host plant tissues specifically to influence biofilm formation and efficient colonization (Gupta et al. 2024). These effector molecule aids in evading identification by the host immune system, avoiding plant immune monitoring, and encouraging microbial colonization or infection by preventing downstream.

4.1.4 Biocontrol Agents

Biocontrol agents play a significant role in defending plants against diseases. Specific metabolites produced by different fungi and bacteria particularly from the genera *Trichoderma*, *Pseudomonas*, and *Bacillus* combat phytopathogenic fungi. Although these biocontrol agents are now effectively employed in the field, they still lack the consistency and efficacy required for widespread commercialization. However, there is room for improvement and as these bacteria advance, they may prove to be a viable substitute for the strong fungicide regimes currently used in agriculture (Arnold et al. 2003).

4.1.5 Fungal Biocontrol Agents

Trichoderma is effective against a wide range of plant pathogens such as *Fusarium* while changing the rhizosphere pH which indirectly aids in nutrient uptake by plants and promotes growth (Alfiky et al. 2021). *Metarhizium acridum* acts as biocontrol agent against locusts and grasshoppers. It also interacts with higher taxa including collembolans, acarians, coleopterans, gryllids, psocopterans, and dipterans (Maute et al. 2017). *Purpureocillium lilacinum*, a rhizosphere inhabitant, acts on various species of plant-parasitic nematodes like root-knot nematode (*Meloidogyn*e spp.) and potato cyst nematode (*Globodera pallida*) which also affects other pests such as *Tetranychus cinnabarinus*, Aphidoidea, and Aleyrodidae. In addition to its biocontrol functions, *P. lilacinum* possesses anticancer, antimicrobial, and insecticidal properties (Chen et al. 2021).

4.1.6 Bacterial Biocontrol Agents

Various *Bacillus* species have demonstrated antagonistic action against a variety of phytopathogenic bacteria in agricultural crops together with rice, maize, and fruit trees (Borriss 2011). For instance, the insecticidal activity of Cry protein produced by *Bacillus thuringiensis* has led researchers to study the ability of several *Bacillus* species, including *Bacillus subtilis*, *Bacillus pumilus*, *Bacillus amyloliquefaciens*, and *Bacillus licheniformis*, to reduce the incidence of diseases important to agriculture. These strains primarily prevent the growth and establishment of phytopathogenic organisms via various strategies: (A) antibiotic excretion, (B) siderophore production, (C) lytic enzyme production, (D) toxin production, and (E) inducing systemic resistance in the plant (ISR) (Villarreal-Delgado et al. 2018).

4.1.7 Viral Biocontrol Agents

Baculoviruses (nucleopolyhedroviruses and granuloviruses) are active in controlling the early larval stages of lepidopteran insects mainly targeting chewing type of insects (Abd-Alla et al. 2020). Granuloviruses are effective against codling moths and caterpillars which extensively infect apple and pear trees worldwide (Lacey et al. 2008). Nuclear polyhedrosis viruses (NPVs) are effective against lepidopteran insects including the American cotton leafworm and beet armyworm (Nawaz et al. 2019).

4.2 Microbe–Plant Molecular Interactions in Response to Different Environmental Stressors

Plant-friendly microbes have produced phytohormones, secondary metabolites, fixing and solubilizing nutrients, and many other beneficial effects, all of which have contributed directly or indirectly to the plant ecosystem (Compant et al. 2019). The plant microbiome has a direct impact on fixation of nitrogen, phosphate solubilization, iron sequestration, and the production and regulation of several phytohormones (auxin, cytokinin, gibberellic acid, and abscisic acid). Concurrently, indirect methods include the synthesis of volatile compounds, antifungal compounds, antibiotics, different types of competition, predators, and immunity mediated by microbes, which has the power to either stimulate or inhibit the host's innate immune responses. Therefore, rhizobia, endophytes, fungi, and plant growth-promoting bacteria (PGPB) are examples of beneficial microbes that show promise for sustainable agriculture and enhancing tolerance of plants to stressor (Singh et al. 2023a, b).

Rhizobium leguminosarum and other nitrogen-fixing bacteria improve soil fertility and plant enhancement by fixing of nitrogen and producing various nodulation factors (Mahapatra et al. 2023). Microbes like *Rhizobium leguminosarum* create phytohormones like auxins and cytokinins, which influence plant development and nutrient uptake by way of receptor signalling (Gómez-Godínez et al. 2023). Various Nod factors from various rhizobia like *Bradyrhizobium japonicum* trigger nodulation signalling and create symbiotic interactions with plants (Yuan et al. 2023). In order to enhance plant immunity and establish a positive relationship, various effector proteins from pathogens like *Pseudomonas syringae* influence host cellular processes (Kalita et al. 2024). To modify their host plants' biological processes, microbes release many molecular effectors through T3SS, T4SS, and T6SS. T6SS genes help rhizobia compete for interbacterial benefits in underground plant environments (De Sousa et al. 2023). Microbes like *Trichoderma harzianum* can potentially cause a systemic reaction in plants through the production of volatile organic compounds (Ahlawat et al. 2024). Microbes employ a variety of tactics, including the formation of exopolysaccharides (EPS), biofilm formation, and enzymatic destruction of components of plant cell walls. Biofilms are formed by certain microorganisms, such as *Pseudomonas aeruginosa* aiding in the adhesion of microbes and shielding them from environmental hazards (Niazi et al. 2023). *Pseudomonas syringae* produces a chemical that mimics hormones, causing a disruption in the balance of hormones and manipulating the development and response of plants (Selwal et al. 2023).

Plant systems have a defense mechanism that can be strengthened and made more aware of by priming in order to react to their constantly shifting surroundings. The priming starts with the sensors detecting stress, which results in a change in form or function of molecules or structures. This sets off a signalling cascade, which in turn prompts an action. Regulation of genes in response to stress and its differential transcription are the results of these sensors' ability to detect reversible physical changes such as modifications in stability of protein structure and membrane fluidity as well as melting of DNA and RNA strands (Westman et al. 2019). Both mitogen-activated protein kinases (MAPKs) and calcium-dependent protein kinases (CDPKs/CPKs) are incited through the binding of multiple known receptors to sensors. These receptors include members of the G protein family and various kinase family (Lohani et al. 2020). These plant defense mechanisms can also be triggered by chemicals, insects, beneficial microorganisms, or environmental cues.

4.3 Factor Affecting the Interaction Between the Plant Microbiome

Soil, host plants, environment, and humans affect plant–microbe interactions (Fig. 1). The host's genotype is a significant factor that determines the composition of the microbiome. Despite growing in close proximity to one another or belonging to the same family, different plant species have distinct microbial assemblages, which can be explained by variations in resource utilization (Samad et al. 2017). Additionally, variations of genotype within plant cultivars have a major influence on microbiological relationships. In addition, the soil's physical, chemical, and biological properties have an impact on the makeup and potential of microbes related to plants. Among other parameters, soil pH, soil carbon, C:N ratio, soil water, air, porosity, and other nutrient amount and cycling are important for preserving the composition of microflora and microfauna (French et al. 2021). Environmental variables like temperature, precipitation, and seasonal dynamics impact the composition and activity of the plant microbiome. This is because the growth and survival of all living things depend on certain conditions. Furthermore, diverse stressors alter plant phenotype according to survival requirements; consequently, the environment can control the relationship with microbes (Koranda et al. 2013; Umar et al. 2021). Ultimately, within a human-altered ecosystem, a modification in the microbiome will occur, leading to adverse effects on the plant. Humans have altered the genetic makeup of plants, used different selection techniques, applied different chemical fertilizers and amendments in excess, and used different microbial inoculants in order to improve traits and reap additional benefits. These actions have an impact on the microbial population and their way of interaction with the plant (Diwan et al. 2022).

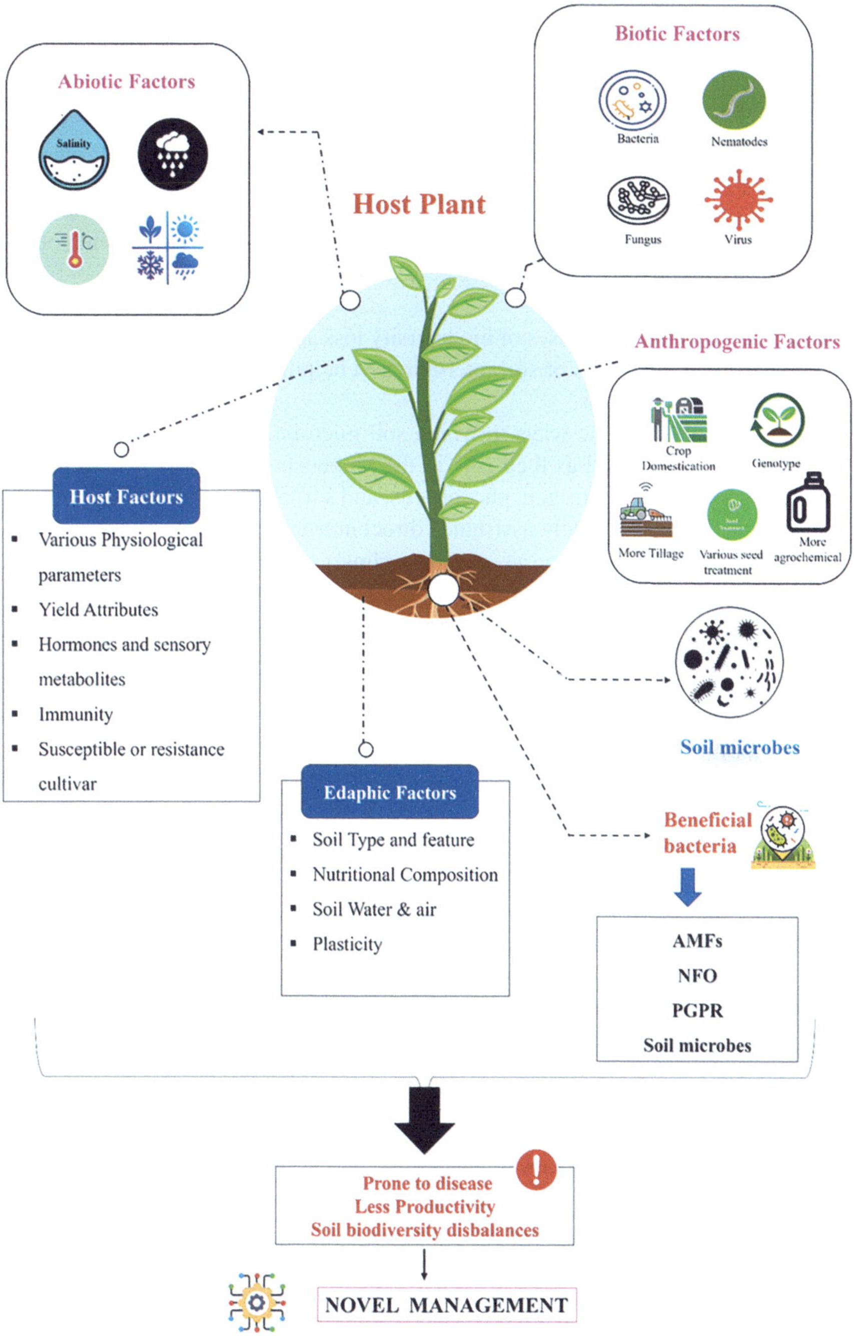

Fig. 1 Overview of factors affecting the relationship between plants and microbes

5 Knowledge Gap About Soil Microbiome

Finding novel solutions to worldwide issues like food insecurity, malnutrition, environmental deterioration, and climate shifting can be facilitated by a deeper understanding of the soil microbiome. Scientific and technological advances in the field of microbiome science have the potential to transform natural resource management, enhance food production, improve ecosystem and human health, and advance a sustainable bio-economy. A more advanced approach can be taken to some of the innovative solutions listed below (Berg et al. 2017).

- Better characterize the causes of biodiversity loss and its possible effects, as well as the diversity and roles of soil microbiota in helping to manage and build sustainable agroecosystems.
- To learn more about the ways in which soil microbiota respond to and control climate change, as well as their role in the biogeochemical cycling of different nutrients like carbon, nitrogen, phosphorus, and sulfur. A better knowing of these mechanisms may ultimately assistance direct nutrient management deftness.
- To take some steps to improve interdisciplinary cooperation in order to learn more about the relationships that exist between different ecosystems.
- To better comprehend the relationships between the soil microbiome and human health, it provides a suitable framework and such an approach.
- To gain knowledge of novel species like soil protista, soil bacteriophages, and the deep soil microbial community that affect the dynamics of nutrients, the preservation of the carbon stock in the agroecosystem, and the occurrence of antibiotic resistance.
- Understanding the effects of micro- and nanoplastics on intricate trophic relationships, ecological processes, and biogeochemical processes mediated by soil microbes.
- Knowing the routes by which antibiotic-resistant components enter agricultural soils, the possibility of exchange between soil and plant resistance machinery, the selection of soil microbial species that degrade plastic and pesticides, and the possible influence on resistant genes in the soil can all contribute to a better gaining of the complex relationship between the soil microbiome and antimicrobial resistance.

Better host performance can be achieved by tailoring the dominant communities in a niche ecosystem through the interaction of the environment and genotype. Therefore, in order to successfully apply microbiome engineering in the field and enhance plant physiological traits under all types of environmental factors while addressing various plant-mediated strategies, it is necessary to understand the ecological hidden mechanisms that control the emergence, persistence, and regional adaptability of plant-associated microbiomes (Lumibao et al. 2020).

6 Exploration of Various Omics Methods

6.1 Genomics

Genomics is an interdisciplinary field of molecular biology focused on the study of genomes including the structure, function, evolution, and interactions of all genes within an organism. The recent strides in this field have revealed how plants adapt to various biotic interactions extending beyond just their beneficial symbionts (Chiquito-Contreras et al. 2024). This shift from phenotypic exploration to molecular unwinding has altered the arena of interaction of plant with microbes. In this domain, genomic techniques like whole genome sequencing (WGS) have altered biological research by providing detailed molecular insights. WGS allows the analysis of complete genome sequences and enzyme annotations, enabling the characterization of genetic variations on a whole genome scale.

Additionally, genome-wide association studies (GWAS) further advance this by linking genetic variation to specific traits which supports the upliftment of improved crop cultivars and beneficial microbial communities. The increased accessibility of next-generation sequencing (NGS) technologies has facilitated deeper unwinding of the genetic machinery underlying critical processes in plant biology. This progress aids in identifying epidemiological factors involved in disease emergence and spread and allows for the comparison of potent candidate genes and the pathways for governing traits such as disease resistance, niche adaptation, and root versus shoot responses (Diwan et al. 2022). Genomics also shows that genome size significantly impacts plant physiology and microbial relationships. Plants with larger genomes often display the paradox effect, developing larger organs and tissues which influences stomata size, CO_2 diffusion, and carbon fixation. This effect can increase the photosynthetic carbon available to root-associated microbes, supporting a large number of microbial symbionts. Changes in genome size can also affect root exudation, stimulating microbial activity and potentially altering microbial diversity (Zhang et al. 2024). Overall, this supports sustainable agriculture by unwinding beneficial and robust microbial-derived products or enabling genetic manipulation to improve crop resilience, optimize microbial interactions, and enhance crop yield, thereby addressing global food security demands (Table 1).

6.2 Metagenomics

Metagenomics is a revolutionary technology in the field of microbiome research that allows for the collection of genetic sequence data from entire microbial communities. This method surpasses the limitations of standard culture-based techniques (Ghosh et al. 2019). This approach enables the direct assembly and analysis of individual genomes from metagenomic data, highlighting the crucial role microbial communities play in the health, growth, and survival of host plants. It allows for the identification and characterization of previously unknown microbial proteins and genetic sequences, providing valuable intuitions into microbial diversity and

Table 1 Overview of omics approaches in plant–microbe interactions and their benefit in sustainable agriculture

Omics approach	Key findings	Tools used	Examples	References
Genomic	Characterized genetic factors influencing symbiosis and nitrogen fixation efficiency Identified plant genes involved in nematode resistance and adaptation mechanisms	Whole genome sequencing and RNA-seq RNA-seq and genome sequencing	Genomic analysis of legume–rhizobium symbiotic interactions Analysis of root-knot nematode (*Meloidogyne incognita*) infection in tomatoes	Garita-Cambronero et al. (2016), Levy et al. (2018)
Metagenomics	Computational methods for analyzing microbial genomes from metagenomic datasets	16S rRNA analysis, shotgun sequencing, MAGs (metagenome-assembled genomes)	Identification of nif genes in diazotrophs, understanding nitrogen fixation and rhizosphere ACC deaminase production Binning of microbial genomes in plant-associated communities	Diwan et al. (2022)
Epigenomics	Focuses on stress responses and environmental adaptation in plants	DNA methylation, histone modifications	Enhancing pathogen resistance, transmission of traits across generations, plant response to *Pseudomonas syringae*	Samantara et al. (2021), Ali et al. (2023), Masenya et al. (2024)
Transcriptomics and meta-transcriptomics	Studies gene expression during plant–microbe interactions	RNA-seq, microarrays	Study of IAA, gibberellins, quorum sensing molecules in plant growth promotion	Fröschel et al. (2021), Morohoshi et al. (2023)
Metabolomics	Reveals metabolite changes during plant–microbe interactions	MS, NMR spectroscopy, MetaboAnalyst	Understanding nutrient uptake in symbiotic relationships, disease detection in grapevines, pathogen resistance	Demiwal et al. (2024), Shafi et al. (2021)

(continued)

Table 1 (continued)

Omics approach	Key findings	Tools used	Examples	References
Proteomics and meta-proteomics	Studies proteins involved in microbial metabolic processes and plant defense responses	2DE, DIGE, MS, Isobaric Tags for Relative and Absolute Quantitation (iTRAQ) proteomics	Understanding *Fusarium oxysporum* pathogenicity, protein expression changes in *Magnaporthe oryzae* in resistant rice plants	Chandok et al. (2022), Priya et al. (2021)
Immunomics	Focuses on the immune responses in plants and the identification of antigens/ epitopes	Genomics, proteomics, transcriptomics integration	Improving plant disease resistance and developing targeted treatments	Dai and Shen (2022)
Microbiomics and culturomics	Characterizes microbial communities in plants and helps understand taxonomic, functional potentials of microbes	Metagenomics, transcriptomics, proteomics, metabolomics	Understanding microbial diversity, plant growth-promoting bacteria roles in disease control	Sarhan et al. (2019), Marchesi and Ravel (2015)
Lipidomics	Investigates cellular lipid profiles and their interactions within cells	GC-MS and LC-MS	Root metabolic profiles with *Azospirillum brasilense* under different phosphorus conditions, salinity tolerance in barley	Gallart-Ayala et al. (2020), Gupta et al. (2021)
Integrated omics approaches	Integrates multi-omics methods to analyze plant–microbe interactions	Exometabolomics, metagenomics, pangenomics, culturomics	Understanding lost traits during domestication, plant pathogen evolution, ancestral microbiome recruitment	Jain et al. (2024), Raaijmakers and Kiers (2022)

composition within their natural environments. It primarily focuses on exploring phylogenetic relationships and functional correlations among microbes, finding new enzyme-encoding genes and sequencing large amounts of environmental materials. By revealing how certain microbes can inhibit plant growth while others enhance plant immunity through pathogen suppression or deterrence, it significantly advances our knowledge about plant–microbe interactions. Additionally, it clarifies the structure of microbial communities and their interactions with both native and invasive pathogens. The adoption of NGS platforms including 16S rRNA analysis, amplicon targeting, and random shotgun sequencing has furthered the identification of novel genes linked with plant microbes (Regalado et al. 2020; Jain et al. 2024). Our knowledge of the molecular mechanisms underpinning plant–microbe interactions and the ways in which plant genotype and environmental factors impact microbiome composition, organization, and evolution has been greatly improved by the availability of large-scale plant–microbe genetic databases (Fig. 2).

6.3 Epigenomic

Epigenomics offers a powerful lens for examining plant–microbe interactions by studying heritable gene expression changes that occur without altering the DNA sequence in both living entities (Samantara et al. 2021). These include DNA methylation, histone alterations (acetylation and methylation), microRNAs, and chromatin remodeling. Such epigenetic changes enable plants and microbes to adapt to their environments, influencing stress responses and potentially transmitting traits across generations (Ali et al. 2023). Scientists may now accomplish precise epigenome changes with the use of certain proteins and gene-editing techniques. Among these instruments are modified CRISPR/Cas9 systems, such as dCas9, zinc finger proteins (ZFP), and transcription activator like effectors (TALEs). The ability to target specific alterations is made feasible by these epi-effectors combining with epigenetic enzyme domains (Doddavarapu et al. 2024). There is growing evidence that epigenomics methods, formerly employed to study gene control, might shed light on the biological functions and linkages between plants and microbes. The application of machine learning in conjunction with epigenomics allows for the localization of differentially methylated regions in plant genomes during interactions with related microbiota (Vigneaud et al. 2023).

6.4 Transcriptomics and Meta-transcriptomics

Through the use of cutting-edge sequencing methods, transcriptomics analyses and quantifies RNA molecules in biological samples, establishing connections between gene activities and particular situations. Methods like microarrays, Sequencing by Oligonucleotide Ligation and Detection-Serial Analysis of Gene Expression (SOLiD-SAGE), and RNA sequencing (RNA-seq) play a crucial role in quantifying all expressed genes. These techniques enable researchers to investigate regulatory changes, mutations, sequence variants, differential gene expression, signalling molecules, and alternative splicing events (Tang et al. 2021).

Similarly, the analysis of total mRNA in environmental samples, known as metatranscriptomics, provides valuable real-time information about gene expression

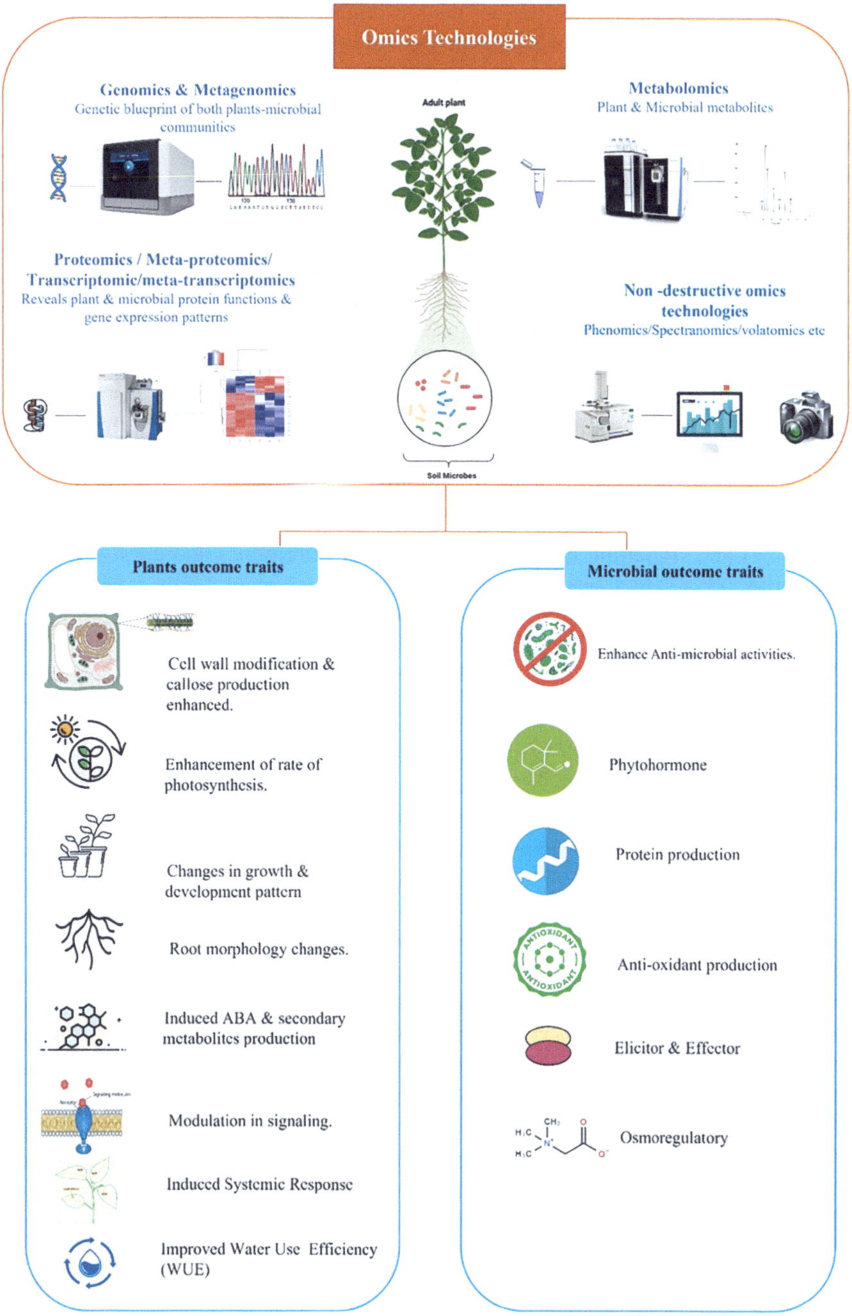

Fig. 2 Different omics techniques for unwinding various traits of plant–microbe relationship

(Dubey et al. 2020). This technique plays a crucial role in pinpointing particular transcribed genes and facilitating the functional analysis of microbial communities associated with plants (Kumar et al. 2021; Gamalero et al. 2022). Studying the expression levels of microorganisms in both clean and polluted samples can greatly improve phytoremediation techniques.

6.5 Metabolomics

Metabolomics is a crucial tool for studying plant–microbe interactions, offering detailed insights into how plants engaged with microbes and revealing significant changes in metabolite profiles. This enhances our understanding of plant responses to both beneficial microbes and pathogens. One of the key mechanisms regulating recruitment, communication, and modulation between host plants and their symbiotic partners is the exchange of metabolites. Both plants and their microbiota use a variety of metabolites, each with specialized functions, to maintain their interactions and relationships. Metabolomics plays a crucial role in identifying these metabolites and signalling molecules, such as phytohormones and quorum sensing compounds (QS), which are essential for understanding plant defenses and interactions with beneficial microbes. Knowledge of the molecular pathways and factors driving these processes is vital for advancing microbiome engineering in sustainable agriculture (Kimotho and Maina 2024). Comprehensive profiling of metabolic alterations is made possible by high-throughput techniques including mass spectrometry (MS), nuclear magnetic resonance (NMR) spectroscopy, and liquid chromatography-mass spectrometry (LC-MS). Tools like XCMS, MZmine, and MetaboAnalyst assist in analyzing metabolic data, providing insights into symbiotic mechanisms, signalling molecules, and plant responses.

6.6 Proteomics and Meta-proteomics

Proteomic studies elucidate the preface of proteins in microbial metabolic processes across various environments. Proteomic analysis is particularly useful for studying unculturable microbes. In plant–microbe interactions, proteins are crucial for maintaining cellular homeostasis; understanding pathogenicity, stress responses, antioxidant mechanisms, and signalling networks; and mediating defense responses. Protein–protein interactions between "Prey" and "Bait" proteins are studied in functional proteomics utilizing the yeast two-hybrid technique (Y2H); when used in conjunction with mass spectrometry, gel-based techniques such as difference gel electrophoresis (DIGE) and two-dimensional electrophoresis (2DE) can identify and quantify proteins and show variations in their expression in different plant tissues.

On the other hand, metaproteomic extends proteomic analysis to entire microbial communities. This methodology is essential for researching nutrient cycle, plant–microbe interactions, soil fertility, and bioremediation. It allows for the screening of proteins involved in key processes such as pathogenicity and host defense modulation. Additionally, meta-proteomics can reveal proteins related to plant diseases, thereby improving our understanding of crop health and disease mechanisms (Chandok et al. 2022) (Fig. 2).

6.7 Lipidomics

Lipidomics is the comprehensive study of cellular lipids, focusing on their identification, quantification, and interactions within cells. This field explores the diverse roles of lipids in cellular processes and their impact on various biological functions. Lipid profiles and compositional changes are frequently analyzed, especially in plant roots, using methods like liquid chromatography-mass spectrometry (LC-MS) and gas chromatography-mass spectrometry (GC-MS). These methods are crucial for understanding lipid metabolism and its role in plant–microbe interactions. Membrane lipids and their derivatives are vital in plant–microbial interactions. Lipids like phospholipids, glycolipids, sphingolipids, and sterol lipids help establish the membrane interface between plants and microbes. Lipid-derived molecules also play essential roles in intracellular signalling and plant–microbial communication. Key signalling lipids include phosphatidic acid, diacylglycerol (DAGs), lysophospholipids, and free fatty acids produced from phospholipase activity. Additionally, apocarotenoids and breakdown products of sphingolipids, such as ceramide, ceramide phosphate, long-chain bases, and their phosphates, are significant. Fatty acids serve as precursors for oxylipins, including jasmonic acid and azelaic acid, which, together with glycerol-3-phosphate, regulate systemic acquired resistance (SAR) (Siebers et al. 2016).

6.8 Microbiomics and Culturomics

Microbiomic explores microbial communities using advanced omics technologies to understand their genetic makeup, gene expressions, protein profiles, and metabolic activities in diverse environment (Kumar 2021). These techniques have been instrumental in characterizing microbial profiles both above- and belowground, revealing taxonomic compositions and functional potentials. For instance, extensive metagenome profiling facilitated by NGS has highlighted the prevalence of biotic factors in plant microbiome with dominant bacterial communities (Marchesi and Ravel 2015). These approaches provide a clearer understanding of various complex relationships, which influence ecosystem dynamics and promote sustainable agriculture.

Culturomics is a high-throughput omics approach that has greatly advanced our understanding of microbial communities, particularly those associated with plants. This method involves cultivating microorganisms from various environments, such as soil and plant tissues, and studying them in controlled laboratory settings. Culturomics generates valuable reference genomes, facilitates gene validation through molecular techniques, andenhances insights into microbial diversity and functional potential (Sarhan et al. 2019).

6.9 Immunomics

Immunomics in plant–microbe interaction involves studying the immune responses of plant using integrated data from genomics, proteomics, and transcriptomics. This approach helps to identify and understand the set of antigens or epitopes that interact with the plant's immune system. By leveraging immunomics, researchers can improve disease resistance, develop targeted treatment, and enhance preventive measures in plants (Dai and Shen 2022).

6.10 Integrated Omic Approaches

Omic techniques like exo-metabolomics, metagenomics, meta-taxanomics, pangenomics, culturomics, meta-proteomics, microbiomics, lipidomics, immunomics, and meta-transcriptomics provide an extensive overview of several biological systems and investigates physiological, biochemical, and molecular aspects of plant–microbe interaction. Integrating these various omics methods adds complexity to the analysis of plant microbiome datasets compared to stand-alone analyses. This integrated approach can identify plant genetic loci involved in recruiting beneficial microbes and reveal molecules released by plants that activate these beneficial microorganisms, potentially improving plant health and breeding practices (Raaijmakers and Kiers 2022).

7 Exploration of Nondestructive Omics Technologies

Noninvasive omics technologies have transformed biological research by enabling the study of complex biological systems without disrupting natural processes. These include advanced imaging techniques such as spectral and fluorescence-based methods which analyze biochemical and physiological changes in plants and microbes nondestructively. The advantages include minimal ecological disturbance, high-throughput data collection, and ethical compliance with sustainable research practices (Großkinsky et al. 2018) (Table 1).

7.1 Phenomics Approach

The selection of phenotypes is crucial for breeding, and modern high-throughput plant phenotyping has transformed this process. This sensor-based phenomics approach provides accurate and precise recording of complex traits related to growth, yield, and stress adaptation from individual organs to whole canopies. It utilizes nondestructive image analysis, data management, and modeling to assess traits consistently over time. The approach measures various structural and functional aspects, including plant biomass, root morphology, leaf characteristics, fruit traits, photosynthetic efficiency, and stress responses. It also evaluates chemical phenotypes such as secondary metabolites involved in plant defense and environmental interactions, including volatile organic compounds (VOCs). Recent advancements include protocols to study the impact of root-associated microbes on different species and genotypes, aiming to identify optimal plant–microbe combinations that enhance performance and resilience (Costa et al. 2019).

7.1.1 Examples of Nondestructive Phenotyping Techniques

7.1.1.1 Volatilomics

It focuses on detecting, characterizing, and quantifying VOCs emitted by plants and associated microbes. These compounds play critical roles in plant defense, communication, and responses to environmental stresses.

7.1.1.2 Shovelomics

It evaluates root system architecture traits such as length, branching, density, and surface area. These traits are essential for improving crop productivity and resilience under varying environmental conditions (Arifuzzaman et al. 2019).

7.1.1.3 Spectranomics

Advances in remote sensing and spectroscopy enable detailed characterization of plant functional traits based on their spectral properties. Spectral data can predict plant growth, metabolism, and defense responses across large scales (Yang et al. 2020) (Table 2).

7.2 Other Noninvasive Techniques and Combinations

7.2.1 Gas Exchange and Transpiration Rate

Simple devices like porometers measure physiological parameters such as gas exchange and transpiration rates. These measurements vary with leaf senescence and provide insights into the physiological status of plants under stress conditions such as water deficit or high temperatures (Lisec et al. 2006).

7.2.2 Nuclear Magnetic Resonance (NMR)

NMR relaxometry allows for the noninvasive measurement of water status at the cellular level. Changes in hydration detected by NMR can correlate with the progression of leaf senescence, providing detailed insights into plant water relations under stress (Kim et al. 2011).

7.2.3 Aboveground Imaging Approach

RGB (red, green, blue) imaging uses digital camera to capture color variations in plant tissues, effectively monitoring stress responses by assessing changes in leaf color and morphology under drought or nutrient stress (Malinowska et al. 2017). On the other hand, spectral imaging employs multispectral or hyperspectral systems to analyze light reflections from plant tissues, providing detailed information on

Table 2 Examples of nondestructive omics technologies and their applications

Technology	Description	Applications	Examples	References
Hyperspectral imaging	Measures reflected light across the electromagnetic spectrum to monitor vegetation health and stress responses	Assessing plant health, stress responses, and biodiversity	Tracking chlorophyll dynamics and photosynthetic activity	Großkinsky et al. (2018)
Fluorescence imaging	Tracks dynamic changes in chlorophyll and photosynthetic activity by detecting emitted light	Monitoring plant stress and physiological changes	Studying chlorophyll degradation under salinity stress	Großkinsky et al. (2018)
Volatilomics	Detects and quantifies VOCs emitted by plants and microbes	Understanding plant defense mechanisms and stress responses	GC-MS analysis of VOC profiles for plant health and stress tolerance	Mhlongo et al. (2018)
Shovelomics	Evaluates root system traits such as length, branching, density, and surface area	Improving crop productivity and resilience	Root system analysis using advanced imaging and tracking systems	Arifuzzaman et al. (2019)
Spectranomics	Uses remote sensing and spectroscopy to characterize plant functional traits	Predicting plant growth, metabolism, and defense responses	Hyperspectral imaging to correlate spectral signatures with biochemical processes	Yang et al. (2020)
Gas exchange and transpiration rate	Measures physiological parameters like gas exchange and transpiration rates	Studying plant physiological status under stress conditions	Porometer measurements for leaf senescence and stress responses	Lisec et al. (2006)
Nuclear magnetic resonance (NMR)	Noninvasive measurement of water status at the cellular level	Monitoring hydration changes and leaf senescence	NMR relaxometry for assessing plant water relations under stress	Kim et al. (2011)
RGB imaging	Captures color variations in plant tissues to monitor stress responses	Assessing changes in leaf color and morphology	Monitoring drought or nutrient stress effects using digital cameras	Malinowska et al. (2017)

morphological parameters and physiological status with the normalized difference vegetation index (NDVI) helping detect traits like stay green in stressed crops. Lastly, fluorescence imaging distinguishes between healthy and senescent tissues based on emitted light, valuable for studying chlorophyll degradation and physiological changes under salinity stress (Großkinsky et al. 2018).

7.2.4 Integrated Imaging Approach

Comprehensive phenomics studies integrate multiple imaging techniques such as RGB, infrared, fluorescence, and 3D imaging. This integration allows researchers to capture a wide range of indices related to physiological changes in plants. For example, combining infrared imaging for temperature sensing with RGB imaging for morphological assessment enables detailed studies of stress responses in crops grown under controlled conditions (Lyu et al. 2017) (Table 2).

8 Advancements in the Omics Technology

8.1 Web-Based Omics Approach

The study of contemporary omics-derived data has significantly benefited from modern web-based tools designed for plant–microbe interactions. These platforms streamline the processing, functional analysis, and visualization of complex biological datasets. Metabolomic studies are supported by platforms like XCMS, MZmine, and MetaboAnalyst, which assist in pathway mapping and in-depth analysis. In metagenomics, tools like QIIME, UPARSE, and MOTHUR facilitate bioinformatics analysis, with MG-RAST and MetaPhlAn2 providing phylogenetic and functional insights. Meta-transcriptomic analysis is enhanced by tools such as HUMAnN2, MetaTrans, and SAMSA, which offer comprehensive gene expression analysis pipelines. For proteomic analysis, tools such as MetaProteomeAnalyzer, SECIMTools, and MetaLab offer efficient data analysis pipelines, while Galaxy-P integrates multi-omics data. The Galaxy Integrated Omics (GIO) platform aids in proteomic identification using genome and transcriptome data. Additional tools like PeptIdent, MultiIdent, MASCOT, SEQUEST, and Sherpa support MS data analysis, while ProFound and MS-Fit improve MALDI-TOF and MS/MS spectra analysis. Gene ontology (GO) enrichment, Kyoto Encyclopedia of Genes and Genomes (KEGG) pathway analysis, and protein–protein interaction (PPI) network analysis further explore gene functions and interactions. Overall, these web-based tools are crucial for advancing omics-driven research and interpreting complex biological data (Jain et al. 2024; Chen et al. 2022) (Table 3).

8.2 Advancement in AI and Machine Learning in Metabolomics for Plant–Microbe Interactions

Artificial intelligence (AI) and machine learning (ML) have become powerful tools in analyzing and interpreting the vast and complex data generated by metabolomics studies in plant–microbe interactions, assisting in data preprocessing, feature selection, metabolite identification, and network analysis (Murmu et al. 2024). AI and ML algorithms automate raw data preprocessing including peak detection, alignment, and normalization that significantly reduce manual efforts and time. These techniques also help identify discriminative metabolites and biomarkers associated

Table 3 Various databases available for efficient storage, management, analysis, and integration of various omics technologies

Database	Aim	Reference
1. RECONSTRUCT	Merging and integrating omics-based in silico model with reconstruction biology in maize breeding to improve the growth and fitness of the crop (https://www.pflanzenforschung.de).	Xie et al. (2019)
2. CUBIC (Complete diallel design plus Unbalanced Breeding like Inter-Cross)	Employed to examine the genetic makeup of numerous essential agronomic characteristics in maize that support interactions between the microbiome and the plant (https://www.ncbi.nlm.nih.gov/bioproject/PRJNA597703).	Liu et al. (2020)
3. GSC (Genomic Standards Consortium)	The GSC aims to make genomic data more accessible and comparable across the international community. It establishes community-driven standards for datasets and ensures the integration and efficient comparison of genomic data (https://www.gensc.org//).	(https://gensc.org/)
4. MIxS Standards (Minimum Information about any (X) Sequence)	MIxS provides a set of minimum standards for describing genomic and marker gene sequence data (https://genomicsstandardsconsortium.github.io/mixs/).	Glass et al. (2014)
5. PAMGO (Plant-Associated Microbe Gene Ontology) Consortium	It focuses on generating and maintaining databases related to the gene ontology of plant-associated microbes (http://pamgo.vbi.vt.edu/about.php).	Torto-Alalibo et al. (2009)
6. STRING (Search Tool for the Retrieval of Interacting Genes/Proteins)	A database of known and predicted protein–protein interactions, including interactions between plant and microbial proteins (https://string-db.org/).	Von Mering et al. (2005)
7. PHI-base (Pathogen-Host Interactions Database)	It's a collection of carefully selected molecular and biological data on genes that have been shown in peer-reviewed scientific publications to influence the course of pathogen–host interactions (http://www.phi-base.org/).	Urban et al. (2020)
8. FungiDB (Fungi database)	A data-mining interface for the functional and comparative genomic data of many species of fungi that are distinct from one another (http://FungiDB.org).	Stajich et al. (2012)
9. Microbial Protein Interaction Database (MPIDB)	To compile and make available every physical microbiological interaction that is known (http://www.jcvi.org/mpidb/).	Goll et al. (2008)
10. PlantGDB (Plant Genome Database)	Assemble a variety of bioinformatics tools that aid in gene prediction and cross-species comparisons to create the foundation for finding gene sets that are specific to a given species or shared by all plants (http://www.plantgdb.org/).	Dong et al. (2004)
11. Planet Microbe	A web-based facilitates the identification of data from selected historical and ongoing oceanic sequencing projects (https://www.planetmicrobe.org/).	Ponsero et al. (2021)

(continued)

Table 3 (continued)

Database	Aim	Reference
12. BAR (Bio-Analytic Resource for Plant Biology)	Utilized to investigate protein–protein interactions, promoter analysis, functional classification enrichment analyses for gene sets, and gene expression and co-expression patterns (https://bar.utoronto.ca/).	Waese et al. (2017)
13. LIS (Legume Information System)	For the legume family, LIS is a genomic data portal (GDP). Access to genetic and genomic data for model legumes and important crops is made possible by LIS (https://www.legumeinfo.org/).	Dash et al. (2016)

with specific plant–microbe interaction outcomes such as disease resistance or symbiotic efficiency. Several ML algorithms, including supervised methods like partial least squares discriminant analysis (PLS-DA) and support vector machines (SVM), predict class membership of new samples based on metabolic profiles, while unsupervised methods like principal component analysis (PCA) and hierarchical clustering identify patterns and relationships among metabolites. Deep learning, a subfield of ML, shows great promise in this field by using artificial neural networks with multiple layers to learn hierarchical representations of metabolomics data. Algorithms such as convolutional neural networks (CNNs) and recurrent neural networks (RNNs) capture complex patterns and relationships among metabolites.

9 Challenges and Limitations of Omics Technologies

Intensive agricultural practices and limited resources create global challenges, impacting crop health and productivity through complex plant–microbe interactions. Despite advances in omics technologies, standardized protocols for validation and analysis are lacking, hindering the thorough exploration of microbial communities. Many microbial species remain inadequately studied, and existing databases are incomplete, complicating technology selection for agricultural issues. Integrating multi-omics data is computationally demanding due to the unique complexities of each field. Developing web-based tools faces challenges in data integration, standardization, scalability, and user-friendliness, with ethical issues around data privacy adding further complexity. Collaborative efforts are crucial to advancing omics methodologies and tools (Kimotho and Maina 2024).

10 Conclusion

The demand for aliment is rising in tandem with the world population. In order to keep up with the present needs for food, agricultural techniques are under constant strain. One of the primary constraints that must be overcome is the presence of unpredictable weather patterns, among other additional issues. Additionally, chemical fertilizers and pesticides can harm human health and the environment. Therefore, it is crucial to understand the intricate relationship between plants and microbes in the agricultural system in order to achieve sustainable agriculture and maintain the health of plants and soil. To overcome these problems, a better analysis of genetic variation, cellular, developmental, and molecular mechanisms affecting crop tolerance toward stress for higher productivity is necessary. Research on these interactions increasingly employs multi-omics platforms to study this abovementioned query. Using experimental and computational methods, multi-omics can improve the interactions with microbes and uncover key microbial players. Exploring these interactions with genomics, transcriptomics, proteomics, and metabolomics is groundbreaking. Choosing the right technology and addressing the limitations of databases are the key obstacle. If we can understand the mechanics and potential of microbes in this type's positive interactions, we can engineer them for biofertilization and biocontrol. As a result, there has been progress in producing crops using different biofertilizers, which may herald a greener method of plant production and the ensuing agricultural revolution. Furthermore, genome editing techniques like CRISPR and omics-based approaches have improved our knowledge regarding these complex interactions by illuminating the role of specific components that contribute to the system's resilience. Despite hurdles like standardization and data integration, omics technologies have great potential for improving agricultural tactics and developing resilient plant–microbe alliances. Focus on improving omics technology resolution and throughput, tackling data integration issues, and promoting interdisciplinary collaborations in future research. Innovations in plant breeding, fertilizer formulations, beneficial microorganisms, and disease management will be benchmarked to increase crop productivity through sustainable agriculture methods, and this chapter highlights the potential for omics techniques to address limited agricultural output, particularly due to environmental stressors, and boost global food productivity.

References

Abd-Alla AM, Meki IK, Demirbas-Uzel G (2020) Insect viruses as biocontrol agents: challenges and opportunities. In: Cottage industry of biocontrol agents and their applications: practical aspects to deal biologically with pests and stresses facing strategic crops. Springer, pp 277–295. https://doi.org/10.1007/978-3-030-33161-0_9

Ahlawat OP, Yadav D, Walia N, Kashyap PL, Sharma P, Tiwari R (2024) Root exudates and their significance in abiotic stress amelioration in plants: a review. J Plant Growth Regul:1–26. https://doi.org/10.1007/s00344-024-11237-7

Alfiky A, Weisskopf L (2021) Deciphering Trichoderma–plant–pathogen interactions for better development of biocontrol applications. J Fungi 7(1):61. https://doi.org/10.3390/jof7010061

Ali MH, Mandal S, Ghorai M, Lal MK, Tiwari RK, Kumar M, Ghosh A, Al-Tawaha AR, Gopalakrishnan AV, Shekhawat MS, Pandey DK (2023) Perspectives of omics and plant microbiome. In: Genomics, transcriptomics, proteomics and metabolomics of crop plants. Academic Press, pp 131–144. https://doi.org/10.1016/B978-0-323-95989-6.00014-0

Arifuzzaman M, Oladzadabbasabadi A, McClean P, Rahman M (2019) Shovelomics for phenotyping root architectural traits of rapeseed/canola (*Brassica napus* L.) and genome-wide association mapping. Mol Gen Genomics 294:985–1000. https://doi.org/10.1007/s00438-019-01563-x

Arnold AE, Mejía LC, Kyllo D, Rojas EI, Maynard Z, Robbins N, Herre EA (2003) Fungal endophytes limit pathogen damage in a tropical tree. Proc Natl Acad Sci 100(26):15649–15654. https://doi.org/10.1073/pnas.2533483100

Astapati AD, Nath S (2023) The complex interplay between plant-microbe and virus interactions in sustainable agriculture: harnessing phytomicrobiomes for enhanced soil health, designer plants, resource use efficiency, and food security. Crop Design 2(1):100028. https://doi.org/10.1016/j.cropd.2023.100028

Bardgett RD, Gibson DJ (2017) Plant ecological solutions to global food security. J Ecol 105(4):859–864. https://doi.org/10.1111/1365-2745.12812

Batool M, Carvalhais LC, Fu B, Schenk PM (2024) Customized plant microbiome engineering for food security. Trends Plant Sci. https://doi.org/10.1016/j.tplants.2023.10.012

Berg G, Rybakova D, Grube M, Köberl M (2016) The plant microbiome explored: implications for experimental botany. J Exp Bot 67(4):995–1002. https://doi.org/10.1093/jxb/erv466

Berg G, Köberl M, Rybakova D, Müller H, Grosch R, Smalla K (2017) Plant microbial diversity is suggested as the key to future biocontrol and health trends. FEMS Microbiol Ecol 93(5):fix050. https://doi.org/10.1093/femsec/fix050

Borriss R (2011) Use of plant-associated Bacillus strains as biofertilizers and biocontrol agents in agriculture. In: Bacteria in agrobiology: plant growth responses. Springer, pp 41–76. https://doi.org/10.1007/978-3-642-20332-9_3

Chandok IK, Afreen H, Afreen R, Haider S, Moharana DP, Hussain T, Aamir M, Shanmugam V, Malik MZ (2022) Functional genomics tools for studying microbe-mediated stress tolerance in plants. In: Mitigation of plant abiotic stress by microorganisms. Academic Press, pp 175–204. https://doi.org/10.1016/B978-0-323-90568-8.00009-2

Chao H, Zhang S, Hu Y, Ni Q, Xin S, Zhao L, Ivanisenko VA, Orlov YL, Chen M (2024) Integrating omics databases for enhanced crop breeding. J Integr Bioinform 20(4):20230012. https://doi.org/10.1515/jib-2023-0012

Chen XL, Sun MC, Chong SL, Si JP, Wu LS (2022) Transcriptomic and metabolomic approaches deepen our knowledge of plant–endophyte interactions. Front Plant Sci 12:700200. https://doi.org/10.3389/fpls.2021.700200

Chen W, Hu Q 2021 'Secondary metabolites of Purpureocillium lilacinum.' Molecules, 27(1), p.18. https://doi.org/10.3390/molecules27010018

Chiquito-Contreras CJ, Meza-Menchaca T, Guzmán-López O, Vásquez EC, Ricaño-Rodríguez J (2024) Molecular insights into plant–microbe interactions: a comprehensive review of key mechanisms. Front Biosci-Elite 16(1):9. https://doi.org/10.31083/j.fbe1601009

Compant S, Samad A, Faist H, Sessitsch A (2019) A review on the plant microbiome: ecology, functions, and emerging trends in microbial application. J Adv Res 19:29–37. https://doi.org/10.1016/j.jare.2019.03.004

Costa C, Schurr U, Loreto F, Menesatti P, Carpentier S (2019) Plant phenotyping research trends, a science mapping approach. Front Plant Sci 9:1933. https://doi.org/10.3389/fpls.2018.01933

Cruz-Hernández MA, Mendoza-Herrera A, Bocanegra-García V, Rivera G (2022) *Azospirillum* spp. from plant growth-promoting bacteria to their use in bioremediation. Microorganisms 10(5):1057. https://doi.org/10.3390/microorganisms10051057

Dai X, Shen L (2022) Advances and trends in omics technology development. Front Med 9:911861. https://doi.org/10.3389/fmed.2022.911861

Dash S, Campbell JD, Cannon EK, Cleary AM, Huang W, Kalberer SR, Karingula V, Rice AG, Singh J, Umale PE, Weeks NT (2016) Legume information system (LegumeInfo. org): a key component of a set of federated data resources for the legume family. Nucleic Acids Res 44(D1):D1181–D1188. https://doi.org/10.1093/nar/gkv1159

De Sousa BFS, Domingo-Serrano L, Salinero-Lanzarote A, Palacios JM, Rey L (2023) The T6SS-dependent effector Re78 of Rhizobium etli Mim1 benefits bacterial competition. Biology 12(5):678. https://doi.org/10.3390/biology12050678

Demiwal P, Tayade S, Yadav SR, Sircar D (2024) A metabolomics perspective on root-derived plant immunity and phytohormone interaction. Physiol Plant 176(1):e14150. https://doi.org/10.1111/ppl.14150

Diwan D, Rashid MM, Vaishnav A (2022) Current understanding of plant-microbe interaction through the lenses of multi-omics approaches and their benefits in sustainable agriculture. Microbiol Res 265:127180. https://doi.org/10.1016/j.micres.2022.127180

Doddavarapu B, Lata C, Shah JM (2024) Epigenetic regulation influenced by soil microbiota and nutrients: paving road to epigenome editing in plants. Biochimica et Biophysica Acta (BBA)-General Subjects:130580. https://doi.org/10.1016/j.bbagen.2024.130580

Dong Q, Schlueter SD, Brendel V (2004) PlantGDB, plant genome database and analysis tools. Nucleic acids research, 32(suppl_1), pp.D354–D359. https://doi.org/10.1093/nar/gkh046

Dubey RK, Tripathi V, Prabha R, Chaurasia R, Singh DP, Rao CS, El-Keblawy A, Abhilash PC, Dubey RK, Tripathi V, Prabha R (2020) Metatranscriptomics and metaproteomics for microbial communities profiling. In: Unravelling the soil microbiome: perspectives for environmental sustainability. Springer, Cham, pp 51–60. https://doi.org/10.1007/978-3-030-15516-2_5

French E, Kaplan I, Iyer-Pascuzzi A, Nakatsu CH, Enders L (2021) Emerging strategies for precision microbiome management in diverse agroecosystems. Nature Plants 7(3):256–267. https://doi.org/10.1038/s41477-020-00830-9

Fröschel C, Komorek J, Attard A, Marsell A, Lopez-Arboleda WA, Le Berre J, Wolf E, Geldner N, Waller F, Korte A, Dröge-Laser W (2021) Plant roots employ cell-layer-specific programs to respond to pathogenic and beneficial microbes. Cell Host Microbe 29(2):299–310. https://doi.org/10.1016/j.chom.2020.11.014

Gallart-Ayala H, Teav T, Ivanisevic J (2020) Metabolomics meets lipidomics: assessing the small molecule component of metabolism. BioEssays 42(12):2000052. https://doi.org/10.1002/bies.202000052

Gamalero E, Bona E, Glick BR (2022) Current techniques to study beneficial plant-microbe interactions. Microorganisms 10(7):1380. https://doi.org/10.3390/microorganisms10071380

Garita-Cambronero J, Palacio-Bielsa A, López MM, Cubero J (2016) Draft genome sequence for virulent and avirulent strains of *Xanthomonas arboricola* isolated from *Prunus spp.* in Spain. Stand Genom Sci 11. https://doi.org/10.1186/s40793-016-0132-3

Ghosh A, Mehta A, Khan AM (2019) Metagenomic analysis and its applications. In: Encyclopedia of bioinformatics and computational biology. Elsevier, pp 184–193. https://doi.org/10.1016/b978-0-12-809633-8.20178-7

Glass EM, Dribinsky Y, Yilmaz P, Levin H, Van Pelt R, Wendel D, Wilke A, Eisen JA, Huse S, Shipanova A, Sogin M (2014) MIxS-BE: a MIxS extension defining a minimum information standard for sequence data from the built environment. ISME J 8(1):1–3. https://doi.org/10.1038/ismej.2013.176

Goll J, Rajagopala SV, Shiau SC, Wu H, Lamb BT, Uetz P (2008) MPIDB: the microbial protein interaction database. Bioinformatics 24(15):1743–1744. https://doi.org/10.1093/bioinformatics/btn285

Gómez-Godínez LJ, Aguirre-Noyola JL, Martínez-Romero E, Arteaga-Garibay RI, Ireta-Moreno J, Ruvalcaba-Gómez JM (2023) A look at plant-growth-promoting bacteria. Plan Theory 12(8):1668. https://doi.org/10.3390/plants12081668

Gomiero T (2016) Soil degradation, land scarcity and food security: reviewing a complex challenge. Sustain For 8(3):281. https://doi.org/10.3390/su8030281

Gopal M, Gupta A (2016) Microbiome selection could spur next-generation plant breeding strategies. Front Microbiol 7:1971. https://doi.org/10.3389/fmicb.2016.01971

Großkinsky DK, Syaifullah SJ, Roitsch T (2018) Integration of multi-omics techniques and physiological phenotyping within a holistic phenomics approach to study senescence in model and crop plants. J Exp Bot 69(4):825–844. https://doi.org/10.1093/jxb/erx333

Gupta S, Smith PM, Boughton BA, Rupasinghe TW, Natera SH, Roessner U (2021) Inoculation of barley with *Trichoderma harzianum* T-22 modifies lipids and metabolites to improve salt tolerance. J Exp Bot 72(20):7229–7246. https://doi.org/10.1093/jxb/erab335

Gupta G, Chauhan PS, Jha PN, Verma RK, Singh S, Yadav VK, Sahoo DK, Patel A (2024) Secretory molecules from secretion systems fine-tune the host-beneficial bacteria (PGPRs) interaction. Front Microbiol 15:1355750. https://doi.org/10.3389/fmicb.2024.1355750

Hardoim PR, van Overbeek LS, van Elsas JD (2008) Properties of bacterial endophytes and their proposed role in plant growth. Trends Microbiol 16(10):463–471. https://doi.org/10.1016/j.tim.2008.07.008

Huang McBeath J, McBeath J, McBeath JH, McBeath J (2010) Plant diseases, pests and food security. In: Environmental change and food security in China, pp 117–156. https://doi.org/10.1007/978-1-4020-9180-3_5

Jain A, Sarsaiya S, Singh R, Gong Q, Wu Q, Shi J (2024) Omics approaches in understanding the benefits of plant-microbe interactions. Front Microbiol 15:1391059. https://doi.org/10.3389/fmicb.2024.1391059

Kalita P, Mohapatra B, Maruthi M (2024) Role of effectors in plant–pathogen interactions. In: Biotechnological advances for disease tolerance in plants. Springer Nature Singapore, Singapore, pp 363–376. https://doi.org/10.1007/978-981-99-8874-7_15

Kim HK, Choi YH, Verpoorte R (2011) NMR-based plant metabolomics: where do we stand, where do we go? Trends Biotechnol 29(6):267–275. https://doi.org/10.1016/j.tibtech.2011.02.001

Kimotho RN, Maina S (2024) Unraveling plant–microbe interactions: can integrated omics approaches offer concrete answers? J Exp Bot 75(5):1289–1313. https://doi.org/10.1093/jxb/erad448

Koranda M, Kaiser C, Fuchslueger L, Kitzler B, Sessitsch A, Zechmeister-Boltenstern S, Richter A (2013) Seasonal variation in functional properties of microbial communities in beech forest soil. Soil Biol Biochem 60:95–104. https://doi.org/10.1016/j.soilbio.2013.01.025

Kumar PS (2021) Microbiomics: were we all wrong before? Periodontol 85(1):8–11. https://doi.org/10.1111/prd.12373

Kumar GC, Chaudhary J, Meena LK, Meena AL, Kumar A (2021) Function-driven microbial genomics for ecofriendly agriculture. In: Microbes in land use change management. Elsevier, pp 389–431. https://doi.org/10.1016/B978-0-12-824448-7.00021-8

Kumari N, Kumawat KC (2024) Microbial ACC-deaminase properties, functions and perspectives in climate stressed agriculture. In: Microbiome drivers of ecosystem function. Academic Press, pp 431–446. https://doi.org/10.1016/B978-0-443-19121-3.00008-9

Lacey LA, Thomson D, Vincent C, Arthurs SP (2008) Codling moth granulovirus: a comprehensive review. Biocontrol Sci Tech 18(7):639–663. https://doi.org/10.1080/09583150802267046

Lesk C, Rowhani P, Ramankutty N (2016) Influence of extreme weather disasters on global crop production. Nature 529(7584):84–87. https://doi.org/10.1038/nature16467

Levy A, Conway JM, Dangl JL, Woyke T (2018) Elucidating bacterial gene functions in the plant microbiome. Cell Host Microbe 24(4):475–485. https://doi.org/10.1016/j.chom.2018.09.005

Li F, Liu M, Li Z, Jiang C, Han F, Che Y (2013) Changes in soil microbial biomass and functional diversity with a nitrogen gradient in soil columns. Appl Soil Ecol 64:1–6. https://doi.org/10.1016/j.apsoil.2012.10.006

Lisec J, Schauer N, Kopka J, Willmitzer L, Fernie AR (2006) Gas chromatography mass spectrometry–based metabolite profiling in plants. Nat Protoc 1(1):387–396. https://doi.org/10.1038/nprot.2006.59

Liu H, Khan MY, Carvalhais LC, Delgado-Baquerizo M, Yan L, Crawford M, Dennis PG, Singh B, Schenk PM (2019) Soil amendments with ethylene precursor alleviate negative impacts of salinity on soil microbial properties and productivity. Sci Rep 9(1):6892. https://doi.org/10.1038/s41598-019-43305-4

Liu HJ, Wang X, Xiao Y, Luo J, Qiao F, Yang W, Zhang R, Meng Y, Sun J, Yan S, Peng Y (2020) CUBIC: an atlas of genetic architecture promises directed maize improvement. Genome Biol 21:1–17. https://doi.org/10.1186/s13059-020-1930-x

Liu H, Qiu Z, Ye J, Verma JP, Li J, Singh BK (2022) Effective colonisation by a bacterial synthetic community promotes plant growth and alters soil microbial community. J Sustain Agric Environ 1(1):30–42. https://doi.org/10.1002/sae2.12008

Lohani N, Jain D, Singh MB, Bhalla PL (2020) Engineering multiple abiotic stress tolerance in canola, *Brassica napus*. Front Plant Sci 11:3. https://doi.org/10.3389/fpls.2020.00003

Lumibao CY, Bernik BM, Formel SK, Kandalepas D, Mighell KL, Pardue J, Van Bael SA, Blum MJ (2020) Rhizosphere microbial communities reflect genotypic and trait variation in a salt marsh ecosystem engineer. Am J Bot 107(6):941–949. https://doi.org/10.1002/ajb2.1497

Lyu JI, Baek SH, Jung S, Chu H, Nam HG, Kim J, Lim PO (2017) High-throughput and computational study of leaf senescence through a phenomic approach. Front Plant Sci 8:250. https://doi.org/10.3389/fpls.2017.00250

Mahapatra R, Mishra P, Patel ZM (2023) The molecular architecture of rhizobium–plant symbiosis in nitrogen fixation. In: The chemical dialogue between plants and beneficial microorganisms. Academic Press, pp 137–144. https://doi.org/10.1016/B978-0-323-91734-6.00006-5

Malinowska M, Donnison IS, Robson PR (2017) Phenomics analysis of drought responses in *Miscanthus* collected from different geographical locations. GCB Bioenergy 9(1):78–91. https://doi.org/10.1111/gcbb.12350

Marchesi JR, Ravel J (2015) The vocabulary of microbiome research: a proposal. Microbiome 3:1–3. https://doi.org/10.1186/s40168-015-0094-5

Masenya K, Manganyi MC, Dikobe TB (2024) Exploring cereal metagenomics: unravelling microbial communities for improved food security. Microorganisms 12(3):510. https://doi.org/10.3390/microorganisms12030510

Maute K, French K, Story P, Bull CM, Hose GC (2017) Short and long-term impacts of ultra-low-volume pesticide and biopesticide applications for locust control on non-target arid zone arthropods. Agric Ecosyst Environ 240:233–243. https://doi.org/10.1016/j.agee.2017.02.024

Mhlongo MI, Piater LA, Madala NE, Labuschagne N, Dubery IA (2018) The chemistry of plant–microbe interactions in the rhizosphere and the potential for metabolomics to reveal signaling related to defense priming and induced systemic resistance. Front Plant Sci 9:112. https://doi.org/10.3389/fpls.2018.00112

Mishra S, Bhattacharjee A, Sharma S (2021) An ecological insight into the multifaceted world of plant-endophyte association. Crit Rev Plant Sci 40(2):127–146. https://doi.org/10.1080/07352689.2021.1901044

Morohoshi T, Arai W, Someya N (2023) N-acylhomoserine lactone-degrading activity of *Trichoderma* species and its application in the inhibition of bacterial quorum sensing. J Microorg Control 28(3):139–143. https://doi.org/10.4265/jmc.28.3_139

Morrissey JP, Dow JM, Mark GL, O'Gara F (2004) Are microbes at the root of a solution to world food production? Rational exploitation of interactions between microbes and plants can help to transform agriculture. EMBO Rep 5(10):922–926. https://doi.org/10.1038/sj.embor.7400263

Murmu S, Sinha D, Chaurasia H, Sharma S, Das R, Jha GK, Archak S (2024) A review of artificial intelligence-assisted omics techniques in plant defense: current trends and future directions. Front Plant Sci 15:1292054. https://doi.org/10.3389/fpls.2024.1292054

Nadarajah K (2021) Metagenomics for improving soil fertility. In: Soil nitrogen ecology. Springer International Publishing, Cham, pp 267–282. https://doi.org/10.1038/s41477-021-00956-4

Nawaz A, Ali H, Sufyan M, Gogi MD, Arif MJ, Ranjha MH, Arshid M, Waseem M, Mustafa T, Qasim M, Rizwan M (2019) Comparative bio-efficacy of nuclear polyhedrosis virus (NPV) and Spinosad against American bollwormm, *Helicoverpa armigera* (Hubner). Revista Brasileira de Entomologia 63(4):277–282. https://doi.org/10.1016/j.rbe.2019.09.001

Niazi P, Monib AW, Ozturk H, Mansoor M, Azizi A, Hassand MH (2023) Review on surface elements and bacterial biofilms in plant-bacterial associations. J Res Appl Sci Biotechnol 2(1):204–214. https://doi.org/10.55544/jrasb.2.1.30

Orrell P, Bennett AE (2013) How can we exploit above–belowground interactions to assist in addressing the challenges of food security? Front Plant Sci 4:432. https://doi.org/10.3389/fpls.2013.00432

Poole P, Ramachandran V, Terpolilli J (2018) Rhizobia: from saprophytes to endosymbionts. Nat Rev Microbiol 16(5):291–303. https://doi.org/10.1038/nrmicro.2017.171

Ponsero AJ, Bomhoff M, Blumberg K, Youens-Clark K, Herz NM, Wood-Charlson EM, Delong EF, Hurwitz, BL 2021 Planet Microbe: a platform for marine microbiology to discover and analyze interconnected 'omics and environmental data. Nucleic acids research, 49(D1), pp.D792–D802.https://doi.org/10.1093/nar/gkh046

Priya P, Aneesh B, Harikrishnan K (2021) Genomics as a potential tool to unravel the rhizosphere microbiome interactions on plant health. J Microbiol Methods 185:106215. https://doi.org/10.1016/j.mimet.2021.106215

Raaijmakers JM, Kiers ET (2022) Rewilding plant microbiomes. Science 378(6620):599–600. https://doi.org/10.1126/science.abn6350

Regalado J, Lundberg DS, Deusch O, Kersten S, Karasov T, Poersch K, Shirsekar G, Weigel D (2020) Combining whole-genome shotgun sequencing and rRNA gene amplicon analyses to improve detection of microbe–microbe interaction networks in plant leaves. ISME J 14(8):2116–2130. https://doi.org/10.1038/s41396-020-0665-8

Rodriguez PA, Rothballer M, Chowdhury SP, Nussbaumer T, Gutjahr C, Falter-Braun P (2019) Systems biology of plant-microbiome interactions. Mol Plant 12(6):804–821. https://doi.org/10.1016/j.molp.2019.05.006

Samad A, Trognitz F, Compant S, Antonielli L, Sessitsch A (2017) Shared and host-specific microbiome diversity and functioning of grapevine and accompanying weed plants. Environ Microbiol 19(4):1407–1424. https://doi.org/10.1111/1462-2920.13618

Samantara K, Shiv A, de Sousa LL, Sandhu KS, Priyadarshini P, Mohapatra SR (2021) A comprehensive review on epigenetic mechanisms and application of epigenetic modifications for crop improvement. Environ Exp Bot 188:104479. https://doi.org/10.1016/j.envexpbot.2021.104479

Sandrini M, Nerva L, Sillo F, Balestrini R, Chitarra W, Zampieri E (2022) Abiotic stress and belowground microbiome: the potential of omics approaches. Int J Mol Sci 23(3):1091. https://doi.org/10.3390/ijms23031091

Sarhan MS, Hamza MA, Youssef HH, Patz S, Becker M, ElSawey H, Nemr R, Daanaa HSA, Mourad EF, Morsi AT, Abdelfadeel MR (2019) Culturomics of the plant prokaryotic microbiome and the dawn of plant-based culture media–a review. J Adv Res 19:15–27. https://doi.org/10.1016/j.jare.2019.04.002

Selwal N, Wani AK, Akhtar N, Kaur M, Jassal PS (2023) Molecular insights of Strigolactone biosynthesis, signalling pathways, regulatory roles, and hormonal crosstalks in plant systems. S Afr J Bot 160:9–22. https://doi.org/10.1016/j.sajb.2023.06.046

Shafi A, Zahoor I, Habib H (2021) Omics technologies to unravel plant-microbe interactions. In: Plant-microbe dynamics. CRC Press, pp 201–220. https://doi.org/10.1201/9781003106784

Siebers M, Brands M, Wewer V, Duan Y, Hölzl G, Dörmann P (2016) Lipids in plant–microbe interactions. Biochimica et Biophysica Acta (BBA)-Mol Cell Biol Lipids 1861(9):1379–1395. https://doi.org/10.1016/j.bbalip.2016.02.021

Singh D, Singh SK, Modi A, Singh PK, Zhimo VY, Kumar A (2020) Impacts of agrochemicals on soil microbiology and food quality. In: Agrochemicals detection, treatment and remediation. Butterworth-Heinemann, pp 101–116. https://doi.org/10.1016/B978-0-08-103017-2.00004-0

Singh G, Agrawal H, Bednarek P (2023a) Specialized metabolites as versatile tools in shaping plant–microbe associations. Mol Plant 16(1):122–144. https://doi.org/10.1016/j.molp.2022.12.006

Singh BK, Delgado-Baquerizo M, Egidi E, Guirado E, Leach JE, Liu H, Trivedi P (2023b) Climate change impacts on plant pathogens, food security and paths forward. Nat Rev Microbiol 21(10):640–656. https://doi.org/10.1038/s41579-023-00900-7

Stajich JE, Harris T, Brunk BP, Brestelli J, Fischer S, Harb OS, Kissinger JC, Li W, Nayak V, Pinney DF, Stoeckert CJ Jr (2012) FungiDB: an integrated functional genomics database for fungi. Nucleic Acids Res 40(D1):D675–D681. https://doi.org/10.1093/nar/gkr918

Summit GB (2020) Expanding the sustainable bioeconomy–vision and way forward. In: Communiqué of the global bioeconomy summit 2020, pp 1–28. Available at https://gbs2015.com/fileadmin/gbs2015/Downloads/Communique_final_neu.pdf

Thoms D, Liang Y, Haney CH (2021) Maintaining symbiotic homeostasis: how do plants engage with beneficial microorganisms while at the same time restricting pathogens? Mol Plant-Microbe Interact 34(5):462–469. https://doi.org/10.1094/MPMI-11-20-0318-FI

Torto-Alalibo T, Collmer CW, Gwinn-Giglio M (2009) The Plant-Associated Microbe Gene Ontology (PAMGO) Consortium: community development of new gene ontology terms describing biological processes involved in microbe-host interactions. BMC Microbiol 9:1–5. https://doi.org/10.1186/1471-2180-9-S1-S1

Trenberth KE, Dai A, Van Der Schrier G, Jones PD, Barichivich J, Briffa KR, Sheffield J (2014) Global warming and changes in drought. Nat Clim Chang 4(1):17–22. https://doi.org/10.1038/nclimate2067

Trivedi P, Delgado-Baquerizo M, Trivedi C, Hu H, Anderson IC, Jeffries TC, Zhou J, Singh BK (2016) Microbial regulation of the soil carbon cycle: evidence from gene–enzyme relationships. ISME J 10(11):2593–2604. https://doi.org/10.1038/ismej.2016.65

Trivedi P, Leach JE, Tringe SG, Sa T, Singh BK (2020) Plant–microbiome interactions: from community assembly to plant health. Nat Rev Microbiol 18(11):607–621. https://doi.org/10.1038/s41579-020-0412-1

Umar OB, Ranti LA, Abdulbaki AS, Bola AL, Abdulhamid AK, Biola MR, Victor KO (2021) Stresses in plants: biotic and abiotic. In: Current trends in wheat research. IntechOpen, pp 1–8. https://doi.org/10.5772/intechopen.87473

Urban M, Cuzick A, Seager J, Wood V, Rutherford K, Venkatesh SY, De Silva N, Martinez MC, Pedro H, Yates AD, Hassani-Pak K (2020) PHI-base: the pathogen–host interactions database. Nucleic Acids Res 48(D1):D613–D620. https://doi.org/10.1093/nar/gkz904

Van der Ent S, Van Wees SC, Pieterse CM (2009) Jasmonate signaling in plant interactions with resistance-inducing beneficial microbes. Phytochemistry 70(13–14):1581–1588. https://doi.org/10.1016/j.phytochem.2009.06.009

Verma R, Annapragada H, Katiyar N, Shrutika N, Das K, Murugesan S (2020) Rhizobium. In: Beneficial microbes in agro-ecology. Academic Press, pp 37–54. https://doi.org/10.1016/B978-0-12-823414-3.00004-6

Vigneaud J, Kohler A, Sow MD, Delaunay A, Fauchery L, Guinet F, Daviaud C, Barry KW, Keymanesh K, Johnson J, Singan V (2023) DNA hypomethylation of the host tree impairs interaction with mutualistic ectomycorrhizal fungus. New Phytol 238(6):2561–2577. https://doi.org/10.1111/nph.18734

Villarreal-Delgado MF, Villa-Rodríguez ED, Cira-Chávez LA, Estrada-Alvarado MI, Parra-Cota FI, Santos-Villalobos SDL (2018) The genus Bacillus as a biological control agent and its implications in the agricultural biosecurity. Revista mexicana de fitopatología 36(1):95–130. https://doi.org/10.18781/r.mex.fit.1706-5

Von Mering C, Jensen LJ, Snel B, Hooper SD, Krupp M, Foglierini M, Jouffre N, Huynen MA, Bork P (2005) STRING: known and predicted protein–protein associations, integrated and transferred across organisms. Nucleic Acids Res 33(suppl_1):D433–D437. https://doi.org/10.1093/nar/gki005

Waese J, Provart NJ (2017) The bio-analytic resource for plant biology. In: Plant genomics databases: methods and protocols. Springer, pp 119–148. https://doi.org/10.1007/978-1-4939-6658-5_6

Westman SM, Kloth KJ, Hanson J, Ohlsson AB, Albrectsen BR (2019) Defence priming in Arabidopsis–a meta-analysis. Sci Rep 9(1):13309. https://doi.org/10.1038/s41598-019-49811-9

Xie K, Guo L, Bai Y, Liu W, Yan J, Bucher M (2019) Microbiomics and plant health: an interdisciplinary and international workshop on the plant microbiome. Mol Plant 12(1):1–3. https://doi.org/10.1016/j.molp.2018.11.004

Yang W, Feng H, Zhang X, Zhang J, Doonan JH, Batchelor WD, Xiong L, Yan J (2020) Crop phenomics and high-throughput phenotyping: past decades, current challenges, and future perspectives. Mol Plant 13(2):187–214. https://doi.org/10.1016/j.molp.2020.01.008

Yang J, Lan L, Jin Y, Yu N, Wang D, Wang E (2022) Mechanisms underlying legume–rhizobium symbioses. J Integr Plant Biol 64(2):244–267. https://doi.org/10.1111/jipb.13207

Yuan S, Ke D, Liu B, Zhang M, Li X, Chen H, Zhang C, Huang Y, Sun S, Shen J, Yang S (2023) The Bax inhibitor GmBI-1α interacts with a Nod factor receptor and plays a dual role in the legume–rhizobia symbiosis. J Exp Bot 74(18):5820–5839. https://doi.org/10.1093/jxb/erad276

Zhang H, Liu H, Han X (2024) Traits-based approach: leveraging genome size in plant–microbe interactions. Trends Microbiol 32(4):333–341. https://doi.org/10.1016/j.tim.2023.10.004

Zhou D, Chen X, Chen X, Xia Y, Liu J, Zhou G (2023) Plant immune receptors interact with hemibiotrophic pathogens to activate plant immunity. Front Microbiol 14:1252039. https://doi.org/10.3389/fmicb.2023.1252039

Metabolomics in Understanding and Mitigating Metal Toxicity

Sutapa Bhowmik, Shravan Kumar, Alok Raj, Dickson Adom, and Reena Saxena

Abstract

Metabolomics, the comprehensive study of metabolites within biological systems, has become a powerful tool for elucidating the mechanisms of metal toxicity and developing mitigation strategies. This study examines recent advancements in the application of metabolomics to understand the biochemical impacts of metal exposure such as arsenic (As), cadmium (Cd), lead (Pb), and mercury (Hg) on various organisms. By analyzing metabolic profiles, researchers can identify specific biomarkers and disrupted pathways associated with metal-induced toxicity. These insights facilitate the development of targeted interventions to mitigate adverse effects. The combination of metabolomics with other omics technologies to increase the accuracy of toxicity assessments and to better understand the mechanisms underlying metal toxicity is discussed here. Furthermore, it highlights the role of metabolomics in identifying potential therapeutic agents and dietary strategies to counteract metal toxicity. Findings from recent studies emphasize the potential of metabolomics to transform various approaches to managing metal toxicity in both environmental and clinical contexts.

S. Bhowmik · R. Saxena (✉)
School of Applied Sciences, Suresh Gyan Vihar University, Gyan Vihar Marg, Jaipur, India

S. Kumar
Department of Biochemical Engineering, Harcourt Butler Technology University, Kanpur, Uttar Pradesh, India

A. Raj
Department of Chemical Engineering, Rashtrakavi Ramdhari Singh Dinkar College of Engineering, Begusarai, Bihar Engineering University, Patna, Bihar, India

D. Adom
Department of Educational Innovations in Science and Technology, Kwame Nkrumah University of Science and Technology, Kumasi, Ghana

S. Kaur et al. (eds.), *Microbial Metabolomics*,
https://doi.org/10.1007/978-981-96-4824-5_15

Keywords

Metabolomics · Metal toxicity · Biochemical impacts · Biomarkers · Environmental contexts

1 Introduction to Metabolomics

Metabolomics is a promising technology that offers a unique perspective on metal-induced alterations in cellular metabolic architecture and can be used to study metal toxicity. The scientific study of chemical reactions involving metabolites, small molecules that are substrates, intermediates, and products of cell metabolism is known as metabolomics. Metabolites, which are typically smaller than 1500 Da, are made up of a range of materials, including steroids, lipids, fatty acids, amino acids, vitamins, minerals, drugs, and carbohydrates (Roberts 2023).

The metabolome is context-dependent and will alter in reaction to external variables, such as illness or exposure to a toxin, just like in related approaches like genomes and proteomics. Rapid readings and cost-effectiveness are two benefits of investigating metabolites (Jones and Cheung 2007). When an organism is under severe stress, tiny alterations in its metabolism can be sensitively reflected using metabolomics. Overall accumulation, which can represent the metabolic changes in organisms both before and after external stimuli, characterizes metabolomics' sensitive reactivity to the environment and external stimuli (Feng et al. 2021). Metabolomics research offers important insights into the qualitative and quantitative metabolite changes brought on by exposure to heavy metals. Any disruption in these pathways can lead to the emergence of progressive illnesses because these metabolites are engaged in a variety of metabolic and biosynthetic pathways (Akash et al. 2023).

1.1 Techniques and Technologies in Metabolomics

A specific technique called "metabolomics" enables assessing changes in a multicellular system's metabolism using targeted and untargeted methods. Studies that are targeted are concentrated on a specific number of identified metabolites, whereas the goal of untargeted metabolomics is to identify every metabolite in a sample (Szeremeta et al. 2021). In metabolomics research, a variety of analytical chemical tools are used; some are used regularly and extensively, while others have specialized uses and are employed less frequently.

The separation, detection, characterization, and quantification of these metabolites and associated metabolic pathways may be made possible by the recent and rapid development of a variety of analytical platforms, such as gas chromatography-mass spectrometry (GC-MS), high-performance liquid chromatography (HPLC), capillary electrophoresis-mass spectrometry (CE-MS), and nuclear magnetic resonance (NMR) spectroscopy.

In non-targeted analysis, GC-MS has been employed as a platform, particularly for hydrophilic metabolites. In general, high-throughput technology is needed for GC-MS-based metabolomics in order to process a large number of samples and accurately identify peaks using mass spectra and standard retention times. A derivatization reaction is necessary for GC-MS to produce volatile compounds for separation on the GC column. The GC-MS analysis fails to identify nonvolatile compounds since these substances are not derivatized. The applicability to metabolomics is thus limited (Tsugawa et al. 2011). It is not necessary to derivatize substances before analysis when using HPLC along with MS. Labile and nonvolatile polar and nonpolar substances in their native form are better analyzed using HPLC separations. Soft ionization has recently been used in the development of LC-MS techniques, which strengthen MS for everyday use. Additionally, LC-MS can offer a list of retention times, m/z (mass by charge) ratios, and an estimate of the relative abundances of identified metabolites that are not actually identified. The application of large-scale metabolomic technologies based on LC-MS in the diagnosis of human diseases is drawing more and more interest (Courant et al. 2009). For charged metabolites, CE-MS is a strong and promising separation method that offers high analyte resolution and primarily provides information on polar or ionic chemicals in biological fluids. In metabolomics, CE-MS is a promising hyphenated micro separation platform because most primary metabolites are polar by nature (Britz-McKibbin 2011). CE first separates metabolites according to size and charge, and then, MS uses ion monitoring across a wide range of m/z values to identify them selectively. Compared to other separation techniques, CE-MS offers a number of significant advantages. The fast analysis time and extremely low sample requirement of CE-MS, with injection volumes ranging from 1 to 20 nanoliter, are two of its major benefits. Because of this characteristic, CE-MS is a highly promising analytical method for high-throughput metabolomics. As a result, both targeted and non-targeted metabolite analyses, including that of inorganic ions, organic acids, amino acids, nucleotides and nucleosides, vitamins, thiols, carbohydrates, and peptides, have been performed using CE-MS (Simó et al. 2011). NMR, one of the most widely used spectroscopic analytical methods, has the ability to simultaneously identify and quantify a large variety of organic molecules in the micromolar range. NMR is simple, mostly automated, and nondestructive, allowing samples to undergo further investigation. Metabolic flux analysis, profiling, and metabolite fingerprinting have all made substantial use of it. The main drawback of NMR for thorough metabolite profiling is its comparatively low sensitivity, which precludes it from being suitable for analyzing a large number of low-abundance metabolites. Because of its rich structural information and capacity to analyze intact biomaterials nondestructively, NMR has always been the method of choice in the field of metabolomics of biofluids. Therefore, a great deal of study and notable advancements have been made in the use of NMR to quantify populations of low-molecular-weight metabolites in biological materials. The ideal method for identifying metabolites in tissue extracts or biofluids may be high-resolution NMR (Soininen et al. 2009). A single analytical instrument cannot be used to detect every metabolite in a biological sample due to the complexity of the metabolome and the variety of metabolite characteristics. The optimal results in metabolomics have been reached by the combined use of contemporary instrumental analytical techniques, which also helps to expand the coverage of discovered metabolites beyond what is possible with single-analysis methods. Thousands of metabolites in a biofluid sample can be detected sensitively and reliably with integrated systems (Zhang et al. 2012).

Comprehensive analysis is a challenging analytical task since metabolites are a diverse group of low-molecular weight compounds that include lipids, amino acids, peptides, nucleic acids, and organic acids. Such chemically varied structures may now be separated, characterized, detected, and quantified a number of analytical platforms based on mass spectrometry (MS) and nuclear magnetic resonance (NMR) that has happened quite quickly recently (Wang et al. 2010). The primary analytical techniques used in metabolomic investigations in the current era are gas chromatography coupled to mass spectrometry (GC-MS), nuclear magnetic resonance spectroscopy (NMR), direct infusion mass spectrometry (DI-MS), two-dimensional GC coupled to MS (GC × GC-MS), capillary electrophoresis coupled to MS (CE-MS), and liquid chromatography coupled to MS (LC-MS) (Begou et al. 2019; Gika et al. 2019).

1.2 Applications of Metabolomics

Metabolomics research has potential for application in many different kinds of circumstances. Since metabolomics is still a relatively new field of study, many of its applications are only now starting to become clear. However, this chapter discusses a few potential current and future applications for metabolomics.

1.2.1 Toxicology

The physiological alterations brought on by a chemical (or combination of chemicals) can be identified by metabolic profiling, particularly on urine or blood plasma samples. Finding biomarkers that show exposure to heavy metals and their harmful effects is made possible by metabolomics. Toxicological assessment may be achieved by metabolic profiling of the blood or urine. The presence of a toxin might cause physiological changes in the physiological sample that can be detected using a variety of approaches (Smith 2019).

There has been significant progress in toxicological research since the introduction of metabolomics. In recent years, metabolomic approaches have yielded a substantial amount of data, leading to the identification of various potential biomarkers linked to the effects, exposure, and/or susceptibility to toxicants. Additionally, these approaches have improved our understanding of the metabolic responses of numerous biological systems (Araújo et al. 2021). One such example of metal toxicity is shown in Table 1.

1.2.2 Functional Genomics

In functional genomics, metabolomics may be particularly helpful in examining potential phenotypes of a given genetic modification. For instance, having enough

Table 1 Studies using metabolomics in toxicological research in terms of metal (Cd) (Araújo et al. 2021)

Toxicant classification	Study model/ matrix	Analytical platform	Study outcomes	Reference
Heavy metal (environmental pollutant): Cadmium (Cd)	Human model: urine	UHPLC-Q Exactive Orbitrap MS	The following metabolic pathways have been severely disrupted: Creatinine pathway Tryptophan metabolism Aminoacyl-tRNA biosynthesis Purine metabolism	Zeng et al. (2021)
	Animal model: urine of wild-type 129/ Sv female mice	UPLC-QTOFMS	Arginine and proline metabolism Alanine, aspartate, and glutamate metabolism Aminoacyl-tRNA biosynthesis Purine metabolism	Sarma et al. (2018)

understanding of how the metabolome behaves could be useful in predicting the phenotypes that would arise from the deletion or insertion of a gene into the genome. There are several real-world applications for the identification of phenotypic alterations. One good example is the ability to look for phenotypic alterations in plant material that has been genetically engineered for human use. Before they are released for public consumption, any alterations that might affect the metabolome of the species that eat the genetically modified material could be anticipated (Smith 2019).

A dynamic, detailed, and accurate picture of the phenotype is provided by the metabolome. Modern high-throughput technologies have made it possible to identify pertinent metabolites that represent a broad range of human phenotypes, including health, illness, drug monitoring, and even aging. Similar to genomics, metabolomics has aided in the knowledge of a wide range of molecular mechanisms and resulted in the development of biomarkers, underscoring its use in precision medicine. Differentiable metabolomic profiles will be identified, which will aid in the development and enhancement of clinical approaches to treat human illness. Metabolomics will be a commonly used technique in the future for medication development, aging, and health and illness diagnosis and monitoring. Utilizing branched-chain amino acids, acylcarnitines, specific phospholipids, and genomics, biomedical applications of metabolomics are already anticipated to track the development of metabolic diseases, such as obesity and diabetes; these can determine the severity of the disease and forecast a possible course of treatment (Gonzalez-Covarrubias et al. 2022).

1.2.3 Nutrigenomics

By bridging the gap between molecular biology and nutrition, nutrigenomics has completely changed the field of nutritional study (Dable-Tupas et al. 2022). Nutrigenomics integrates human nutrition concepts with insights from the fields of genomics, transcriptomics, proteomics, and metabolomics. Every bodily fluid has a metabolome that is influenced by a number of exogenous variables including nutrition, other food ingredients, and drugs, as well as endogenous factors like the person's age, gender, body composition, genetic susceptibilities, and coexisting medical disorders. The application of metabolomics in nutrigenomics can be utilized to ascertain an individual's metabolic fingerprint, which illustrates the impact of both endogenous and exogenous influences on the body's metabolism (Smith 2019). Specifically, metabolomics has been essential in the creation of food consumption indicators to improve the precision of dietary evaluations (Brennan and De Roos 2021).

1.2.4 Health and Medical

Metabolomics has potential for improving our existing knowledge of, ability to diagnose, and ability to treat a number of diseases, including cancer and endocrine disorders. The field can assist in determining the disease's pathophysiological processes and the mechanisms that can be targeted for disease management. Metabolomics is being utilized more and more to track therapy outcomes, find novel drug targets, diagnose diseases, and comprehend disease mechanisms (Wishart 2016).

The impact of medications, poisons, and medical conditions on metabolites have all been investigated using mass spectrometry (MS)-based metabolomics techniques. Numerous diseases, including breast, kidney, heart, bladder, esophageal, and gastric cancers, as well as natural metabolic mistakes, toxicological effects, and nutrition, have been studied using MS-based metabolomics techniques (Zhang et al. 2013a, b; Feng et al. 2018). Finding diagnostic biomarkers is a major emphasis of cancer metabolomics research, but the field is also being used to gain deeper mechanistic understanding of cancer and carcinogenesis. In kidney disorders, metabolomics is crucial because it understands disrupted metabolic pathways, identifies early biomarkers, and monitors therapy outcomes (Ilyas et al. 2024).

Metabolomics is a widely used technique in biomedicine to investigate the molecular pathways underlying the development and progression of a wide range of disorders, as well as to identify novel biomarkers for biomonitoring and diagnosis (López-López et al. 2018).

1.2.5 Environmental Metabolomics

Characterizing an organism's interactions with its surroundings is another use for metabolomics. It is possible to learn important information about how the environment affects an organism's health by examining these environmental interactions and evaluating an organism's function and health at the molecular level. This can also be used to collect data for a larger population in order to supply ecology and other related fields of study (Smith 2019).

Metabolomics has been utilized more and more in environmental research in the past few years to examine how different environmental exposures, mostly chemical pollutants, affect an organism's or human's metabolism. As a result, metabolomics is now a crucial tool in exposome research, helping to investigate the molecular effects that xenobiotics have on biological processes. In exposome studies and the development of adverse outcome pathways (AOPs), the combination of metabolomics with information about the presence of xenobiotics in biological samples (xeno-metabolome) and other molecular information obtained from omics technologies (e.g., transcriptomics, proteomics) is crucial to understanding the biological mechanisms triggered by specific chemical compounds (Bedia 2022).

Metabolomics is used in environmental science research to describe how biotic and abiotic stressors affect ecosystems and human health (Bedia et al. 2018). The majority of the study on abiotic variables affecting metabolomics is concerned with common environmental chemical contaminants and their combinations. In this context, the mechanism of action and effects of these substances in people and biota are studied using metabolomics as an ecotoxicological approach (Bedia et al. 2018; Zhang et al. 2021). Owing to the distinct chemical makeup of lipid structures and their significance in cell biology processes, lipidomics—a subfield of metabolomics—has arisen in this field, with its own extraction and analytical procedures (Wei et al. 2019). Numerous environmental studies have focused on identifying lipid alterations in various biological matrices as a response to environmental stresses (Koelmel et al. 2020).

2 Metal Toxicity

The worldwide community is concerned about the hazardous effects of heavy metals (HMs) as manganese (Mn), mercury (Hg), lead (Pb), arsenic (As), and cadmium (Cd). Despite being naturally occurring elements in the earth's crust, metal mining, industry, agriculture, and other anthropogenic activities have caused the concentrations of certain metals to rise above levels found in nature in the soil, air, and water. These metals make up most of the contaminants in the environment and are used in many different production processes. These pollutants are chemically changed, which modifies their toxicity and bioavailability, but they cannot be eliminated easily once released into the environment. Ultimately, drinking water, soil, groundwater, air, and food chains can all get contaminated by metals (Wallace et al. 2020).

2.1 Types of Toxic Metals

Heavy metals are divided into two categories: essential and nonessential; per research by Kim et al. (2019), crucial at low quantities, heavy metals are less harmful and function as coenzymes in biological processes. For instance, iron is found in hemoglobin and myoglobin, whereas cobalt is found in vitamin B12. Nonessential heavy metals are extremely hazardous to living things, even at very low quantities, and they are not biodegradable (Kim et al. 2019):

1. Heavy metals are classified on the basis of their impact on the human body. Lead (Pb), mercury (Hg), arsenic (As), nickel (Ni), chromium (Cr), and cadmium (Cd) are extremely poisonous and carcinogenic and able to harm a number of organs. These heavy metals have greater stability and least biodegradability.
2. Tin (Sn) and aluminum (Al) are less harmful.
3. Barium (Ba) and lithium (Li) are not necessary for biological and physiological processes.
4. Copper (Cu), zinc (Zn), nickel (Ni), manganese (Mn), molybdenum (Mo), cobalt (Co), selenium (Se), chromium (Cr), and iron (Fe) are vital to biological and physiological processes.

2.2 Sources of Metal Contamination

The term "heavy metal contamination of the environment" refers to the rise in naturally occurring levels of hazardous heavy metals brought on by human activity in urban surface sediment (Fig. 1). There are three different sizes of particles that make up urban surface sediment: dust, fine sand, and coarse sand. Certain heavy metals, such as chromium (Cr), lead (Pb), cadmium (Cd), and arsenic (As), as well as zinc (Zn), copper (Cu), nickel (Ni), tin (Sn), and vanadium (V), are found in substantial amounts in the surface sediment of metropolitan areas (Hanfi et al. 2020).

As the world's population grows, industrialization and development advance quickly, becoming the primary causes of environmental contaminations. The increased concentration of heavy metals and metalloids in water is caused by a variety of factors, including the release of suddenly mine tailings, the disposal of wastes with metals, industrialization, the use of gasoline and paints, fertilizers, animal

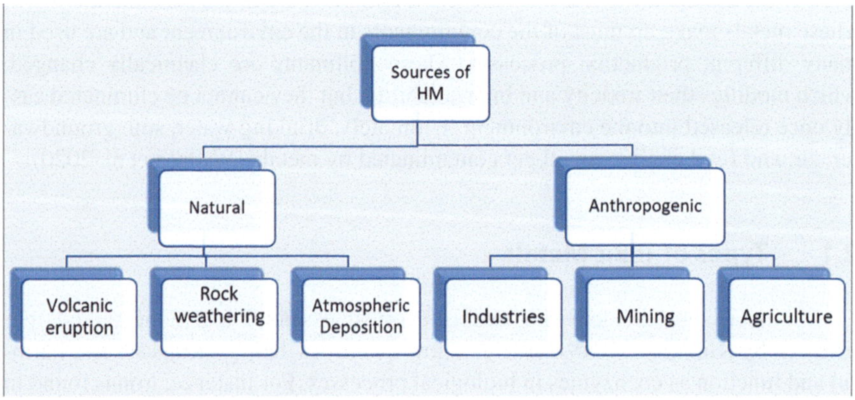

Fig. 1 Main sources of heavy metals

Table 2 Main sources of metal toxicity (Rama Jyothi 2021)

Toxic metals	Sources	References
Chromium (Cr)	Steel fabrication Electroplating Textile Tanneries	Mohammad Ali et al. (2021)
Lead (Pb)	Batteries Coal combustion Paint industry	Mahuta (2020)
Arsenic (As)	Atmospheric deposition Mining pesticides	Huq et al. (2020)
Mercury (Hg)	Coal combustion Fish Mining Paint industry Paper industry Volcanic eruption	Driscoll et al. (2013)
Cadmium (Cd)	Plastic Fertilizers pesticides	Zhang and Reynolds (2019)
Copper (Cu)	Copper polishing Plating Printing	Yabe et al. (2010)
Nickel (Ni)	Electroplating Iron and steel manufacture Fertilizers Mining and metallurgy	Mohammad Ali et al. (2021)

manures, wastewater, sewage, sludge, pesticides, coal, and others (Singh et al. 2022) (Table 2).

2.3 Mechanisms of Metal Toxicity

Poisons, either acute or chronic, can result from exposure to food, drink, or air. These heavy metals bioaccumulate and have a wide range of harmful effects on different body tissues and organs. Apoptosis, damage-repairing mechanisms, growth, proliferation, differentiation, and other cellular functions are all hampered by heavy metals. When the methods of action of these metals are compared, they are found to have comparable mechanisms of action, which include oxidative stress, reactive oxygen species (ROS) production, antioxidant defense weakness, and enzyme inactivation (Fig. 2). However, some of them have the ability to attach selectively to particular macromolecules. Genomic instability is brought on by a few hazardous metals, such as arsenic, cadmium, and chromium. Their carcinogenicity has been attributed to defects in DNA repair that occur after the three metals induce oxidative stress and damage DNA (Balali-Mood et al. 2021).

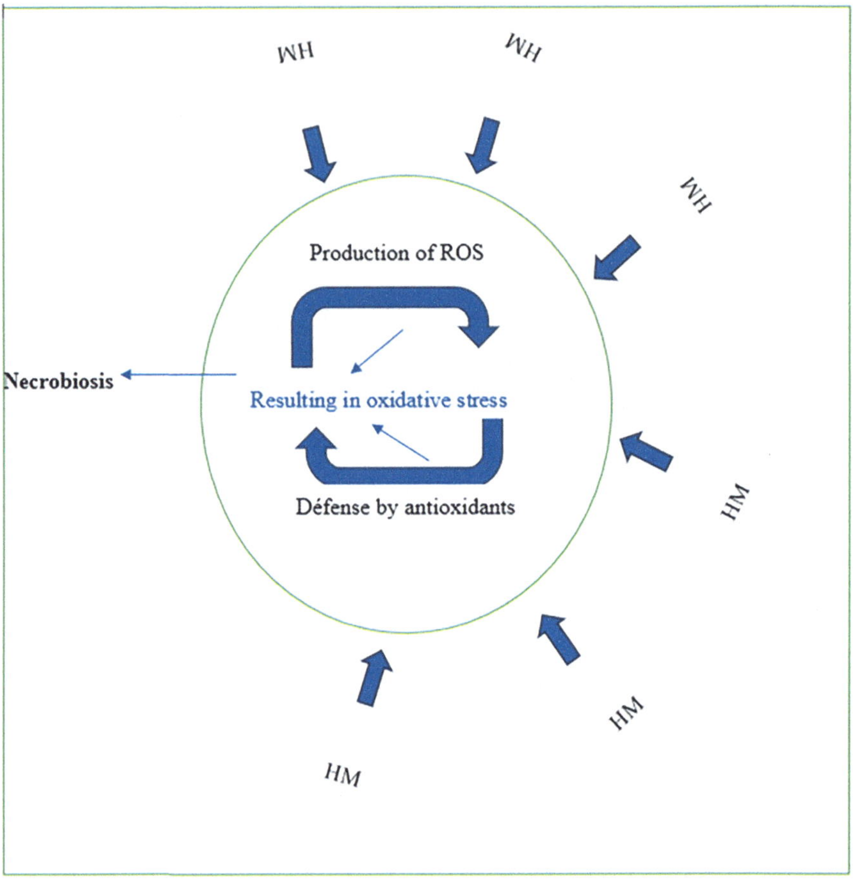

Fig. 2 Schematic representation of heavy metal-induced toxicity mechanism. Heavy metals attack on the cell

A metal that reacts with biotic systems loses its valent electrons, which are responsible for its oxidation property, and creates cations (Jayamurali et al. 2021). ROS such as superoxide, anions, hydrogen peroxides, hydroxyl ions, etc. are the first to cause toxicity. Oxidative stress is reportedly brought on by ROS and poses a risk to cells (Fig. 2). It is also known that the metal poisons cause the body's antioxidant system to become weaker. By attacking cell molecules with nucleophilicity, heavy metals such as mercury combine to produce stable compounds. The complexes that develop cause oxidative stress in the cells, which leads to the final depletion of lipids, nucleic acids, and enzymes. Although arsenic compounds can also produce ROS, their mechanisms of action involve damaging DNA, inhibiting DNA repair mechanisms, and blocking signal transduction pathways. According to reports, the metal may be genotoxic (Jayamurali et al. 2021). ROS and bonding with sulfhydryl groups are two of the ways that heavy metals (HMs) induce toxicity to biological systems. Due to this, important macromolecules become inactive, which causes oxidative stress and glutathione depletion to appear. When harmful metals are exposed to the body and then enter it, a number of things happen, such as the inhibition or involvement of specific metabolic pathways (Rehman et al. 2021). According to Hubaux et al. (2013), arsenic can cause cancer through posttranslational modifications and epigenetic alterations. Histone modifications and changes in DNA methylation are part of the mechanism. The harmful metal may be the cause of the epigenetic changes that underlie cancer.

It has been reported that heavy metals cause cell apoptosis. Arresting the cell cycle is the mechanism by which the cell enters apoptosis. Additionally, when mitochondria are harmed by oxidative phosphorylation after being exposed to ROS, apoptosis occurs (Tchounwou et al. 2012).

2.4 Health and Environmental Impacts

Industrial activities expose people to heavy metals (HMs), which can be absorbed through the skin or by ingestion and inhalation (Ali et al. 2019). Excessive concentrations of necessary or nonessential metals can cause genetic alterations, physiological problems, or morphological abnormalities (Mitra et al. 2022).

Toxic heavy metal poisoning of aquatic and terrestrial ecosystems is a serious environmental issue that affects public health. The majority of heavy metals can be fatal to humans and pollute the ecosystem and atmosphere. By combining with various environmental components including water, soil, and air, heavy metals can become extremely hazardous, and humans and other living things may be exposed to them through the food chain (Mitra et al. 2022).

The buildup of heavy metals from the surrounding water and soil can contaminate food supplies, including fish, shellfish, grains, vegetables, and fruits. In humans, heavy metals can also lead to a number of very serious health issues that impact the kidney, liver, brain system, and respiratory system. The majority of heavy metals have high carcinogenicity. Moreover, HMs can cause bio-regulatory system disruption that results in neurodegenerative diseases like Parkinson's and Alzheimer's diseases as well as functional or psychosomatic problems like chronic fatigue syndrome. They can also cause delays in human growth and development (Sall et al. 2020).

3 Role of Metabolomics in Toxicology

Examining the toxicity of the increasing number of pollutants in the environment is necessary from the perspective of chemical safety and risk assessment. Adverse outcome pathway (AOP) is a useful framework in this regard for assessing the biomolecular impacts of chemicals taking into account the various levels of organization, including biological and molecular data. An efficient method for providing physiological information on AOPs pertinent to a specific pollutant exposure is metabolomics (Bedia 2022).

3.1 Principles of Toxicometabolomics

Over the last 20 years, metabolomics has become a highly promising substitute and/or additional instrument for conventional toxicological tests. It involves the qualitative and quantitative analysis of small molecules (less than 1500 Da) present in biological samples or organisms to determine the important metabolites connected to different biological functions (Bernhard et al. 2023). The highly controlled surroundings of toxicological models offer suitable platforms for metabolomics research with a range of applications, since environmental factors can affect metabolic abnormalities in addition to diet, sex, and illness (da Silva et al. 2021). A branch of metabolomics called environmental toxic metabolomics seeks to understand how chemical pollutants and environmental stresses affect an organism's metabolic pathways, changing physiological responses and perhaps having negative impacts (Bernhard et al. 2023). The most accurate representation of an organism's or biological system's phenotype is provided by metabolomics, which records the byproducts of cellular activities and offers direct information on metabolic reactions and physiological condition. Comprehensive and in-depth molecular understanding of impacted metabolic processes can be obtained by metabolomics. When metabolomics and environmental toxicology are combined, it can be easier to identify biomarkers to measure exposure and/or effects, as well as to comprehend the processes and/or modes of action that underlie the toxicity of environmental pollutants and provide insight into pathways leading to unfavorable outcomes (Bernhard et al. 2023).

3.2 Biomarkers of Metal Exposure

In epidemiological research, biomarkers are frequently utilized to evaluate exposure to metals and determine internal dose since they incorporate many exposure sources, such as food, water, and air. Metals can be exposed through both common and uncommon sources and routes, as well as through biotransformation and elimination processes. The interpretation of metal biomarkers requires an understanding of the metabolic biotransformation and removal of metals, as not all biomarkers provide sufficient levels of exposure. Since not all metal biomarkers give high enough levels of exposure, interpreting them necessitates knowledge of the metabolic biotransformation and elimination of metals (Martinez-Morata et al. 2023).

A biomarker, an acronym for biological markers, is a quantifiable biological marker associated with several states or specific conditions (Table 3). Biological reactions to the level of exposure and the harmful consequences of environmental pollutants are also referred to as biomarkers. As an early indicator of biological xenobiotic effects at the molecular, cellular, tissue, and organ levels, biomarkers serve as the last ecotoxicological test to determine subchronic impacts on organisms (Kadim and Risjani 2022).

The exposure window of interest must be taken into account while choosing biomarkers.Although many biomarkers represent recent exposures, chronic

Table 3 Summary of the most commonly used biomarkers for toxic metals

Metals	Biomarkers	References
Arsenic	Blood Urine Toenail Hair Fingernail	Signes et al. (2021)
Cadmium	Blood Urine Feces Placenta Soft tissues like liver, kidneys	Zhao et al. (2023)
Lead	Blood Bone Teeth Nails Hair Urine	Hu et al. (2007)
Mercury	Urine Hair Blood Feces Human milk Nails Placenta	Risher et al. (1999)
Ni	Urine Feces Serum Hair Plasma	Abadin et al. (2005)

exposure to metals is usually most relevant, particularly for people exposed in the low-to-moderate range. Biomarkers reflecting integrated exposures over longer durations may be preferred if metal exposures change over time (Martinez-Morata et al. 2023). In this sense, toenails are appealing and appear to be trustworthy indicators of exposure to As, Hg, Mn, and Se (Signes-Pastor et al. 2021; Salcedo-Bellido et al. 2021). But for other elements like Cd, Pb, or Ni, toenails don't seem to be sufficient or require more research.

4 Metabolomic Profiling in Metal Toxicity

By providing insight into the metabolic state, metabolomics makes it easier to analyze biochemical pathways, find biomarkers, and comprehend how organisms function in response to various stimuli or circumstances (Qiu et al. 2023). The first stage is designing the experiment, which is based on the goals of the study and the characteristics of the sample (de Sá et al. 2023). Biofluids, cell/tissue-based models, and human/animal specimens are among the options. To preserve consistency and reduce the chance of contamination, it is essential to adhere to the established sample collecting protocols (Kim et al. 2016). The sample integrity is maintained through meticulous handling and specific approaches. In order to ensure reliable analysis, sample preparation removes any interfering chemicals and extracts the metabolites. Concentration or derivatization could improve the detectability (Akash et al. 2023). Using specialized equipment, mainly nuclear magnetic resonance (NMR) and mass spectrometry (MS) techniques, the features of the metabolite are retrieved after sample processing. While mass spectrometry (MS), which includes gas chromatography-mass spectrometry (GC-MS) and liquid chromatography-mass spectrometry (LC-MS), offers broad metabolite detection and high sensitivity, nondestructive analysis and structural insights with great reproducibility are provided by NMR. While GC-MS works well for volatile chemicals, LC-MS offers complementing advantages by covering a large spectrum of metabolomes without derivatization. Collectively, they offer an all-encompassing strategy suited to the particular needs of metabolomics studies (Ilyas et al. 2024).

The two primary categories of metabolomics methodologies are complementary untargeted metabolomics and targeted metabolomics, which are frequently applied concurrently or consecutively (Fig. 3) (Di Minno et al. 2021; González-Peña and Brennan 2019). Untargeted metabolomics is an all-inclusive that measures every metabolite in a sample, even those with unknown targets. Targeted metabolomics, on the other hand, involves measuring a specific group of identified and biochemically annotated analytes (Lopes et al. 2017).

The metabolomics approach begins with a precise and uncomplicated description of the biological issue that has to be solved. The metabolomics approach type (targeted vs. untargeted), sample type (biological fluids, tissues, cells, and/or intact organisms), sample size (number of specimens to be assessed), experimental conditions to which samples will be submitted, frequency of sample collection, metabolic quenching to interrupt enzymatic activity (addition of organic solvents and/or immediate freezing of samples by the use of dry ice or liquid nitrogen), storage conditions (−80 °C is usually preferred for long-term storage of biological fluids), analytical platforms to be used, and sample preparation strategies must all be defined at this point because they are all related (Klassen et al. 2017).

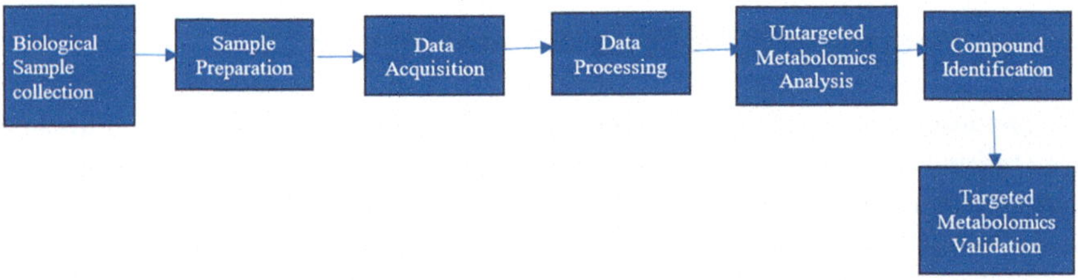

Fig. 3 Schematic representation of a general metabolomics workflow

4.1 Analytical Techniques (e.g., NMR, LC-MS, GC-MS)

Because of the unique properties of metabolomics data, several bioinformatics tools must be used in accordance with a predetermined workflow. The initial stage in the metabolomics workflow is to identify and characterize several groupings of metabolites using various techniques. Nuclear magnetic resonance (NMR) spectroscopy and mass spectrometry (MS) are the two primary platforms used in metabolomics study. Each has benefits and drawbacks of its own (Beisken et al. 2015). In order to simplify the biological sample and enable MS analysis of various groups of molecules at various times, a separation phase is typically conducted prior to MS-based metabolomics. MS obtains spectrum data in the form of the ionized compound's relative intensity and mass-to-charge ratio (m/z) (Chen et al. 2022).

In the field of MS technology, liquid chromatography (LC) and gas chromatography (GC) columns (referred to as LC-MS and GC-MS, respectively) are the most widely used separation techniques. The accurate identification of metabolites is made possible by MS. MS has a better ability to resolve issues, particularly when utilized in conjunction with chromatographic separation techniques. Selected qualitative and quantitative studies are also possible with MS due to its brief analysis cycle, which lasts between 5 and 140 minutes. The primary drawbacks of mass spectrometry (MS) are the high instrument costs and the need to separate or purify the material before loading it into the mass spectrometer. The basis of nuclear magnetic resonance (NMR) is the idea that atomic nuclei absorb and reemit energy in response to changes in an external magnetic field. The concentration and chemical structure of metabolites can be described and quantified using the spectral data produced by NMR. NMR has the advantages of being a highly reproducible, nondestructive method that doesn't require a lot of sample preparation (Chen et al. 2022). NMR in particular has the capacity to deliver a high level of structural information quickly. However, because NMR has a poorer sensitivity, smaller amounts of chemicals that may be significant may be undetected because they are obscured by larger peaks. Metabolite identification and quantification are difficult tasks that are difficult to automate. Therefore, in order to extract relevant and trustworthy information from these profiles, thorough data processing and statistical analysis are needed. For the detection of particular types of substances, such as fatty acids, alcohols, phenols, vitamins, organic acids, polyamines, nucleotides, polyphenols, terpenes, flavonoids, lipids, and other compounds, LC-MS is more appropriate (Chaleckis et al. 2019). The fact that GC-MS can only identify volatile substances or molecules that can be converted into volatiles is an intrinsic restriction of the technique. A variety of materials can be found using GC-MS, including sugars, polyols, amines, fatty acids, organic acids, sugar phosphates, and other compounds (Wei et al. 2021; Lai and Fiehn 2018). It is important to remember that derivatization is necessary for GC-MS measurement of water-soluble compounds because direct GC-MS analysis is only possible if the sample is volatile and stable at room temperature. The chemical and physical characteristics of each sample or the hypothetical target compounds, as well as the type of analysis to be performed (untargeted or targeted), are taken into

Table 4 Few studies on environmental metabolomics under heavy metal stress based on animal (Zhang et al. 2021)

Heavy metals	Instrument used	Organism	Metabolic pathways affected by metal stress	Reference
As	UPLC-MS/MS	Sprague–Dawley (SD) rat	Phenylalanine metabolism; aminoacyl-tRNA, phenylalanine, tyrosine, tryptophan, ubiquinone, and terpenoid–quinone biosynthesis	Huang et al. (2016)
Cu	GC-MS	Mussels (*Perna canaliculus*)	Oxidative stress and apoptosis involved in the transsulfuration pathway; glutathione and taurine metabolism	Nguyen et al. (2018)
Metal (loid): V, Cr, Mn, Fe, Co, Ni, Cu, Zn, As, Se, Cd, Sb, and Pb	NMR	Mosquitofish (*Gambusia holbrooki*)	Aspartate, histidine, myo-inositol, taurine, and choline	Melvin et al. (2018)

consideration when selecting a separation method, as each has a unique resolution and sensitivity in the identification of metabolites (Zhang et al. 2012) (Table 4).

4.2 Data Analysis and Interpretation

Once metabolites (biomarkers) have been identified, it is intriguing to investigate the connection between exposure to metals and disruption of metabolic pathways. If the incorrect extraction process and analytical technique are applied, the biological interpretation of metabolomic data may be deceptive. Furthermore, because metabolism is intricately linked and changes in one molecule can impact many metabolic pathways, interpreting changes in metabolites is a challenging undertaking. Furthermore, it is crucial to comprehend the relationship between metabolites, which calls for cooperation among experts in statistics, biology, and chemistry (García-Sevillano et al. 2015).

The two most popular categories of statistical analytic techniques in the field of metabolomics are univariate and multivariate techniques. Every approach provides a different perspective on the data structure. Using a matrix of variables, multivariate analysis exposes features based on how one variable is related to the others. Univariate analysis produces results that are differentially weighted since it only considers one variable.

The classification and prediction of sample properties via the creation of models that encapsulate the data found in data matrices is the aim of statistical analysis. Principal component analysis (PCA) and partial least squares (PLS) are well-established techniques for multivariate analysis of metabolomics data, without delving too deeply into statistics. By reducing the dimensionality of our data to inferred variables, PCA is a technique that aids in the identification of significant patterns and features. Exercises involving prediction, regression, and classification can make advantage of the dimensionality reduction techniques. The procedures utilized for preprocessing, scaling, and normalization of the data have a major impact on the quality of the statistical models we infer. Thus, rigorous examination of several models for consensus building is necessary for good data analysis.

The metabolomic experiments yielded complex data due to the large number of observed factors, necessitating a thorough statistical analysis of the data. For a trustworthy analysis, metabolomic studies typically employ multivariate statistical methods such as partial least square-discriminant analysis (PLSDA), hierarchical cluster analysis (HCA), and classical statistical methods such as t-tests, ANOVA, and a nonparametric Mann–Whitney–Wilcoxon test (Bartel et al. 2013). The false discovery rate (FDR) and Bonferroni correction techniques are required to limit false positive data due to the multiple-testing issue, particularly in untargeted metabolomics (Bartroff and Song 2020). The following analyses can be performed on raw data for biomarker search and pathway enrichment analysis using free web-based platforms like MetaboAnalyst (Pang et al. 2021), XCMS Online (Domingo-Almenara and Siuzdak 2020), or PAIRUP-MS (Hsu et al. 2019), which are designed for mass spectrometry-based metabolomics data processing.

4.3 Integration with Other Omics Data

The study of numerous, intricate chemical reactions occurring inside living things and how they are impacted by internal or external disturbances is known as metabolomics. It is the closest layer to the phenome and the underlying biochemical layer that reflects information expressed by the transcriptome, proteome, and genome. This makes it the foundation of omics profiling methods. The integration of metabolomics data with genomics, transcriptomics, and proteomics information presents so far unseen opportunities to advance our understanding of biological functions, clarify their underlying mechanisms, and reveal latent relationships among omics variables (Jendoubi 2021) (Fig. 4). Each layer of the so-called omics cascade, which

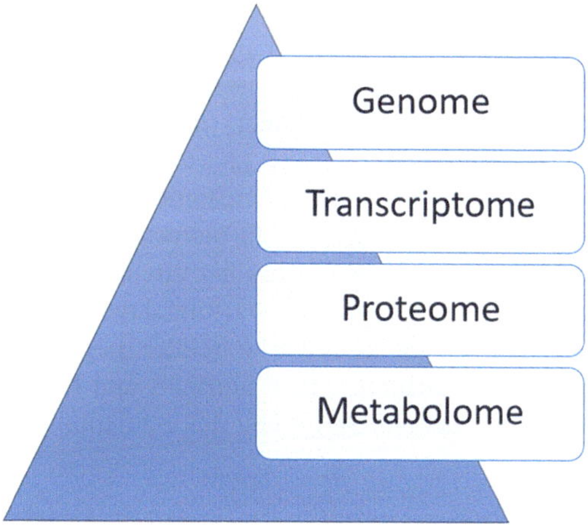

Fig. 4 The most downstream position of the metabolome is reflected in the hierarchical dimension of the omics, which establishes a direct correlation between genotype and sickle cell phenotype

consists of the genome, metabolome, proteome, and transcriptome, characterizes a biosystem or an organism at several biomolecular levels (Haukaas et al. 2017).

5 Metabolomics in Mitigating Metal Toxicity

Metabolomics offers distinct view points on alterations in metabolic patterns and functions brought about by cytotoxicity. The pathophysiology of heavy metal poisoning is mostly dependent on highly reactive free radicals (ROS), which are excessively accumulated as a result of heavy metals impairing antioxidant defense. By precisely detecting, measuring, and characterizing minute metabolites, metabolomics provides unique knowledge about critical metabolic alterations influencing pathological processes in a broad range of biological systems (Akash et al. 2023).

5.1 Identifying Protective Metabolites

Antioxidants: Enzymes that neutralize ROS and shield cells from oxidative stress include glutathione, vitamin C, and vitamin E. Metabolites that can bind to metals and aid in their excretion are known as chelating agents.

Anti-inflammatory metabolites: Substances that lessen inflammation brought on by metal exposure, such as flavonoids (such as quercetin) and polyphenols. Proteins and metabolites that have the ability to sequester or transport metals to secure areas within cells are known as metal transporters or binding proteins.

5.2 Dietary and Pharmacological Interventions

Worldwide, exposure to harmful environmental factors is linked to a multitude of acute and chronic illnesses, such as cancer, neurological and developmental problems, cardiovascular disease, and numerous others. These exposures, whether acute or chronic, are linked to metabolic alterations that may have a favorable or unfavorable impact on an individual's health and well-being. Exposure reduction is the main strategy for guarding against harmful exposures, although it isn't always practical for everyone. If successful, creating interventions to counteract detrimental exposures (such through diet) might be a useful backup plan in these situations. To make an intervention more pleasant and boost adherence, dietary changes or nutritional supplements can be customized to a person's preferences and are generally safe to use. Similar metabolic pathways are modulated by nutrients and other dietary components, which may in some circumstances neutralize or even increase the positive effects of exposures. Nutritional therapies can be tailored to these targets using this knowledge in order to reduce negative impacts or increase good effects. Precision nutrition, where nutrients and food are utilized to target gene–environment interactions and other subpopulation characteristics, may benefit from this strategy. Creating these "nutrient cocktails" offers a chance to change how exposures work in order to improve public health (Rushing et al. 2023).

5.3 Bioremediation Strategies

In bioremediation, which uses biological forms to extract hazardous metal ions through a variety of processes, toxic contaminants are detoxified from bacteria in a natural and affordable way. There are two categories of remediation methods that bacteria use: active/non-metabolic and passive/metabolic. Complexation and ion-exchange reactions are examples of the passive mechanism, whereas precipitation, the efflux mechanism, and accumulation processes are examples of the active mechanism (Lata et al. 2023).

Microbes can acquire metal ions through two distinct processes: adsorption and absorption. While absorption involves the entire mass of materials, adsorption happens at the surface. Because of the mucus layer, hazardous heavy metals are absorbed into the cell during the adsorption process. At the outer layer of the cell, functional groups such as oxygen, nitrogen, phosphorus, and sulfur can form complexes with metal ions. Furthermore, the surfaces of heavy metals have phosphoric acid anions, carboxyl anionic groups, and cationic groups that enable the heavy metal ions to pass through the cell membrane (Lata et al. 2023).

5.4 Role of Antioxidants and Chelating Agents

Antioxidants are compounds that either prevent or delay oxidative stress-related damage (Gulcin 2020). Antioxidants control free radical generation in physiological settings (Parveen et al. 2016). Antioxidants can be separated into two groups: endo- and exogenic—based on where they are found. Superoxide dismutase, glutathione peroxidase, and catalase are in the first category. Among the several exogenous antioxidants are flavonoids, vitamins, and minerals (Bouayed and Bohn 2010).

By binding metal ions to chelating chemicals, chelating therapy creates "chelates," which are ring-shaped structures that help remove metal ions from the body. Antioxidants and chelating agents are two pharmacological treatments for heavy metal toxicity. The pharmaceutical treatment of heavy metal contamination in the human body faces a number of difficulties at the moment. Heavy metal poisoning is frequently treated with chelation therapy; nevertheless, the gastrointestinal, renal, and hematological adverse effects of conventional chelating drugs can be severe. It is a continuous challenge to develop a chelator that efficiently binds to heavy metals to promote their elimination while limiting detrimental side effects. Several chemical antagonists have been found to be safe for use, including glutathione (GSH), trisodium citrate dihydrate (TCD), and epigallocatechin-3-gallate (EGCG). One of the primary constituents of catechins, EGCG has the ability to scavenge intracellular ROS and chelate metal ions to provide antioxidative effects (Zhou et al. 2024).

The most researched antioxidant in green tea, EGCG, has been demonstrated to alter a number of molecular pathways connected to oxidative stress, inflammation, and cell death. The pathophysiology of several diseases, including cancer, neurological disorders, and cardiovascular diseases, has been linked to the deregulation of these molecular pathways, which are essential for controlling immune responses, redox balance, and programmed cell death (Zhou et al. 2024).

6 Environmental Applications of Metabolomics

A number of research has been done on the harmful consequences of environmental contaminants using metabolomics. Currently, the toxicological impacts of environmental contaminants are being evaluated using metabolomics technologies. By examining the changes in environmental pollutant exposure on the endogenous metabolic profile of organisms, the toxicological consequences of pollutants and their internal mechanisms are investigated. The final result of changes in the expression of all connected genes or related enzyme activity in the metabolic pathway is an overall change in endogenous metabolites in organisms. Furthermore, there exists a clear correlation between the physiological state of an organism and the overall alterations in its endogenous metabolites (Liu et al. 2022).

6.1 Monitoring Metal Pollution

As a result of its extreme toxicity and durability, HM has emerged as one of the most hazardous environmental contaminants (Table 5). The most common environmental contaminants containing heavy metals are arsenic (As), mercury (Hg), lead (Pb), cadmium (Cd), and so forth (Rahman and Singh 2019).

6.2 Impact Assessment on Wildlife

Heavy metal exposure occurs in animals mostly due to anthropogenic environmental contamination, which contaminates their food and water. Heavy metals stay in the body for a long time after consumption, and depending on the exposure, the body may experience acute, chronic, clinical, subclinical, or mild toxicities. Because of the harmful effects of metals, oxidative stress is brought on by free radicals, which upset cellular homeostasis. Typical clinical signs may appear in cases of acute heavy metal poisoning, which may result in animal mortality with similar results from necropsies. Animals with chronic toxicities have decreased body condition scores overall and have less potential for output (Afzal and Mahreen 2024).

The purpose of environmental impact assessments (EIAs) is to evaluate the anticipated impacts of upcoming projects on the environment, public health, and the region's economic potential. Ecological, cultural, and financial implications arise from accurately evaluating projects' impact on wildlife populations. In addition to

Table 5 Harmful effects of heavy metals using metabolomics

Heavy metal	Organisms	Instrument used	Sample	Alterations in metabolic profiles	References
Cadmium (Cd)	Male Sprague–Dawley rats	^{1}H NMR	Urine	Alterations to the metabolism of amino acids, glutathione production, and mitochondrial energy.	Lee et al. (2014)
	Woman	LC-Q-ToF-MS and GC-MS	Urine	Modifications to the metabolism of amino acids, galactose, purines, and the route leading to the manufacture of steroid hormones.	Gao et al. (2014)
	Chlamys farreri	UPLC-TOF/MS	Gill	Alterations in signal transduction, autophagy activation, and lipid phosphorylation.	Han et al. (2021)
Lead (Pd)	Earthworms (*Eisenia fetida*)	^{1}H NMR	Earthworms (*Eisenia fetida*)	Alterations in the metabolism of energy, amino acids, and oxidative stress.	Chen et al. (2016)
	Male rats	GC-MS	Urine	Rat blood urine nitrogen, urea, uric acid, and creatinine levels all significantly arose.	Soussi et al. (2018)
Mercury (Hg)	Earthworms (*Eisenia fetida*)	^{1}H–^{13}C NMR	Earthworms (*Eisenia fetida*)	Alterations in energy metabolism, amino acid metabolism, and osmoregulation.	Tang et al. (2018)
Arsenic (As)	C57BL/6 mice	GC/MS	Brain tissue and serum	Alterations in the tricarboxylic acid cycle and amino acid metabolism.	Dai et al. (2016)

offering a framework for characterizing, measuring, and tracking wildlife health in upcoming EIAs, wildlife health measures are employed in EIA procedures. Integrating wildlife health into environmental impact assessments (EIAs) by incorporating community-based sampling, participatory epidemiology, context-specific wildlife health indicators, and local ecological knowledge to enhance ecological understanding and decision-making.The wildlife sections of environmental impact assessments (EIAs) typically concentrate on three primary areas: (i) establishing baselines for pertinent wildlife species before the project is implemented, (ii) projecting how the project's implementation will alter baseline conditions, and (iii) suggesting monitoring and mitigation strategies to offset any projected negative effects of the project (Aleuy et al. 2022).

6.3 Ecosystem Health Indicators

The vital indications of our world are ecosystem health indicators, which give crucial details regarding the state and functioning of ecosystems. Scientists and decision-makers can monitor these indicators to evaluate the general health of an ecosystem, identify possible risks, and implement conservation and restoration plans into action.

The health of the environment is drastically declining as a result of human activity, endangering economic activity, animal and human health, and the life-support system. Thus, keeping an eye on the well-being of ecosystems and wildlife populations is essential to figuring out ecological dynamics and evaluating management initiatives (Ribas et al. 2023). An ecosystem's general health can be assessed using ecosystem health indicators.

7 Future Perspectives and Challenges

The fast-moving field of metabolomics has great promise for improving our knowledge of human health and illness. The identification of novel biomarkers and pathways linked with a broad spectrum of diseases has been made possible by recent advancements in analytical technologies and the growing availability of large-scale datasets. Precision medicine, personalized nutrition, multi-omics integration, the use of AI and machine learning in metabolomics data analytics, the use of metabolomics in translational biology, and its connection to drug development and advanced wearable technology for faster disease detection and monitoring are some of the future paths in metabolomics research (Soni et al. 2023). Targeted metabolomics relies on quantifying the known metabolites of certain pathways, making it a quantitative approach to metabolomics. Targeted metabolomics is a field that is always developing; hence, there are several challenges in its widespread application. As of right now, no one analytical technique in targeted metabolomics can provide information beyond that of a single metabolite. Even with the combination use of tools and techniques, the following constraints persisted: the information gathered focused mostly on a subset of metabolites.

The methods themselves have some particular drawbacks. Metabolism has many benefits and is used in many areas of clinical medicine and the life sciences, but some obstacles and worries prevent its complete implementation. Firstly, precise and highly repeatable metabolite measurement is a prerequisite for applying metabolomics to biomarker and medication discovery. Owing to the complexity of analytical techniques, meeting such a need is frequently challenging. Secondly, appropriate sample size and appropriate sample quality should always be monitored in order to determine the best possible pharmacological response. Thirdly, choosing the right statistical techniques is crucial to prevent data misinterpretation because metabolomics data frequently contains a large number of variables with a small sample size (Meng et al. 2024).

8 Conclusion

To sum up, the utilization of metabolomics in disease diagnostics signifies noteworthy advancements in the medical domain. Metabolomics holds promise for bettering patient quality of life, managing diseases better, and opening up new avenues for medical research.

The resilience and great potential of metabolomics as a state-of-the-art tool to explore heavy metal toxicities at the cellular, tissue, and organism levels are only now becoming apparent, when compared to other comparably more established "omics" such as transcriptomics. Particular viewpoints on cytotoxicity-induced modifications to metabolic patterns and activities are offered by metabolomics. The underlying cause of heavy metal poisoning is mostly dependent on highly reactive free radicals (ROS), which are excessively accumulated as a result of heavy metals impairing antioxidant defense. Metabolomics uses highly precise identification, quantification, and characterization of tiny metabolites to provide unique information about important metabolic changes affecting pathological processes in a wide range of living systems. Metabolomics is a promising approach for finding new detoxifying processes in a range of organisms for metal toxicity investigations. Furthermore, it might be applied to detect modifications in metabolic pathways that other "omics" methods would overlook. Protein and mRNA levels in the cell cannot measure changes in metabolic flow and may not always correlate with modifications in metabolic pathways. Metabolomics holds the potential to bridge this gap by offering data that cannot be obtained through other methods.

References

Abadin H, Fay M, Wilbur SB (2005) Toxicological profile for Nickel

Afzal A, Mahreen N (2024) Emerging insights into the impacts of heavy metals exposure on health, reproductive and productive performance of livestock. Front Pharmacol 15. Frontiers Media SA. https://doi.org/10.3389/fphar.2024.1375137

Akash MSH, Yaqoob A, Rehman K, Imran M, Assiri MA, Al-Rashed F, Al-Mulla F, Ahmad R, Sindhu S (2023) Metabolomics: a promising tool for deciphering metabolic impairment in heavy metal toxicities. Front Mol Biosci 10. Frontiers Media SA. https://doi.org/10.3389/fmolb.2023.1218497

Aleuy OA, Kutz S, Mallory ML, Provencher JF (2022) Wildlife health in environmental impact assessments: are we missing a key metric? Environ Rev 31(2):348–359. https://doi.org/10.1139/er-2022-0023

Ali H, Khan E, Ilahi I (2019) Environmental chemistry and ecotoxicology of hazardous heavy metals: environmental persistence, toxicity, and bioaccumulation. J Chem 2019. Hindawi Limited. https://doi.org/10.1155/2019/6730305

Araújo AM, Carvalho F, De Pinho PG, Carvalho M (2021) Toxicometabolomics: small molecules to answer big toxicological questions. Metabolites 11(10). MDPI. https://doi.org/10.3390/metabo11100692

Balali-Mood M, Naseri K, Tahergorabi Z, Khazdair MR, Sadeghi M (2021) Toxic mechanisms of five heavy metals: mercury, lead, chromium, cadmium, and arsenic. Front Pharmacol 12. Frontiers Media S.A. https://doi.org/10.3389/fphar.2021.643972

Bartel J, Krumsiek J, Theis FJ (2013) Statistical methods for the analysis of high-throughput metabolomics data. Comput Struct Biotechnol J 4(5):e201301009. Elsevier B.V. https://doi.org/10.5936/csbj.201301009

Bartroff J, Song J (2020) Sequential tests of multiple hypotheses controlling false discovery and nondiscovery rates. Seq Anal 39(1):65–91. https://doi.org/10.1080/07474946.2020.1726686

Bedia C (2022) Metabolomics in environmental toxicology: applications and challenges. In: Trends in environmental analytical chemistry, vol 34. Elsevier B.V. https://doi.org/10.1016/j.teac.2022.e00161

Bedia C, Cardoso P, Dalmau N, Garreta-Lara E, Gómez-Canela C, Gorrochategui E, Navarro-Reig M, Ortiz-Villanueva E, Puig-Castellví F, Tauler R (2018) Applications of metabolomics analysis in environmental research. In: Comprehensive analytical chemistry, vol 82, pp 533–582. https://doi.org/10.1016/bs.coac.2018.07.006

Begou O, Deda O, Agapiou A, Taitzoglou I, Gika H, Theodoridis G (2019) Urine and fecal samples targeted metabolomics of carobs treated rats. J Chromatogr B Analyt Technol Biomed Life Sci 1114–1115:76–85. https://doi.org/10.1016/j.jchromb.2019.03.028

Beisken S, Eiden M, Salek RM (2015) Getting the right answers: understanding metabolomics challenges. In: Expert review of molecular diagnostics, vol 15, issue 1. Expert Reviews Ltd. pp 97–109. https://doi.org/10.1586/14737159.2015.974562

Bernhard A, Poulsen R, Brun Hansen AM, Hansen M (2023) Toxicometabolomics as a tool for next generation environmental risk assessment. EFSA J 21(S1). https://doi.org/10.2903/j.efsa.2023.e211005

Bouayed J, Bohn T (2010) Exogenous antioxidants - Double-edged swords in cellular redox state: health beneficial effects at physiologic doses versus deleterious effects at high doses. In: Oxidative medicine and cellular longevity vol 3, issue 4. pp 228–237. https://doi.org/10.4161/oxim.3.4.12858

Brennan L, De Roos B (2021) Nutrigenomics: lessons learned and future perspectives. Am J Clin Nutr 113(3):503–516. https://doi.org/10.1093/ajcn/nqaa366

Britz-McKibbin P (2011) Capillary Electrophoresis–Electrospray Ionization-Mass Spectrometry (CE–ESI-MS)-based metabolomics. In: Methods in molecular biology, vol 708. Humana Press Inc, pp 229–246. https://doi.org/10.1007/978-1-61737-985-7_14

Chaleckis R, Meister I, Zhang P, Wheelock CE (2019) Challenges, progress and promises of metabolite annotation for LC–MS-based metabolomics. Curr Opin Biotechnol 55:44–50. https://doi.org/10.1016/j.copbio.2018.07.010

Chen T, Liu Y, Li M-H, Xu H-D, Sheng J-Y, Zhang L, Wang J-S (2016) Integrated 1H NMR-based metabolomics analysis of earthworm responses to sub-lethal Pb exposure. Environ Chem 13(5):792. https://doi.org/10.1071/EN15192

Chen Y, Li EM, Xu LY (2022) Guide to metabolomics analysis: a bioinformatics workflow. In: Metabolites vol 12, issue 4. MDPI. https://doi.org/10.3390/metabo12040357

Courant F, Pinel G, Bichon E, Monteau F, Antignac JP, Le Bizec B (2009) Development of a metabolomic approach based on liquid chromatography-high resolution mass spectrometry to screen for clenbuterol abuse in calves. Analyst 134(8):1637–1646. https://doi.org/10.1039/b901813a

Dable-Tupas G, Capirig CJ, Roy M, Pathak A (2022) Nutrigenomics research: methods and applications. In: Role of nutrigenomics in modern-day healthcare and drug discovery. Elsevier, pp 35–82. https://doi.org/10.1016/B978-0-12-824412-8.00004-7

Dai H, Xia Y-Y, Ting-Li Han T-LH, Philip N Baker PNB, Tang X, Zhang R-Y, Du H, Cai T-J, Cheng S-Q (2016) Effect of chronic arsenic exposure on mouse brain tissue and serum metabolomics Nan Fang Yi Ke Da Xue Xue Bao = J Southern Med Univ 36(9):1192–1197

Di Minno A, Gelzo M, Stornaiuolo M, Ruoppolo M, Castaldo G (2021) The evolving landscape of untargeted metabolomics. In: Nutrition, metabolism and cardiovascular diseases vol 31, issue 6. Elsevier B.V. pp 1645–1652. https://doi.org/10.1016/j.numecd.2021.01.008

Domingo-Almenara X, Siuzdak G (2020) Metabolomics data processing using XCMS. In: Computational methods and data analysis for metabolomics, pp 11–24. https://doi.org/10.1007/978-1-0716-0239-3_2

Driscoll CT, Mason RP, Chan HM, Jacob DJ, Pirrone N (2013) Mercury as a global pollutant: sources, pathways, and effects. In: Environmental science and technology, vol 47, issue 10. pp 4967–4983. https://doi.org/10.1021/es305071v

Feng C-C, Zhang A-H, Miao J-H, Sun H, Han Y, Yan G-L, Wu F-F, Wang X-J, Com P (2018) Recent advances in understanding cross-talk between bile acids and gut microbiota. Open J Proteomics Genomics. https://doi.org/10.17352/ojpg

Feng Z, Ji S, Ping J, Cui D (2021) Recent advances in metabolomics for studying heavy metal stress in plants. In: TrAC - Trends in Analytical Chemistry, vol 143. Elsevier B.V. https://doi.org/10.1016/j.trac.2021.116402

Gao Y, Lu Y, Huang S, Gao L, Liang X, Wu Y, Wang J, Huang Q, Tang L, Wang G, Yang F, Hu S, Chen Z, Wang P, Jiang Q, Huang R, Xu Y, Yang X, Ong CN (2014) Identifying early urinary metabolic changes with long-term environmental exposure to cadmium by mass-spectrometry-based metabolomics. Environ Sci Technol 48(11):6409–6418. https://doi.org/10.1021/es500750w

García-Sevillano MÁ, García-Barrera T, Gómez-Ariza JL (2015) Environmental metabolomics: biological markers for metal toxicity. In: Electrophoresis, vol 36, issue 18. Wiley-VCH Verlag. pp 2348–2365. https://doi.org/10.1002/elps.201500052

Gika H, Virgiliou C, Theodoridis G, Plumb RS, Wilson ID (2019) Untargeted LC/MS-based metabolic phenotyping (metabonomics/metabolomics): the state of the art. J Chromatogr B: Analyt Technol Biomed Life Sci 1117:136–147. Elsevier B.V. https://doi.org/10.1016/j.jchromb.2019.04.009

Gonzalez-Covarrubias V, Martínez-Martínez E, Del Bosque-Plata L (2022) The potential of metabolomics in biomedical applications. In: Metabolites, vol 12, issue 2. MDPI. https://doi.org/10.3390/metabo12020194

González-Peña D, Brennan L (2019) Recent advances in the application of metabolomics for nutrition and health. Annu Rev Food Sci Technol 10:479 519. https://doi.org/10.1146/annurev-food-032818

Gulcin İ (2020) Antioxidants and antioxidant methods: an updated overview. In: Archives of toxicology, vol 94, issue 3. Springer. pp 651–715. https://doi.org/10.1007/s00204-020-02689-3

Hanfi MY, Mostafa MYA, Zhukovsky MV (2020) Heavy metal contamination in urban surface sediments: sources, distribution, contamination control, and remediation. In: Environmental monitoring and assessment, vol 192, issue 1. Springer. https://doi.org/10.1007/s10661-019-7947-5

Haukaas TH, Euceda LR, Giskeødegård GF, Bathen TF (2017) Metabolic portraits of breast cancer by HR MAS MR spectroscopy of intact tissue samples. In: Metabolites, vol 7, issue 2. MDPI AG. https://doi.org/10.3390/metabo7020018

Hsu YHH, Churchhouse C, Pers TH, Mercader JM, Metspalu A, Fischer K, Fortney K, Morgen EK, Gonzalez C, Gonzalez ME, Esko T, Hirschhorn JN (2019) PAIRUP-MS: pathway analysis and imputation to relate unknowns in profiles from mass spectrometry-based metabolite data. PLoS Comput Biol 15(1). https://doi.org/10.1371/journal.pcbi.1006734

Hu H, Shih R, Rothenberg S, Schwartz BS (2007) The epidemiology of lead toxicity in adults: measuring dose and consideration of other methodologic issues. In: Environmental health perspectives, vol 115, issue 3. pp 455–462. https://doi.org/10.1289/ehp.9783

Huang Q, Luo L, Alamdar A, Zhang J, Liu L, Tian M, Eqani SAMAS, Shen H (2016) Integrated proteomics and metabolomics analysis of rat testis: mechanism of arsenic-induced male reproductive toxicity. Sci Rep 6. https://doi.org/10.1038/srep32518

Hubaux R, Becker-Santos DD, Enfield KS, Rowbotham D, Lam S, Lam WL, Martinez VD (2013) Molecular features in arsenic-induced lung tumors. Mol Cancer 12(1):20. https://doi.org/10.1186/1476-4598-12-20

Huq ME, Fahad S, Shao Z, Sarven MS, Khan IA, Alam M, Saeed M, Ullah H, Adnan M, Saud S, Cheng Q, Ali S, Wahid F, Zamin M, Raza MA, Saeed B, Riaz M, Khan WU (2020) Arsenic in a groundwater environment in Bangladesh: occurrence and mobilization. J Environ Manag 262. https://doi.org/10.1016/j.jenvman.2020.110318

Ilyas K, Iqbal H, Akash MSH, Rehman K, Hussain A (2024) Heavy metal exposure and metabolomics analysis: an emerging frontier in environmental health. In: Environmental science and pollution research, vol 31, issue 26. Springer. pp 37963–37987. https://doi.org/10.1007/s11356-024-33735-7

Jayamurali D, Varier KM, Liu W, Raman J, Ben-David Y, Shen X, Gajendran B (2021) An overview of heavy metal toxicity. In: Metal, metal oxides and metal sulphides for biomedical applications, pp 323–342. https://doi.org/10.1007/978-3-030-56413-1_12

Jendoubi T (2021) Approaches to integrating metabolomics and multi-omics data: a primer. In: Metabolites, vol 11, issue 3. MDPI AG. https://doi.org/10.3390/metabo11030184

Jones OA, Cheung VL (2007) Comparative medicine an introduction to metabolomics and its potential application in veterinary science. Comp Med 57(5):436–442. https://www.mpimp-golm.mpg.de/

Kadim MK, Risjani Y (2022) Biomarker for monitoring heavy metal pollution in aquatic environment: an overview toward molecular perspectives. In: Emerging contaminants, vol 8. KeAi Communications Co, pp 195–205. https://doi.org/10.1016/j.emcon.2022.02.003

Kim SJ, Kim SH, Kim JH, Hwang S, Yoo HJ (2016) Understanding metabolomics in biomedical research. In: Endocrinology and metabolism, vol 31, issue 1. Korean Endocrine Society. pp 7–16. https://doi.org/10.3803/EnM.2016.31.1.7

Kim JJ, Kim YS, Kumar V (2019) Heavy metal toxicity: an update of chelating therapeutic strategies. J Trace Elem Med Biol 54:226–231. Elsevier GmbH. https://doi.org/10.1016/j.jtemb.2019.05.003

Klassen A, Faccio AT, Canuto GAB, da Cruz PLR, Ribeiro HC, Tavares MFM, Sussulini A (2017) Metabolomics: definitions and significance in systems biology. In: Advances in experimental medicine and biology, vol 965. Springer, New York LLC, pp 3–17. https://doi.org/10.1007/978-3-319-47656-8_1

Koelmel JP, Napolitano MP, Ulmer CZ, Vasiliou V, Garrett TJ, Yost RA, Prasad MNV, Godri Pollitt KJ, Bowden JA (2020) Environmental lipidomics: understanding the response of organisms and ecosystems to a changing world. In: Metabolomics, vol 16, issue 5. Springer. https://doi.org/10.1007/s11306-020-01665-3

Lai Z, Fiehn O (2018) Mass spectral fragmentation of trimethylsilylated small molecules. In: Mass spectrometry reviews, vol 37, issue 3. Wiley. pp 245–257. https://doi.org/10.1002/mas.21518

Lata S, Sharma S, Kaur S (2023) OMICS approaches in mitigating metal toxicity in comparison to conventional techniques used in cadmium bioremediation. In: Water, air, and soil pollution, vol 234, issue 3. Institute for Ionics. https://doi.org/10.1007/s11270-023-06145-7

Lee YK, Park EY, Kim S, Son JY, Kim TH, Kang WG, Jeong TC, Kim KB, Kwack SJ, Lee J, Kim S, Lee BM, Kim HS (2014) Evaluation of cadmium-induced nephrotoxicity using urinary metabolomic profiles in Sprague–Dawley male rats. J Toxicol Environ Health - Part A: Current Issues 77:1384–1398. https://doi.org/10.1080/15287394.2014.951755

Liu, H., Li, H., Han, D. Zhang, X., Gong, X., , Zhang, H., Tian, X., & Xu, Y. (2021). Metabolomics comparison of metabolites and functional pathways in the gills of *Chlamys farreri* under cadmium exposure. Environ Toxicol Pharmacol, 86. https://doi.org/10.1016/j.etap.2021.103683

Liu L, Wu Q, Miao X, Fan T, Meng Z, Chen X, Zhu W (2022) Study on toxicity effects of environmental pollutants based on metabolomics: a review. In: Chemosphere, vol 286. Elsevier Ltd. https://doi.org/10.1016/j.chemosphere.2021.131815

Lopes AS, Cruz ECS, Sussulini A, Klassen A (2017) Metabolomic strategies involving mass spectrometry combined with liquid and gas chromatography. In: Metabolomics: from fundamentals to clinical applications, pp 77–98. https://doi.org/10.1007/978-3-319-47656-8_4

López-López Á, López-Gonzálvez Á, Barker-Tejeda TC, Barbas C (2018) A review of validated biomarkers obtained through metabolomics. In: Expert review of molecular diagnostics, vol 18, issue 6. Taylor and Francis Ltd. pp 557–575. https://doi.org/10.1080/14737159.2018.1481391

Mahuta MA (2020) An overlook of sources and strategies of minimising lead poisoning in Nigeria. JHSS 20(6):2020. https://www.hummingpubng.com

Martinez-Morata I, Sobel M, Tellez-Plaza M, Navas-Acien A, Howe CG, Sanchez TR (2023) A state-of-the-science review on metal biomarkers. In: Current environmental health reports, vol 10, issue 3. Springer Science and Business Media Deutschland GmbH. pp 215–249. https://doi.org/10.1007/s40572-023-00402-x

Melvin SD, Lanctôt CM, Doriean NJC, Carroll AR, Bennett WW (2018) Untargeted NMR-based metabolomics for field-scale monitoring: Temporal reproducibility and biomarker discovery in mosquitofish (*Gambusia holbrooki*) from a metal(loid)-contaminated wetland. Environ Pollut 243:1096–1105. https://doi.org/10.1016/j.envpol.2018.09.071

Meng X, Liu Y, Xu S, Yang L, Yin R (2024) Review on analytical technologies and applications in metabolomics. In: Biocell, vol 48, issue 1. Tech Science Press. pp 65–78. https://doi.org/10.32604/biocell.2023.045986

Mitra S, Chakraborty AJ, Tareq AM, Emran TB, Nainu F, Khusro A, Idris AM, Khandaker MU, Osman H, Alhumaydhi FA, Simal-Gandara J (2022) Impact of heavy metals on the environment and human health: Novel therapeutic insights to counter the toxicity. J King Saud Univ - Sci 34(3). https://doi.org/10.1016/j.jksus.2022.101865

Mohammad Ali M, Hossain D, Al-Imran SK, Begum M, Hasan Osman M (2021) Environmental pollution with heavy metals: a public health concern. In: Heavy metals - their environmental impacts and mitigation. IntechOpen. https://doi.org/10.5772/intechopen.96805

Nguyen TV, Alfaro AC, Merien F, Lulijwa R, Young T (2018) Copper-induced immunomodulation in mussel (*Perna canaliculus*) haemocytes. Metallomics 10(7):965–978. https://doi.org/10.1039/c8mt00092a

Pang Z, Chong J, Zhou G, De Lima Morais DA, Chang L, Barrette M, Gauthier C, Jacques PÉ, Li S, Xia J (2021) MetaboAnalyst 5.0: narrowing the gap between raw spectra and functional insights. Nucleic Acids Res 49(W1):W388–W396. https://doi.org/10.1093/nar/gkab382

Parveen A, Sajid M, Akash H, Rehman K, Kyunn WW (2016) Recent investigations for discovery of natural antioxidants: a comprehensive review. In: Critical reviews™ in eukaryotic gene expression, vol 26, issue 2. https://www.begellhouse.com

Qiu S, Cai Y, Yao H, Lin C, Xie Y, Tang S, Zhang A (2023) Small molecule metabolites: discovery of biomarkers and therapeutic targets. In: Signal transduction and targeted therapy, vol 8, issue 1. Springer Nature. https://doi.org/10.1038/s41392-023-01399-3

Rahman Z, Singh VP (2019) The relative impact of toxic heavy metals (THMs) (arsenic (As), cadmium (Cd), chromium (Cr)(VI), mercury (Hg), and lead (Pb)) on the total environment: an overview. In: Environmental monitoring and assessment, vol 191, issue 7. Springer International Publishing. https://doi.org/10.1007/s10661-019-7528-7

Rama Jyothi N (2021) Heavy metal sources and their effects on human health. In: Heavy metals - their environmental impacts and mitigation. IntechOpen. https://doi.org/10.5772/intechopen.95370

Rehman AU, Nazir S, Irshad R, Tahir K, Rehman K, Islam RU, Wahab Z (2021) Toxicity of heavy metals in plants and animals and their uptake by magnetic iron oxide nanoparticles. J Mol Liq 321. Elsevier BV. https://doi.org/10.1016/j.molliq.2020.114455

Ribas MP, García-Ulloa M, Espunyes J, Cabezón O (2023) Improving the assessment of ecosystem and wildlife health: microbiome as an early indicator. In: Current opinion in biotechnology, vol 81. Elsevier Ltd. https://doi.org/10.1016/j.copbio.2023.102923

Risher JF, De Rosa CT, Jones DE, Murray HE (1999) Summary report for the expert panel review of the toxicological profile for mercury. In: Toxicology and industrial health, vol 15. https://doi.org/10.1177/074823379901500504

Roberts R (2023) An introduction to metabolomics what is metabolomics? how is metabolomics data generated and analyzed?

Rushing BR, Thessen AE, Soliman GA, Ramesh A, Sumner SCJ (2023) The exposome and nutritional pharmacology and toxicology: a new application for metabolomics. Exp Dermatol 3(1). https://doi.org/10.1093/exposome/osad008

de Sá HC, Dos Santos EKP, Dos Santos SC, Nunes RS, Canuto GAB (2023) Fundamentals and analytical strategies for metabolomics workflow: an overview and microbial applications. Brazilian J Anal Chem 10(40):35–53. Visao Fokka Communication Agency. https://doi.org/10.30744/brjac.2179-3425.RV-136-2022

Salcedo-Bellido I, Gutiérrez-González E, García-Esquinas E, Fernández de Larrea-Baz N, Navas-Acien A, Téllez-Plaza M, Pastor-Barriuso R, Lope V, Gómez-Ariza JL, García-Barrera T, Pollán M, Jiménez Moleón JJ, Pérez-Gómez B (2021) Toxic metals in toenails as biomarkers of exposure: a review. In: Environmental research, vol 197. Academic Press Inc. https://doi.org/10.1016/j.envres.2021.111028

Sall ML, Karim A, Diaw D, Gningue-Sall D, Aaron SE, Aaron J-J (2020) Toxic heavy metals: impact on the environment and human health, and treatment with conducting organic polymers, a review. Environ Sci Pollut Res 27:29927–29942. https://doi.org/10.1007/s11356-020-09354-3/Published

Sarma SN, Saleem A, Lee J-Y, Tokumoto M, Hwang G-W, Man Chan H, Satoh M (2018) Effects of long-term cadmium exposure on urinary metabolite profiles in mice. J Toxicol Sci 43(2):89–100. https://doi.org/10.2131/jts.43.89

Signes-Pastor AJ, Gutiérrez-González E, García-Villarino M, Rodríguez-Cabrera FD, López-Moreno JJ, Varea-Jiménez E, Pastor-Barriuso R, Pollán M, Navas-Acien A, Pérez-Gómez B, Karagas MR (2021) Toenails as a biomarker of exposure to arsenic: a review. In: Environmental research, vol 195. Academic Press Inc. https://doi.org/10.1016/j.envres.2020.110286

Signes-Pastor AJ, Martinez-Camblor P, Baker E, Madan J, Guill MF, Karagas MR (2021) Prenatal exposure to arsenic and lung function in children from the New Hampshire Birth Cohort Study. Environment International, 155

da Silva KM, Iturrospe E, Bars C, Knapen D, Van Cruchten S, Covaci A, van Nuijs ALN (2021) Mass spectrometry-based zebrafish toxicometabolomics: a review of analytical and data quality challenges. In: Metabolites, vol 11, issue 9. MDPI. https://doi.org/10.3390/metabo11090635

Simó C, Ibáñez C, Gómez-Martínez Á, Ferragut JA, Cifuentes A (2011) Is metabolomics reachable? Different purification strategies of human colon cancer cells provide different CE-MS metabolite profiles. Electrophoresis 32(13):1765–1777. https://doi.org/10.1002/elps.201100019

Singh A, Sharma A, Verma K, Chopade RL, Pandit P, Nagar V, Aseri V, Choudhary SK, Awasthi G, Awasthi KK, Sankhla SM (2022) Heavy metal contamination of water and their toxic effect on living organisms. In: The toxicity of environmental pollutants. IntechOpen. https://doi.org/10.5772/intechopen.105075

Smith Y (2019) Applications of metabolomics. https://www.news-medical.net/life-sciences/Applications-of-Metabolomics.aspx

Soininen P, Kangas AJ, Würtz P, Tukiainen T, Tynkkynen T, Laatikainen R, Järvelin MR, Kähönen M, Lehtimäki T, Viikari J, Raitakari OT, Savolainen MJ, Ala-Korpela M (2009) High-throughput serum NMR metabonomics for cost-effective holistic studies on systemic metabolism. Analyst 134(9):1781–1785. https://doi.org/10.1039/b910205a

Soni V, Bartelo N, Schweickart A, Chawla Y, Dutta A, Jain S (2023) Future perspectives of metabolomics: gaps, planning, and recommendations. In: Metabolomics. Springer International Publishing, pp 479–512. https://doi.org/10.1007/978-3-031-39094-4_14

Soussi A, Gargouri M, Akrouti A, El Feki A (2018) Antioxidant and nephro-protective effect of Juglans Regia vegetable oil against lead-induced nephrotoxicity in rats and its characterization by GC-MS. EXCLI J 17:492–504. https://doi.org/10.17179/excli2018-1235

Szeremeta M, Pietrowska K, Niemcunowicz-Janica A, Kretowski A, Ciborowski M (2021) Applications of metabolomics in forensic toxicology and forensic medicine. Int J Mol Sci 22(6):1–16. MDPI AG. https://doi.org/10.3390/ijms22063010

Tang R, Ding C, Dang F, Ma Y, Wang J, Zhang T, Wang X (2018) NMR-based metabolic toxicity of low-level Hg exposure to earthworms. Environ Pollut 239:428–437. https://doi.org/10.1016/j.envpol.2018.04.027

Tchounwou PB, Yedjou CG, Patlolla AK, Sutton DJ (2012) Heavy metal toxicity and the environment. In: EXS, vol 101, pp 133–164. https://doi.org/10.1007/978-3-7643-8340-4_6

Tsugawa H, Bamba T, Shinohara M, Nishiumi S, Yoshida M, Fukusaki E (2011) Practical non-targeted gas chromatography/mass spectrometry-based metabolomics platform for metabolic phenotype analysis. J Biosci Bioeng 112(3):292–298. https://doi.org/10.1016/j.jbiosc.2011.05.001

Wallace DR, Taalab YM, Heinze S, Tariba Lovaković B, Pizent A, Renieri E, Tsatsakis A, Farooqi AA, Javorac D, Andjelkovic M, Bulat Z, Antonijević B, Buha Djordjevic A (2020) Toxic-metal-induced alteration in miRNA expression profile as a proposed mechanism for disease development. Cells 9(4). NLM (Medline). https://doi.org/10.3390/cells9040901

Wang JH, Byun J, Pennathur S (2010) Analytical approaches to metabolomics and applications to systems biology. Semin Nephrol 30(5):500–511. https://doi.org/10.1016/j.semnephrol.2010.07.007

Wei F, Lamichhane S, Orešič M, Hyötyläinen T (2019) Lipidomes in health and disease: analytical strategies and considerations. In: TrAC - Trends in analytical chemistry, vol 120. Elsevier B.V. https://doi.org/10.1016/j.trac.2019.115664

Wei J, Xiang L, Cai Z (2021) Emerging environmental pollutants hydroxylated polybrominated diphenyl ethers: from analytical methods to toxicology research. Mass Spectrom Rev 40(3):255–279. Wiley. https://doi.org/10.1002/mas.21640

Wishart DS (2016) Emerging applications of metabolomics in drug discovery and precision medicine. In: Nature reviews drug discovery, vol 15, issue 7. Nature Publishing Group. pp 473–484. https://doi.org/10.1038/nrd.2016.32

Yabe J, Ishizuka M, Umemura T (2010) Current levels of heavy metal pollution in Africa. J Vet Med Sci 72(10)

Zeng T, Liang Y, Chen J, Cao G, Yang Z, Zhao X, Tian J, Xin X, Lei B, Cai Z (2021) Urinary metabolic characterization with nephrotoxicity for residents under cadmium exposure. Environ Int 154. https://doi.org/10.1016/j.envint.2021.106646

Zhang H, Reynolds M (2019) Cadmium exposure in living organisms: a short review. In: Science of the total environment, vol 678. Elsevier B.V, pp 761–767. https://doi.org/10.1016/j.scitotenv.2019.04.395

Zhang A, Sun H, Wang P, Han Y, Wang X (2012) Modern analytical techniques in metabolomics analysis. Analyst 137(2):293–300. https://doi.org/10.1039/C1AN15605E

Zhang AH, Sun H, Qiu S, Wang XJ (2013a) Metabolomics in noninvasive breast cancer. In: Clinica Chimica Acta, vol 424, pp 3–7. https://doi.org/10.1016/j.cca.2013.05.003

Zhang AH, Sun H, Qiu S, Wang XJ (2013b) NMR-based metabolomics coupled with pattern recognition methods in biomarker discovery and disease diagnosis. Magn Reson Chem 51(9):549–556. https://doi.org/10.1002/mrc.3985

Zhang LJ, Qian L, Ding LY, Wang L, Wong MH, Tao HC (2021) Ecological and toxicological assessments of anthropogenic contaminants based on environmental metabolomics. In: Environmental science and ecotechnology, vol 5. Elsevier B.V. https://doi.org/10.1016/j.ese.2021.100081

Zhao D, Wang P, Zhao FJ (2023) Dietary cadmium exposure, risks to human health and mitigation strategies. In: Critical reviews in environmental science and technology, vol 53, issue 8. Taylor and Francis Ltd. pp 939–963. https://doi.org/10.1080/10643389.2022.2099192

Zhou J, Hong H, Zhao J, Fang R, Chen S, Tang C (2024) Metabolome analysis to investigate the effect of heavy metal exposure and chemoprevention agents on toxic injury caused by a multi-heavy metal mixture in rats. Sci Total Environ 906:167513. https://doi.org/10.1016/j.scitotenv.2023.167513

Artificial Intelligence and Machine Learning in Microbial Degradation of Pollutants and Toxins

Payal Trivedi, Ashwani Kumar, Namrata Gupta, and Chirag N. Patel

Abstract

Pollution is a critical global issue, adversely affecting the environment, human health, plants, and animals. Microbial degradation of pollutants and toxins offers a promising strategy for environmental remediation. Optimizing these processes requires a thorough understanding of complex microbial interactions and environmental factors. This chapter delves into the sources and impacts of pollutants and toxins, alongside their microbial degradation using advanced techniques such as artificial intelligence (AI) and machine learning (ML). It is valuable for environmental scientists, biogeochemists, and industrial practitioners aiming to manage polluted sites and mitigate toxin presence. Further, it provides a succinct overview of how AI and ML are revolutionizing microbial degradation studies to enhance pollutant removal. AI and ML provide robust tools for predictive modelling and process optimization in microbial degradation. ML algorithms identify optimal conditions for microbial activity, thereby improving bioremediation efficiency. Furthermore, AI-driven decision support systems enable dynamic adjustments to environmental conditions, maximizing pollutant removal. In conclusion, integrating AI and ML with microbial degradation holds great potential for advancing environmental remediation. Through computational modelling and

P. Trivedi · A. Kumar · N. Gupta (✉)
Department of Biotechnology, University Institute of Biotechnology, Chandigarh University, Mohali, Punjab, India

C. N. Patel
Biotechnology Research Center, Technology Innovation Institute, Abu Dhabi, United Arab Emirates

Drug Design & Development Section, Translational Gerontology Branch, Intramural Research Program National Institute on Aging, NIH, Baltimore, MD, USA

S. Kaur et al. (eds.), *Microbial Metabolomics*,
https://doi.org/10.1007/978-981-96-4824-5_16

data-driven insights, researchers can optimize microbial degradation strategies to address complex pollutants, contributing to sustainable environmental management and ecosystem restoration.

Keywords

Artificial intelligence · Machine learning · Microbial metabolomics · Pollutants · Toxins · Bioremediation

1 Introduction

Environmental pollution by industrial activities such as hydrocarbons, toxic wastes, agricultural practices, and urbanization poses a severe threat to ecosystems and human health. Traditional methods of pollutant degradation, while effective to an extent, often face limitations in scalability, efficiency, and adaptability. Microbial degradation, utilizing bacteria, fungi, and other microorganisms to break down pollutants, offers a promising solution due to its sustainability and eco-friendliness. However, optimizing these microbial processes remains a challenge. The integration of artificial intelligence (AI) and machine learning (ML) into environmental biotechnology, particularly in the microbial degradation of pollutants and toxins, represents a groundbreaking advancement. This chapter explores how AI and ML techniques can be applied to enhance the efficiency and effectiveness of microbial degradation of pollutants and toxins. We review current methodologies, highlight significant case studies, and discuss future perspectives in leveraging AI and ML to address environmental pollution through microbial means.

Bacteria, algae, protozoa, fungi, and archaea are found in nearly every aquatic environment, playing a crucial role in contaminant reduction through degradation processes. Microbial degradation, a key factor in environmental transformation, often determines the fate of pollutants (Fenner et al. 2021). The biodegradation of pharmaceutically active compounds (PhACs) during biological processes is linked to both primary metabolic and co-metabolic activities of microorganisms in mixed liquor (Gupta et al. 2024). Environmental factors such as pH, temperature, dissolved oxygen, and nutrients significantly influence microbial metabolism, making optimal conditions essential for microbial growth. The extensive use and disposal of pesticides, antibiotics, nonsteroidal anti-inflammatory drug, and other chemicals have raised concerns due to their ecotoxicological effects, leading to increased concentrations in water, sediments, and soil. Growing awareness of the risks posed by organic pollutants has driven the development of strategies to detoxify and remediate these contaminants (Mbachu et al. 2020).

Microbial degradation, or biodegradation, involves using microorganisms to break down, detoxify, or transform environmental pollutants. A wide range of microorganisms, including bacteria, fungi, and protozoa, play a crucial role in the biological decomposition of diverse substrates (Rozalska and Iwanicka-Nowicka 2016). Degradation processes encompass both abiotic and biotic mechanisms, where organic pollutants are either transformed into less harmful byproducts or fully mineralized to simpler carbon forms. Interest in microbial degradation has grown significantly due to advancements in genomic, metagenomic, bioinformatic, and high-throughput analyses that have enhanced our understanding of environmentally relevant microorganisms (Diaz 2008; Koukkou 2011). Bioremediation is an increasingly popular and effective method for removing toxic waste from polluted environments. This approach offers a cost-effective and practical solution for eliminating environmental contaminants (Tripathi et al. 2021). Numerous researchers have applied bioremediation technologies to address both organic and inorganic pollutants (Tripathi and Garg 2014; Sonawane et al. 2022). The process relies on microbial enzymes to degrade hydrocarbons and other harmful substances into less toxic compounds (Singh et al. 2022). By utilizing the enzymatic pathways of bacteria, algae, fungi, and plants, bioremediation accelerates biochemical reactions that degrade contaminants, offering a practical solution for addressing both organic and inorganic pollution (Khalid et al. 2021).

While bioremediation is a clean and cost-effective method for remediating contaminated sites, it often faces challenges such as prolonged treatment times and reduced efficiency when microbes or plants struggle to tolerate harsh environmental conditions (Cecchin et al. 2017). To address these limitations, it is crucial to advance bioremediation techniques or develop integrated approaches for more efficient hydrocarbon removal. In recent years, nanotechnology has garnered significant attention due to its unique properties and superior efficiency (Guerra et al. 2018). However, despite the growing applications and reputation of nanoparticles (NPs), many are chemically synthesized using reducing or capping agents that can have harmful environmental effects (Verma 2018).

Artificial intelligence (AI), another major tool, is increasingly recognized for its precision, speed, and accuracy across various fields. As an advanced and sophisticated technique, AI leverages digital algorithms or software to perform complex tasks, offering the potential to revolutionize bioremediation by developing more efficient, faster, and robust methods for rehabilitating oil-contaminated systems. Advanced machine learning methods, such as artificial neural networks (ANN) and support vector machines (SVM), are promising tools that provide deep insights and assist in interpreting nonlinear data in environmental science (Shadrin et al. 2020). AI-based models can predict the correlation between the origin and spatial distribution of contaminants, their concentration, and properties in specific locations, thus aiding environmentalists in the effective planning and execution of remediation efforts (Ebrahimi et al. 2019). This chapter also explores the potential of AI to enhance the accuracy and efficiency of bioremediation processes for pollutant removal. It discusses various research efforts that have integrated AI into pollutant remediation, as documented in internationally recognized scientific journals and databases like PubMed Central, ScienceDirect, and Scopus. AI is also being used to monitor environmental quality and improve the precision of remediation efforts. As such, it is anticipated that AI-driven bioremediation research will play a crucial role in addressing environmental challenges. Moving forward, the focus should be on developing highly efficient algorithms and integrating them into remediation studies to achieve superior outcomes (Patowary et al. 2023).

1.1 Importance of Microbial Degradation

Microorganisms are highly effective in degrading a wide range of organic pollutants due to their versatile metabolic machinery and ability to adapt to challenging environmental conditions. As such, they play a crucial role in site remediation. However, their effectiveness in this process is influenced by various factors, including the chemical nature and concentration of pollutants, bioavailability, and the physicochemical properties of the environment (Tahri et al. 2013). Bioremediation, which involves the use of microorganisms like bacteria, fungi, algae, yeast, etc., is a process designed to remove contaminants from the environment (Enerijiofi 2021; Kour et al. 2022) (Fig. 1, Table 1). As key components of ecosystems, microbial systems play a crucial role in maintaining ecological balance and are highly effective in degrading harmful pollutants (Bavumiragira et al. 2022). Microbial degradation is a biochemical process in which microorganisms break down pollutants through metabolic and co-metabolic pathways (Eheneden et al. 2023). These microbes exhibit remarkable adaptability to extreme temperatures, ranging from −196 °F to

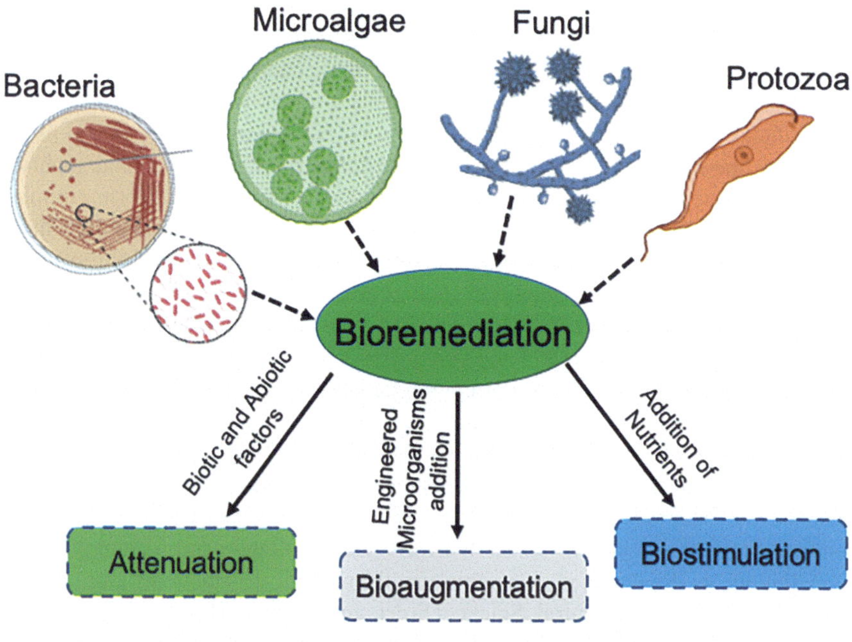

Fig. 1 Illustration of bioremediation utilizing various microbial taxa. Common microbial-based bioremediation approaches involve either stimulating native microbial communities through the addition of specific compounds (biostimulation) or introducing specific microbial species with the ability to degrade or detoxify contaminants (bioaugmentation) (Adams et al. 2015; Brar et al. 2017)

Table 1 Various microorganisms used in bioremediation

Pollutants	Species	Organisms	References
Crude oil	*Pseudomonas aeruginosa*, *Penicillium chrysogenum*, *Ralstonia pickettii*	Bacteria	Mukjang et al. (2022), Purnomo et al. (2019)
PAHs	*Pseudomonas* sp., *Cycloclasticus* sp.		Messina et al. (2015)
Phenanthrene	*Alcaligenes aquatilis*, *Bacillus*, *Burkholderia*		Mahjoubi et al. (2019)
Hydrocarbons and chromium	*Agarivorans* sp., *Pseudomonas aeruginosa*, *Spirulina* sp.		Dell'Anno et al. (2021), Gaur et al. (2022)
Phenol, pyrene	*Chlorella*, *Chlamydomonas*	Microalgae	Wu et al. (2022)
Crude oil	*Chlorella vulgaris*		Kalhor et al. (2017)
Fluoranthene	*Nitzschia* sp.		Hong et al. (2008)
Arsenic and cadmium	*Microcystis aeruginosa*		Deng et al. (2020)
Pyrene	*Aspergillus sclerotiorum*, *Ganoderma lucidum*	Fungi	Agrawal et al. (2018)
Hexadecane	*Aspergillus* sp.		Al-Hawash et al. (2018)
Chlorophenols and aliphatic hydrocarbons	Aspergillus sp., *Penicillium* sp., *Aspergillus* sp., and *Cladosporium* sp.		Li et al. (2020)

1200 °F, making them particularly suitable for environmental remediation efforts (Hussain et al. 2022). This biochemical breakdown of pollutants by microorganisms is essential for the natural detoxification of environments contaminated with organic and inorganic substances.

Various microbial species, including *Achromobacter*, *Alcaligenes*, *Xanthobacter*, *Arthrobacter*, *Pseudomonas*, *Bacillus*, *Mycobacterium*, *Corynebacterium*, *Flavobacterium*, and *Nitrosomonas*, are known to participate in the degradation and purification of pollutants (Sharma et al. 2021). The effectiveness of bioremediation is influenced by numerous factors, including physical and chemical properties of the environment, the type of soil, available carbon and nitrogen sources, and whether the process involves a single microorganism or a microbial consortium (Garg et al. 2012). Microbial consortia, which consist of multiple species working synergistically to optimize substrate utilization, often exhibit enhanced multifunctionality and resilience, leading to greater bioremediation efficiency compared to single-species approaches (Abatenh et al. 2017). Among the essential nutrients, carbon plays a pivotal role in supporting microbial activity, thereby accelerating the bioremediation process and promoting the breakdown of existing pollutants in situ.

Recent advancements have significantly enhanced the use of microbes in bioremediation. Microbial glycoconjugates, for instance, have been found to reduce surface tension, increase bioavailability, and create a solvent interface for organic pollutants, thereby improving their removal from the environment (Bhatt et al. 2021). Additionally, microbial biofilms—composed of polysaccharides, extracellular DNA, and proteins—are increasingly being utilized in the bioremediation of stubborn pollutants, especially recalcitrant organic contaminants (Sonawane et al. 2022). Another promising technology is the bioelectrochemical system, which integrates biological and electrochemical methods to effectively manage pollutants, particularly petroleum hydrocarbons (Ambaye et al. 2023).

Genetic and metabolic engineering is also a rapidly evolving field aimed at producing novel microbes with specific, desired traits. Typically, genetically modified organisms (GMOs) are created by altering bacterial genetic codes through recombinant DNA techniques. These engineered microbes can develop new metabolic pathways for pharmaceutical degradation, reduce the accumulation of harmful metabolites, and enhance metabolic catabolism. Moreover, microbial enzymes play a crucial role in the biodegradation of environmental contaminants. A focused study on these enzymes, along with techniques to modify them using nanoparticles, could further enhance their efficiency in remediation processes (Xue et al. 2022; Patowary et al. 2023).

In 2019, Wang et al. employed an AI system known as an integrated extended short-term memory network (LSTM), which leverages cross-correlation and association rules (Apriori) to identify pollutant point sources and trace industrial contaminants in water bodies. This approach enabled the creation of water quality cross-correlation maps, facilitating the tracking of contaminant sources. Similarly, Shadrin et al. (2020) utilized machine learning models, including artificial neural networks (ANN) and support vector machines (SVM), to predict the phytotoxicity of total petroleum hydrocarbons (TPH). AI has also been applied to interpret rheological properties that impact the transport and refining of petroleum. For instance, Stratiev et al. (2023) developed an ANN-based model to predict the viscosity of crude oil. These examples illustrate that AI can be effectively used to pinpoint oil pollution sites and improve the precision and efficiency of remediation processes. The integration of machine learning algorithms with conventional remediation technologies thus offers a more accurate, faster, and cost-effective approach to environmental restoration.

1.2 Challenges in Microbial Degradation Processes

One of the major challenges in microbial degradation is the absence of standardized methodologies for evaluating and comparing the degradation capabilities of different microbial species. Existing biodegradation analysis methods often lack accuracy, necessitating further investigation. Additionally, the presence of multiple pollutants within a system can hinder the degradation process. For instance, the impact of antibiotics on the transport behavior of pharmaceuticals is crucial for effective pollutant removal, yet this area remains poorly understood and needs further exploration. While the mineralization process generally breaks down pharmaceutical substances, some microorganisms biotransform these compounds, generating intermediates that can be more hazardous than the original pollutants. Consequently, employing a mixed microbial consortium for complete pollutant mineralization could be more effective and preferable than relying on a single pure culture, which may not have the same potential. Microbial degradation has predominantly focused on organic substrates, often neglecting persistent inorganic pollutants. Therefore, intensified research is needed to identify microbes capable of degrading inorganic pollutants effectively.

In the realm of waste management, bioremediation stands out as an effective technique that uses organisms to consume or neutralize pollutants, particularly in contaminated areas (Giri et al. 2020). The process involves leveraging data from various biological databases—including those detailing chemical structures, RNA/protein expression, organic compounds, catalytic enzymes, microbial degradation pathways, and comparative genomics—to decipher the specific mechanisms by which an organism degrades a particular pollutant (Yergeau et al. 2012). To enhance the effectiveness of bioremediation, a range of bioinformatics tools are employed to analyze these data sources, ultimately contributing to the development of more advanced environmental remediation technologies. However, there is a notable lack of comprehensive data on the factors that influence the growth and metabolism of microbes with bioremediation potential. This gap has limited the widespread application of bioremediation techniques (Zheng et al. 2018). Addressing this scarcity of information is crucial for expanding the use of bioremediation in diverse environmental contexts.

The integration of statistical analysis and bioinformatics plays a critical role in optimizing process conditions for bioremediation. Furthermore, extensive research on nanobioremediation (NBR) processes, coupled with their optimization and scaling, integrated with advanced AI, modern biotechnological techniques, enzyme technology, and genetic recombination, can lead to the development of sustainable solutions for effectively addressing environmental challenges). This multidisciplinary approach holds the potential to significantly enhance the efficiency and scalability of bioremediation efforts.

2 Significance of Artificial Intelligence and Machine Learning

2.1 Introduction to AI and ML Technologies

In the modern era of technology, few fields have gathered as much attention and transformative potential as artificial intelligence (AI) and machine learning (ML). These domains are not merely technical disciplines but are pivotal to understanding the future trajectory of technology and its impact on society. This chapter delves into the fundamental concepts of AI and ML, exploring their evolution, core principles, applications, and implications for various sectors. It aims to provide a comprehensive overview that sets the stage for deeper exploration in subsequent areas (Silver et al. 2016; Thrun 2019).

(a) *The Emergence of Artificial Intelligence*

Artificial intelligence, as a field, traces its roots to the mid-twentieth century, emerging from the intersection of computer science, cognitive psychology, and mathematical logic. Early AI research was driven by the ambition to simulate human cognitive processes and create machines capable of performing tasks that required intelligence. The term "artificial intelligence" itself was coined in 1956 by John McCarthy during the Dartmouth Conference, which is considered the founding event of AI as a research discipline. In its early stages, AI research focused on symbolic methods, such as logic- and rule-based systems, aimed at solving well-defined problems through explicit instructions. However, the limitations of these approaches became evident as the complexity of real-world problems grew. This period of enthusiasm, followed by a phase of disillusionment known as the "AI winter," led researchers to reassess their strategies and seek new methodologies (Russell and Norvig 2016).

The resurgence of AI in recent decades has been largely fueled by advancements in computational power, the availability of large datasets, and innovative algorithms. The introduction of neural networks and deep learning has marked a significant shift, enabling machines to perform tasks with unprecedented accuracy. This modern era of AI is characterized by its ability to learn from data, adapt to new information, and make predictions or decisions with minimal human intervention (Murphy 2012).

(b) *Machine Learning: A Core Component of AI*

Machine learning, a subset of AI, is centered around the concept that systems can learn from data, identify patterns, and make decisions with minimal human intervention. Unlike traditional AI approaches that rely heavily on explicitly programmed rules, ML algorithms improve their performance by analyzing and learning from data. This paradigm shift has led to the development of sophisticated models that can handle vast amounts of information and deliver insights previously unattainable through conventional methods.

ML encompasses various approaches, including supervised learning, unsupervised learning, and reinforcement learning. Supervised learning involves training algorithms on labeled data, where the input–output pairs are known, allowing the model to learn the mapping function. Unsupervised learning, on the other hand, deals with unlabeled data and aims to uncover hidden patterns or structures within the data. Reinforcement learning is inspired by behavioral psychology and focuses on training models to make sequences of decisions by rewarding desired outcomes and penalizing undesirable ones (Silver et al. 2016; Thrun 2019).

The rapid advancement in ML techniques has led to significant breakthroughs in diverse applications such as natural language processing, computer vision, and autonomous systems. These innovations have not only enhanced technological capabilities but have also brought about profound changes in how we interact with technology and the world around us (LeCun et al. 2015).

(c) *Implications of AI and ML in Different Fields*

The practical applicability of AI and ML is very extensive and transformative. In healthcare, AI-driven diagnostic tools and predictive models are revolutionizing patient care and personalized medicine (Goodfellow et al. 2016). In finance, algorithms are employed for risk assessment, fraud detection, and algorithmic trading. The retail sector benefits from AI through enhanced customer experiences, inventory management, and personalized recommendations. Additionally, AI-powered systems are making strides in autonomous vehicles, smart cities, and environmental monitoring, showcasing the versatility and potential of these technologies (Noviati et al., 2024). However, the widespread adoption of AI and ML also brings forth important ethical, social, and economic considerations. Issues related to data privacy, algorithmic bias, and the impact on employment are critical areas of concern that necessitate thoughtful deliberation and regulatory frameworks. As AI and ML technologies continue to evolve, it is essential to address these challenges to ensure that their benefits are equitably distributed and their risks mitigated (Jain et al. 1999).

Further, the future of AI and ML holds both exciting possibilities and significant challenges. The ongoing research in areas such as general artificial intelligence, quantum computing, and ethical AI promises to push the boundaries of what is currently achievable. The collaboration of AI with other emerging technologies like IoT is likely to open new frontiers and create innovative solutions to complex problems (Sutton and Barto 2018). Nonetheless, the path forward is fraught with challenges that include technical hurdles, ethical dilemmas, and the need for interdisciplinary collaboration. Ensuring the robustness, fairness, and transparency of AI systems, as well as addressing the societal impacts of these technologies, will be crucial in shaping a future where AI and ML contribute positively to humanity (Hassabis et al. 2017). Understanding the fundamental principles and current advancements leads in evolving the landscape of AI and ML and contributes to its responsible development and application.

2.2 Applications of AI and ML in Environmental Science

The interplay between artificial intelligence (AI) and machine learning (ML) and environmental science is rapidly transforming how we understand, monitor, and address environmental issues. The integration of these technologies offers novel approaches to managing complex environmental challenges, from climate change to biodiversity conservation. This note explores the diverse applications of AI and ML in environmental science, highlighting their impact on research, policy, and practical solutions.

(a) *Climate Change Modeling and Prediction*

Climate change poses one of the most pressing global challenges, and AI/ML technologies are playing a crucial role in enhancing climate models and predictions.

Traditional climate models rely on complex mathematical equations and extensive data, which can be computationally intensive. AI and ML can improve these models by identifying patterns and making predictions based on large datasets. ML algorithms are used to assimilate data from various sources, such as satellite imagery and weather stations, to refine climate models. For example, deep learning techniques can enhance weather forecasts by analyzing historical weather patterns and current atmospheric conditions (Rasp et al. 2018).

AI models have been employed to predict extreme weather events, such as hurricanes and heatwaves, with greater accuracy. By analyzing vast amounts of historical data and real-time inputs, ML algorithms can forecast the intensity and trajectory of these events, aiding in disaster preparedness and response (Koster et al. 2020).

(b) *Remote Sensing and Environmental Monitoring*

Remote sensing technologies, which use satellite and aerial imagery to monitor environmental changes, benefit significantly from AI and ML. These technologies help in tracking deforestation, land use changes, and water quality.

- *Land Cover Classification*: ML algorithms are used to classify land cover types from satellite images. Convolutional neural networks (CNNs), a type of deep learning model, have demonstrated high accuracy in distinguishing between different land cover types, such as forests, wetlands, and urban areas (Zhu et al. 2017).
- *Biodiversity Monitoring*: AI techniques are applied to analyze camera trap images and acoustic data for wildlife monitoring. For instance, computer vision algorithms can identify and count animal species in camera trap images, aiding in biodiversity assessments and conservation efforts (Yuan et al. 2020).

(c) *Pollution Management*

Pollution management is another critical area where AI and ML contribute to environmental science. These technologies assist in tracking pollution sources, predicting pollution levels, and devising strategies to mitigate pollution.

- *Air Quality Prediction*: ML models predict air quality by analyzing data from sensors, weather conditions, and historical pollution levels. These models help in forecasting pollution events and assessing the impact of various pollutants on public health (Cheng et al. 2020).
- *Water Quality Monitoring*: AI is used to monitor water quality in real time by analyzing sensor data. Machine learning algorithms can detect anomalies and predict water quality issues, enabling timely interventions to address contamination and maintain safe water supplies (Jin et al. 2021).

(d) *Conservation and Resource Management*

Conservation efforts and resource management are greatly enhanced by AI and ML technologies. These tools support the sustainable management of natural resources and help in protecting endangered species.

- *Habitat Modeling*: ML algorithms are used to model and predict habitat suitability for various species. By analyzing environmental variables and species distribution data, these models assist in identifying critical habitats and planning conservation strategies (Elith et al. 2006).
- *Resource Optimization*: AI is applied in managing natural resources, such as forests and fisheries. For example, ML models optimize forest management practices by predicting growth patterns and assessing the impact of harvesting activities (Asner et al. 2017).

(e) *Environmental Policy and Decision-Making*

Environmental policy and decision-making benefit from AI and ML through improved data analysis and predictive capabilities. These technologies support evidence-based policy development and facilitate more effective environmental management.

- *Policy Impact Assessment*: AI models assess the potential impacts of environmental policies by simulating various scenarios and analyzing outcomes. This approach helps policymakers understand the consequences of different strategies and make informed decisions (Liu et al. 2018).
- *Public Engagement*: AI-driven tools enhance public engagement in environmental issues by providing interactive platforms for data visualization and feedback. For example, AI-powered apps can offer real-time information on air quality and suggest actions to reduce exposure to pollutants (Wang et al. 2021).

2.3 Potential Contributions of AI and ML to Microbial Degradation Studies

Microbial degradation is a crucial process for the bioremediation of environmental pollutants and involves the breakdown of organic substances by microorganisms. It is essential for the cleanup of contaminated sites and the recycling of organic matter in ecosystems. Recent advancements in artificial intelligence (AI) and machine learning (ML) offer transformative potential for enhancing our understanding and optimization of microbial degradation processes. This note explores how AI and ML can contribute to microbial degradation studies, highlighting their applications, benefits, and future prospects (Fig. 2).

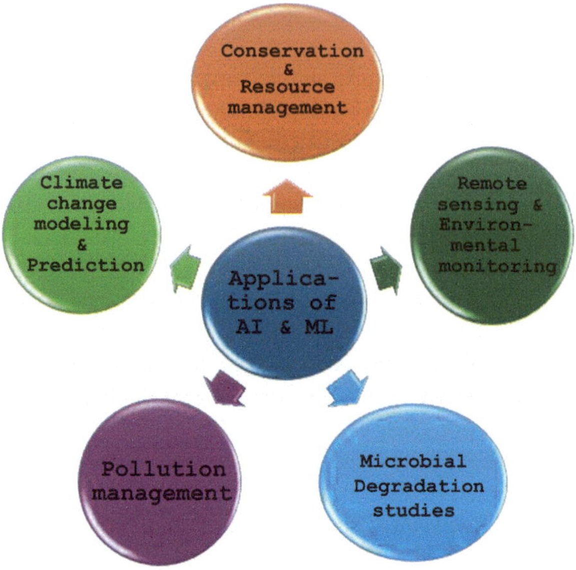

Fig. 2 Applications of artificial intelligence and machine learning in environmental science

(a) *Enhancing Microbial Community Analysis*

Understanding microbial communities involved in degradation processes is fundamental to optimizing bioremediation strategies. Traditional methods for analyzing microbial communities, such as culture-based techniques and 16S rRNA gene sequencing, provide valuable insights but can be limited by their resolution and scope. AI and ML offer advanced tools for improving microbial community analysis.

- *High-Throughput Sequencing Analysis*: ML algorithms are increasingly used to analyze high-throughput sequencing data, which provides detailed profiles of microbial communities. Techniques such as clustering and classification algorithms help in identifying and characterizing diverse microbial species and their functional roles in degradation processes (Hao et al. 2020). For instance, deep learning models can analyze metagenomic data to uncover patterns related to microbial diversity and function.
- *Predictive Modeling*: AI models, including neural networks and ensemble methods, can predict microbial community dynamics based on environmental variables and degradation pathways. These models help in understanding how changes in environmental conditions affect microbial community structure and function, enabling more effective management of bioremediation processes (Karkman et al. 2019).

(b) *Optimizing Degradation Pathways*

Microbial degradation involves complex biochemical pathways that vary across different microorganisms and pollutants. AI and ML can enhance the optimization and understanding of these pathways by analyzing large datasets and identifying key factors influencing degradation efficiency.

- *Pathway Prediction*: ML algorithms can predict metabolic pathways involved in the degradation of specific pollutants. By analyzing genomic and proteomic data, these models identify enzymes and metabolic routes essential for breaking down organic compounds. For example, AI-driven approaches have been used to predict the degradation pathways of pollutants such as hydrocarbons and pesticides (Cheng et al. 2021).
- *Enzyme Function Prediction*: AI models can predict the function of uncharacterized enzymes involved in microbial degradation. Techniques such as protein structure modeling and functional annotation using deep learning algorithms provide insights into the catalytic activities of enzymes and their potential applications in bioremediation (Lee et al. 2022).

(c) *Improving Bioreactor Design and Performance*

Bioreactors are systems used to cultivate microorganisms for degradation processes. AI and ML contribute to the optimization and control of bioreactor conditions, enhancing the efficiency of microbial degradation.

- *Real-Time Monitoring*: AI-driven sensors and control systems enable real-time monitoring of bioreactor parameters such as temperature, pH, and nutrient levels. ML algorithms analyze sensor data to optimize these conditions, ensuring optimal microbial activity and degradation rates (Vargas et al. 2021).
- *Predictive Maintenance*: ML models can predict equipment failures and maintenance needs based on historical data and real-time monitoring. This predictive maintenance approach minimizes downtime and ensures continuous operation of bioreactors, improving the overall efficiency of degradation processes (Zhao et al. 2020).

(d) *Enhancing Data Integration and Interpretation*

Microbial degradation studies generate vast amounts of data from various sources, including genomics, proteomics, metabolomics, and environmental monitoring. AI and ML offer powerful tools for integrating and interpreting this complex data.

- *Data Integration*: AI techniques, such as data fusion and integrative modeling, combine data from different sources to provide a comprehensive understanding of degradation processes. For example, integrative models can combine genomic data with environmental factors to predict microbial behavior and degradation outcomes (Li et al. 2019).
- *Pattern Recognition*: ML algorithms excel at identifying patterns and correlations in large datasets. In microbial degradation studies, these algorithms can uncover relationships between microbial activity, environmental conditions, and degradation efficiency, leading to more informed decision-making (Xie et al. 2020).

(e) *Facilitating Field Applications and Scale-Up*

The application of AI and ML extends beyond laboratory studies to field applications and large-scale bioremediation projects. These technologies help in scaling up microbial degradation processes and implementing them effectively in real-world scenarios.

- *Field Monitoring*: AI-driven remote sensing and data analytics tools facilitate the monitoring of bioremediation processes in the field. For instance, AI algorithms can analyze satellite imagery and sensor data to assess the progress of bioremediation efforts and detect potential issues (Kumar et al. 2021).
- *Scale-Up Optimization*: ML models assist in scaling up microbial degradation processes from laboratory to industrial scales. These models predict how microbial processes will perform under different conditions and guide the design of large-scale bioreactors and treatment systems (Gao et al. 2022).

The integration of AI and ML into microbial degradation studies represents a significant advancement in environmental science and biotechnology. These technologies offer enhanced capabilities for analyzing microbial communities, optimizing degradation pathways, improving bioreactor performance, and interpreting complex data. As AI and ML continue to evolve, their applications in microbial degradation are likely to expand, leading to more effective and efficient bioremediation strategies.

3 Case Studies and Applications

3.1 Case Study 1: Functional Profiling and Pathway Prediction of Genomic and Metagenomic Data by Using AI and ML

The integration of artificial intelligence (AI) and machine learning (ML) with genomic and 22 metagenomic data of microbial communities has shown tremendous potential in addressing environmental pollution. This case study explores how these technologies are leveraged to enhance the understanding and utilization of microbial communities for the degradation of pollutants and toxins. Microbial communities play a crucial role in bioremediation, the process of using microorganisms to degrade environmental contaminants (Ayilara and Babalola 2023) Genomic and metagenomic data provide insights into the genetic potential and functional capabilities of these communities (Vikram et al. 2016). However, the complexity and vastness of the data pose challenges in interpretation and application (Lawson et al. 2021). AI and ML offer powerful tools to overcome these challenges by enabling data-driven insights and predictive modelling. Machine learning models can predict the metabolic pathways involved in pollutant degradation by analyzing metagenomic sequences (Wijaya et al. 2024). Functional profiling using AI approaches enables the identification of key enzymes and microbial taxa responsible for specific degradation processes. This information is crucial for optimizing bioremediation strategies (Haque et al. 2022). AI-driven predictive models can forecast the behavior of microbial communities under different environmental conditions. These models help in predicting the efficiency of pollutant degradation and in designing microbial consortia with enhanced bioremediation capabilities. For instance, machine learning has been used to analyze the human microbiome to identify microbial signatures linked to disease states, showcasing the potential for similar applications in environmental microbiomes (Sharma and Shukla 2020).

3.2 Case Study 2: AI in Biomarker Identification

AI technologies, including machine learning (ML) and deep learning (DL), are employed to process vast datasets generated from microbial studies. These technologies facilitate the identification of specific biomarkers like genes, proteins, or metabolites that indicate the presence and activity of microorganisms capable of degrading pollutants.

High-throughput sequencing technologies generate extensive genomic, transcriptomic, and metabolomic data. AI algorithms preprocess this data, handling noise and variability to ensure accuracy (Li et al. 2022). Machine learning models analyze the preprocessed data to extract relevant features. Techniques like random forests, support vector machines (SVM), and neural networks identify potential biomarkers by recognizing patterns and associations between microbial activity and pollutant degradation (Zhong et al. 2021). AI models are trained on annotated datasets to predict microbial degradation capabilities based on the identified biomarkers. These models can then be used to screen and predict the degradation potential of new microbial strains or communities (Marcos-Zambrano et al. 2021). The identified biomarkers undergo experimental validation to confirm their roles in degradation pathways. AI aids in interpreting complex interactions and pathways, providing insights into microbial ecology and functional dynamics (Wijaya et al. 2024).

3.3 Case Study 3: AI-Driven Microbial Degradation of Polycyclic Aromatic Hydrocarbons (PAHs)

Polycyclic aromatic hydrocarbons (PAHs) are persistent environmental pollutants with significant health risks. A recent study applied AI-driven approaches to identify biomarkers for PAH degradation in soil microbiomes. Soil samples contaminated with PAHs were subjected to metagenomic sequencing. The resulting data included microbial community compositions and functional gene profiles. Machine learning models, including random forests and neural networks, processed the metagenomic data. The models identified key genes and metabolic pathways associated with PAH degradation. The AI models highlighted several biomarkers, such as specific dioxygenase genes, which are crucial for the initial oxidative breakdown of PAHs. Experimental assays confirmed the presence and activity of these biomarkers in degrading PAHs, demonstrating the efficacy of AI-driven biomarker identification.

3.4 Case Study 4: Screening Methods for Isolation of High-Performance Degraders

The increasing prevalence of environmental pollutants has driven the need for efficient bioremediation strategies. High-performance microbial degraders are pivotal in this context, capable of breaking down complex pollutants into nontoxic forms. This case study explores the methodologies employed in isolating these potent microbial degraders.

Isolation and Screening Techniques

1. *Enrichment Culture Technique*
   - Objective: To selectively enrich and isolate microorganisms capable of degrading specific pollutants.
   - Method: Environmental samples (soil, water, sludge) are incubated in a medium containing the target pollutant as the sole carbon and energy source. Over successive subcultures, microorganisms that thrive in these conditions are isolated.
   - Outcome: This technique enhances the abundance of pollutant-degrading microbes, making isolation easier (Wu et al. 2013).
2. *Biochemical Assays*
   - Objective: To screen isolated strains for their degradative abilities.
   - Method: Isolates are grown in the presence of the pollutant, and various assays (e.g., CO_2 evolution, oxygen uptake, and specific enzyme activity) are conducted to quantify degradation.
   - Outcome: These assays provide a preliminary assessment of the microbial degradation potential (Tian et al. 2018).

3. *Molecular Techniques*
   - Objective: To identify genetic markers associated with pollutant degradation.
   - Method: Techniques like PCR, qPCR, and metagenomics are employed to detect and quantify genes encoding degradative enzymes.
   - Outcome: These techniques offer precise identification and quantification of high-performance degraders at the genetic level (Widada et al. 2002).
4. *Omics Approaches*
   - Objective: To comprehensively analyze the metabolic and genetic capabilities of microbial communities.
   - Method: Metagenomics, meta-transcriptomics, and proteomics are used to understand the functional potential and active metabolic pathways in microbial communities exposed to pollutants.
   - Outcome: Omics approaches provide a holistic view of the microbial community's potential to degrade pollutants and help identify key players in the process (Scholes et al. 2021).

3.5 Case Study 5: Isolation of Plastic Degraders

Plastic pollution is a significant environmental issue, and isolating microorganisms capable of degrading plastics is critical for bioremediation efforts (Cai et al. 2023).

Methodology

1. Sample Collection: Soil samples from plastic-contaminated sites were collected.
2. Enrichment Culture: Samples were incubated in a medium with plastic polymers (e.g., polyethylene, polystyrene) as the sole carbon source.
3. Isolation: Microorganisms showing growth were isolated and purified.
4. Screening: Biochemical assays (e.g., weight loss of plastic films, CO_2 evolution) were conducted to assess degradation.
5. Molecular Analysis: PCR was used to detect genes related to plastic degradation, such as alkB (alkane hydroxylase) and laccase genes.
6. Omics Analysis: Metagenomic sequencing provided insights into the community structure and functional genes involved in plastic degradation.

4 Applications, Challenges, and Future Prospects of Artificial Intelligence and Machine Learning in Microbial Degradation of Pollutants and Toxins

Artificial intelligence (AI) and machine learning (ML) are revolutionizing various scientific fields, including environmental biotechnology. The application of AI and ML in microbial degradation of pollutants and toxins offers innovative solutions to environmental challenges.

4.1 Applications of AI and ML in Microbial Degradation

(i) *Predictive Modeling*: AI and ML algorithms are used to predict the biodegradation potential of various microbial strains. By analyzing large datasets of microbial genomes and environmental conditions, AI models can identify the most effective microbes for degrading specific pollutants, such as toxic industrial chemicals and oil spills (Mishra et al. 2021).
(ii) *Optimization of Degradation Pathways*: ML techniques help in optimizing the metabolic pathways involved in microbial degradation. This involves adjusting environmental parameters to enhance microbial activity and efficiency in breaking down pollutants. For example, machine learning models have been used to optimize conditions for degrading organic pollutants in wastewater treatment (Dale et al. 2010).
(iii) *Monitoring and Control Systems*: AI-powered sensors and control systems can continuously monitor microbial degradation processes, ensuring optimal conditions are maintained. These systems use real-time data to adjust parameters such as temperature, pH, and nutrient levels, thereby improving the efficiency and effectiveness of bioremediation efforts (Venkateswaran et al. 2023).

4.2 Challenges in Implementing AI and ML

(i) Data Quality and Availability: High-quality, comprehensive datasets are crucial for training AI and ML models. Incomplete or biased data can lead to inaccurate predictions and suboptimal performance. There is a need for standardized data collection and sharing practices across research institutions (Aldoseri et al. 2023).
(ii) Computational Complexity: Developing and deploying AI and ML models require significant computational resources. This can be a barrier for many research institutions, particularly those in developing regions. Efficient algorithms and cloud-based solutions are needed to overcome these limitations.
(iii) Interdisciplinary Collaboration: Successful implementation of AI in microbial degradation requires collaboration between microbiologists, data scientists, and environmental engineers. Bridging these diverse fields can be challenging due to differences in terminology, methodologies, and objectives (Asnicar et al. 2024).

4.3 Future Prospects

(i) *Enhanced Predictive Capabilities*: Future advancements in AI and ML are expected to provide even more accurate predictive models, enabling the identification of novel microbial strains and consortia with superior degradation capabilities.

(ii) *Integration with Synthetic Biology*: The integration of AI with synthetic biology could lead to the design of engineered microbes tailored for specific bioremediation tasks. AI algorithms can assist in designing genetic modifications to enhance microbial degradation pathways.
(iii) *Scalable Solutions*: AI-driven bioremediation solutions can be scaled to address large-scale environmental pollution. This includes the development of autonomous systems that can operate in diverse and challenging environments, from industrial sites to marine ecosystems.
(iv) *Policy and Regulation Support*: AI can assist policymakers by providing data-driven insights and predictive models to formulate effective environmental regulations and standards for pollution control.

5 Conclusion

In conclusion, the integration of microbial degradation with advanced technologies such as artificial intelligence (AI) and machine learning (ML) represents a transformative approach to tackling environmental pollution. While traditional methods of pollutant degradation have served their purpose, they often fall short in addressing the complex and ever-evolving nature of contaminants in our ecosystems. Microbial degradation, with its sustainability and eco-friendliness, has emerged as a powerful solution, yet it is not without challenges. The use of AI and ML to optimize these microbial processes marks a significant advancement, allowing for more precise, scalable, and adaptable remediation strategies. By harnessing the potential of AI and ML, we can enhance the efficiency of bioremediation, making it a more viable option for a broader range of pollutants, including those that are currently difficult to degrade. Moreover, the chapter underscores the importance of ongoing research and development in environmental biotechnology. It is evident that the future of pollution management lies in interdisciplinary collaboration, where AI, nanotechnology, and microbial sciences converge to create innovative solutions. As we move forward, it is crucial to continue exploring the synergies between these fields, refining our methodologies, and expanding our understanding of microbial communities and their interactions with pollutants. The promise of AI-driven microbial degradation lies not only in its potential to remediate existing contaminants but also in its ability to prevent future pollution by predicting and mitigating environmental risks. This chapter paves the way for a new era in environmental sustainability, where technology and nature work hand in hand to protect our planet and ensure the health of future generations.

References

Abatenh E, Gizaw B, Tsegaye Z, Wassie M (2017) The role of microorganisms in bioremediation-a review. Open J Environ Biol 2:038–046. https://doi.org/10.17352/ojeb.000007
Adams GO, Fufeyin PT, Okoro SE, Ehinomen I, Biology E (2015) Bioremediation, biostimulation and bioaugmention: a review. Int J Environ Bioremediat Biodegrad 3:28–39

Agrawal N, Verma P, Shahi SK (2018) Degradation of polycyclic aromatic hydrocarbons (phenanthrene and pyrene) by the ligninolytic fungi *Ganoderma lucidum* isolated from the hardwood stump. Bioresour Bioprocess 5(1). Available from: https://doi.org/10.1186/s40643-018-0197-5

Aldoseri A, Al-Khalifa KN, Hamouda AM (2023) Re-thinking data strategy and integration for artificial intelligence: concepts, opportunities, and challenges. Appl Sci 13(12):7082

Al-Hawash AB, Zhang J, Li S, Liu J, Ghalib HB, Zhang X et al (2018) Biodegradation of n-hexadecane by *Aspergillus* sp. RFC-1 and its mechanism. Ecotoxicol Environ Saf 164:398–408. Available from: https://doi.org/10.1016/j.ecoenv.2018.08.049

Ambaye TG, Vaccari M, Franzetti A, Prasad S, Formicola F, Rosatelli A et al (2023) Microbial electrochemical bioremediation of petroleum hydrocarbons (PHCs) pollution: recent advances and outlook. Chem Eng J 452:139372. https://doi.org/10.1016/j.cej.2022.139372

Asner GP, Knapp DE, Broadbent EN (2017) Selective logging in the Brazilian Amazon. Science 313(5787):206–210

Asnicar F, Thomas AM, Passerini A, Waldron L, Segata N (2024) Machine learning for microbiologists. Nat Rev Microbiol 22(4):191–205

Ayilara MS, Babalola OO (2023) Bioremediation of environmental wastes: the role of microorganisms. Front Agron 5:1183691

Bavumiragira JP, Ge J, Yin H (2022) Fate and transport of pharmaceuticals in water systems: a processes review. Sci Total Environ 823:153635. https://doi.org/10.1016/j.scitotenv.2022.153635

Noviati ND, Putra FE, Sadan S, Ahsanitaqwim R, Septiani N, Santoso NPL (2024) Artificial intelligence in autonomous vehicles: current innovations and future trends. International Journal of Cyber and IT Service Management 4(2):97–104. https://doi.org/10.34306/ijcitsm.v4i2.16

Bhatt P, Verma A, Gangola S, Bhandari G, Chen S (2021) Microbial glycoconjugates in organic pollutant bioremediation: recent advances and applications. Microb Cell Factories 20. https://doi.org/10.1186/s12934-021-01556-9

Brar A, Kumar M, Vivekanand V, Pareek N (2017) Photoautotrophic microorganisms and bioremediation of industrial effluents: current status and future prospects. 3 Biotech 7:18

Cai Z, Li M, Zhu Z, Wang X, Huang Y, Li T, Gong H, Yan M (2023) Biological degradation of plastics and microplastics: a recent perspective on associated mechanisms and influencing factors. Microorganisms 11(7):1661. https://doi.org/10.3390/microorganisms11071661

Cecchin I, Reddy KR, Thomé A, Tessaro EF, Schnaid F (2017) Nanobioremediation: integration of nanoparticles and bioremediation for sustainable remediation of chlorinated organic contaminants in soils. Int Biodeterior Biodegradation 119:419–428. https://doi.org/10.1016/j.ibiod.2016.09.027

Cheng J, Yang M, Luo X (2020) Air quality prediction using machine learning. J Environ Manag 254:109810

Cheng J, Liu Q, Zhang W (2021) Machine learning-based prediction of microbial degradation pathways. J Hazard Mater 408:124491

Dale JM, Popescu L, Karp PD (2010) Machine learning methods for metabolic pathway prediction. BMC Bioinformatics 11:1–14

Dell'Anno F, Rastelli E, Tangherlini M, Corinaldesi C, Sansone C, Brunet C et al (2021) Highly contaminated marine sediments can host rare bacterial taxa potentially useful for bioremediation. Front Microbiol 12:584850. Available from: https://doi.org/10.3389/fmicb.2021.584850

Deng J, Fu D, Hu W, Lu X, Wu Y, Bryan H (2020) Physiological responses and accumulation ability of *Microcystis aeruginosa* to zinc and cadmium: implications for bioremediation of heavy metal pollution. Bioresour Technol 303:122963. Available from: https://doi.org/10.1016/j.biortech.2020.122963

Diaz E (2008) Microbial biodegradation: genomics and molecular biology, 1st edn. Caister Academic Press, p 402

Ebrahimi M, Sarikhani MR, Sinegani AAS, Ahmadi A, Keesstra S (2019) Estimating the soil respiration under different land uses using artificial neural network and linear regression models. Catena 174:371–382. https://doi.org/10.1016/j.catena.2018.11.035

Eheneden I, Wang R, Zhao J (2023) Antibiotic removal by microalgae-bacteria consortium: metabolic pathways and microbial responses. Sci Total Environ 891:164489. https://doi.org/10.1016/j.scitotenv.2023.164489

Elith J, Leathwick JR, Hastie T (2006) A working guide to boosted regression trees. J Anim Ecol 77(4):802–813

Enerijiofi KE (2021) Bioremediation of environmental contaminants: a sustainable alternative to environmental management. Elsevier eBooks, pp 461–480. https://doi.org/10.1016/b978-0-12-820524-2.00019-5

Fenner K, Elsner M, Lueders T, McLachlan MS, Wackett LP, Zimmermann M et al (2021) Methodological advances to study contaminant biotransformation: new prospects for understanding and reducing environmental persistence? ACS ES&T Water 1:1541–1554. https://doi.org/10.1021/acsestwater.1c00025

Gao J, Zhang Y, Xu X (2022) Scaling up microbial bioremediation processes using machine learning. Bioresour Technol 344:126130

Garg SK, Tripathi M, Srinath T (2012) Strategies for chromium bioremediation of tannery effluent. Rev Environ Contam Toxicol:75–140. https://doi.org/10.1007/978-1-4614-2329-4_2

Gaur VK, Tripathi V, Manickam N (2022) Bacterial- and fungal-mediated biodegradation of petroleum hydrocarbons in soil. In: Development in wastewater treatment research and processes. Elsevier eBooks, pp 407–427. Available from: https://doi.org/10.1016/b978-0-323-85839-7.00008-6

Giri BS, Geed S, Vikrant K, Lee SS, Kim K-H, Kailasa SK et al (2020) Progress in bioremediation of pesticide residues in the environment. Environ Eng Res 26:200446. https://doi.org/10.4491/eer.2020.446

Goodfellow I, Bengio Y, Courville A (2016) Deep learning. MIT Press

Guerra F, Attia M, Whitehead D, Alexis F (2018) Nanotechnology for environmental remediation: materials and applications. Molecules 23:1760. https://doi.org/10.3390/molecules23071760

Gupta A, Kumar S, Bajpai Y, Chaturvedi K, Johri P, Tiwari RK et al (2024) Pharmaceutically active micropollutants: origin, hazards and removal. Front Microbiol 15. https://doi.org/10.3389/fmicb.2024.1339469

Hao Q, Liu L, Wu X (2020) Deep learning approaches for high-throughput microbial community analysis. Bioinformatics 36(4):1157–1163

Haque S, Srivastava N, Pal DB, Alkhanani MF, Almalki AH, Areeshi MY, Gupta VK (2022) Functional microbiome strategies for the bioremediation of petroleum-hydrocarbon and heavy metal contaminated soils: a review. Sci Total Environ 833:155222

Hassabis D, Silver D, Salisbury M (2017) Neurosymbolic AI. Nature 547:511–518

Hong YW, Yuan DX, Lin QM, Yang TL (2008) Accumulation and biodegradation of phenanthrene and fluoranthene by the algae enriched from a mangrove aquatic ecosystem. Mar Pollut Bull 56(8):1400–1405. Available from: https://doi.org/10.1016/j.marpolbul.2008.05.003

Hussain A, Rehman F, Rafeeq H, Waqas M, Asghar A, Afsheen N et al (2022) In-situ, Ex-situ, and nano-remediation strategies to treat polluted soil, water, and air—A review. Chemosphere 289:133252. https://doi.org/10.1016/j.chemosphere.2021.133252

Jain AK, Murty MN, Flynn PJ (1999) Data clustering: a review. ACM Computing Surveys (CSUR) 31(3):264–323

Jin X, Zhou Z, Li H (2021) Real-time water quality monitoring and forecasting using machine learning. Environ Sci Technol 55(5):3076–3084

Kalhor AX, Movafeghi A, Mohammadi-Nassab AD, Abedi E, Bahrami A (2017) Potential of the green alga *Chlorella vulgaris* for biodegradation of crude oil hydrocarbons. Mar Pollut Bull 123(1–2):286–290. Available from: https://doi.org/10.1016/j.marpolbul.2017.08.045

Karkman A, Pärssinen K, Larsson DGJ (2019) Machine learning-based predictions of microbial community dynamics. Environ Microbiol Rep 11(4):482–490

Khalid F, Hashmi MZ, Jamil N, Qadir A, Ali MI (2021) Microbial and enzymatic degradation of PCBs from e-waste-contaminated sites: a review. Environ Sci Pollut Res 28:10474–10487. https://doi.org/10.1007/s11356-020-11996-2

Koster RD, Mahanama SP, Kim H (2020) Predicting extreme weather events with machine learning. J Clim 33(16):6750–6771
Koukkou A (2011) Microbial bioremediation of non-metals: current research. Caister Academic Press, Poole
Kour D, Khan SS, Kour H, Kaur T, Devi R, Judy C et al (2022) Microbe-mediated bioremediation: current research and future challenges. J Appl Biol Biotechnol:6–24. https://doi.org/10.7324/jabb.2022.10s202
Kumar A, Singh V, Sharma M (2021) AI-driven remote sensing for monitoring bioremediation processes. Environ Monit Assess 193(2):32
Lawson CE, Martí JM, Radivojevic T, Jonnalagadda SVR, Gentz R, Hillson NJ, Martin HG (2021) Machine learning for metabolic engineering: a review. Metab Eng 63:34–60
LeCun Y, Bengio Y, Hinton G (2015) Deep learning. Nature 521:436–444
Lee S, Park J, Choi J (2022) Predicting enzyme functions in microbial degradation using deep learning. Comput Biol Chem 93:107654
Li X, Zhang M, Wang X (2019) Integrative modeling approaches for microbial degradation studies. Front Microbiol 10:2234
Li Q, Liu J, Gadd GM (2020) Fungal bioremediation of soil co-contaminated with petroleum hydrocarbons and toxic metals. Appl Microbiol Biotechnol 104(21):8999–9008. Available from: https://doi.org/10.1007/s00253-020-10854-y
Li R, Li L, Xu Y, Yang J (2022) Machine learning meets omics: applications and perspectives. Brief Bioinform 23(1):bbab460
Liu J, Wang X, Wu J (2018) Integrating machine learning with policy analysis. Environmental Policy & Governance 28(4):265–278
Mahjoubi M, Aliyu H, Cappello S, Naifer M, Souissi Y, Cowan DA et al (2019) The genome of *Alcaligenes aquatilis* strain BU33N: insights into hydrocarbon degradation capacity. PLoS One 14(9):e0221574. Available from: https://doi.org/10.1371/journal.pone.0221574
Marcos-Zambrano LJ, Karaduzovic-Hadziabdic K, Loncar Turukalo T, Przymus P, Trajkovik V, Aasmets O et al (2021) Applications of machine learning in human microbiome studies: a review on feature selection, biomarker identification, disease prediction and treatment. Front Microbiol 12:634511
Mbachu AE, Chukwura EI, Mbachu NA (2020) Role of microorganisms in the degradation of organic pollutants: a review. Energy Environ Eng 7(1):11. https://doi.org/10.13189/eee.2020.070101
Messina E, Denaro R, Crisafi F, Smedile F, Cappello S, Genovese M et al (2015) Genome sequence of obligate marine polycyclic aromatic hydrocarbons-degrading bacterium *Cycloclasticus* sp. 78-ME, isolated from petroleum deposits of the sunken tanker Amoco Milford Haven, Mediterranean Sea. Mar Genomics 25:11–13. Available from: https://doi.org/10.1016/j.margen.2015.10.006
Mishra B, Varjani S, Kumar G, Awasthi MK, Awasthi SK, Sindhu R et al (2021) Microbial approaches for remediation of pollutants: innovations, future outlook, and challenges. Energy Environ 32(6):1029–1058
Mukjang N, Chitov T, Mhuantong W, Champreda V, Pathom-Aree W, Sattayawat P et al (2022) Bacterial communities associated with crude oil bioremediation through composting approaches with indigenous bacterial isolate. Life 12(11):1712. Available from: https://doi.org/10.3390/life12111712
Murphy KP (2012) Machine learning: a probabilistic perspective. MIT Press
Patowary R, Devi A, Mukherjee AK (2023) Advanced bioremediation by an amalgamation of nanotechnology and modern artificial intelligence for efficient restoration of crude petroleum oil-contaminated sites: a prospective study. Environ Sci Pollut Res 30:74459–74484. https://doi.org/10.1007/s11356-023-27698-4
Purnomo AS, Rizqi HD, Harmelia L, Anggraeni SD, Melati RE, Damayanti ZH et al (2019) Biodegradation of crude oil by *Ralstonia pickettii* under high salinity medium. Malays J Fundam Appl Sci 15(3):377–380. Available from: https://doi.org/10.11113/mjfas.v15n3.1181
Rasp S, Neufeld AT (2018) Deep learning for climate prediction. Nat Commun 9:4324

Rozalska S, Iwanicka-Nowicka R (2016) In: Dlugonski J (ed) Organic pollutants degradation by microorganisms. Genomics, metagenomics and metatranstriptomics: backgrounds. Microbial biodegradation, from omics to function and application. Caister Academic Press, pp 1–12

Russell SJ, Norvig P (2016) Artificial intelligence: a modern approach. Pearson

Scholes NS, Mayor-Ruiz C, Winter GE (2021) Identification and selectivity profiling of small-molecule degraders via multi-omics approaches. Cell Chem Biol 28(7):1048–1060

Shadrin D, Pukalchik M, Kovaleva E, Fedorov M (2020) Artificial intelligence models to predict acute phytotoxicity in petroleum contaminated soils. Ecotoxicol Environ Saf 194:110410. https://doi.org/10.1016/j.ecoenv.2020.110410

Sharma B, Shukla P (2020) Designing synthetic microbial communities for effectual bioremediation: a review. Biocatal Biotransformation 38(6):405–414

Sharma P, Pandey AK, Kim S-H, Singh SP, Chaturvedi P, Varjani S (2021) Critical review on microbial community during in-situ bioremediation of heavy metals from industrial wastewater. Environ Technol Innov 24:101826. https://doi.org/10.1016/j.eti.2021.101826

Silver D, Huang A, Maddison CJ (2016) Mastering the game of Go with deep neural networks and tree search. Nature 529:484–489

Singh S, Singh S, Kushwaha R (2022) Bioremediation of hydrocarbons and xenobiotic compound. BENTHAM SCIENCE PUBLISHERS eBooks, pp 1–48. https://doi.org/10.2174/9789815036039122010004

Sonawane JM, Rai AK, Sharma M, Tripathi M, Prasad R (2022) Microbial biofilms: recent advances and progress in environmental bioremediation. Sci Total Environ 824:153843. https://doi.org/10.1016/j.scitotenv.2022.153843

Stratiev D, Shishkova I, Dinkov R, Nenov S, Sotirov S, Sotirova E et al (2023) Prediction of petroleum viscosity from molecular weight and density. Fuel 331:125679. https://doi.org/10.1016/j.fuel.2022.125679

Sutton RS, Barto AG (2018) Reinforcement learning: an introduction. MIT Press

Tahri N, Bahafid W, Sayel H, El Ghachtouli N (2013) Biodegradation: involved microorganisms and genetically engineered microorganisms. In: Biodegradation - life of science. InTech. https://doi.org/10.5772/56194

Thrun S (2019) Artificial intelligence in games. Cambridge University Press

Tian X, Wang X, Peng S, Wang Z, Zhou R, Tian H (2018) Isolation, screening, and crude oil degradation characteristics of hydrocarbons-degrading bacteria for treatment of oily wastewater. Water Sci Technol 78(12):2626–2638

Tripathi M, Garg SK (2014) Dechlorination of chloroorganics, decolorization, and simultaneous bioremediation of Cr6+ from real tannery effluent employing indigenous *Bacillus cereus* isolate. Environ Sci Pollut Res 21:5227–5241. https://doi.org/10.1007/s11356-013-2479-y

Tripathi M, Singh DN, Prasad N, Gaur R (2021) Advanced bioremediation strategies for mitigation of chromium and organics pollution in tannery. Springer eBooks, pp 195–215. https://doi.org/10.1007/978-981-16-0602-1_10

Vargas R, Rivera C, Ramírez M (2021) AI-based real-time monitoring and optimization of bioreactors. Chem Eng J 414:128733

Venkateswaran N, Kumar SS, Diwakar G, Gnanasangeetha D, Boopathi S (2023) Synthetic biology for waste water to energy conversion: IOT and AI approaches. In: Applications of synthetic biology in health, energy, and environment, pp 360–384

Verma N (2018) A green synthetic approach for size tunable nanoporous gold nanoparticles and its glucose sensing application. Appl Surf Sci 462:753–759. https://doi.org/10.1016/j.apsusc.2018.08.175

Vikram S, Guerrero LD, Makhalanyane TP, Le PT, Seely M, Cowan DA (2016) Metagenomic analysis provides insights into functional capacity in a hyperarid desert soil niche community. Environ Microbiol 18(6):1875–1888

Wang J, Zhang J, Hu J (2021) AI-driven environmental public engagement. Int J Environ Res Public Health 18(2):480

Widada J, Nojiri H, Omori T (2002) Recent developments in molecular techniques for identification and monitoring of xenobiotic-degrading bacteria and their catabolic genes in bioremediation. Appl Microbiol Biotechnol 60:45–59

Wijaya J, Park J, Yang Y, Siddiqui SI, Oh S (2024) A metagenome-derived artificial intelligence modeling framework advances the predictive diagnosis and interpretation of petroleum-polluted groundwater. J Hazard Mater 472:134513

Wu M, Chen L, Tian Y, Ding Y, Dick WA (2013) Degradation of polycyclic aromatic hydrocarbons by microbial consortia enriched from three soils using two different culture media. Environ Pollut 178:152–158

Wu P, Zhang Z, Luo Y, Bai Y, Fan J (2022) Bioremediation of phenolic pollutants by algae - current status and challenges. Bioresour Technol 350:126930. Available from: https://doi.org/10.1016/j.biortech.2022.126930

Xie Z, Chen Y, Xu S (2020) Pattern recognition in microbial degradation datasets using machine learning. J Comput Biol 27(5):745–758

Xue Y, Du P, Shendi AAI, Yu B (2022) Mercury bioremediation in aquatic environment by genetically modified bacteria with self-controlled biosecurity circuit. J Clean Prod 337:130524. https://doi.org/10.1016/j.jclepro.2022.130524

Yergeau E, Sanschagrin S, Beaumier D, Greer CW (2012) Metagenomic analysis of the bioremediation of diesel-contaminated Canadian high arctic soils. PLoS One 7:e30058. https://doi.org/10.1371/journal.pone.0030058

Yuan Y, Wu H, Liu Q (2020) Computer vision for wildlife monitoring. Eco Inform 58:101071

Zhao J, Wang L, Zhang H (2020) Predictive maintenance in bioreactor systems using machine learning. J Process Control 92:121–133

Zheng Y, Li Y, Long H, Zhao X, Jia K, Li J et al (2018) bifA regulates biofilm development of *Pseudomonas putida* MnB1 as a primary response to H_2O_2 and Mn^{2+}. Front Microbiol 9. https://doi.org/10.3389/fmicb.2018.01490

Zhong S, Zhang K, Bagheri M, Burken JG, Gu A, Li B, Zhang H (2021) Machine learning: new ideas and tools in environmental science and engineering. Environ Sci Technol 55(19):12741–12754

Zhu X, Liu X, Liu Y (2017) Land cover classification with deep learning. Remote Sens Environ 202:266–278

Ethical Implications and Regulatory Frameworks for Microbial Products and Processes: Microbial Contributions to a Circular Economy

Deepak Kumar, Ritik Kumar, Sakshi, and Anil Kumar

Abstract

Microbial products have become integral to various industries, playing a crucial role in advancing sustainability and innovation. This chapter offers a comprehensive review of several microbiological products, including their applications within the circular economy, as well as ethical considerations involved. Microbial products derived from organic waste, such as bioethanol and biogas, support resource recovery and waste reduction while providing sustainable energy sources. The microbial synthesis of bioethanol, biodiesel, and biohydrogen is a component of biorenewable energy. Microorganisms are essential to the pharmaceutical industry, producing hormones, antibiotics, vaccines, and therapeutic enzymes, that advances healthcare and medicine. Microbial products are used in agriculture as biofertilizers and plant growth promoters, thereby improving crop yield and soil fertility in an eco-friendly manner. Probiotics and other microbial products like biopolymers have several advantages in both health and materials science. The ethical implications of microbial products and processes raise concerns about biosafety, biosecurity, intellectual property, and access equity. Ensuring that microbial technologies are safe, secure, and fairly distributed is essential for their advancement and responsible implementation. To reduce risks and uphold public confidence, microbial product creation is regulated by stringent safety and quality standards, environmental monitoring, and ethical reviews. These microbial products play a vital role within the circular economy, reducing environmental impact by facilitating the conversion of waste into useful

D. Kumar (✉) · R. Kumar · Sakshi
Department of Biotechnology, Chandigarh University Gharuan, Mohali, India
e-mail: deepak.e12689@cumail.in

A. Kumar
School of Chemical and Metallurgical Engineering, University of the Witwatersrand, Johannesburg, South Africa

S. Kaur et al. (eds.), *Microbial Metabolomics*, https://doi.org/10.1007/978-981-96-4824-5_17

resources, providing sustainable fossil fuel substitutes, and promoting pharmaceutical innovation and accessibility. Additionally, they improve resource efficiency and environmental sustainability in other domains, such as bioremediation and material manufacturing. Overall, microbial products are at the forefront of tackling global issues, providing long-term solutions, and advancing numerous industries while also raising serious ethical and legal concerns that require careful consideration.

Keywords

Microbial products · Biowaste · Enzymes · Biofertilizer

1 Introduction

Microbial products have revolutionized a diverse array of industries by harnessing the capabilities of microorganisms to synthesize valuable compounds. These products span multiple domains, including agriculture, energy, pharmaceuticals, and environmental management, reflecting the profound impact of microbial technology on modern life (Meena and Mohanty 2020). The field of industrial microbiology leverages the metabolic activities of microbes to produce economically significant substances through processes such as fermentation, which underpins the large-scale generation of these products (Abdel-Aziz et al. 2017).

The significance of microbial products is underscored by their broad applications. From biofuels and biowaste-derived products to pharmaceuticals and agricultural biofertilizers, microorganisms contribute to sustainable solutions and innovative technologies (Chander et al. 2023). Microbes possess the remarkable ability to convert organic waste into valuable resources and to produce compounds that are difficult or costly to synthesize chemically, highlighting their indispensable role in addressing global challenges (Zhou et al. 2022).

In agriculture, microbial products like biofertilizers and biopesticides enhance soil health and boost crop productivity while reducing environmental impact compared to traditional chemical inputs (De Palma et al. 2022; Kumar et al. 2024). Similarly, in energy sector, microbes offer sustainable alternatives to fossil fuels by enabling the production of bioethanol, biodiesel, and other bioenergy forms (Shahbazi and Rahimpour 2023). Pharmaceuticals industry gets benefits significantly from microbial products, with antibiotics, antivirals, and other therapeutic agents being derived from microbial sources (Kapoor et al. 2020). Furthermore, microbial processes are leveraged biowaste management and environmental remediation to tackle pollution and waste challenges (Mekala et al. 2023).

Despite the benefits, the use of microbial products and processes raises ethical considerations that must be carefully managed. The principles of beneficence, nonmaleficence, respect for individuals, and justice serve as ethical framework for evaluating of microbial technologies (Khan et al. 2015). Ensuring that these innovations contribute positively to society while minimizing potential risks and inequalities is crucial for the responsible advancement of this field (Nora 2013). By exploring the multifaceted impact of microbial products, we gain insight into how these tiny organisms are shaping the future of technology and sustainability.

2 Microbial Products

Microorganisms are very vital in synthesizing a variety of products useful to humans (Paul et al. 2019). Microorganisms like fungi, bacteria, and viruses have become integral to biotechnology, impacting industries such as agriculture, food production, and pharmaceuticals. In agriculture, microbes are harnessed for pest control and soil health, while in biomanufacturing, they act as biological factories producing amino acids, enzymes, medicines, and food additives (Shin et al. 2010). Their role in traditional fermentation processes and large-scale medical applications highlights the growing significance of microbial products in the food and pharmaceutical sectors. Industrial microbiology refers to the large-scale utilization of microbes to synthesize commercially valuable compounds known as microbial products for a wide range of applications (Abdel-Aziz et al. 2017). These microbial products include beverages, antibiotics, food additives, human and animal health treatment products, and biofuels. Furthermore, microbes have the power to make chemical substances that have not been produced in a laboratory or can only be made with extreme difficulty. Additionally, the activity of microbes on inexpensive carbohydrate sources can produce some chemical products at a lower cost than other chemical molecules (Paul et al. 2019). Significant investments have been made in this field, and expectations are high due to the enormous potential applications of microbial products and the emergence of useful technologies (Sessitsch et al. 2018). The number of start-ups creating and selling microbiological products has greatly expanded in the past few years (Batista and Singh 2021) (Fig. 1).

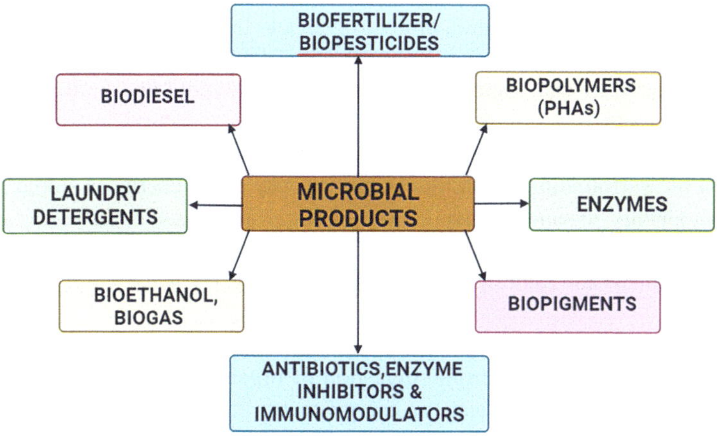

Fig. 1 List of some microbial product

3 Microbial Products Can Broadly be Categorized as Follows

3.1 Microbial Products Derived from Biowaste

Global population growth and industrialization are driving an increase in demand for energy and chemicals (Kang et al. 2023). Biowaste accounts for a sizable amount of solid waste and often contains organic waste such as domestic, agricultural, municipal, food, paper, and other items (Visco et al. 2022). Although organic substances, such as proteins, polysaccharides, lipids, fiber, or water, make up the majority of these biowastes, their composition varies greatly (Jiménez-Rosado et al. 2023). The organic component of these wastes can be easily metabolized by microorganisms to yield value-added substances. Microbial fermentation techniques can help create waste biorefinery techniques (Zhou et al. 2022). Microbes with genetic machinery capable of producing cellulases and β-glucosidase may efficiently breakdown complex organic materials. Microbial technologies are addressing future demands in organic waste management, production of clean energy/bioethanol, biofertilizer, biopigments, and some enzymes (Chander et al. 2023). Biofertilizers are environmentally friendly products made up of microbes that boost soil health (Kumar et al. 2024). They can be made from food waste through anaerobic digestion, aerobic decomposition, and chemical hydrolysis. Agricultural wastes, such as wheat straw, can also be directly delivered to soil (Du et al. 2018). Used coffee grounds could potentially be valorized and reused by appropriate management (e.g., composting, digestion, pyrolysis), providing useful products to agriculture and boosting the soil's composition and fertility (Stylianou et al. 2018).

3.2 Bioenergy

Microalgae have caught researchers' interest as an alternate substrate for bioenergy production, including bioethanol, biohydrogen, and biodiesel (Awasthi et al. 2019). Furthermore, switching to a circular economy (CE) is critical for creating a low-carbon, sustainable, and resource-efficient culture (Leipold and Petit-Boix 2018). The usage of bioenergy is based on photosynthesis, a natural process in which plants receive sunlight and convert it into chemical energy by retaining carbon dioxide from the surrounding environment. In this way, solar energy is stored by biomass. Bioenergy systems convert stored energy into usable forms for several applications, providing a renewable and environmentally friendly substitute to fossil fuels (Shahbazi and Rahimpour 2023). There are several major types of bioenergy:

Biogas: It is produced by anaerobically digesting organic material such as livestock manure and sewage. Anaerobic digestion of organic materials, such as animal dung, wastewater, food waste, or energy crops, produces biogas. Methane and carbon dioxide are emitted as organic components decompose during this process. Biogas may be used to produce power, heat buildings, and fuel automobiles (Kougias and Angelidaki 2018; Mishra et al. 2021).

Bioethanol: It is produced by fermenting and distilling high-carbohydrate biomass sources such as sugarcane, corn, or wheat. It can be used in vehicles as an independent fuel or in combination with gasoline. Bioethanol production is widely established, and it is frequently used as a sustainable fuel additive around the world (Melendez et al. 2022; Moodley et al. 2020). Biofuel production has steadily increased in comparison to fossil fuels as greenhouse gas emissions have been significantly decreased. Bioethanol is currently the most widely used biofuel in the world, and its production has increased significantly. Furthermore, a review of commercial bioethanol production reveals that the United States and Brazil have major businesses that are heavily driven by their respective national energy policies (Moreira et al. 2023).

Biodiesel: The method of transesterification uses fatty acids from vegetable or animal fats to create sustainable diesel fuel. Without requiring engine changes, biodiesel may be used directly to replace petroleum-based diesel in compression–ignition engines (Balat and Balat 2010; Wang et al. 2021a).

3.3 Pharmaceutical

Microbial pharmaceuticals include antibiotics, anticancer medications, immunomodulators, and enzyme inhibitors. Coccidiostatic agents, nematicides, and insecticides are some other bacterial-derived bioactive compounds. Furthermore, *Escherichia coli*, the pioneer of molecular biology, is one of the most essential hosts for the creation of therapeutic recombinant proteins (Lancini and Demain 2013). Carefully considered research can be done on the pharmacological components of microbial diversity in order to promote and safeguard human health norms. Numerous investigations have uncovered amazing breakthroughs that aim to treat pathogenic diseases and various metabolic abnormalities, consequently extending and improving human health and supporting human society showed in Table 1. Pharmaceutical microbiology aims to provide knowledge and understanding of the presence of bacteria, yeasts, molds, viruses, and toxins in various pharmaceutical raw materials, products, intermediates, and environments supporting therapeutic construction. It also considers the role of microbes in regulating medicinal harvests, manufacturing settings, and human health.

3.4 Agriculture

One of the biggest problems approaching humanity in agriculture will be sustainable intensification that reduces the ecological imprint. Environment friendly crop management is currently receiving more attention due to the increased demand to minimize the usage of chemicals (De Palma et al. 2022). Researchers, businesses, and regulators have all shown a great deal of interest in using microbial tools to

Table 1 List of important pharmaceutical products

Microbial products	Source	Role
Diastaphenazines	*Streptomyces diastaticus*	It has antibacterial and antifungal activity (Yu et al. 2011).
Diketopiperazines	*Streptomyces* sp.	It has anti H1N1 activity (Wang et al. 2013).
β-carboline	*Microbispora* sp. LGMB259	It has anticancer activity (Savi et al. 2015).
Arisugacin K	*P. echinulatum*	It can treat infections that are caused by *E. coli* (Wang et al. 2021b).
7-isoprenylindole-3-carboxylic acid	*Streptomyces* sp. neau-D50	It has antifungal and cytotoxic activities (Zhang et al. 2014).
Helveticin J	*Lactobacillus helveticus*	It alters the membrane potential, resulting in ATP outflow (Negash and Tsehai 2020).
Fengycin	*Bacillus subtilis*	It serves as immunomodulators; toxins; enzyme inhibitors; antibacterial, antiviral, and anticancer agents; and bioagents with hypocholesterolemic properties (Sur et al. 2018).
Pyrrolopyrazines	*Streptomyces* sp. VITMK1	Antimicrobial properties (Manimaran et al. 2017)

boost agricultural productivity in a sustainable way (Batista and Singh 2021). Biofertilizers are environmentally friendly products made of living organisms that strengthen plant nutrition intake, protect plants from phytopathogens, and improve soil health (Kumar et al. 2024). microorganisms that are helpful for plants to partially replace chemicals and minimize the negative environmental effects of agriculture. Single strains or a collection of well-characterized, biosafe live microorganisms are included in prepared microbial products, also known as inoculants, for use in agriculture. These microbes can support a plant host's growth, health, and development (Sanjuán et al. 2023).

3.5 Other Products

Food safety, manufacturing, processing, the preservation of food, and storage all depend on microbiology. Microorganisms like bacteria, molds, and yeasts are used to produce food and ingredients, including dairy products, wine, beer, and baked goods (Lorenzo et al. 2018). Since ancient times, microorganisms have been employed in food fermentation, and many food items are still prepared using these techniques (Soccol et al. 2006). Since microbial enzymes are more stable than those found in plants and animals, they are used extensively in the food industry. Their great consistency makes process modification and optimization simple, and their production through fermentation processes can be done in an economical manner by reducing the time and space required (Gurung et al. 2013).

Numerous industrial sectors employ these enzymes extensively; for example, amylolytic enzymes are used in the food, laundry detergent, paper, and textile industries (Pandey et al. 2000). They are employed in the manufacturing of maltose syrups, high fructose corn syrups, crystalline glucose, and glucose syrups. They are also added to detergents as an aid to get rid of stains composed of starch. They are employed in the paper industry to lower the viscosity of starch so that paper can be coated accurately. Amylases are employed for warp sizing of textile fibers in the textile industry (Souza and Magalhães 2010). Enzymes such as lipases, xylanases, and proteases are also widely used in the food industry (Raveendran et al. 2018).

The current state of the world has a significant impact on the development of biopolymers originating from microorganisms because of their numerous advantages and universal demand, which aid in the replacement of their artificial and dangerous substitutes. The amount of hazardous and toxic pollutants in the air continues to increase due to growing urbanization and industrialization. Extracellular polymeric substances (EPS) and polyhydroxyalkanoates (PHAs), which are produced by a wide variety of microbial taxa, are examples of microbial biopolymers that could be used as alternatives to help waste bioremediation work more effectively (Gupta et al. 2020). An extensive range of extracellular and intracellular biopolymers, including polyamides, polysaccharides, polyphosphates, polyesters, proteinaceous compounds, and extracellular DNA, may be efficiently produced by bacteria, which are regarded as the primary cell manufacturers (Ghosh et al. 2021).

4 Ethical Implications of Microbial Products and Processes

The use of microbiological goods or procedures has a number of ethical considerations. The four major tenets of ethics are beneficence, non-maleficence, respect for individuals, and fairness (Gupta et al. 2017). The concept of non-maleficence highlights the significance of averting short- and long-term harm, such as contamination of the environment, genetic alterations in microbial populations, or unanticipated effects on the environment and human health. Microbial products, such as biofertilizers and biopesticides, must be beneficent if they provide definite benefits, such as enhanced sustainability, health, and environmental protection, while avoiding damage. But it's important to carefully consider any potential hazards, such unanticipated health effects and environmental disruptions (Khan et al., 2015).

When threats to ecosystems, public health, or vulnerable groups are not completely recognized or handled or when the benefits are not clear, ethical problems arise. Microbial products can support sustainable agriculture, but they also run the risk of upsetting regional ecosystems and destroying biodiversity, which goes against the beneficence principle. To reduce these dangers, long-term research and regulation are crucial. Many religious traditions place a great value on the harmony of creation and the natural order. One may argue that changing or introducing microorganisms especially genetically modified ones interfere with this balance, raising ethical and theological concerns. The concept of non-maleficence must not be violated by microbial products, such as probiotics, biofertilizers, and microbial medications, if they result in poisoning, allergic responses, infections, or other negative health impacts in humans or animals. It is also necessary to control the discharge of these products, such biofertilizers and bioremediation bacteria, in order to prevent damaging biodiversity, uprooting local species, or upsetting ecosystems (Thakur et al. 2023) (Varkey 2021). In order to be just, the advantages of microbial products like probiotics, biofertilizers, and microbial medications must be shared equitably by all groups, especially underprivileged, rural, and low-income groups. Unfair disparities result from limiting the use of these technology to large-scale farms or wealthy countries. It is important to work on making microbial goods available and reasonably priced for all (Nora 2013). Justice also includes the needs of future generations, stressing the need to prevent long-term environmental degradation, resource depletion, or irreversible ecological harm such as overusing biofertilizers or introducing genetically altered bacteria with unknown effects that might unjustly burden them.

5 Regulation of Microbial Product Formulations

Microbial systems are being used in a wide range of industries and are expected to transform a number of industries, including environmental remediation, paper, textile, food and agriculture, and healthcare. The controversy over the social, ethical, and legal ramifications of genetically modified organisms is growing in tandem with the rapid improvements in technology for their development. These innovative technologies must be protected by intellectual property rights (IPR) regimes in order for them to be commercialized and form the foundation of prosperous corporate endeavors. The innovation must not only be protected legally but also adhere to the regulatory framework while taking societal concerns into consideration (Meena and Mohanty 2020).

Regulations are essential for ensuring the safety and quality of microbiological products. These frameworks address several crucial topics, including risk assessment, to find possible dangers like environmental impact, pathogenicity, and allergenicity. In order to guarantee consistency in product formulation, claims, and labeling, they also use standardization and labeling standards. To verify the product efficacy, guidelines also require thorough safety and efficacy testing through clinical or field trials. In order to identify and mitigate any risks that may occur after market entry, ongoing post-market surveillance is equally important (Lahlali et al. 2022).

Using criteria, one can determine if a product is compliant or defective by evaluating the analytical data. A microbiological criterion is a set of standards that establishes the acceptability of a food product, food batch, or process. These standards are based on the quantity of toxins or metabolites produced by microorganisms per unit(s) of mass, volume, area, or batch, as well as the presence, absence, or number of microorganisms (Hildebrandt 2014).

6 Role of Microbial Products in Circular Economy

Take, make, and dispose is a common way to sum up linear economy. In other words, use the resources you require, create the things that will be sold and generate profit, and then get rid of anything you don't need, including products that are approaching the end of their useful lives. When a product reaches the end of its life cycle, it should be retained in the economy as a resource that may be utilized repeatedly to generate additional value (Sariatli 2017). Politically and industrially, the idea of the circular economy (CE) has gained popularity as a potentially revolutionary new production structure that blends environmental sustainability with economic growth (Kirchherr et al. 2017; Pansera et al. 2024). The goal of the CE model is to establish a closed-loop system where materials are recycled, mended, and reused instead of being discarded after their initial use (Genovese et al. 2017). According to the UN, the circular bioeconomy which emphasizes sustainable and renewable energy sources has emerged as a critical worldwide concern. The goal of generating value by extracting energy from organic materials previously considered waste has united nations worldwide.

A huge amount of biowastes has been produced as a result of the unprecedented growth in the human population. These biowastes are made up of essential elements like potassium, nitrogen, and phosphorous, which can be very helpful in protecting the farming industry from the negative impacts of synthetic fertilizers (Upadhyay et al. 2024). In order to promote a circular economy, reusing biowaste materials derived from sewage sludge, animal waste, and agricultural waste can help recover nutrients and create valuable goods (Kurniawati et al. 2023). Biological systems and microbes provide sustainable bioconversion processes to transform biowaste into useful bioproducts. Utilizing bio-based fertilizers made from biowaste can boost agricultural output while reducing the harm that synthetic fertilizers do to the environment (Upadhyay et al. 2024).

6.1 Role of Microbial Products in Medicine/Pharmaceuticals

One of most important microbial products is antibiotics which provide us protection from various diseases. *Actinomycetes* bacteria are a main source of microbial-based bioactive products (Abdel-Razek et al. 2020). A wide range of antibiotics, including anthracyclines, angucyclinones, phenazines, piercidins, macrolides, anthraquinones, polyenes, tetracyclines, beta-lactams, etc., are produced by *actinomycetes*. Tetracycline antibiotics are widely recognized for their broad-spectrum effectiveness, targeting a diverse array of Gram-positive and Gram-negative bacteria, spirochetes, intracellular bacteria, and even protozoan parasites, and it is produced by *actinomycetes* and *Streptomyces aureofaciens* (Grossman 2016). *Streptomycetes* are a prolific source of bioactive compounds, producing notable macrolide polyene antifungals like natamycin from *Streptomyces natalensis* and nystatin from *Streptomyces noursei* as well as key antibacterial agents such as kanamycin from *Streptomyces kanamyceticus* streptomycin from *Streptomyces griseus*

(Abdel-Razek et al. 2020). Streptomycin, the first discovered aminoglycoside, is used to treat gram-negative bacterial infections like brucellosis, plague, and tuberculosis (Quinn et al. 2020). Erythromycin, a wide-spectrum macrolide antibiotic originally from *Saccharopolyspora erythraea*, is effective against both gram-positive and gram-negative pathogens, including *Corynebacterium diphtheriae*, *Staphylococcus aureus*, and *Mycoplasma pneumoniae*. Metabolic disorders like nonalcoholic fatty liver disease, obesity, diabetes, insulin resistance, and cardiovascular diseases are increasingly prevalent in developing countries due to sedentary lifestyles and poor diets (Jangra and Pothuraju 2020). Trillions of microorganisms in the human colon, known as the gut microbiota, play crucial roles in nutrient metabolism, vitamin production, immune system modulation, and pathogen resistance (Jandhyala et al. 2015). Research groups globally have documented the beneficial effects of probiotics like *lactobacilli* and *bifidobacteria* on obesity and gut microbiota in both rodents and humans, noting that different probiotics combat obesity through various mechanisms, such as *Lactobacillus rhamnosus* PL60 and *Lactobacillus plantarum* PL62, which produce conjugated linoleic acid (Thakur et al. 2024) (Lee et al. 2007). Fungal metabolites have been used to create a variety of medicinal products, such as griseofulvin, mycophenolic acid, fusidic acid, and cyclosporine. Significantly, cyclosporine has produced the clinical agent Debio 025, and statins such as lovastatin and mevastatin—which are derived from *Aspergillus terreus* and *Penicillium citrinum*, respectively—are essential antilipidemic medications for the treatment of cardiovascular disease (Flisiak et al. 2008). A vaccine is a biological preparation derived from disease-causing microorganisms or substances resembling them, designed to confer immunity against the disease caused by that organism. These preparations are typically made from killed or weakened forms of microbes or from surface proteins or toxins of the pathogen. Vaccination is generally safe, with minimal adverse reactions. Through routine immunization programs, health departments have successfully protected millions of children from numerous opportunistic infections and diseases that once caused significant mortality. In these vaccines, microorganisms are killed using chemicals, antibiotics, radiation, or heat. Vaccines produced in this way are used to protect against diseases like rabies, polio, hepatitis A, and influenza. Some vaccines, however, contain live microorganisms that have been rendered non-virulent. This is achieved by culturing them under specific conditions that disable their ability to cause disease. Most attenuated vaccines are derived from viruses and bacteria and are commonly used to prevent diseases such as typhoid, mumps, measles, yellow fever, and rubella (Lartigue et al. 2019).

6.2 Role of Microbial Products in Agriculture

Biofertilizers are the important microbial product which provides many benefits to agriculture as well as environment because they are not toxic to environment as compared to chemical fertilizers. Biofertilizers, also known as microbial inoculants, are organic products made up of specific microorganisms that originate from plant roots and the rhizosphere, which facilitate plant growth by enhancing nutrient availability to the soil, seeds, or plants (Daniel et al. 2022). Microbial biofertilizers, serving as a sustainable biotechnological alternative to chemical fertilizers, are vital for enhancing soil fertility, improving plant–water relations, and protecting against drought, diseases caused by mycotoxin-producing fungi, and insect pests while also boosting crop yields, reducing production costs, minimizing environmental impact, and supporting plant growth and development through nutrient solubilization, enhanced water uptake via root hair formation, and the generation of phytohormones, all without harmful side effects (Kumar et al. 2024). The conversion of phosphorus from insoluble form to soluble form is carried out by bacteria like *Agrobacterium* and *Flavobacterium* and fungi such as *Trichoderma* spp. and *Penicillium* spp. (Nosheen et al. 2021). *Rhizobium*, *Erwinia*, *Arthrobacter Agrobacterium*, *Bacillus*, and *Pseudomonas* spp. are rhizobacteria that promote the growth of plants by producing hormones which increase growth of root growth and supply nutrients (Nosheen et al. 2021). Another major microbial product is biopesticide, or microbial pesticide. Chemical pesticides are used to boost agricultural output, but they have several downsides, including the fact that they are poisonous in nature, lack specificity in action, and are expensive (Anani et al. 2020). Microbial pesticides are particular in action, can be easily obtained without the use of costly chemicals, and are environmentally sustainable with no residual effects (Ayilara et al. 2023). The major categories of microbes used as biopesticides include fungal genera *Metarhizium*, *Paecilomyces*, *Verticillium*, *Beauveria*, *Hirsutella*, and *Lecanicillium*; bacteria genera *Chromobacterium*, *Pseudomonas*, and *Yersinia*; and nematodes belonging to the genera *Heterorhabditis* and *Steinernema* (Adeleke et al. 2022).

6.3 Role of Microbial Products in Energy

Various microbial community play great role in production of energy like bioethanol, biodiesel, and microbial fuel cell. Microbial biotechnology serves as a vital strategy for eco-friendly bioprocesses, utilizing microorganisms and their enzymes to convert carbohydrates, lignins, and glycerols into various renewable resources, such as bioenergy production (Ramamurthy et al. 2021). Being essential providers of enzymes, microorganisms are diverse entities that aid in the consumption and bioconversion of biomass (Kilbane 2016). Microorganism-based biofuel production has the potential to contribute significantly to energy provision, and countries lacking fossil fuel resources will strongly support the bioprocessing of renewable energy, and using *Trichoderma reesei* to produce bioethanol from lignocellulosic biomass is a very efficient method that does not require pretreatment with NaClO or acid hydrolysis (Fatma et al. 2021). To increase yields, a microbial community may be formed by mixing *Trichoderma reesei* for enzyme synthesis with *Saccharomyces cerevisiae* and *Scheffersomyces stipitis* to effectively use both hexose and pentose sugars (Rastogi and Shrivastava 2017). In addition, the inclusion of *Penicillium echinulatum* and *Anoxybacillus flavithermus* promotes cellulase and xylanase activity (Todero Ritter et al. 2013), while *Aspergillus* spp. aids in optimizing carbohydrate biosynthesis via exo- and endo-inulinases, making biofuel production from renewable resources even more effective (Yeoman et al. 2010). Microbial fuel cells (MFCs) represent an advanced technology with significant potential for sustainable energy generation. These systems harness microorganisms to transform organic substances into electrical power (Sonawane et al. 2024). By combining abiotic electrochemical mechanisms with biocatalytic redox reactions, MFCs emerge as a compelling and fast-growing technology. The presence of microorganisms enhances the catalytic efficiency of electrochemical processes, leading to redox reactions that produce energy. Within the anode, electro-active bacteria actively interact with the MFCs to facilitate this energy generation (Logan and Elimelech 2012). Microbial fuel cells (MFCs) effectively treat a wide range of contaminants from various wastewater sources, such as municipal, textile, tannery, distillery, and petroleum industries, while also advancing wastewater treatment and bioenergy production (Moinee and Sanzida 2021). Through a process like transesterification, biodiesel, alternative fuel, is made from oils and fats. Crops such as rapeseed and soybeans help to reduce greenhouse gases, while microbial oils and genetically modified bacteria have the potential to improve production efficiency. For biodiesel production, microalgae are thought to be extremely advantageous due to their high fatty acid content.

6.4 Others

Microorganism or their products play a major role in degrading pollutants or treatment of waste. Many pollutants such as naphthalene, fluorine, and phenols and its derivative, 2,4-dichlorophenoxyacetic acid, are degraded by some endophytic microorganisms like *Pseudomonas putid*a PD1, *Alcaligenes* sp., and *Pseudomonas putida* strain VM144 (Sakshi et al. 2023). Various microorganism contains carbon reserves such as polyhydroxyalkanoate, which is a biopolymer and best alternative to synthetic plastic. This will reduce the harmful effects that are caused by synthetic plastics, and also, polyhydroxyalkanoate doesn't cause any pollution and is degraded easily (Zhou et al. 2022).

Fruit pulp, pea pod powder, whey, molasses, maize steep liquor, bran, and a variety of other waste products from agro-industries have been considered as possible supplies of carbon, nitrogen, and minerals for the microbial pigment production process (Panesar et al. 2015). Industries involved in the processing of fish produce a lot of by-products. One growing issue for the environment and human health is the disposal of these wastes. Heads, chitinous material, viscera, wastewater, and other fish wastes were prepared and tested as growth substrates for the synthesis of microbial enzymes, including lipase, protease, and chitinolytic and ligninolytic enzymes (Ben Rebah and Miled 2013).

Researchers at *ACS Central Science* have discovered a plastic-eating *Escherichia coli* that can effectively convert polyethylene terephthalate (PET) plastic waste into adipic acid, which is used to make nylon materials, medications, and fragrances. This is the first upcycling of polyethylene terephthalate plastic waste into the widely used petrochemical and nylon precursor adipic acid (Valenzuela-Ortega et al. 2023). With their abundant contents, cheese whey, leftover milk, and other by-products of the dairy industry have been thought of as low-cost resources and assessed as a substitute substrate for microbial production processes (such as biomass, enzymes, organic acids, bioenergy, and biopolymers) (Sar et al. 2022).

7 Conclusion

In conclusion, microbial products are leading the way in promoting sustainability in a variety of industries, such as energy, pharmaceuticals, agriculture, and environmental management. These microbes are essential for turning biowaste into useful resources, creating biofuels that are sustainable like bioethanol and biodiesel, increasing agricultural output by using biofertilizers and biopesticides, and developing groundbreaking medical breakthroughs. Their capacity to transform organic waste into valuable by-products promotes the development of renewable energy sources and reduces environmental effect, which makes a substantial contribution to the circular economy. However, the use of microbial technology brings with it a number of legal and ethical issues that need to be properly handled. Ensuring the safety of microbial products is paramount, as their widespread use can pose risks of environmental contamination and unintended health impacts. Furthermore, navigating intellectual property concerns is necessary to safeguard breakthroughs and advance equal access to these technology. Particularly in underdeveloped regions, equal access to microbial technologies is essential to prevent inequalities and guaranteeing that the benefits are shared fairly. As microbial technologies advance, it is crucial to maintain a balance between leveraging their potential to drive sustainability and addressing these ethical and regulatory concerns. This involves rigorous safety assessments, robust regulatory frameworks, and ongoing dialogue with stakeholders to uphold ethical standards. By doing so, we can ensure that microbial innovations continue to contribute positively to global sustainability goals while fostering a fair and just distribution of their benefits. The continued evolution of microbial technologies offers significant promise for a more sustainable and equitable future, provided that their development is managed with care and responsibility.

References

Abdel-Aziz SM, Abo Elsoud MM, Anise AAH (2017) Microbial biosynthesis: a repertory of vital natural products. In: Food biosynthesis. Elsevier, pp 25–54. https://doi.org/10.1016/B978-0-12-811372-1.00003-8

Abdel-Razek AS, El-Naggar ME, Allam A, Morsy OM, Othman SI (2020) Microbial natural products in drug discovery. Processes 8(4):470. https://doi.org/10.3390/pr8040470

Adeleke BS, Ayangbenro AS, Babalola OO (2022) In vitro screening of sunflower associated endophytic bacteria with plant growth-promoting traits. Front Sustain Food Syst 6. https://doi.org/10.3389/fsufs.2022.903114

Anani OA, Mishra RR, Mishra P, Enuneku AA, Anani GA, Adetunji CO (2020) Effects of toxicant from pesticides on food security: current developments. In: Innovations in food technology. Springer, Singapore, pp 313–321. https://doi.org/10.1007/978-981-15-6121-4_22

Awasthi MK, Sarsaiya S, Wainaina S, Rajendran K, Kumar S, Quan W, Duan Y, Awasthi SK, Chen H, Pandey A, Zhang Z, Jain A, Taherzadeh MJ (2019) A critical review of organic manure biorefinery models toward sustainable circular bioeconomy: technological challenges, advancements, innovations, and future perspectives. Renew Sust Energ Rev 111:115–131. https://doi.org/10.1016/j.rser.2019.05.017

Ayilara MS, Adeleke BS, Akinola SA, Fayose CA, Adeyemi UT, Gbadegesin LA, Omole RK, Johnson RM, Uthman QO, Babalola OO (2023) Biopesticides as a promising alternative to synthetic pesticides: a case for microbial pesticides, phytopesticides, and nanobiopesticides. Front Microbiol 14. https://doi.org/10.3389/fmicb.2023.1040901

Balat M, Balat H (2010) Progress in biodiesel processing. Appl Energy 87(6):1815–1835. https://doi.org/10.1016/j.apenergy.2010.01.012

Batista BD, Singh BK (2021) Realities and hopes in the application of microbial tools in agriculture. Microb Biotechnol 14(4):1258–1268. https://doi.org/10.1111/1751-7915.13866

Ben Rebah F, Miled N (2013) Fish processing wastes for microbial enzyme production: a review. 3 Biotech 3:255–265. https://doi.org/10.1007/s13205-012-0099-8

Chander AM, Singh NK, Venkateswaran K (2023) Microbial technologies in waste management, energy generation and climate change: implications on earth and space. J Indian Inst Sci 103(3):833–838. https://doi.org/10.1007/s41745-023-00388-3

Daniel AI, Fadaka AO, Gokul A, Bakare OO, Aina O, Fisher S, Burt AF, Mavumengwana V, Keyster M, Klein A (2022) Biofertilizer: the future of food security and food safety. Microorganisms 10(6):1220. https://doi.org/10.3390/microorganisms10061220

De Palma M, Scotti R, D'Agostino N, Zaccardelli M, Tucci M (2022) Phyto-friendly soil bacteria and fungi provide beneficial outcomes in the host plant by differently modulating its responses through (In)direct mechanisms. Plan Theory 11(20):2672. https://doi.org/10.3390/plants11202672

Du C, Abdullah JJ, Greetham D, Fu D, Yu M, Ren L, Li S, Lu D (2018) Valorization of food waste into biofertiliser and its field application. J Clean Prod 187:273–284. https://doi.org/10.1016/j.jclepro.2018.03.211

Fatma S, Saleem A, Tabassum R (2021) Wheat straw hydrolysis by using co-cultures of *Trichoderma reesei* and *Monascus purpureus* toward enhanced biodegradation of the lignocellulosic biomass in bioethanol biorefinery. Biomass Convers Biorefinery 11(3):743–754. https://doi.org/10.1007/s13399-020-00652-x

Flisiak R, Horban A, Gallay P, Bobardt M, Selvarajah S, Wiercinska-Drapalo A, Siwak E, Cielniak I, Higersberger J, Kierkus J, Aeschlimann C, Grosgurin P, Nicolas-Métral V, Dumont J-M, Porchet H, Crabbé R, Scalfaro P (2008) The cyclophilin inhibitor Debio-025 shows potent anti-hepatitis C effect in patients coinfected with hepatitis C and human immunodeficiency virus. Hepatology 47(3):817–826. https://doi.org/10.1002/hep.22131

Genovese A, Acquaye AA, Figueroa A, Koh SCL (2017) Sustainable supply chain management and the transition towards a circular economy: evidence and some applications. Omega 66:344–357. https://doi.org/10.1016/j.omega.2015.05.015

Ghosh S, Lahiri D, Nag M, Dey A, Sarkar T, Pathak SK, Atan Edinur H, Pati S, Ray RR (2021) Bacterial biopolymer: its role in pathogenesis to effective biomaterials. Polymers 13(8):1242. https://doi.org/10.3390/polym13081242

Grossman TH (2016) Tetracycline antibiotics and resistance. Cold Spring Harb Perspect Med 6(4):a025387. https://doi.org/10.1101/cshperspect.a025387

Gupta V, Sengupta M, Prakash J, Tripathy BC (2017) Biosafety and bioethics. In: Basic and applied aspects of biotechnology. Springer, Singapore, pp 503–520. https://doi.org/10.1007/978-981-10-0875-7_24

Gupta J, Rathour R, Medhi K, Tyagi B, Thakur IS (2020) Microbial-derived natural bioproducts for a sustainable environment: a bioprospective for waste to wealth. In: Refining biomass residues for sustainable energy and bioproducts. Elsevier, pp 51–85. https://doi.org/10.1016/B978-0-12-818996-2.00003-X

Gurung N, Ray S, Bose S, Rai V (2013) A broader view: microbial enzymes and their relevance in industries, medicine, and beyond. Biomed Res Int 2013:1–18. https://doi.org/10.1155/2013/329121

Hildebrandt G (2014) Sampling plans on microbiological criteria. In: Encyclopedia of food microbiology. Elsevier, pp 353–359. https://doi.org/10.1016/B978-0-12-384730-0.00302-5

Jandhyala SM, Talukdar R, Subramanyam C, Vuyyuru H, Sasikala M, Nageshwar Reddy D (2015) Role of the normal gut microbiota. World J Gastroenterol 21(29):8787–8803. https://doi.org/10.3748/wjg.v21.i29.8787

Jangra S, Pothuraju R (2020) Living medicines for health and disease management. In: Microbial diversity, interventions and scope. Springer, Singapore, pp 321–333. https://doi.org/10.1007/978-981-15-4099-8_18

Jiménez-Rosado M, Perez-Puyana V, Romero A (2023) The use of biowaste for the production of biodegradable superabsorbent materials. Curr Opin Food Sci 49:100975. https://doi.org/10.1016/j.cofs.2022.100975

Kang B-J, Jeon J-M, Bhatia SK, Kim D-H, Yang Y-H, Jung S, Yoon J-J (2023) Two-stage biohydrogen and polyhydroxyalkanoate production: upcycling of spent coffee grounds. Polymers 15(3):681. https://doi.org/10.3390/polym15030681

Kapoor D, Sharma P, Sharma MMM, Kumari A, Kumar R (2020) Microbes in Pharmaceutical Industry. In: Microbial diversity, interventions and scope. Springer, Singapore, pp 259–299. https://doi.org/10.1007/978-981-15-4099-8_16

Khan MS, Gorea RK, Qamar S, Mustafa G, Gorea A (2015) Some ethical perspectives in the discipline of microbiology. Int J Ethics Trauma Victimol 1(02):45–50. https://doi.org/10.18099/ijetv.v1i2.6817

Kilbane JJ (2016) Future applications of biotechnology to the energy industry. Front Microbiol 7. https://doi.org/10.3389/fmicb.2016.00086

Kirchherr J, Reike D, Hekkert M (2017) Conceptualizing the circular economy: an analysis of 114 definitions. Resour Conserv Recycl 127:221–232. https://doi.org/10.1016/j.resconrec.2017.09.005

Kougias PG, Angelidaki I (2018) Biogas and its opportunities—a review. Front Environ Sci Eng 12(3):14. https://doi.org/10.1007/s11783-018-1037-8

Kumar D, Kumar R, Sakshi D, Kumar A (2024) Commercialization and market perspectives of biofertilizers through advanced approaches. In: Metabolomics, proteomics and gene editing approaches in biofertilizer industry. Springer Nature, Singapore, pp 339–357. https://doi.org/10.1007/978-981-97-2910-4_17

Kurniawati A, Stankovics P, Hilmi YS, Toth G, Smol M, Toth Z (2023) Understanding the future of bio-based fertilisers: the EU's policy and implementation. Sustain Chem Clim Action 3:100033. https://doi.org/10.1016/j.scca.2023.100033

Lahlali R, Ezrari S, Radouane N, Kenfaoui J, Esmaeel Q, El Hamss H, Belabess Z, Barka EA (2022) Biological control of plant pathogens: a global perspective. Microorganisms 10(3):596. https://doi.org/10.3390/microorganisms10030596

Lancini G, Demain AL (2013) Bacterial pharmaceutical products. In: The prokaryotes. Springer, Berlin/Heidelberg, pp 257–280. https://doi.org/10.1007/978-3-642-31331-8_28

Lartigue C, Valverde Timana Y, Labroussaa F, Schieck E, Liljander A, Sacchini F, Posthaus H, Batailler B, Sirand-Pugnet P, Vashee S, Jores J, Blanchard A (2019) Attenuation of a pathogenic *Mycoplasma* strain by modification of the *obg* gene by using synthetic biology approaches. MSphere 4(3). https://doi.org/10.1128/mSphere.00030-19

Lee K, Paek K, Lee HY, Park JH, Lee Y (2007) Antiobesity effect of *trans* −10, *cis* −12-conjugated linoleic acid-producing *Lactobacillus plantarum* PL62 on diet-induced obese mice. J Appl Microbiol 103(4):1140–1146. https://doi.org/10.1111/j.1365-2672.2007.03336.x

Leipold S, Petit-Boix A (2018) The circular economy and the bio-based sector – perspectives of European and German stakeholders. J Clean Prod 201:1125–1137. https://doi.org/10.1016/j.jclepro.2018.08.019

Logan BE, Elimelech M (2012) Membrane-based processes for sustainable power generation using water. Nature 488(7411):313–319. https://doi.org/10.1038/nature11477

Lorenzo JM, Munekata PE, Dominguez R, Pateiro M, Saraiva JA, Franco D (2018) Main groups of microorganisms of relevance for food safety and stability. In: Innovative technologies for food preservation. Elsevier, pp 53–107. https://doi.org/10.1016/B978-0-12-811031-7.00003-0

Manimaran M, Gopal JV, Kannabiran K (2017) Antibacterial activity of *Streptomyces* sp. VITMK1 isolated from mangrove soil of Pichavaram, Tamil Nadu, India. Proc Natl Acad Sci India Sect B Biol Sci 87(2):499–506. https://doi.org/10.1007/s40011-015-0619-5

Meena SS, Mohanty A (2020) Ethical, patent, and regulatory issues in microbial engineering. In: Engineering of microbial biosynthetic pathways. Springer, Singapore, pp 133–142. https://doi.org/10.1007/978-981-15-2604-6_8

Mekala KPR, Dinesan A, Serva Peddha M, Dhale MA (2023) Valorization of biowastes as fermentative substrate for production of *Exiguobacterium* sp. GM010 pigment and toxicity effect in rats. Food Chem 407:135131. https://doi.org/10.1016/j.foodchem.2022.135131

Melendez JR, Mátyás B, Hena S, Lowy DA, El Salous A (2022) Perspectives in the production of bioethanol: a review of sustainable methods, technologies, and bioprocesses. Renew Sust Energ Rev 160:112260. https://doi.org/10.1016/j.rser.2022.112260

Mishra A, Kumar M, Bolan NS, Kapley A, Kumar R, Singh L (2021) Multidimensional approaches of biogas production and up-gradation: opportunities and challenges. Bioresour Technol 338:125514. https://doi.org/10.1016/j.biortech.2021.125514

Moinee AA, Sanzida N (2021) Performance analysis of different anode materials of a double chambered microbial fuel cell. Chem Eng Res Bull:26–31. https://doi.org/10.3329/cerb.v22i1.54295

Moodley P, Sewsynker-Sukai Y, Gueguim Kana EB (2020) Progress in the development of alkali and metal salt catalysed lignocellulosic pretreatment regimes: potential for bioethanol production. Bioresour Technol 310:123372. https://doi.org/10.1016/j.biortech.2020.123372

Moreira BP, Sganzerla WG, Torres-Mayanga PC, Ruiz HA, Lachos-Perez D (2023) Overview of Commercial Bioethanol Production Plants. In: Betiku E, Ishola MM. (eds) Bioethanol: A Green Energy Substitute for Fossil Fuels. Green Energy and Technology. Springer, Cham. https://doi.org/10.1007/978-3-031-36542-3_11

Negash AW, Tsehai BA (2020) Current applications of bacteriocin. Int J Microbiol 2020:1–7. https://doi.org/10.1155/2020/4374891

Nora LM (2013) Law, ethics, and the clinical neurologist. Handb Clin Neurol 118:63–78. https://doi.org/10.1016/B978-0-444-53501-6.00005-6

Nosheen S, Ajmal I, Song Y (2021) Microbes as biofertilizers, a potential approach for sustainable crop production. Sustain For 13(4):1868. https://doi.org/10.3390/su13041868

Pandey A, Nigam P, Soccol CR, Soccol VT, Singh D, Mohan R (2000) Advances in microbial amylases. Biotechnol Appl Biochem 31(2):135. https://doi.org/10.1042/BA19990073

Panesar R, Kaur S, Panesar PS (2015) Production of microbial pigments utilizing agro-industrial waste: a review. Curr Opin Food Sci 1:70–76. https://doi.org/10.1016/j.cofs.2014.12.002

Pansera M, Barca S, Martinez Alvarez B, Leonardi E, D'Alisa G, Meira T, Guillibert P (2024) Toward a just circular economy: conceptualizing environmental labor and gender justice in circularity studies. Sustain Sci Pract Policy 20(1). https://doi.org/10.1080/15487733.2024.2338592

Paul PEV, Sangeetha V, Deepika RG (2019) Emerging trends in the industrial production of chemical products by microorganisms. In: Recent developments in applied microbiology and biochemistry. Elsevier, pp 107–125. https://doi.org/10.1016/B978-0-12-816328-3.00009-X

Quinn GA, Banat AM, Abdelhameed AM, Banat IM (2020) Streptomyces from traditional medicine: sources of new innovations in antibiotic discovery. J Med Microbiol 69(8):1040–1048. https://doi.org/10.1099/jmm.0.001232

Ramamurthy PC, Singh S, Kapoor D, Parihar P, Samuel J, Prasad R, Kumar A, Singh J (2021) Microbial biotechnological approaches: renewable bioprocessing for the future energy systems. Microb Cell Factories 20(1):55. https://doi.org/10.1186/s12934-021-01547-w

Rastogi M, Shrivastava S (2017) Recent advances in second generation bioethanol production: an insight to pretreatment, saccharification and fermentation processes. Renew Sust Energ Rev 80:330–340. https://doi.org/10.1016/j.rser.2017.05.225

Raveendran S, Parameswaran B, Ummalyma SB, Abraham A, Mathew AK, Madhavan A, Rebello S, Pandey A (2018) Applications of microbial enzymes in food industry. Food Technol Biotechnol 56(1):10.17113/ftb.56.01.18.5491

Sakshi, Kumar R, Kaur H, Kumar D, Chauhan S (2023) Microbial endophytes: a novel approach for emerging pollutants. In: Management and mitigation of emerging pollutants. Springer International Publishing, pp 249–272. https://doi.org/10.1007/978-3-031-41005-5_9

Sanjuán J, Nápoles MC, Pérez-Mendoza D, Lorite MJ, Rodríguez-Navarro DN (2023) Microbials for Agriculture: Why Do They Call Them Biostimulants When They Mean Probiotics? Microorganisms, 11(1),153. https://doi.org/10.3390/microorganisms11010153

Sar T, Harirchi S, Ramezani M, Bulkan G, Akbas MY, Pandey A, Taherzadeh MJ (2022) Potential utilization of dairy industries by-products and wastes through microbial processes: a critical review. Sci Total Environ 810:152253. https://doi.org/10.1016/j.scitotenv.2021.152253

Sariatli F (2017) Linear economy versus circular economy: a comparative and analyzer study for optimization of economy for sustainability. Visegrad J Bioecon Sustain Dev 6(1):31–34. https://doi.org/10.1515/vjbsd-2017-0005

Savi DC, Shaaban KA, Vargas N, Ponomareva LV, Possiede YM, Thorson JS, Glienke C, Rohr J (2015) *Microbispora* sp. LGMB259 endophytic actinomycete isolated from *Vochysia divergens* (Pantanal, Brazil) producing β-Carbolines and indoles with biological activity. Curr Microbiol 70(3):345–354. https://doi.org/10.1007/s00284-014-0724-3

Sessitsch A, Brader G, Pfaffenbichler N, Gusenbauer D, Mitter B (2018) The contribution of plant microbiota to economy growth. Microb Biotechnol 11(5):801–805. https://doi.org/10.1111/1751-7915.13290

Shahbazi MJ, Rahimpour MR (2023) Economic assessment of bioenergy utilization. In: Reference module in earth systems and environmental sciences. Elsevier. https://doi.org/10.1016/B978-0-323-93940-9.00097-9

Shin SG, Han G, Lim J, Lee C, Hwang S (2010) A comprehensive microbial insight into two-stage anaerobic digestion of food waste-recycling wastewater. Water Res 44(17):4838–4849. https://doi.org/10.1016/j.watres.2010.07.019

Soccol CR, Rojan PJ, Patel AK, Woiciechowski AL, Vandenberghe LP, Pandey A (2006) Glucoamylase. In: Enzyme technology. Springer, New York, pp 221–237. https://doi.org/10.1007/978-0-387-35141-4_11

Sonawane AV, Rikame S, Sonawane SH, Gaikwad M, Bhanvase B, Sonawane SS, Mungray AK, Gaikwad R (2024) A review of microbial fuel cell and its diversification in the development of green energy technology. Chemosphere 350:141127. https://doi.org/10.1016/j.chemosphere.2024.141127

Souza PM d, Magalhães P d O e (2010) Application of microbial α-amylase in industry – a review. Braz J Microbiol 41(4):850–861. https://doi.org/10.1590/S1517-83822010000400004

Stylianou M, Agapiou A, Omirou M, Vyrides I, Ioannides IM, Maratheftis G, Fasoula D (2018) Converting environmental risks to benefits by using spent coffee grounds (SCG) as a valuable resource. Environ Sci Pollut Res 25(36):35776–35790. https://doi.org/10.1007/s11356-018-2359-6

Sur S, Romo TD, Grossfield A (2018) Selectivity and mechanism of fengycin, an antimicrobial lipopeptide, from molecular dynamics. J Phys Chem B 122(8):2219–2226. https://doi.org/10.1021/acs.jpcb.7b11889

Thakur B, Kaur S, Tripathi M, Upadhyay SK (2023) Exploring the potential of lactic acid bacteria and its molecular mechanism of action in the development of biosurfactants: current finding and future outlook. Biotechnol Genet Eng Rev 00(00):1–32. https://doi.org/10.1080/02648725.2023.2216421

Thakur B, Kaur S, Dwibedi V, Albadrani GM, Al-Ghadi MQ, Abdel-Daim MM (2024) Unveiling the antimicrobial and antibiofilm potential of biosurfactant produced by newly isolated *Lactiplantibacillus plantarum* strain 1625. Front Microbiol 15:1459388

Todero Ritter CE, Camassola M, Zampieri D, Silveira MM, Dillon AJP (2013) Cellulase and Xylanase production by *Penicillium echinulatum* in submerged media containing cellulose amended with sorbitol. Enzyme Res 2013:1–9. https://doi.org/10.1155/2013/240219

Upadhyay SK, Singh G, Rani N, Rajput VD, Seth CS, Dwivedi P, Minkina T, Wong MH, Show PL, Khoo KS (2024) Transforming bio-waste into value-added products mediated microbes for enhancing soil health and crop production: perspective views on circular economy. Environ Technol Innov 34:103573. https://doi.org/10.1016/j.eti.2024.103573

Valenzuela-Ortega M, Suitor JT, White MF, Hinchcliffe T, Wallace S (2023) Microbial upcycling of waste PET to adipic acid. ACS Central Sci 9(11):2057–2063. https://doi.org/10.1021/acscentsci.3c00414

Varkey B (2021) Principles of clinical ethics and their application to practice. Med Princ Pract 30(1):17–28. https://doi.org/10.1159/000509119

Visco A, Scolaro C, Facchin M, Brahimi S, Belhamdi H, Gatto V, Beghetto V (2022) Agri-food wastes for bioplastics: European prospective on possible applications in their second life for a circular economy. Polymers 14(13):2752. https://doi.org/10.3390/polym14132752

Wang P, Xi L, Liu P, Wang Y, Wang W, Huang Y, Zhu W (2013) Diketopiperazine derivatives from the marine-derived Actinomycete *Streptomyces* sp. FXJ7.328. Mar Drugs 11(4):1035–1049. https://doi.org/10.3390/md11041035

Wang H, Peng X, Zhang H, Yang S, Li H (2021a) Microorganisms-promoted biodiesel production from biomass: a review. Energy Conver Manag X 12:100137. https://doi.org/10.1016/j.ecmx.2021.100137

Wang C, Tang S, Cao S (2021b) Antimicrobial compounds from marine fungi. Phytochem Rev 20(1):85–117. https://doi.org/10.1007/s11101-020-09705-5

Yeoman CJ, Han Y, Dodd D, Schroeder CM, Mackie RI, Cann IKO (2010) Thermostable enzymes as biocatalysts in the biofuel industry. Adv Appl Microbiol 70:1–55. https://doi.org/10.1016/S0065-2164(10)70001-0

Yu Z, Zhao L-X, Jiang C-L, Duan Y, Wong L, Carver KC, Schuler LA, Shen B (2011) Bafilomycins produced by an endophytic actinomycete *Streptomyces* sp. YIM56209. J Antibiot 64(1):159–162. https://doi.org/10.1038/ja.2010.147

Zhang J, Wang J-D, Liu C-X, Yuan J-H, Wang X-J, Xiang W-S (2014) A new prenylated indole derivative from endophytic actinobacteria *Streptomyces* sp. neau-D50. Nat Prod Res 28(7):431–437. https://doi.org/10.1080/14786419.2013.871546

Zhou Y, Kumar V, Harirchi S, Vigneswaran VS, Rajendran K, Sharma P, Wah Tong Y, Binod P, Sindhu R, Sarsaiya S, Balakrishnan D, Mofijur M, Zhang Z, Taherzadeh MJ, Kumar Awasthi M (2022) Recovery of value-added products from biowaste: a review. Bioresour Technol 360:127565. https://doi.org/10.1016/j.biortech.2022.127565

Microbial Contributions to a Circular Economy

Gajendra B. Singh, Shubhankhi Dey, and Gaurav Mudgal

Abstract

Microbial involvement in the circular economy is undoubtedly revolutionizing the future of sustainability. The role of microbes is essential for waste reduction and resource regeneration. The ability of microorganisms to convert organic waste into biofertilizers, biofuels, and bioplastics promotes sustainability in agriculture and industry sectors. In agriculture, microbes are known to improve the soil health, assist nutrient cycling, and encourage sustainable farming approaches, eventually minimizing the need for toxic chemical fertilizers. Developments in microbial biotechnology have resulted in the development of microbial factories in the industrial sector. These include organisms engineered to produce biodegradable chemicals and materials, which are environmentally friendly replacements for nonrenewable resources. Through bioremediation or biodetoxification, microbes can transform harmful substances into useful or less toxic forms. Apart from removing pollutants from the environment, these tiny organisms also recycle them for safe reuse. Apart from reducing adverse effects on the environment, microorganisms also solve global challenges like plastic pollution by creating biodegradable materials. With the advancement in microbial technology, these organisms are offering promising solutions to challenges such as waste management, climate change, resource depletion, and many more. Using microbial potential, a sustainable system can be developed which can help us to regenerate essential resources and ensure a secure future for coming generations. This chapter covers the different ways that microbes could promote a circular economy, with a focus on waste reduction, recycling, and environmental remediation. It also emphasizes the most recent developments in microbial biotechnology which have the potential to help with environmentally friendly management practices.

G. B. Singh (✉) · S. Dey · G. Mudgal
University Institute of Biotechnology, Chandigarh University, Mohali, Punjab, India
e-mail: gajendra.singh@cumail.in

S. Kaur et al. (eds.), *Microbial Metabolomics*,
https://doi.org/10.1007/978-981-96-4824-5_18

Keywords

Circular economy · Microbial biotechnology · Waste conversion · Waste to energy · Sustainable agriculture · Bioremediation

1 Introduction

An economic model where the materials are part of a continuous circulation, without becoming a waste, is recognized as an ideal circular economy. Processes like reuse, refurbishment, recycle, and remanufacturing distinctively separate it from the linear economy which operates on the traditional manufacturing, using, and discarding (Elisha 2020). This not only promotes economic sustainability but also ensures minimized waste generation by maintaining the utilization of the manufactured goods, thereby offering a critical alternative to render less damage to the environment and reduced depletion of resources.

Where the physical alternatives are constantly at play, biologically, microorganisms have been established to play a pivotal role in strengthening the circular economy at core. Over the years, different species of microorganisms have also established a strong foothold in both agricultural and industrial sectors, which directly and indirectly creates a multifaceted, eco-friendly, and stable circular economy (Kee et al. 2021; Upadhyay et al. 2024). Along with this, various microorganisms have evolved with definitive solutions resulting in breakdown of complex organic and inorganic compounds which caters to proper waste management. According to Chavan et al. (2022), a newer reported approach, microbial bioconversion, has also been reported, which involves the conversion of organic waste into valuable products which contributes positively to the economy and environmental sustainability.

The above claim was also established by Kee et al. (2021) who had reported that agricultural waste or crop residues can be biologically converted into biofertilizers, thereby enhancing soil fertility. This was strengthened by Priya et al. (2023) who ensured that the microorganisms are not only contributing in generating biofertilizers and biopesticides, but the process has also resulted in better crop yields and reduced dependency on chemical pesticides and fertilizers which would otherwise cause adverse effects on both the environment and humans.

Microbes have given strong alternatives not only in the agricultural sectors but also in the food industry. With the implementation of probiotics as well as prebiotics, enhancement of gut health has also been observed. Thus, the inclusion of microbes in both the agricultural and food industries can sustainably contribute to the circular economy by not only reducing waste generation but also modifying resources for efficient reusage.

Similarly, Lemes et al. (2022) had also claimed in one of his studies that the by-products generated by the action of certain microorganisms can result into the production of bioactive compounds which can be implemented in the pharmaceutical industry, followed by cosmetics and dermatology, along with enhanced production of biofuels and biogas which eventually offer an alternative to fossil fuels.

In addition to the abovementioned contributions, microbes have also played a key role in bioremediation which detoxifies polluted environments. Microbes, via the process of bioremediation, scours up oil spills, degrades heavy metals, and also successfully neutralizes hazardous compounds, thereby converting them into either comparatively lesser harmful by-products or other beneficial compounds (Imam et al. 2021).

The most astounding role of microorganisms that directly contribute to the circular economy is the ability to recycle. Through recycling and bioremediation, if incorporated into proper waste management as well as industrial practices, a balanced and sustained environment can be established. Such approaches can be better understood and evaluated in microbial biotechnology, which opens new doors to enhance the potential of microorganisms in expanding the circular economy more.

The extent to which microbial bioengineering has progressed has enabled the generation of engineered microbes which have the capability of producing newer products with better efficiency thereby implemented in pharmaceutical, biofuel, and chemical-based industries. Such integration leading to advancements in various sectors has had profound effects on the modulation of circular economy. Thus, with the help of microbial engineering, the establishment of closed-loop systems where the manufactured products or energy sources can be reutilized efficiently can lead to the formation of a sustained environment (Kumar et al. 2023).

These systems not only minimize waste but also reduce the environmental impact of production processes, contributing to a more resilient economy. While understanding the potential and prospects of microbial biotechnology, a brief discussion is necessary about the implementations of microbe-based techniques in both the food and agriculture sectors.

In the gradual establishment of a circular economy, the role played by microbes have encompassed waste management, recycling of materials, and sustainability in various sectors of agriculture, pharmacy, and various industries. With the advancements in the application of microbe-based techniques, a gradual balance is being established between a sustainable environment and socioeconomic well-being. Thus, the following paper focuses on discussing each of the factors in detail while exploring the potential of microbial biotechnology and its application in different sectors.

2 Microbial Roles in Waste Management

With the rise in necessity of proper waste management, microbes have emerged as a reliable solution, which not only helps in reducing piling of unmanaged waste but also contributes in generating valuable products through recycling. Thus, via processes like decomposition as well as biological degradation which also helps in nutrient cycling, microbes have successfully contributed to a healthy ecosystem and to proper establishment of a circular economy.

2.1 Decomposition and Nutrient Cycling

2.1.1 Microbial Decomposition Processes

Microbial decomposition applies usage of enzymatic or nonenzyme-based processes which breaks down complex organic matter into simple molecules. This process is furnished by a number of bacterial and fungal species as well as actinomycetes, which use extracellular enzymes for the degradation of organic components like proteins, lipids, or carbohydrates into more stable compounds (Rodríguez-Espinosa et al. 2023).

Fungal species of Penicillium and Aspergillus are recognized for their ability to decompose and degrade lignocellulosic molecules, which are present in wood and straw, with the help of enzymes that are capable of degrading lignin abundantly available in the cell walls of plants (Mehta and Chelike 2024; Gupte et al. 2023). Actinomycetes, as mentioned before, also helps in the degradation of chitin and cellulose converting them into more soluble forms (Bhatti et al. 2017).

2.1.2 Nutrient Cycling and Soil Fertility

The breakdown of complex organic matter in simple forms contributes to efficient nutrient cycling, as the simpler molecules or elements are now easily absorbed by the soil which also enhances the soil fertility.

A fascinating example of this is observed during nitrogen cycling where organic nitrogen is converted into inorganic nitrogen derivatives such as nitrate (NO_3^-) and ammonium (NH_4^+), which makes nitrogen easily available in the soil for the plants to absorb from (Rodríguez-Espinosa et al. 2023).

A similar trend can be observed in case of potassium and phosphorus cycling. In case of potassium cycling, plant residues are degraded by microbes for the release of potassium which is absorbed by the plants for restoring and maintaining physiological functions (Olaniyan et al. 2022). Along with this, phosphorus cycling by bacterial species such as Bacillus and Pseudomonas also contribute in nutrient cycling, thereby making it available in easily solubilizing form. Thereafter, mycorrhizal fungi assist in the uptake and absorption of phosphorus by plants (Rawat et al. 2021).

2.1.3 Formation of Humus and Soil Structure

Organic matter is broken down microbially, and the stable organic product that results from microbial decomposition of this matter is known as humus. Humus is the unifying product of microbial degradation of plant residues and animal manure, composed of organic material inherited from decomposed bodies. This transformed humus provides a complex mixture of biomolecules that form an important matrix in the soil, regulating soil structure, water-holding capacity, and nutrient availability (Guo et al. 2019).

It creates soil structure by holding the mineral particles together in granules, giving better aeration and dispersion of the minerals throughout the soil. By doing so, you aggregate the soil, which opens up the pore space in the soil to allow for quicker water infiltration and deeper root growth. Additionally, humus is high in cation exchange capacity (CEC), which enables it to either hold or release valuable nutrients and thereby feed plants. In addition, humus helps maintain the moisture in the soil, reducing irrigation frequency. Humus help fertilize the ground and reduce demands for synthetic fertilizers so it helps us farm better sustainably.

2.2 Biodegradation of Organic Waste

With the help of biodegradation, microbes break down organic waste and helps to recycle the compounds, thereby creating a circular economy approach toward waste management which will support environmental sustainability. This is an essential process because it helps in decreasing the waste volume, reduces our need for landfills, and recycles nutrients back into the ecosystem. A wide variety of microbes are involved in biodegradation of organic waste, which includes leftover food, garden waste, agricultural by-products, as well as toxic compounds, along with solid-phase fermentation using over two dozen genera of bacterial species and reports on numerous enzymatic classes other than fungi (Kaur et al. 2023; Wainaina et al. 2020; Bhatnagar et al. 2016; Mehndiratta et al. 2015; Singh et al. 2013, 2011a, b, c). In composting, organic waste is digested by aerobic bacteria and fungi in conditions that are controlled and tailored for the conversion of organic waste into compost: a soil amendment. Compost decreases the size of what goes into a landfill and creates nutritious compost that contributes to enhanced soil fertility.

Anaerobic bacteria also play a critical role in the decomposition of organic matter, producing biogas in the form of methane dioxide, when breaking down organic waste under anaerobic conditions (e.g., landfills or oxygen-free digesters). The biogas can then be captured and used as a renewable source of energy, with the remaining digestate suitable for use as a soil amendment or further treated to harvest nutrients. Anaerobic digestion is an environment-friendly alternative to landfilling, and it also aligns with the principles of a circular economy, converting waste into energy and useful materials.

Along with this, biodegradation is also a very significant component of waste management, especially in terms of agricultural as well as industrial wastes. It is capable of accommodating agricultural and industrial waste which can later be turned into natural fertilizer. In the food area, biological processes convert waste apparatus such as fruit skins and vegetable scraps into compost or bioenergy, thereby augmenting sustainability (Szyba and Mikulik 2022). In agriculture, microbes treat livestock manure by converting it to compost or biogas, reducing environmental impact and providing a nutrient recycling process back to the soil.

3 Microbial Bioconversion

Microbial bioconversion has emerged as a reliable alternative in creating not only a sustainable environment but also a circular economy in the long run. By the process, microorganisms including bacteria, fungi, as well as algae are able to successfully convert waste substrates into products that yield definitive value. This potentially resolves the rising issue of waste management, thereby creating feasible resources which contributes in establishing a circular economy.

3.1 Conversion of Waste into Valuable Products

Utilizing the ability of microorganisms in decomposition of organic waste material into generation of much simple compounds for myriad processes like efficient nutrient cycling is known as microbial bioconversion. Microbes use several biochemical pathways for breaking down complex compounds generated through food waste, industrial remnants, plant residue, as well as agricultural by-products, which yield high-value products such as biofertilizers as well as biopesticides (Kaur et al. 2022). To furnish this, where fungal species of *Aspergillus*, *Trichoderma*, and *Penicillium* have been used for the decomposition of lignocellulosic materials, bacterial species such as *Bacillus* and *Pseudomonas* contribute in making phosphorus available in soluble form, thereby contributing to efficient nutrient cycling (Gunjal et al. 2020).

Along with this, the bioconversion of waste generated from different industries such as food and beverage industry into bioactive compounds which have found significant applications in the pharmaceutical industry as well as biotechnological fields stands as a testament to the potential of microbe-based conversion in the establishment of circular economy.

3.2 Production of Biogas and Biofuels

Biogas and biofuels have made a significant impact on human civilization and have risen as an extreme necessity. Especially with the depletion of fossil fuels, these have emerged as a sustainable alternative in terms of energy source. Biogas, being a renewable energy, is produced by bacterial species such as *Methanobacterium* and *Methanococcus* as well as *Clostridium* sp., through anaerobic digestion by the breakdown of organic compounds (Ottoni et al. 2022). The methane thus produced is procured as a clean as well as renewable energy for different household applications such as generation of electricity, heating, as well as fuel for automobile. The substrate for the production of biogas is garnered from waste materials of food and agricultural manure, and it involves four distinct steps of hydrolysis, acidogenesis, acetogenesis, and methanogenesis. The entirety of these steps involves breaking of organic materials, conversion into volatile fatty acids and alcoholic compounds, and utilization by methanogenic bacteria for the production of carbon dioxide and methane.

Biofuels, such as ethanol and biodiesel, are another product of microbial bioconversion. Ethanol is primarily produced through the fermentation of carbohydrates by yeast species like *Saccharomyces cerevisiae* (Verma and Kumar 2021). This process involves the conversion of sugars into ethanol and carbon dioxide, and the resulting ethanol is used as a renewable fuel additive in gasoline. Biodiesel, on the other hand, is produced through the transesterification of vegetable oils or animal fats with methanol or ethanol, utilizing microbial lipases as catalysts. Biodiesel serves as an environmentally friendly alternative to fossil fuels, reducing greenhouse gas emissions and reliance on nonrenewable resources (Christopher et al. 2014).

3.3 Bioconversion in Agriculture and Industry

Microbial bioconversion has offered lucrative solutions in the agricultural and industrial sectors as well. Agriculture has seen immense transformation leading to better fertility of soil which had directly reflected in crop productivity. Further inclusion of microbial inoculants, as well as biological fertilizers and pesticides, replacing chemical alternatives has also contributed to overall crop health and has also reduced the amount of negative impact on the environment.

Further, in the industries, microbial bioconversion with the help of enzymatic reactions has resulted in massive advancements in textile, detergent, as well as paper manufacturing. Enzymes produced by fungal species, such as cellulases and xylanases which are capable of breaking down cellulose and hemicellulose in either industry of paper or textile, have not only enhanced the efficiency but also improved the quality resulting in better turnover (Chukwuma et al. 2020). Along with this, a similar trend has been observed in the pharmaceutical industry where the production of bio-based drugs has garnered significant attention toward the field of industrial biotechnology, thereby yielding high-value products with lower dependency on petrochemical feedstocks and enhanced sustainability on bio-based production.

4 Microbes in Recycling

With the rising necessity of recycling, microbes have emerged as a reliable alternative to chemical recycling. The ability of the microorganisms to break down complex materials into more simple and reusable compounds has contributed immensely to waste management (Table 1).

4.1 Role in Recycling Processes

Microbial recycling has extended to myriad applications, from organic waste management to the recovery of valuable resources from various types of waste. In the recycling process, the primary function is the decomposition of organic waste, where microorganisms break down biodegradable components into simpler compounds. The process results in transformation of the waste component into compost which becomes a nutrition for the soil and enhances the fertility of soil, thereby promoting plant growth. The breakdown of complex compounds into such simple molecules results in reintegration into the natural environmental cycling. The

Table 1 Selected examples of microbial contributions to circular economy processes

Microbial process	Role in circular economy	Examples of applications	Environmental benefits	References
Bioconversion of waste	Converts organic and industrial waste into biofuels, bioplastics, and biofertilizers	Production of ethanol, biogas, and biodiesel from agricultural waste	Reduces reliance on fossil fuels and synthetic fertilizers	Singh et al. (2024a) and Alazaiza et al. (2022)
Biodegradation and recycling	Breaks down organic pollutants and recycles waste components	Microbial treatment of wastewater, plastics, and oil spills	Minimizes environmental pollution, promotes resource recovery	Vinayak and Singh (2022a) and Bhardwaj et al. (2021a)
Microbial factories	Produces bio-based chemicals and high-value materials	Synthesis of biopolymers, antibiotics, and enzymes	Reduces environmental footprint of traditional chemical industries	Vinayak et al. (2022)
Probiotics and prebiotics in agriculture	Enhances soil microbiota, improves crop yield and plant health	Use of *rhizobium* for nitrogen fixation, *bacillus* for root health	Reduces the need for chemical pesticides and fertilizers	Vinayak et al. (2021) and Prasad et al. (2015)
Microbial fertilizers	Increases soil nutrient availability and promotes nutrient cycling	Use of phosphate-solubilizing and nitrogen-fixing bacteria	Enhances soil fertility, decreases dependency on synthetic inputs	Dhar et al. (2024) and Parashar et al. (2023)
Bioremediation of pollutants	Cleans up contaminated environments by degrading toxic substances	Treatment of heavy metal contamination, oil spills, and pesticides	Restores polluted ecosystems, contributes to soil and water health	Singh et al. (2011a), Bhardwaj et al. (2021b), and Vinayak and Singh (2022b)
Microbial fuel cells (MFCs)	Converts organic compounds directly into electricity	Microbial fuel cells generating power from wastewater treatment	Provides sustainable energy while treating waste	Mohyudin et al. (2022)
Plastic biodegradation	Breaks down synthetic plastics into environmentally friendly by-products	*Ideonella sakaiensis* for degrading PET plastic	Reduces plastic waste accumulation and microplastic pollution	Ali et al. (2021)

(continued)

Table 1 (continued)

Microbial process	Role in circular economy	Examples of applications	Environmental benefits	References
Microbial mining (bioleaching)	Extracts valuable metals from ores using microbial processes	*Acidithiobacillus ferrooxidans* for copper bioleaching	Reduces harmful chemical use in mining, promotes resource recovery	Shi et al. (2023)
Pharmaceutical production	Microbes engineered to produce drugs, vaccines, and biopharmaceuticals	Use of yeast and bacteria to produce insulin, antibiotics, and vaccines	Sustainable drug production, minimizes waste from chemical synthesis	Jana et al. (2024)
Textile waste treatment	Breaks down textile dyes and pollutants	Use of fungi and bacteria to degrade toxic dyes from textile waste	Reduces toxic waste in water bodies from textile industries	Singh et al. (2024b) and Vinayak and Singh (2022c)
Food waste management	Converts food waste into bioenergy and organic fertilizers	Microbial composting and anaerobic digestion of food waste	Reduces landfill waste, produces energy and organic fertilizers	Vinayak et al. (2020)
Carbon sequestration	Captures and stores atmospheric CO_2 through microbial activity	Use of cyanobacteria and microalgae to absorb CO_2	Mitigates climate change by reducing atmospheric carbon levels	Vinayak et al. (2023) and Nayak et al. (2022)
Microbial desalination	Desalination of seawater using microbial electrolysis cells	Removal of salt from water by microbial electrolysis	Provides sustainable freshwater production	Zahid et al. (2022)

process of efficient recycling can be understood by the working of *Aspergillus niger*, a fungal species, as well as *Penicillium* which are proficient in degradation of lignocellulosic compounds which have been very difficult to be decomposed with the help of chemical processes (Saini and Sharma 2021). The same is observed in case of bacterial species such as Bacillus and Pseudomonas which breakdown organic agricultural and industrial waste products.

Microbial recycling is not limited to organic waste products but also in the case of inorganic waste materials. It had been observed that certain microbial species can furnish metal recovery with the help of a process known as bioleaching, from by-products generated by mining and e-waste. Species like *Thiobacillus ferrooxidans* and *Leptospirillum ferrooxidans* are also known for their capability of oxidizing iron and sulfur compounds as well as for solubilizing metal elements such as zinc, copper, and gold which can not only be recovered but also reused (Dave and Tipre 2019).

4.2 Bioremediation of Pollutants and Contaminants

Bioremediation—the strategy of using living organisms to clear up the environment—involves the application of microbiology for the degradation or transformation of pollutants and contaminants. The implementation of bioremediation in such cases is furnished with the help of microbes which degrade contaminants such as hydrocarbons, heavy metals, as well as synthetic chemicals and metabolize them into generation of simple molecules of water and carbon dioxide which decreases the requirement of chemical-based treatment options. Microbial species such as *Alcanivorax borkumensis* and *Pseudomonas putida* are currently being implemented in cleaning of oil spills in various places across the globe (Rahmati et al. 2022).

Microbial bioremediation is not only limited to decontamination of water bodies and conversion of hydrocarbons but also have extended to recycling of waste generated from pharmaceutical and industrial solvents. Further, *Sphingomonas* and *Burkholderia* species have been shown to degrade persistent organic pollutants such as chlorinated solvents and polychlorinated biphenyls (PCBs), contributing to the cleanup of industrial sites and polluted water sources (Bhandari 2024).

4.3 Applications in Water and Soil Remediation

Microbes have largely contributed in the establishment of stable ecosystem with the help of processes like nutrient cycling and preservation or less utilization of natural resources. This is possible because of a number of advancements, among which bioremediation has played a crucial role. For efficient remediation of water bodies, several bioreactors have been constructed which uses different species of microbes, as required, for the degradation of organic contaminants. This, successfully restores the quality of water by reducing issues of waste in the most cost-effective way possible.

Soils accumulate heavy metals from different waste sources which contaminates it and reduces its fertility. To this, the use of chemical pesticides also contributes to it, which raises an issue to deal with. Thus, hydrocarbon-degrading bacteria have been implemented for this purpose which breaks down these complex particles and converts them into simple absorbable forms which enhances the quality and fertility of the soils.

Along with this, the usage of biochar, which is a carbon-rich source obtained from biomass and mixed with bacterial inoculants, has offered a lucrative approach toward dealing with soil remediation (Xiang et al. 2022). It has successfully provided with a stable matrix for microbial colonization which enhances the degradation and absorption of pollutants. This approach supports the recycling of organic waste into valuable soil amendments while addressing soil contamination issues. Figure 1 illustrates the few of the diverse applications of microbial processes across key sectors, highlighting their role in promoting sustainability through waste recycling, resource regeneration, and environmental remediation.

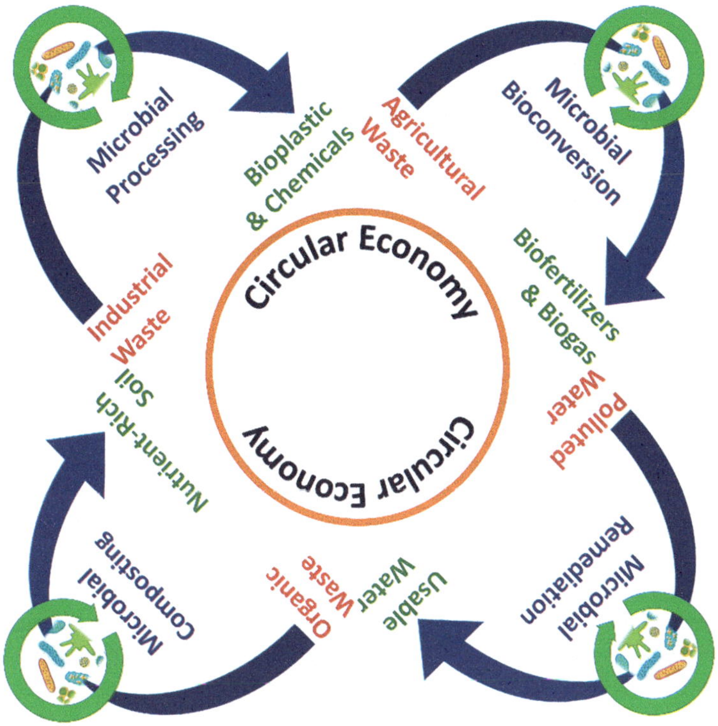

Fig. 1 Microbial contributions to the circular economy

5 Microbial Biotechnology

With reference to the myriad implementations of microorganisms, microbial biotechnology has emerged as a reliable niche which provides better scope for exploring the unique capabilities and different prospects in this field. Along with synthetic biology, microbial biotechnology has advanced to a great extent and has provided alternative solutions to a number of existing problems. To understand the scope of microbial biotechnology and its contributions to the circular economy, the following sections now highlights the role, applications, advancements, and scope in brief.

5.1 Advances in Microbial Engineering

With the emergence of genetic modification techniques, microbial engineering has successfully been ventured newer sectors and opened fresh opportunities in different industries. A striking example of such advancements is the advent of CRISPR-Cas 9, which has become revolutionary in case of gene editing (Hemalatha et al. 2023). The technique involves alterations of microbial genomes which gives new directions to existing problems. For instance, the well-known discovery of modified yeast strains with altered genes involved in carbohydrate metabolism has resulted in enhanced production of biochemical.

Further, genetically engineered strains of *E. coli* have the capability of producing drugs like insulin which has helped millions of people in dealing with diabetes. Along with this, microbial biotechnology has also contributed in the production of antibiotics and vaccines which have been revolutionary in the medical and pharmaceutical industry.

Besides, metabolic engineering has also been used to study certain strains and design them artificially to seep in and modify metabolic intermediates. This can also be observed in the production of bio-based chemical such as lactic acid as well as succinic acid which have been utilized in the manufacturing of biodegradable plastic materials.

5.2 Synthetic Biology and Microbial Factories

To offer a more structured solution in the field of microbial biotechnology, synthetic biology has emerged with the possibility of construction and designing of newer biological formulations and for the creation of microorganisms with novel attributes; the field of synthetic biology thereby brings about a one-stop platform by including biological principles and computer science and presents itself as an amalgamation to the field of science. With the help of this, understanding the functioning and modifying the microorganisms have not only become easy but have also opened newer possibilities. One of the prime concepts of synthetic biology is the production of microbial factories, where microorganisms are designed to result in the production of valuable materials which can be utilized in different industries, standardized biological tools such as BioBricks, as well as genetic circuits, which can be used for assembling different factors to produce a properly functioning biological system (Rautray et al. 2023). As per research, engineered *Saccharomyces cerevisiae*, which is otherwise known as baker's yeast, is now capable of producing cannabidiol after the integration of genes isolated from cannabis plants into the yeast genomes (Qiu et al. 2022). This stands as a living proof of the extent of cost-effective prospects which can be included in cultivation of newer plants.

Along with this, polyhydroxyalkanoates (PHAs), a bio-based polymer, which is also a biodegradable plastic form, can also be produced with the help of synthetic biology with the help of microorganisms (Saravanan et al. 2022). By introducing synthetic pathways into bacterial strains, scientists have created microorganisms that can produce PHAs from renewable feedstocks, contributing to the reduction of plastic waste and the promotion of a circular economy.

6 Microbial Processes in Food and Agriculture

Microbial processes are integral to modern food and agricultural practices, offering solutions that enhance food quality, promote soil health, and support sustainable farming. Microorganisms play multifaceted roles in these areas, contributing to the principles of a circular economy by recycling nutrients, improving resource efficiency, and minimizing environmental impact.

6.1 Probiotics and Prebiotics

One of the most prominent contributions of microorganisms in the food and health sector is the emergence of probiotics and prebiotics. These products adhere to lowering of pH and inhibition of pathogenic bacterial growth. Where the former are living microorganisms administered in sufficient proportions that contribute to health benefits of the consumer, the latter consists of nondigestible components of food which enhances the growth of good bacteria in the gut, thereby promoting gut health of the individual. Microbial species such as *Lactobacillus*, *Saccharomyces boulardii*, as well as *Bifidobacterium* which are frequently found in sauerkraut, yogurt, and kefir have promoted healthy gut microbiota, reducing gastrointestinal anomalies, thereby enhancing overall digestive health and immunity (Vinayak et al. 2021).

Prebiotics serve as a potential source of supplements which promote the growth of such bacteria which help in restoration of gut health. Naturally available prebiotic components include galactooligosaccharides, inulin, as well as fructooligosaccharides in vegetables like onion and garlic, and fruits such as banana also support healthy gut microbiome, thereby contributing to better metabolic function and gut health (Kaur et al. 2021).

6.2 Microbial Fertilizers and Soil Health

The use of microbe-based fertilizers has also contributed in the enhancement of the soil fertility and better crop yield in the agricultural sectors. Further, rhizobial inoculants have also emerged as a reliable source of biofertilizers (Kaur et al. 2022). The inoculant, in this case, utilizes the ability of the rhizobia of fixing the atmospheric nitrogen to enhance soil health and reduce the need for artificial nitrogen-based fertilizers.

Along with this, mycorrhizal fungi have also emerged as a suitable microbial fertilizer. These form mutual associations with the roots of the plants and by extending hyphae in the soil increase the available surface area for the absorption of nutrient and water. This caters to improvement of phosphorus uptake, increases tolerance of the plants toward drought, and maintains soil integrity.

6.3 Sustainable Agriculture Practices

Microbial processes are central to sustainable agriculture, which aims to balance productivity with environmental stewardship. Sustainable practices that harness microbial capabilities contribute to the principles of a circular economy by promoting resource efficiency, reducing waste, and enhancing ecosystem health. One key aspect of sustainable agriculture is the use of microbial-based soil amendments. In this, the process of composting by bacterial species such as Actinobacteria has resulted in reduced requirement of synthetic or chemical-based fertilizers for efficient restoration and enhancement of the soil quality.

A different but significant form of agricultural practice known as cover cropping has also been a widely accepted processes included in microbe-based processes. The process includes usage of cover crops such as clover and vetch which are planted in the fields amid different sessions of vegetation. This not only enhances the soil health but also prevents the soil from getting eroded. The role of microbes begins during the decomposition and degradation of the cover crops by decomposing bacteria which enriches the soil with the necessary nutrients, thereby increasing the chances of having a more resilient agricultural system.

A popular practice of IPM or integrated pest management also includes the usage of microbes instead of chemicals while dealing with pesticides. Microbe-based pesticides which specifically help in eradicating pathogens without destroying the soil or the crop are emerging as a reliable source of pesticides as an alternative to chemical-based ones. This can also be understood via the functioning of *Bacillus thuringiensis* which produces toxins that are effective against certain insect pests while being safe for humans and animals (Heckel 2020).

Techniques such as anaerobic digestion utilize microorganisms to decompose organic waste, producing biogas and digestate. Biogas can be used as a renewable energy source, while digestate serves as a nutrient-rich soil amendment. This approach not only reduces waste but also recycles nutrients, aligning with the principles of a circular economy.

7 Challenges and Future Directions

As we advance our understanding and application of microbial contributions to a circular economy, several challenges and limitations must be addressed. These challenges span technological, policy, and research domains, each impacting the effective deployment and scalability of microbial technologies. Addressing these issues is crucial for optimizing microbial processes in waste management, resource recovery, and sustainable practices.

7.1 Technological Barriers and Limitations

One of the primary technological barriers to the widespread implementation of microbial technologies is the complexity of microbial systems. The interaction between different microbial species and their environments can be intricate and difficult to control. For example, in bioreactors used for waste treatment or biogas production, maintaining optimal conditions for microbial activity—such as temperature, pH, and nutrient availability—can be challenging. Fluctuations in these conditions can affect microbial performance and reduce the efficiency of processes like biodegradation or biogas production.

Another technological challenge is the scalability of microbial processes. While laboratory-scale experiments often demonstrate the potential of microbial technologies, scaling these processes to industrial levels presents significant hurdles. Factors such as reactor design, process control, and cost efficiency must be optimized to ensure that microbial technologies are economically viable and practical for large-scale applications. Additionally, the integration of microbial processes into existing infrastructure, such as wastewater treatment plants or agricultural systems, requires careful planning and adaptation.

The genetic engineering of microbes also presents challenges. While advances in synthetic biology and genetic modification have expanded the capabilities of microorganisms, there are concerns about the stability and safety of genetically modified organisms (GMOs). Ensuring that engineered microbes do not have unintended environmental or health impacts is crucial for their successful deployment. This includes addressing potential issues such as horizontal gene transfer, where genetic material from engineered microbes could be transferred to other organisms in the environment.

7.2 Future Research and Innovation in Microbial Contributions

Looking ahead, future research and innovation are crucial for advancing microbial contributions to the circular economy. Several areas of research hold promise for enhancing the effectiveness and applicability of microbial technologies.

One key area of research is the development of novel microbial strains with enhanced capabilities. Advances in synthetic biology and genome editing techniques, such as CRISPR-Cas9, offer opportunities to engineer microorganisms with specific traits, such as improved efficiency in waste degradation or enhanced production of biofuels. Research focused on exploring the genetic and metabolic pathways of microbes can lead to the discovery of new microbial functions and applications.

Another promising area is the integration of microbial technologies with other emerging technologies. For instance, combining microbial processes with advanced materials, such as nanomaterials or smart sensors, can create innovative solutions for environmental monitoring and resource recovery. Research into hybrid systems that combine biological and chemical processes could enhance the efficiency of waste treatment and resource recovery.

The exploration of microbial diversity in natural environments, such as extreme or underexplored habitats, can also lead to the discovery of novel microorganisms with unique capabilities. These microorganisms may possess specialized enzymes or metabolic pathways that can be harnessed for applications in waste management, bioremediation, or sustainable agriculture.

Furthermore, interdisciplinary research that combines microbiology with fields such as engineering, environmental science, and economics can provide a holistic approach to addressing challenges in microbial technologies. Collaborative efforts

Table 2 Challenges and future directions for microbial technologies in a circular economy

Challenges	Technological barriers	Potential solutions	Research and innovation areas
Scalability of microbial processes	Difficulty in scaling from lab to industrial levels	Developing more efficient, large-scale bioreactors, continuous monitoring systems	Designing reactors that maintain optimal conditions for microbes
Genetic engineering constraints	Issues with the stability of engineered microbes in natural environments	Advanced gene editing tools like CRISPR for precision modification	Exploring microbial diversity to naturally find robust strains
Cost and economic viability	High cost of developing and maintaining microbial systems	Cost-effective production technologies, public–private partnerships	Economies of scale, using locally available feedstocks to reduce costs
Regulatory and policy issues	Lack of standardized global regulations for microbial GMOs	International collaboration on regulatory frameworks	Risk assessments to address concerns over GMOs and environmental safety
Public perception and acceptance	Misconceptions or skepticism about microbial biotechnology	Public education campaigns, transparent risk communication	Demonstrating clear benefits through case studies and local projects
Environmental impact concerns	Potential risk of uncontrolled microbial growth or horizontal gene transfer	Rigorous environmental monitoring, containment measures for GMOs	Research into microbial interactions with natural ecosystems
Innovation and research gaps	Limited understanding of unexplored microbial capabilities	Expanding research on extremophiles, underexplored microbial communities	Bioprospecting for new microbial species in extreme environments (deep sea, hot springs)

between researchers, industry, and policymakers can drive innovation and accelerate the adoption of microbial solutions in the circular economy. Table 2 outlines the key challenges and future directions for microbial technologies within the circular economy, highlighting technological barriers, potential solutions, and areas for research and innovation. It emphasizes the need for scalable processes, improved genetic engineering methods, economic viability, regulatory frameworks, public acceptance, environmental safety, and a deeper understanding of microbial capabilities to enhance the effectiveness of these technologies in sustainable practices.

8 Conclusion

In the intricate tapestry of sustainable development, microbial contributions to a circular economy emerge as a pivotal force, weaving together the principles of waste reduction, resource regeneration, and environmental resilience. The detailed exploration of microbial roles in waste management, bioconversion, recycling, biotechnology, and agricultural processes underscores their transformative impact on achieving circular economic goals. Microbes, despite their microscopic size, play monumental roles in converting waste into valuable products, enhancing soil health, and mitigating environmental pollutants.

Microbial processes facilitate the decomposition and nutrient cycling essential for maintaining soil fertility and supporting agricultural sustainability. By breaking down complex organic materials into simpler, nutrient-rich compounds, microbes enhance soil structure and nutrient availability, thereby reducing reliance on synthetic inputs and promoting sustainable farming practices. The biodegradation of organic waste, driven by microbial activity, further exemplifies their critical role in reducing waste volumes, recovering energy, and recycling nutrients, all of which align with the objectives of a circular economy.

The advancements in microbial biotechnology, including synthetic biology and microbial factories, represent significant strides toward a sustainable future. Engineered microorganisms with enhanced capabilities offer innovative solutions for producing biofuels, bioplastics, and other bio-based materials, providing alternatives to nonrenewable resources and reducing environmental impact. The applications of microbial biotechnology extend to various sectors, including agriculture and industry, where microbial processes contribute to resource efficiency and sustainability.

Despite these advancements, challenges remain in harnessing the full potential of microbial technologies. Technological barriers, such as process scalability and system complexity, must be addressed to optimize microbial applications for large-scale implementation. Policy implications and regulatory considerations also play a critical role in shaping the adoption and integration of microbial technologies, necessitating balanced approaches that ensure safety while fostering innovation. Continued research and development are essential for overcoming these challenges and exploring new opportunities in microbial science.

Future research holds promise for further enhancing microbial contributions to a circular economy. Advances in synthetic biology, genome editing, and microbial diversity exploration offer avenues for developing novel strains and processes with improved efficiency and functionality. Interdisciplinary collaborations and integration with emerging technologies can drive innovation and accelerate the deployment of microbial solutions across various sectors.

In conclusion, microbial contributions are integral to the principles of a circular economy, offering solutions for waste management, resource recovery, and environmental sustainability. As we continue to advance our understanding and application of microbial technologies, it is crucial to address existing challenges and embrace ongoing research and innovation. By harnessing the power of microbes, we can build a more sustainable future, where resources are efficiently recycled, environmental impacts are minimized, and the circular economy thrives.

References

Alazaiza MY, Albahnasawi A, Ahmad Z, Bashir MJ, Al-Wahaibi T, Abujazar MSS, Amr SSA, Nassani DE (2022) Potential use of algae for the bioremediation of different types of wastewater and contaminants: production of bioproducts and biofuel for green circular economy. J Environ Manag 324:116415

Ali SS, Elsamahy T, Al-Tohamy R, Zhu D, Mahmoud YAG, Koutra E, Metwally MA, Kornaros M, Sun J (2021) Plastic wastes biodegradation: mechanisms, challenges and future prospects. Sci Total Environ 780:146590

Bhandari S (2024) Environmental pollution caused by organic pollutants, their harmful impacts, and treatment through a microbiological approach. In: Nano-phytoremediation and Environmental pollution. CRC Press, pp 1–36

Bhardwaj A, Sharma RK, Singh GB (2021a) Contagious progression and distribution of arsenic in India: a key towards bioremediation. Nat Environ Pollut Technol 20(2)

Bhardwaj A, Sharma RK, Singh GB (2021b) Identification and characterization of arsenic transforming *bacillus* species from abandoned mining regions of Madhya Pradesh and Jharkhand. J Pure Appl Microbiol 15(1):175

Bhatnagar Y, Singh GB, Mathur A, Srivastava S, Gupta S, Gupta N (2016) Biodegradation of carbazole by *Pseudomonas* sp. GBS. 5 immobilized in polyvinyl alcohol beads. J Biochem Technol 6(3):1003–1007

Bhatti AA, Haq S, Bhat RA (2017) Actinomycetes benefaction role in soil and plant health. Microb Pathog 111:458–467

Chavan S, Yadav B, Atmakuri A, Tyagi RD, Wong JW, Drogui P (2022) Bioconversion of organic wastes into value-added products: a review. Bioresour Technol 344:126398

Christopher LP, Kumar H, Zambare VP (2014) Enzymatic biodiesel: challenges and opportunities. Appl Energy 119:497–520

Chukwuma OB, Rafatullah M, Tajarudin HA, Ismail N (2020) Lignocellulolytic enzymes in biotechnological and industrial processes: a review. Sustain For 12(18):7282

Dave SR, Tipre DR (2019) Diversity of iron and Sulphur oxidizers in sulphide mine leachates. In: Microbial diversity in ecosystem sustainability and biotechnological applications: volume 1. Microbial diversity in normal & extreme environments. Springer Nature, Singapore, pp 293–317

Dhar SK, Kaur J, Singh GB, Chauhan A, Tamang J, Lakhara N, Asyakina L, Atuchin V, Mudgal G, Abdi G (2024) Novel *Bacillus* and *Prestia* isolates from Dwarf century plant enhance crop yield and salinity tolerance. Sci Rep 14(1):14645

Elisha OD (2020) Moving beyond take-make-dispose to take-make-use for sustainable economy. Int J Sci Res Educ 13(3):497–516

Gunjal AB, Patil NN, Shinde SS (2020) Enzymes in degradation of the lignocellulosic wastes. Springer International Publishing

Guo XX, Liu HT, Wu SB (2019) Humic substances developed during organic waste composting: formation mechanisms, structural properties, and agronomic functions. Sci Total Environ 662:501–510

Gupte A, Prajapati D, Bhatt A, Pandya S, Raghunathan M, Gupte S (2023) Agro-industrial residues: an eco-friendly and inexpensive substrate for fungi in the development of white biotechnology. In: Fungi and fungal products in human welfare and biotechnology. Springer Nature, Singapore, pp 571–603

Heckel DG (2020) How do toxins from *Bacillus thuringiensis* kill insects? An evolutionary perspective. Arch Insect Biochem Physiol 104(2):e21673

Hemalatha P, Abda EM, Shah S, Prabhu SV, Jayakumar M, Karmegam N, Kim W, Govarthanan M (2023) Multi-faceted CRISPR-Cas9 strategy to reduce plant based food loss and waste for sustainable bio-economy–a review. J Environ Manag 332:117382

Imam A, Kanaujia PK, Ray A, Suman SK (2021) Removal of petroleum contaminants through bioremediation with integrated concepts of resource recovery: a review. Indian J Microbiol 61:250–261

Jana S, Gupta BD, Kar A, Ghosh A, Pervin M, Bala A, Matsabisa MG, Das R, Ghosh S, Mukherjee PK, Haldar PK (2024) Production of biopharmaceuticals on genetically modified organisms. In: Concepts in pharmaceutical biotechnology and drug development. Springer Nature, Singapore, pp 91–101

Kaur AP, Bhardwaj S, Dhanjal DS, Nepovimova E, Cruz-Martins N, Kuča K, Chopra C, Singh R, Kumar H, Şen F, Kumar D (2021) Plant prebiotics and their role in the amelioration of diseases. Biomol Ther 11(3):440

Kaur J, Mudgal G, Chand K, Singh GB, Perveen K, Bukhari NA, Debnath S, Mohan TC, Charukesi R, Singh G (2022) An exopolysaccharide-producing novel *Agrobacterium pusense* strain JAS1 isolated from snake plant enhances plant growth and soil water retention. Sci Rep 12(1):21330

Kaur J, Mudgal G, Negi A, Tamang J, Singh S, Singh GB, Bose KJC, Debnath S, Wadaan MA, Farooq Khan M, Ruokolainen J, Kesari KK (2023) Reactive black-5, Congo red and methyl orange: chemical degradation of azo-dyes by agrobacterium. Water 15(9):1664

Kee SH, Chiongson JBV, Saludes JP, Vigneswari S, Ramakrishna S, Bhubalan K (2021) Bioconversion of agro-industry sourced biowaste into biomaterials via microbial factories–a viable domain of circular economy. Environ Pollut 271:116311

Kumar V, Vangnai AS, Sharma N, Kaur K, Chakraborty P, Umesh M, Singhal B, Utreja D, Carrasco EU, Andler R, Taherzadeh MJ (2023) Bioengineering of biowaste to recover bioproducts and bioenergy: a circular economy approach towards sustainable zero-waste environment. Chemosphere 319:138005

Lemes AC, Egea MB, Oliveira Filho JGD, Gautério GV, Ribeiro BD, Coelho MAZ (2022) Biological approaches for extraction of bioactive compounds from agro-industrial by-products: a review. Front Bioeng Biotechnol 9:802543

Mehndiratta P, Jain A, Singh GB, Sharma S, Srivastava S, Gupta S, Gupta N (2015) Magnetite nanoparticle aided immobilization of *Pseudomonas* sp. GBS. 5 for carbazole degradation. J Biochem Technol 5(4):823–825

Mehta P, Chelike DK (2024) Utilizing fungal biodegradation for valorisation of lignocellulosic waste biomass and its diverse applications. Appl Res 3(4):e202300119

Mohyudin S, Farooq R, Jubeen F, Rasheed T, Fatima M, Sher F (2022) Microbial fuel cells a state-of-the-art technology for wastewater treatment and bioelectricity generation. Environ Res 204:112387

Nayak N, Mehrotra R, Mehrotra S (2022) Carbon biosequestration strategies: a review. Carbon Capt Sci Technol 4:100065

Olaniyan FT, Alori ET, Adekiya AO, Ayorinde BB, Daramola FY, Osemwegie OO, Babalola OO (2022) The use of soil microbial potassium solubilizers in potassium nutrient availability in soil and its dynamics. Ann Microbiol 72(1):45

Ottoni JR, Bernal SPF, Marteres TJ, Luiz FN, Dos Santos VP, Mari ÂG, Somer JG, de Oliveira VM, Passarini MRZ (2022) Cultured and uncultured microbial community associated with biogas production in anaerobic digestion processes. Arch Microbiol 204(6):340

Parashar M, Dhar SK, Kaur J, Chauhan A, Tamang J, Singh GB, Lyudmila A, Perveen K, Khan F, Bukhari NA, Mudgal G (2023) Two novel plant-growth-promoting *Lelliottia amnigena* isolates from *Euphorbia prostrata* Aiton enhance the overall productivity of wheat and tomato. Plan Theory 12(17):3081

Prasad R, Kumar M, Varma A (2015) Role of PGPR in soil fertility and plant health. In: Plant-growth-promoting rhizobacteria (PGPR) and medicinal plants. Springer, Cham, pp 247–260

Priya AK, Alagumalai A, Balaji D, Song H (2023) Bio-based agricultural products: a sustainable alternative to agrochemicals for promoting a circular economy. RSC Sustain 1(4):746–762

Qiu J, Hou K, Li Q, Chen J, Li X, Hou H, Wang L, Liu J, Xue Q, Wang C (2022) Boosting the cannabidiol production in engineered *Saccharomyces cerevisiae* by harnessing the vacuolar transporter BPT1. J Agric Food Chem 70(38):12055–12064

Rahmati F, Asgari Lajayer B, Shadfar N, van Bodegom PM, van Hullebusch ED (2022) A review on biotechnological approaches applied for marine hydrocarbon spills remediation. Microorganisms 10(7):1289

Rautray P, Roy A, Eisenbart B (2023) Application of bio-bricks & its benefits. Proc Des Soc 3:603–612

Rawat P, Das S, Shankhdhar D, Shankhdhar SC (2021) Phosphate-solubilizing microorganisms: mechanism and their role in phosphate solubilization and uptake. J Soil Sci Plant Nutr 21(1):49–68

Rodríguez-Espinosa T, Papamichael I, Voukkali I, Gimeno AP, Candel MBA, Navarro-Pedreño J, Zorpas AA, Lucas IG (2023) Nitrogen management in farming systems under the use of agricultural wastes and circular economy. Sci Total Environ 876:162666

Saini S, Sharma KK (2021) Fungal lignocellulolytic enzymes and lignocellulose: a critical review on their contribution to multiproduct biorefinery and global biofuel research. Int J Biol Macromol 193:2304–2319

Saravanan K, Umesh M, Kathirvel P (2022) Microbial polyhydroxyalkanoates (PHAs): a review on biosynthesis, properties, fermentation strategies and its prospective applications for sustainable future. J Polym Environ 30(12):4903–4935

Shi S, Pan J, Dong B, Zhou W, Zhou C (2023) Bioleaching of rare earth elements: perspectives from mineral characteristics and microbial species. Fortschr Mineral 13(9):1186

Singh GB, Gupta S, Srivastava S, Gupta N (2011a) Biodegradation of carbazole by newly isolated *Acinetobacter* spp. Bull Environ Contam Toxicol 87:522–526

Singh GB, Srivastava S, Gupta S, Gupta N (2011b) Evaluation of carbazole degradation by *Enterobacter* sp. isolated from hydrocarbon contaminated soil. Recent Res Sci Technol 3(11)

Singh GB, Srivastava A, Saigal A, Aggarwal S, Bisht S, Gupta S, Srivastava S, Gupta N (2011c) Biodegradation of carbazole and dibenzothiophene by bacteria isolated from petroleum-contaminated sites. Biorem J 15(4):189–195

Singh GB, Gupta S, Gupta N (2013) Carbazole degradation and biosurfactant production by newly isolated *Pseudomonas* sp. strain GBS. 5. Int Biodeterior Biodegradation 84:35–43

Singh GB, Sharma A, Thapa J, Nidhi. (2024a) Foliar-based nanoformulations: leads and flaws. In: Metabolomics, proteomics and gene editing approaches in biofertilizer industry: volume II. Springer Nature, Singapore, pp 223–245

Singh GB, Vinayak A, Mudgal G, Kesari KK (2024b) Azo dye bioremediation: an interdisciplinary path to sustainable fashion. Environ Technol Innov 103832:103832

Szyba M, Mikulik J (2022) Energy production from biodegradable waste as an example of the circular economy. Energies 15(4):1269

Upadhyay SK, Singh G, Rani N, Rajput VD, Seth CS, Dwivedi P, Minkina T, Wong MH, Show PL, Khoo KS (2024) Transforming bio-waste into value-added products mediated microbes for enhancing soil health and crop production: perspective views on circular economy. Environ Technol Innov 103573:103573

Verma N, Kumar V (2021) Microbial conversion of waste biomass into bioethanol: current challenges and future prospects. Biomass Convers Biorefinery:1–38

Vinayak A, Singh GB (2022a) Biodecolorization of reactive black 5 using magnetite nanoparticles coated Bacillus sp. RA5. Mater Today Proc 48:1523–1526

Vinayak A, Singh GB (2022b) Synthetic azo dye bio-decolorization by *Priestia* sp. RA1: process optimization and phytotoxicity assessment. Arch Microbiol 204(6):318

Vinayak A, Singh GB (2022c) Decolorization of reactive black 5 by immobilized bacterial cells. ECS Trans 107(1):11105

Vinayak A, Mudgal G, Singh G (2020) Waste to energy: sustainable approach for renewable energy. Plant Cell Biotechnol Mol Biol 21(15–16):292–302

Vinayak A, Mudgal G, Sharma S, Singh GB (2021) Prebiotics for probiotics. In: Advances in probiotics for sustainable food and medicine. Springer, Singapore, pp 63–82

Vinayak A, Sharma S, Singh GB (2022) Biopolymers from industrial waste. In: Biopolymers: recent updates, challenges and opportunities. Springer International Publishing, Cham, pp 129–149

Vinayak A, Sharma S, Singh GB (2023) Bioinspired materials for CO_2 capture and conversion. In: CO_2-Philic polymers, nanocomposites and chemical solvents. Elsevier, pp 57–76

Wainaina S, Awasthi MK, Sarsaiya S, Chen H, Singh E, Kumar A, Ravindran B, Awasthi SK, Liu T, Duan Y, Taherzadeh MJ (2020) Resource recovery and circular economy from organic solid waste using aerobic and anaerobic digestion technologies. Bioresour Technol 301:122778

Xiang L, Harindintwali JD, Wang F, Redmile-Gordon M, Chang SX, Fu Y, He C, Muhoza B, Brahushi F, Bolan N, Xing B (2022) Integrating biochar, bacteria, and plants for sustainable remediation of soils contaminated with organic pollutants. Environ Sci Technol 56(23):16546–16566

Zahid M, Savla N, Pandit S, Thakur VK, Jung SP, Gupta PK, Prasad R, Marsili E (2022) Microbial desalination cell: desalination through conserving energy. Desalination 521:115381

Clinical Application of Metabolomics in Infectious Diseases and Future Perspectives

Prakash Bharti, Ramesh Chandra Tripathi, and Anshul Singh

Abstract

Metabolomics is an advanced and emerging field in the biological sciences that provides a comprehensive understanding of biological systems' small molecules (metabolites). This technique has been advanced over the last decades and has drawn much interest from the medical community due to its unique ability to identify biomarkers associated with specific diseases and elucidate the biochemical basis of several diseases . Metabolomics techniques seem to be heading toward significant experimental and practical development, especially in the field of diagnostics. Nuclear magnetic resonance (NMR), mass spectrometry (MS), and chromatography methods identify metabolites in cells, tissues, feces, saliva, and urine samples. Most importantly, metabolomics makes it easier to analyze metabolites in huge amounts, which makes early and precise disease detection possible. Its use in clinical settings is crucial for detecting metabolic trends associated with a diversity of diseases, enabling early diagnosis and personalized treatment. Through the identification of associated biomarkers that indicate the presence and variability of disease, this method is crucial to the diagnosis of several diseases. In this chapter, we discuss the clinical application of metabolomics approaches in infectious diseases, including viral (AIDS, COVID-19, hepatitis B, and hepatitis C) and bacterial (tuberculosis and *Clostridium difficile*), along with future perspectives.

P. Bharti (✉) · R. C. Tripathi
Department of Biological Sciences, Mahatma Gandhi Chitrakoot Gramodaya Vishwavidyalaya Chitrakoot, Chitrakoot, Madhya Pradesh, India

A. Singh
School of Allied Health Sciences, Sharda University, Greater Noida, Uttar Pradesh, India

S. Kaur et al. (eds.), *Microbial Metabolomics*,
https://doi.org/10.1007/978-981-96-4824-5_19

Keywords

Metabolomics · Disease biomarker · Metabolites · Infectious diseases · COVID-19 biomarkers

1 Introduction

Metabolomics is a comprehensive study of metabolites inside biological systems; In recent years, it has become an emerging field in infectious diseases. Metabolomics provides a complete understanding of the biochemical variations caused by infections through the analysis of the distinct biochemical fingerprints left by cellular actions. In addition, to enhance our understanding of pathogen–host interactions, this all-inclusive approach has the potential to significantly change infectious disease diagnosis and treatment.

Metabolomics is an analytical clinical procedure with multiple applications in infectious diseases. With its help, various biomarkers can be identified and analyzed and used as diagnostic tools that can accurately differentiate between many diseases and their associated pathways. This approach plays an more important role where early identification of infection is difficult using conventional methods (Tounta et al. 2021). Moreover, by allowing insight into the metabolic pathways that viruses hijack, it offers novel targets for therapeutic intervention. In a time where antibiotic resistance is an increasing problem, such information is advantageous in the development of novel antimicrobial drugs targeting drug resistance (Rahman et al. 2022). Additional, remarkable use of metabolomics is the tracking of disease progress and the effectiveness of treatment. Throughout time, metabolic variations can be monitored by doctors to assess patient response to treatment and make well-informed decisions regarding therapeutic changes. Enhancing patient health, these dynamic monitoring abilities can add customized and effective treatment strategies (Cheng et al. 2017). Additionally, metabolomics can provide insight into the metabolic effects of infectious diseases, supporting the understanding of the long-term complications on patient health and directing the management of postinfections (Rahman and Schellhorn 2023).

The use of metabolomics in infectious diseases is promising, as advances in analytical techniques make metabolomics studies more accurate, reliable, and sensitive. The detailed information on infectious disease is expected through the integration of metabolomics with other omics techniques such as proteomics and genomics (Rahman and Schellhorn 2023). This technology can elucidate the complex systems of interaction between the virus and the host, opening up new possibilities for infectious disease prevention. In this chapter, we discuss the clinical applications of metabolomics in infectious diseases, some recent studies, and future perspectives.

2 Metabolomics Technology

The field of metabolomics approach originated with pioneering work by Pauling and coworkers in 1971 (Cui et al. 2018; Pauling et al. 1971). They quantitatively detected hundreds of chemicals in different biological samples to evaluate the functioning condition of biological systems. In 1999, Jeremy Nicholson introduced the idea of metabolomics (Nicholson et al. 1999). The scientific discipline of metabolomics is concerned with analyzing alterations in small molecules, or metabolites, found in biological tissues. The number of identified metabolites, sample preparation, accuracy and precision of the data acquired, and the quantitative level of the study aims are some of the differences between non-targeted and targeted analytic approaches in the field of research (Dunn et al. 2011). It is feasible to investigate the association between a range of metabolites from various pathways using non-targeted metabolomics without having to make any assumptions beforehand. While trying to detect metabolites that were previously unknown presents obstacles for experimental activities, it also results in the identification of novel metabolites and creates huge data sets. Conversely, targeted metabolomics entails the measurement or quantification of metabolites with established chemical structures and biological activities. A more precise explanation of the molecules associated with various illness scenarios is made possible by targeted metabolomics (Roberts et al. 2012). The traditional workflow of metabolomics study includes sample collection (including plasma, serum, urine, saliva, cells and tissues, and feces), sample pretreatment, metabolite measurement, data processing, and statistical analysis to identify biologically significant small molecules closely associated with metabolic traits (Roca et al. 2021). In metabolomics, nuclear magnetic resonance (NMR), gas chromatography-mass spectrometry (GC-MS), and liquid chromatography-mass spectrometry (LC-MS) are the most often utilized measurement methods (Carbone et al. 2021; Edison et al. 2021). The benefits of the NMR for metabolomics techniques include easy sample preparation, small sample quantity requirement, rapid measurement times, nondestructive samples, excellent repeatability, and nonselective metabolite detection. Nevertheless, NMR sensitivity is inferior to that of the mass spectrometry (MS) method, and it might not be able to identify some biochemicals in trace amounts. Rapid and effective detection is made possible by the effective preliminary processing of the LC-MS technique, which makes it not essential for the measured components to be volatile or thermally stable. Without a complete reference spectrum library, the identification of metabolites becomes more complicated and necessitates the use of standard materials, which is a tough way. The GC-MS technique provides excellent sensitivity and resolution and a trustworthy reference spectrum library for metabolite characterization. However, for samples with low volatility, derivation processing is required; also, the first processing is difficult and may cause some chemical signals to be lost. The complexity and diversity of metabolites make it more complicated for a single technology to measure and identify them all simultaneously. To have a more comprehensive understanding of metabolite details, it is necessary to integrate data from many analytical platforms (Pomyen

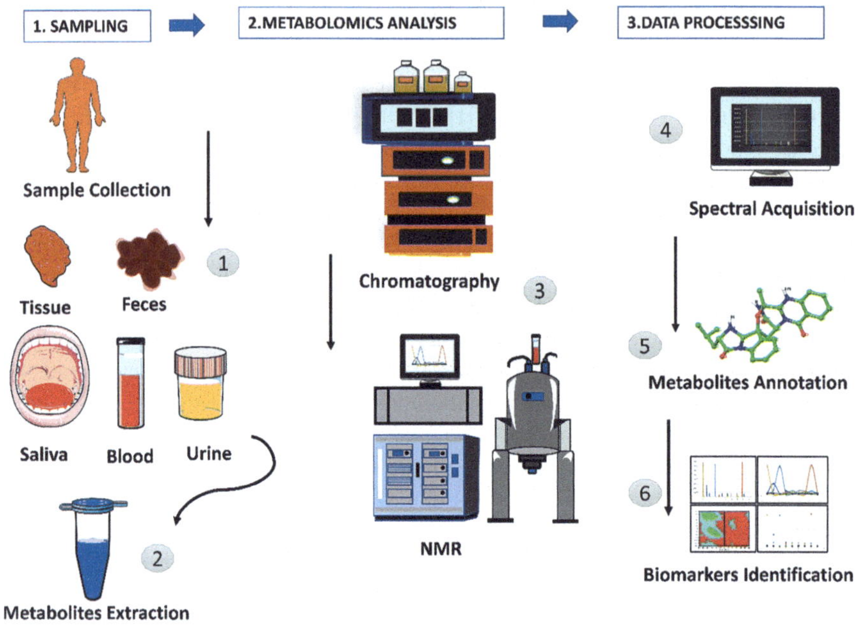

Fig. 1 Graphical workflow for metabolomics study usually includes investigational design. (1) Sample preparation: study design, sample collection, and metabolite extraction. (2) Metabolomics analysis: GC-MS, LC-M, and NMR. (3) Data processing: spectral acquisition, metabolite annotation, and biomarker identification

et al. 2020; Spina et al. 2021). Analyzing metabolic profiles in biological materials accurately allows one to identify important metabolic pathways and associated enzymes that are involved in the development of disease. According to Schmidt et al. (2021), this procedure also provides regulatory information on upstream proteins and genes. Through individual metabolomics research, several biological processes inside the biomolecular network during an illness are defined (Alarcon-Barrera et al. 2022). In terms of disease diagnosis, prognosis, medication effects, and comprehension of disease association and development, metabolomics technology represents a distinct medical benefit (Fig. 1).

3 Metabolomics and Viral Infections

Over the past few decades, extensive studies have been carried out on the innate immune response of the host to various viral infections. Most studies conducted are the immune-metabolic reaction to the *Herpesviridae* virus family. However, several investigations have also been carried out on other viruses, such as the rhinovirus, hepatitis B and C, influenza virus, human immunodeficiency virus (HIV), and severe acute respiratory syndrome-CoV-2 (SARS-CoV-2) (Overmyer et al. 2021). It is generally recognized that during viral infections ,this is accomplished by directing the host cells metabolism away from energy-producing processes like aerobic glycolysis and toward more efficient ones like anaerobic glycolysis (Mayer et al. 2019). Recent advancements have improved our knowledge of the roles played by lipids, metabolites, and inflammation in viral infectious illnesses. In a thorough investigation of dengue virus infection, blood samples from 44 dengue-infected patients and 50 individuals were compared for their metabolome and lipidome (Cui et al. 2013). Studies revealed the downregulation of many pathways, including lipolysis, glycolysis, fatty acid β-oxidation, and polyunsaturated phospholipid hydrolysis, by weeks 3–4. Their pathways had been elevated during the acute stages of dengue fever. The researchers identified several polyunsaturated phospholipids that both increased and decreased inflammation. This finding suggested that the host cell needed to regulate the level of inflammation throughout the infection in order to manage the virus and avoid causing damage to the host. Furthermore, T cells are the focus of several persistent viral infections, which frequently control and inhibit the adaptive immune response (Mayer et al. 2019).

3.1 Biomarkers of Viral Infections

Metabolites, as biomarkers, showed dissimilarities in abundance that might point to certain disease conditions or phases. These indicators play a vital role in both the diagnosis and treatment of viral infections. The host immunological response and the infection trouble of basic metabolic pathways generally result in the difference in their levels. This section focuses on four viral infectious diseases : COVID-19, acquired immunodeficiency syndrome (AIDS), hepatitis C, and hepatitis B. These diseases have been the subject of several metabolomics studies and are most related.

3.2 Metabolomics Application to COVID-19

About 100 million confirmed COVID-19 cases had been reported as of March 2021 globally, and approximately 2.5 million people had died as a result of the unpredicted global pandemic that began in late December 2019 and was brought on by the emergence and rapid spread of SARS-CoV-2. The absence of proven treatments emphasizes the urgent need for rapid, accurate diagnosis methods. Metabolomics is a potentially effective and promising approach for the diagnosis and treatment of COVID-19 as it can efficiently screen large-scale data sets and detect molecules to detect biomarkers in the prediction of disease prognosis and severity. Its use in the study of COVID-19 will lead to excellent efficiency. By analyzing the metabolic profile of individuals infected with SARS-CoV-2, metabolomics enables the identification of specific metabolites associated with a viral infection and disease progression (Kimhofer et al. 2020). Some recent research has shown that COVID-19 is linked to many other metabolic diseases, including diabetes; this suggests that immune metabolism is a key factor in the COVID-19 disease (Mauvais-Jarvis

2020). However, only a limited number of studies have directly demonstrated these results.

In one study, using LC-MS and NMR analytical techniques, investigators analyzed the whole human blood metabolites from 17 samples from SARS-CoV-2-positive confirmed individuals and 25 samples from a negative control individual (Kimhofer et al. 2020). On the basis of the findings, blood samples from SARS-CoV-2-positive confirmed individuals had greater levels of tryptophan ratio, raised triglyceride, and higher alpha-1-cis-gylcoprotein signal A, among other inflammatory biomarker molecules. They also detect variations in protein–lipid profiles, including high-density lipoprotein (HDL) and low-density lipoprotein (LDL). In another study, researchers compared total blood samples from 102 SARS-CoV-2-positive patients with 26 negative SARS-CoV-2 control group. This study combines various omics data sets (including metabolomics through LC-MS/GC-MS and lipidomics through LC-MS). It was discovered that COVID-19 patients had an increased tryptophan ratio (Overmyer et al. 2021). Ultimately, these findings allow for the identification of molecules that drive inflammation and other comorbidities in COVID-19. Several findings of this study indicate that COVID-19 triggers a severe inflammatory response in the body, making COVID-19 a life-threatening cause in patients with diabetes or cardiovascular disease (Cui et al. 2013; Nishiga et al. 2020).

In addition to immune-metabolomics investigations on the virus, several studies have revealed unique host biomarkers associated with COVID-19. Recent research suggests that plasma samples that determine positive COVID-19 testing might be distinguished from the healthy control group with accuracy by calculating the arginine/kynurenine ratio. Investigators use 1H-NMR and direct injection LC/MS to look at 183 metabolites in plasma samples from positive SARS-CoV-2 patients in the ICU. Once verified, these samples were categorized as either positive SARS-CoV-2 or negative. According to the study, changes in arginine, sarcosine, creatinine, kynurenine, and lysophosphatidylcholine (LPCs) were indicative of a particular biochemical signal for positive SARS-CoV-2. The kynurenine route, which is initiated by tryptophan breakdown, culminates in the production of NAD+ for cellular energy. Elevated levels are linked to enhanced tryptophan catabolism and enhanced T cell activation-induced synthesis of interferon gamma. The documented decrease in arginine might be a significant indicator of the illness's severity (Fraser et al. 2020).

In 2024, NMR spectroscopy techniques were used to examine the metabolomic and lipidomic profiles of 263 (training cohort) symptomatic and 135 (validation cohort) asymptomatic patients, who were hospitalized after SARS-CoV-2 positive test. In addition, the researchers created profiles for 280 individuals collected before the start of the COVID-19 pandemic. Principal component analysis (PCA) showed how effectively the groups were differentiated and highlighted the substantial metabolic changes induced by the infection. A pathogenic shift in the size and composition of lipoprotein particles was found by lipidomic analysis, indicating a higher risk of atherosclerosis. Similarly, a metabolomic analysis study identified unusually higher levels of the ketone bodies acetone carboxylic acid, beta-hydroxybutyric acid, dimethyl ketone, and 2-hydroxybutyric, an outcome of hepatic glutathione synthesis and a biomarker of oxidative stress (Whiley et al. 2024).

3.3 Metabolomics Application to Hepatitis B

Hepatitis B virus (HBV) is the cause of chronic hepatitis. The lack of biomarkers has long been a barrier to clinical diagnosis (Zhang et al. 2013). Metabolomics technology has great potential for transforming HBV detection. It provides a sensitive and efficient method to diagnose and determine the stage of the disease without demanding the use of risky techniques like biopsies and histology (El-Serag 2012). Acute HBV infection advances to chronic infection, which increases the chance of developing cirrhosis and hepatic fibrosis as well as hepatocellular carcinoma (HCC).

A non-targeted study, gas chromatography–time-of-flight–mass spectrometry (GC-TOF-MS) metabolomics techniques to find disease-associated metabolites biomarkers that might help with initial HCC diagnosis by differentiating between HBV stages (Gao et al. 2015). In this study, researchers analyzed 201 blood serum samples from the healthy control groups and patients who suffered from different stages of diseases such as liver cirrhosis, HBV, and HCC (Gao et al. 2015). In conclusion, investigators identified 15 metabolites that were closely associated with the procedure. As potential biomarkers unique to hepatic disease, phenylalanine, 2-hydroxybutanedioic acid, and 5-methoxytryptamine were examined for HBV vs. healthy control group, hexadecanoic acid for liver cirrhosis vs. HBV, and also asparagine and β-glutamate for HCC vs. liver cirrhosis. These all markers revealed dominant discriminant performance. Another one non-targeted metabolomics study in which, researchers collected saliva fluid from 16 healthy control groups and 20 samples from chronic HBV patients were investigated by NMR technique (Gilany et al. 2019). Subsequently, metabolites that might distinguish between two groups were identified using statistical analysis. Ultimately, researchers identified a series of metabolites including propionic acid, putrescine, acetic acid, succinic acid, tyrosine, lactic acid, butyric acid, pyruvic acid, pyridoxine, and p-hydroxybenzoic acid, which used might accurately discriminate patients from healthy control groups.

Zhang et al. (2013) used ultraperformance liquid chromatography (UPLC/TOF) techniques for determining metabolite change analysis in HBV patients. By using the MetPA tool to analyze metabolic pathways, this study identified a total 11 (tyrosinamide, alpha-N-phenylacetyl-L-glutamine, biotin sulfone, hexanoic acid, 5-oxo-heneicosanoic acid, D-glucosaminide, 1-amino naphthalene, phenyl sulfate, 7-dehydrocholesterol, azelaic acid, and 2-methylhippuric acid) metabolites that were involved in the development of HBV (Zhang et al. 2013). In terms of biomarker potential for disease diagnosis, some metabolites have promising results. Conspicuously, in terms of precision, sensitivity, and specificity, especially four metabolites–biomarkers showed potent efficacy in the diagnosis of HBV. Metabolomics is a non-invasive procedure for biomarkers tracking in the early diagnosis of HBV. This study may progress methods for diagnosing, treating, and preventing HBV and offer insightful data on its pathogenesis. This study may progress methods for diagnosing, treating, and preventing HBV and offer insightful data on its pathogenesis.

3.4 Metabolomics Application to AIDS

AIDS is a deadly infectious disorder caused by the retrovirus known as HIV. This pathogen leads to life-threatening infection and mortality worldwide (Waymack and Sundareshan 2024). HIV shows variations in clinical symptoms and indications; its key characteristics consist of high a mortality rate and rapid transmission (Ghannoum et al. 2013). HIV is commonly transmitted through infected blood transfusions, unprotected sexual intercourse, and infected needles (Hewer et al. 2006). It rapidly targets and undermines the immune system after infecting the body, especially affecting $CD4^+$ T lymphocytes, which results in the destruction or malfunction of the immune cells, leading to reduced immune response to infections and other immune-related diseases and even advancement of malignancies (Ghannoum et al. 2013).

Metabolomics techniques are an emerging and crucial tool in HIV research, especially in the field of viral vaccine development and early disease diagnostics. Scientists' primary objective of studies has been to identify biomarkers present in bio-samples, including blood plasma, to evaluate the protective effects of potential immunizations. Early research showed that verifying HIV patient's blood metabolic profile can be a beneficial implement for discrimination between HIV-positive and HIV-negative individuals (Hewer et al. 2006). In several studies, researchers have found significant differences between HIV-positive people who received antiretroviral therapy (ART+) and HIV-negative people (Hewer et al. 2006; Philippeos et al. 2009).

A study with a targeted metabolomics standard method attempted to determine the molecules related to the progress of atherosclerosis and endothelial dysfunction by analyzing the metabolomic profile in a sample of HIV-negative, ART+, and HIV-positive patients. Scientists recruited 144 participants and were categorized into 3 groups: HIV negative, HIV positive, ART+, and HIV positive/ART-, which were targeted metabolomic analyses via standard operating procedures (SOPs) (Frias et al. 2020). The fibromuscular dysplasia (FMD) was used to measure endothelial dysfunction, while coronary intima-media thickness (CIMT) was used to measure the advancement of atherosclerosis, and Meso Scale Discovery (MSD) was also used to measure inflammatory biomarkers. Ultimately, researchers identified 257 metabolites, 8 metabolites associated with FMD, and 11 associated with CIMT. Cytosine showed a negative correlation with FMD and a significant correlation with CIMT. Eighteen metabolites identified in ART+ patients showed the increased quantity of kynurenine and 3–2-hydroxyphenyl propanoate, along with the reduced level of itaconate compared to ART- patients. Also comparing HIV-negative individuals and HIV-positive patients showed a significant proatherogenic profile.

GC-MS and LC-MS techniques were used for metabolite characterization in HIV patients, oral samples from 12 HIV-positive patients were collected, and 12 healthy individuals were analyzed by metabolomics (Ghannoum et al. 2013). After the process, researchers identified 198 detectable and 85 nondetectable metabolites, of which 27 (12 increased, e.g., p-cresol sulfate, allantoin, and tryptophan, and 15 decreased, e.g., fucose, fumarate, and N-acetylglucosamine) detectable metabolites were identified in HIV-positive patients. When compared to HIV-negative patients, HIV-positive patients on ART+ had the highest quantity of statistically metabolite correlation changes. Curiously, HIV-positive people who had not started ART showed a higher oral phenylalanine ratio than both control individuals and those who were using ART.

3.5 Metabolomics Application to Hepatitis C

According to world health organization(WHO), the hepatitis C virus (HCV) is a major infectious disease that, in 70% of cases, progresses to chronic illness. The most widely used method to measure the advancement of liver fibrosis in HCV patients is the Meta-analysis of Histological data in Viral Hepatitis (METAVIR) grading system, which categorized fibrosis stages from F0 (no fibrosis) to F4 (cirrhosis). There is an increased chance that advanced-stage fibrosis may progress to HCC (Bedossa and Poynard 1996). The field of metabolomics, which fills the gap between genotype and phenotype, has an emerging and crucial tool for the diagnosis of HCV, enhancing our knowledge of the disease progression and assisting in the development of targeted treatments. Metabolomics allows a sensitive and noninvasive procedure to analyze phases of the disease, and several effective studies have been done in this area, but there are limited studies available on metabolic changes in hepatic damage. This research compared a group of patients who tested positive for HCV with and without direct-acting antiviral interferon (DAA-IFN-free regimens). Using 1H-NMR, their findings clearly determine that HCV patients' metabolomic profiles differed before and after receiving successful DAA treatment. Observable metabolite level changes in treated individuals suggest modifications to many metabolic pathways (Meoni et al. 2019).

Nomair et al. (2019) examined the differences in plasma metabolites between patients with HCV-induced cirrhosis and those with HCC, utilizing a metabolomics approach. Thirty-four different metabolites were found using GC/MS after the investigations; 22 HCC and 22 HCC patients were included (Nomair et al. 2019). Five metabolites glycine, oxalic acid, decanoic acid, oleic acid, and octanoic acid showed substantial discriminating effects. These findings suggest that metabolomics may be useful in comparing the features of individuals with HCV-induced cirrhosis and HCC, signifying that treatment intervention may focus on metabolic pathways linked to fatty acids.

4 Metabolomics of Bacterial Infections

The metabolomics approach makes it possible to comprehensively explore the metabolic changes that are carried by adjacent pathogens, offering vital insights on the relations between pathogens and hosts as well as the significant processes of bacterial infection. The field of metabolomics has been employed to quantify and qualitatively distinguish between metabolites to identify differences in the metabolic pathways associated with bacterial infections. As an outcome, several biomarkers related to pathogenic conditions have been identified; these can be key molecules for early diagnosis and the making of personalized treatments (Luu et al. 2023). In this section, we discuss the application of metabolomics to bacterial infectious diseases such as *Clostridium difficile* and tuberculosis.

4.1 Metabolomics Application to Tuberculosis

Mycobacterium tuberculosis (Mt), hereafter referred to as Mtb, is the bacterium that causes tuberculosis (TB), a serious infectious disease. Background TB continues to be the leading single infectious agent cause of death, despite being curable. Estimates currently suggest that approximately 15% of the world's population is carrying this pathogen. However, Only 10% of these patients develop symptomatic TB, while the rest continue to have latent disease. People with HIV have a lifetime risk of 5–15%, compared to those without HIV who only have the risk during their first year after infection. The most common combination drug for treating tuberculosis is a 6-month course of multiple antibiotics. The global epidemic of multidrug-resistant (MDR) and extensively drug-resistant (XDR) TB now presents a major obstacle to effective management. Metabolomics has been proposed as an influential tool to deepen TB research by characterizing the metabolic changes caused by *Mycobacterium tuberculosis* infection (Howard and Khader 2020). It helps in uncovering novel biomarkers for early diagnostics disease monitoring (Table 1). Metabolomic studies have identified unique metabolic signatures in TB patients that provide insight into the host–pathogen interaction and new opportunities for targeted intervention or improved diagnostics (Yu et al. 2023). There are several metabolomics studies completed for disease diagnosis and disease-associated biomarker identification; here, we discuss some latest metabolomics studies. One study used the UPLC-MS for TB-associated biomarker identification. Researchers collected blood samples from healthy individuals, TB patients, TB-cured peoples, and pneumonia (non-TB) patients and further performed amino acid profiling analysis. Outcomes of the degree of lung injury and participants (TB patients, healthy individuals, TB cured patients, and pneumonia non-TB patients) were shown to be significantly correlated with amino acid levels. Phenylalanine had an indicated respectable efficiency in TB diagnosis. The ratio of plasma amino acid substance to total blood amino acid substance was used to create a prediction model, and it performed well in TB diagnosis, showing a sensitivity of 84% and specificity of 97%. Moreover, a robust negative correlation between plasma glutamine levels and illness severity was found by examining the correlation between high-resolution computerized tomography (HRCT) scores and amino acid levels. Notably, after cured TB groups, plasma had lower amounts of ornithine, and the ratio of four amino acids significantly changed (Sun et al. 2023).

Table 1 Diagnostic biomarkers for tuberculosis

Types of samples	Disease	Techniques	Identified biomarkers	References
Blood plasma	Tuberculosis and household contacts (HHCs)	LC-MS and MS	Phosphatidylglycerol, Lys- phosphatidylinositol, and acyl-phosphatidylinositol mannoside	Collins et al. (2018)
Blood plasma	Tuberculosis, DM, tuberculosis–DM, and HCs	1H-NMR	Histidine, phenylalanine, cholesterol, and sphingomyelin	Vrieling et al. (2018)
Blood plasma	Pulmonary tuberculosis and HHCs	LC-MS and MS	Glycolipids (trehalose-6-mycolate, phosphatidylinositol)	Frediani et al. (2014)
Urine	Symptomatic TB and asymptomatic control groups	HPLC-MS	Diacetylspermine, neopterin, hydroxybenzoic acid, and N-acetylhexosamine	Isa et al. (2018)
Urine	TB, non-TB, and HCs	1H-NMR	Variations in metabolic patterns noticed in urine samples	Comella-del-Barrio et al. (2021)
Serum	TB and non-TB	UPLC-MS/MS	Has-microRNA-215-5p and γ-glutamylthreonine	Krishnan et al. (2021)
Plasma	LTBI (treated and untreated), DS-TB, HCs, MDR-TB, HHCs, and HCs	MS/MS	Tryptophan and kynurenine/tryptophan	Collins et al. (2020)
Mtb strains and plasma sputum	TB and HCs	UPLC-MS/MS	Cholestenone in plasma sputum	Chandra et al. (2022)

4.1.1 Tuberculosis Diagnostic Biomarkers

Several host biomarker signs have been identified via extensive studies on the metabolomics of the host immune response, which has assisted with the diagnosis, prognosis, and condition of bacterial infections (Fernández-García et al. 2020). These markers have been found using a diversity of metabolomics approaches in combination with a diversity of host samples, including blood plasma, urine, saliva, and tissues (Table 1). The development of diagnostic tests precise to bacterial diseases might be advanced by these potential indicators.

4.2 Metabolomics Application to *Clostridium difficile*

Clostridioides difficile, formerly known as *Clostridium difficile*, is a gram-positive spore-forming bacillus bacterium. This bacterium causes an infection called *C. difficile* infection also known as (CDI), which is generally spread by spores from toxic variants of the bacterium; these infections can cause serious diseases including toxic megacolon and colitis, and in rare situations, they can even be fatal (Abt et al. 2016; Burdon et al. 1978). The high risk of CDI development is due to factors involving antibiotics usage, overage, and underlying poor health conditions like inflammatory bowel disease (IBD). The variation of the normal gut flora is a key factor in this improved sensitivity. In recent, there has been an increase in the frequency of complications, severity of the disease, and incidence of CDI (Chitnis et al. 2013; Ma and Lewis 2018). Metabolomics enables the precise identification of CDI-associated biomarkers, which help in the treatment of infection. Several studies have been conducted that help in CDI management.

Zhou et al. (2018) conducted a metabolome analysis using the UPLC-MS technology. Fecal samples were obtained from diarrhea patients, CDI-positive patients, CDI-negative individuals, and a healthy control group. In the PCA plots, the samples from the sick patients and the healthy individuals formed separate clusters. There was an even greater difference between the CDI patients and the non-CDI control group according to the partial least squares discriminant analysis (PLS-DA) model. Diarrhea patients who tested positive for CDI had relatively high levels of sphingosine, chenodeoxycholic acid, phenylalanine, lysophosphatidylcholine, and propylene glycol stearate. In contrast, the relative abundances of fatty amide, tyrosine, chenodeoxycholylglycine, and N-acyl-4-sphingenyl-1-o-phosphorylcholine were lower than those of the other sets. In conclusion, the analysis of fecal metabolome CDI diarrhea patients' samples provided strong evidence for the detection of infection. This study might be a helpful diagnostic method for CDI-associated diarrhea.

5 Conclusion and Future Perspectives

Advancements in technology and an improved understanding of metabolic pathways suggest promising future horizons for the clinical applicability of metabolomics-based analyses to infectious diseases. The analysis of the full set of metabolites in a biological system is termed as "metabolomics." Metabolomics provides multiple advantages for infectious disease research. Therefore, systemic metabolic changes introduced by the host–pathogen interplay during an early phase of infection offer insights toward the development of a prognostic marker to enable rapid and reliable diagnosis leading to timely intervention in controlling infection outbreaks (Rahman and Schellhorn 2023). Distinction of patients based on metabolic profile corresponding to disease severity and progression would enable individualized treatment pathways improving cost-effective use of resources. Metabolomics is helpful in identifying new therapeutic targets because of its value in showing the metabolic changes after pathogen infection.

References

Abt MC, McKenney PT, Pamer EG (2016) *Clostridium difficile* colitis: pathogenesis and host defence. Nat Rev Microbiol 14(10):609–620. https://doi.org/10.1038/nrmicro.2016.108

Alarcon-Barrera JC, Kostidis S, Ondo-Mendez A, Giera M (2022) Recent advances in metabolomics analysis for early drug development. Drug Discov Today 27(6):1763–1773. https://doi.org/10.1016/j.drudis.2022.02.018

Bedossa P, Poynard T (1996) An algorithm for the grading of activity in chronic hepatitis C. Hepatology 24(2):289–293. https://doi.org/10.1002/hep.510240201

Burdon DW, Brown JD, George RH, Arabi Y, Alexander-Williams J, Keighley MR (1978) Pseudomembranous colitis caused by Clostridia. N Engl J Med 299(1):48–49

Carbone F, Bruzzaniti S, Fusco C, Colamatteo A, Micillo T, De Candia P, Bonacina F, Norata GD, Matarese G (2021) Metabolomics, lipidomics, and immunometabolism. In: Annunziato F, Maggi L, Mazzoni A (eds) T-helper cells, vol 2285. Springer, pp 319–328. https://doi.org/10.1007/978-1-0716-1311-5_24

Chandra P, Coullon H, Agarwal M, Goss CW, Philips JA (2022) Macrophage global metabolomics identifies cholestenone as host/pathogen cometabolite present in human *Mycobacterium tuberculosis* infection. J Clin Invest 132(3):e152509. https://doi.org/10.1172/JCI152509

Cheng S, Shah SH, Corwin EJ, Fiehn O, Fitzgerald RL, Gerszten RE, Illig T, Rhee EP, Srinivas PR, Wang TJ, Jain M (2017) Potential impact and study considerations of metabolomics in cardiovascular health and disease: a scientific statement from the American Heart Association. Circ Cardiovasc Genetics 10(2):e000032. https://doi.org/10.1161/HCG.0000000000000032

Chitnis AS, Holzbauer SM, Belflower RM, Winston LG, Bamberg WM, Lyons C, Farley MM, Dumyati GK, Wilson LE, Beldavs ZG, Dunn JR, Gould LH, MacCannell DR, Gerding DN, McDonald LC, Lessa FC (2013) Epidemiology of community-associated *Clostridium difficile* infection, 2009 through 2011. JAMA Intern Med 173(14):1359. https://doi.org/10.1001/jamainternmed.2013.7056

Collins JM, Walker DI, Jones DP, Tukvadze N, Liu KH, Tran VT, Uppal K, Frediani JK, Easley KA, Shenvi N, Khadka M, Ortlund EA, Kempker RR, Blumberg HM, Ziegler TR (2018) High-resolution plasma metabolomics analysis to detect *Mycobacterium tuberculosis*-associated metabolites that distinguish active pulmonary tuberculosis in humans. PLoS One 13(10):e0205398. https://doi.org/10.1371/journal.pone.0205398

Collins JM, Siddiqa A, Jones DP, Liu K, Kempker RR, Nizam A, Shah NS, Ismail N, Ouma SG, Tukvadze N, Li S, Day CL, Rengarajan J, Brust JCM, Gandhi NR, Ernst JD, Blumberg HM, Ziegler TR (2020) Tryptophan catabolism reflects disease activity in human tuberculosis. JCI Insight 5(10):e137131. https://doi.org/10.1172/jci.insight.137131

Comella-del-Barrio P, Izquierdo-Garcia JL, Gautier J, Doresca MJC, Campos-Olivas R, Santiveri CM, Muriel-Moreno B, Prat-Aymerich C, Abellana R, Pérez-Porcuna TM, Cuevas LE, Ruiz-Cabello J, Domínguez J (2021) Urine NMR-based TB metabolic fingerprinting for the diagnosis of TB in children. Sci Rep 11(1):12006. https://doi.org/10.1038/s41598-021-91545-0

Cui L, Lee YH, Kumar Y, Xu F, Lu K, Ooi EE, Tannenbaum SR, Ong CN (2013) Serum metabolome and lipidome changes in adult patients with primary dengue infection. PLoS Negl Trop Dis 7(8):e2373. https://doi.org/10.1371/journal.pntd.0002373

Cui L, Lu H, Lee YH (2018) Challenges and emergent solutions for LC-MS/MS based untargeted metabolomics in diseases. Mass Spectrom Rev 37(6):772–792. https://doi.org/10.1002/mas.21562

Dunn WB, Broadhurst DI, Atherton HJ, Goodacre R, Griffin JL (2011) Systems level studies of mammalian metabolomes: the roles of mass spectrometry and nuclear magnetic resonance spectroscopy. Chem Soc Rev 40(1):387–426. https://doi.org/10.1039/B906712B

Edison AS, Colonna M, Gouveia GJ, Holderman NR, Judge MT, Shen X, Zhang S (2021) NMR: unique strengths that enhance modern metabolomics research. Anal Chem 93(1):478–499. https://doi.org/10.1021/acs.analchem.0c04414

El-Serag HB (2012) Epidemiology of viral hepatitis and hepatocellular carcinoma. Gastroenterology 142(6):1264–1273.e1. https://doi.org/10.1053/j.gastro.2011.12.061

Fernández-García M, Rey-Stolle F, Boccard J, Reddy VP, García A, Cumming BM, Steyn AJC, Rudaz S, Barbas C (2020) Comprehensive examination of the mouse lung metabolome following *Mycobacterium tuberculosis* infection using a multiplatform mass spectrometry approach. J Proteome Res 19(5):2053–2070. https://doi.org/10.1021/acs.jproteome.9b00868

Fraser DD, Slessarev M, Martin CM, Daley M, Patel MA, Miller MR, Patterson EK, O'Gorman DB, Gill SE, Wishart DS, Mandal R, Cepinskas G (2020) Metabolomics profiling of critically ill coronavirus disease 2019 patients: identification of diagnostic and prognostic biomarkers. Crit Care Explor 2(10):e0272. https://doi.org/10.1097/CCE.0000000000000272

Frediani JK, Jones DP, Tukvadze N, Uppal K, Sanikidze E, Kipiani M, Tran VT, Hebbar G, Walker DI, Kempker RR, Kurani SS, Colas RA, Dalli J, Tangpricha V, Serhan CN, Blumberg HM, Ziegler TR (2014) Plasma metabolomics in human pulmonary tuberculosis disease: a pilot study. PLoS One 9(10):e108854. https://doi.org/10.1371/journal.pone.0108854

Frias M, Berarpour N, Pagano S, Sidibe J, Kamau F, Lecour S, Strijdom H, Thomas A, Vuilleumier N (2020) Evaluation of antiretroviral therapy on metabolomics and atherogenic markers in HIV patients. Eur Heart J 41(Supplement_2):ehaa946.3307. https://doi.org/10.1093/ehjci/ehaa946.3307

Gao R, Cheng J, Fan C, Shi X, Cao Y, Sun B, Ding H, Hu C, Dong F, Yan X (2015) Serum metabolomics to identify the liver disease-specific biomarkers for the progression of hepatitis to hepatocellular carcinoma. Sci Rep 5(1):18175. https://doi.org/10.1038/srep18175

Ghannoum MA, Mukherjee PK, Jurevic RJ, Retuerto M, Brown RE, Sikaroodi M, Webster-Cyriaque J, Gillevet PM (2013) Metabolomics reveals differential levels of oral metabolites in HIV-infected patients: toward novel diagnostic targets. OMICS J Integr Biol 17(1):5–15. https://doi.org/10.1089/omi.2011.0035

Gilany K, Mohamadkhani A, Chashmniam S, Shahnazari P, Amini M, Arjmand B, Malekzadeh R, Nobakht Motlagh Ghoochani BF (2019) Metabolomics analysis of the saliva in patients with chronic hepatitis B using nuclear magnetic resonance: a pilot study. Iran J Basic Med Sci 22(9):1044–1049. https://doi.org/10.22038/ijbms.2019.36669.8733

Hewer R, Vorster J, Steffens FE, Meyer D (2006) Applying biofluid 1H NMR-based metabonomic techniques to distinguish between HIV-1 positive/AIDS patients on antiretroviral treatment and HIV-1 negative individuals. J Pharm Biomed Anal 41(4):1442–1446. https://doi.org/10.1016/j.jpba.2006.03.006

Howard NC, Khader SA (2020) Immunometabolism during *Mycobacterium tuberculosis* infection. Trends Microbiol 28(10):832–850. https://doi.org/10.1016/j.tim.2020.04.010

Isa F, Collins S, Lee MH, Decome D, Dorvil N, Joseph P, Smith L, Salerno S, Wells MT, Fischer S, Bean JM, Pape JW, Johnson WD, Fitzgerald DW, Rhee KY (2018) Mass spectrometric identification of urinary biomarkers of pulmonary tuberculosis. EBioMedicine 31:157–165. https://doi.org/10.1016/j.ebiom.2018.04.014

Kimhofer T, Lodge S, Whiley L, Gray N, Loo RL, Lawler NG, Nitschke P, Bong S-H, Morrison DL, Begum S, Richards T, Yeap BB, Smith C, Smith KGC, Holmes E, Nicholson JK (2020) Integrative modeling of quantitative plasma lipoprotein, metabolic, and amino acid data reveals a multiorgan pathological signature of SARS-CoV-2 infection. J Proteome Res 19(11):4442–4454. https://doi.org/10.1021/acs.jproteome.0c00519

Krishnan S, Queiroz ATL, Gupta A, Gupte N, Bisson GP, Kumwenda J, Naidoo K, Mohapi L, Mave V, Mngqibisa R, Lama JR, Hosseinipour MC, Andrade BB, Karakousis PC (2021) Integrative multi-omics reveals serum markers of tuberculosis in advanced HIV. Front Immunol 12:676980. https://doi.org/10.3389/fimmu.2021.676980

Luu GT, Little JC, Pierce EC, Morin M, Ertekin CA, Wolfe BE, Baars O, Dutton RJ, Sanchez LM (2023) Metabolomics of bacterial–fungal pairwise interactions reveal conserved molecular mechanisms. Analyst 148(13):3002–3018. https://doi.org/10.1039/D3AN00408B

Ma GK, Lewis JD (2018) Increasing incidence of multiply recurrent *Clostridium difficile* infection. Ann Intern Med 168(4):308. https://doi.org/10.7326/L17-0602

Mauvais-Jarvis F (2020) Aging, male sex, obesity, and metabolic inflammation create the perfect storm for COVID-19. Diabetes 69(9):1857–1863. https://doi.org/10.2337/dbi19-0023

Mayer KA, Stöckl J, Zlabinger GJ, Gualdoni GA (2019) Hijacking the supplies: metabolism as a novel facet of virus-host interaction. Front Immunol 10:1533. https://doi.org/10.3389/fimmu.2019.01533

Meoni G, Lorini S, Monti M, Madia F, Corti G, Luchinat C, Zignego AL, Tenori L, Gragnani L (2019) The metabolic fingerprints of HCV and HBV infections studied by nuclear magnetic resonance spectroscopy. Sci Rep 9(1):4128. https://doi.org/10.1038/s41598-019-40028-4

Nicholson JK, Lindon JC, Holmes E (1999) "Metabonomics": understanding the metabolic responses of living systems to pathophysiological stimuli via multivariate statistical analysis of biological NMR spectroscopic data. Xenobiotica 29(11):1181–1189. https://doi.org/10.1080/004982599238047

Nishiga M, Wang DW, Han Y, Lewis DB, Wu JC (2020) COVID-19 and cardiovascular disease: from basic mechanisms to clinical perspectives. Nat Rev Cardiol 17(9):543–558. https://doi.org/10.1038/s41569-020-0413-9

Nomair A, Madkour M, Shamseya M, Elsheredy H, Shokr A (2019) Profiling of plasma metabolomics in patients with hepatitis C-related liver cirrhosis and hepatocellular carcinoma. Clin Exp Hepatol 5(4):317–326. https://doi.org/10.5114/ceh.2019.89478

Overmyer KA, Shishkova E, Miller IJ, Balnis J, Bernstein MN, Peters-Clarke TM, Meyer JG, Quan Q, Muehlbauer LK, Trujillo EA, He Y, Chopra A, Chieng HC, Tiwari A, Judson MA, Paulson B, Brademan DR, Zhu Y, Serrano LR et al (2021) Large-scale multi-omic analysis of COVID-19 severity. Cell Syst 12(1):23–40.e7. https://doi.org/10.1016/j.cels.2020.10.003

Pauling L, Robinson AB, Teranishi R, Cary P (1971) Quantitative analysis of urine vapor and breath by gas-liquid partition chromatography. Proc Natl Acad Sci USA 68(10):2374–2376. https://doi.org/10.1073/pnas.68.10.2374

Philippeos C, Steffens FE, Meyer D (2009) Comparative 1H NMR-based metabonomic analysis of HIV-1 sera. J Biomol NMR 44(3):127–137. https://doi.org/10.1007/s10858-009-9329-8

Pomyen Y, Wanichthanarak K, Poungsombat P, Fahrmann J, Grapov D, Khoomrung S (2020) Deep metabolome: applications of deep learning in metabolomics. Comput Struct Biotechnol J 18:2818–2825. https://doi.org/10.1016/j.csbj.2020.09.033

Rahman M, Schellhorn HE (2023) Metabolomics of infectious diseases in the era of personalized medicine. Front Mol Biosci 10:1120376. https://doi.org/10.3389/fmolb.2023.1120376

Rahman M, Schellhorn H, Jithesh PV, Rahman MM (2022) Editorial: metabolomics in infectious diseases. Front Genet 13:875835. https://doi.org/10.3389/fgene.2022.875835

Roberts LD, Souza AL, Gerszten RE, Clish CB (2012) Targeted metabolomics. Curr Protoc Mol Biol 98(1). https://doi.org/10.1002/0471142727.mb3002s98, Unit 30.2.1

Roca M, Alcoriza MI, Garcia-Cañaveras JC, Lahoz A (2021) Reviewing the metabolome coverage provided by LC-MS: focus on sample preparation and chromatography-A tutorial. Anal Chim Acta 1147:38–55. https://doi.org/10.1016/j.aca.2020.12.025

Schmidt DR, Patel R, Kirsch DG, Lewis CA, Vander Heiden MG, Locasale JW (2021) Metabolomics in cancer research and emerging applications in clinical oncology. CA Cancer J Clin 71(4):333-358. https://doi.org/10.3322/caac.21670.

Spina R, Saliba S, Dupire F, Ptak A, Hehn A, Piutti S, Poinsignon S, Leclerc S, Bouguet-Bonnet S, Laurain-Mattar D (2021) Molecular identification of endophytic bacteria in *Leucojum aestivum* in vitro culture, NMR-based metabolomics study and LC-MS analysis leading to potential amaryllidaceae alkaloid production. Int J Mol Sci 22(4):1773. https://doi.org/10.3390/ijms22041773

Sun Y, Liao Y, Xiong N, He X, Zhang H, Chen X, Xiao G, Wang Z, Rao W, Zhang G (2023) Amino acid profiling as a screening and prognostic biomarker in active tuberculosis patients. Clin Chim Acta 548:117523. https://doi.org/10.1016/j.cca.2023.117523

Tounta V, Liu Y, Cheyne A, Larrouy-Maumus G (2021) Metabolomics in infectious diseases and drug discovery. Mol Omics 17(3):376–393. https://doi.org/10.1039/D1MO00017A

Vrieling F, Ronacher K, Kleynhans L, Van Den Akker E, Walzl G, Ottenhoff THM, Joosten SA (2018) Patients with concurrent tuberculosis and diabetes have a pro-atherogenic plasma lipid profile. EBioMedicine 32:192–200. https://doi.org/10.1016/j.ebiom.2018.05.011

Waymack JR, Sundareshan V (2024) Acquired immune deficiency syndrome. In: StatPearls. StatPearls Publishing. http://www.ncbi.nlm.nih.gov/books/NBK537293/

Whiley L, Lawler NG, Zeng AX, Lee A, Chin S-T, Bizkarguenaga M, Bruzzone C, Embade N, Wist J, Holmes E, Millet O, Nicholson JK, Gray N (2024) Cross-validation of metabolic phenotypes in SARS-CoV-2 infected subpopulations using targeted liquid chromatography–mass spectrometry (LC-MS). J Proteome Res 23(4):1313–1327. https://doi.org/10.1021/acs.jproteome.3c00797

Yu Y, Jiang X-X, Li J-C (2023) Biomarker discovery for tuberculosis using metabolomics. Front Mol Biosci 10:1099654. https://doi.org/10.3389/fmolb.2023.1099654

Zhang A, Sun H, Han Y, Yan G, Wang X (2013) Urinary metabolic biomarker and pathway study of hepatitis B virus infected patients based on UPLC-MS system. PLoS One 8(5):e64381. https://doi.org/10.1371/journal.pone.0064381

Zhou P, Zhou N, Shao L, Li J, Liu S, Meng, Duan J, Xiong X, Huang X, Chen Y, Fan X, Zheng Y, Ma S, Li C, Wu A (2018) Diagnosis of Clostridium difficile infection using an UPLC–MS based metabolomics method. Metabolomics 14(8). https://doi.org/10.1007/s11306-018-1397-x

Advancing Fish Nutrition Research Through Metabolomics: Unveiling and Optimising Nutritional Pathways

Nitesh Kumar Yadav, Sourabh Debbarma, Parvind Kumar, Jaya Angom, Yilbong Yirang, and Arun Bhai Patel

Abstract

Aquaculture production is expanding rapidly, necessitating the development of economically and environmentally sustainable alternative feedstuffs to reduce reliance on fish meal and fish oil-based diets. Efforts to enhance diet formulation for sustainable aquaculture have primarily involved substituting marine resources with plant-based alternatives. However, there is a rising interest in exploring additional alternative feed options. Technologies like metabolomics can expand our understanding of fish nutrition and the effects of feeds and nutrients on fish metabolism. Nonetheless, data from growth trials and body composition assessments remain crucial for evaluating fish nutritional health. In this chapter, we highlight findings from studies investigating the application of metabolomics in fish nutrition. We explore the concept of metabolomics and discuss specific technical considerations related to sample types, sampling techniques, and sample preparation methods. Our chapter demonstrates how metabolomics studies contribute to elucidating the impact of nutrition on fish metabolism.

Keywords

Aquaculture · Fish nutrition · Metabolomics · Fish metabolism · Feeds

N. K. Yadav (✉) · P. Kumar · J. Angom · A. B. Patel (✉)
Department of Aquaculture, College of Fisheries, Central Agricultural University (Imphal), Lembucherra, Tripura, India

S. Debbarma
Fisheries College and Research Institute, Tamil Nadu Dr. J. Jayalalithaa Fisheries University, Thoothukudi, Tamil Nadu, India

Y. Yirang
ICAR—Central Institute of Fisheries Education, Mumbai, Maharashtra, India

S. Kaur et al. (eds.), *Microbial Metabolomics*,
https://doi.org/10.1007/978-981-96-4824-5_20

1 Introduction

Metabolomics is the comprehensive chemical analysis of metabolic products (metabolites) and external pollutants (xenobiotics) within a biological system, such as cells, biofluids, or tissues. This approach involves high-throughput detection of small molecules (with molecular weights less than 1500 Da) and their absolute or relative quantification, whether or not these molecules have been identified (Wishart 2008; Chin and Slupsky 2013). Metabolites encompass a wide range of substances, including but not limited to organic acids, carbohydrates, amino acids, fatty acids, nucleosides and nucleotides, steroids and their derivatives, terpenoids, carotenoids, and flavonoids. Unlike targeted methods, the holistic approach of metabolomics allows for a more comprehensive understanding of the complex responses of a biological system when exposed to external factors, such as nutrition (Brennan et al. 2015; Sébédio 2017; Alfaro and Young 2018). By detecting small molecule metabolites, metabolomics provides insights into how nutrient metabolism is altered, offering a holistic view of fish metabolism and health (Roques et al. 2019; Debbarma et al. 2024). This approach focuses not only on metabolite identification and quantification but also on understanding various biological processes. Moreover, metabolomics facilitates the exploration of metabolic networks and pathways that regulate different physiological functions in fish. By mapping these pathways and their interconnections, researchers can identify novel indicators of nutritional status, metabolic health, and stress responses in fish populations. The data gathered is essential for enhancing feed efficiency, refining dietary formulations, and improving the overall health of both farmed and wild fish species.

Typically employing nuclear magnetic resonance (NMR) and mass spectrometry (MS), fingerprinting and profiling are two widely used metabolomic techniques that can extract valuable information from biological systems (Goodacre et al. 2004; Dunn 2005). Fingerprinting involves the high-throughput analysis of numerous subjects to develop a model that incorporates all the analytical variables generated by metabolite detection. This method is primarily used to assess the impact of xenobiotics or environmental stressors, such as farming-related activities, and is also crucial for verifying the authenticity of food or feed. Since this approach focuses on the classification of individuals, signal identification is not required. On the other hand, profiling is based on the identification and/or quantification of analytical variables derived from the nonselective study of the biological system.

The identification of metabolites offers qualitative insights into metabolic pathways affected by internal or external influences (Yadav et al. 2024a, b). For instance, in prison grouper (*Epinephelus septemfasciatus*), metabolite profiling has been utilised to characterise gut content following the transition from a natural diet to a synthetic diet (Asakura et al. 2014). Profiling and fingerprinting are complementary methods for understanding how environmental factors impact metabolism. For example, Gil-Solsona et al. (2017) integrated these techniques to identify nutritional indicators and associated metabolic adaptations in fasting gilthead sea bream (*Sparus aurata*). Metabolomics in fish nutrition research primarily utilises NMR and MS, which, despite their complementary strengths in selectivity, sensitivity, and structural information, are rarely combined. This chapter aims to explore how metabolomics enhances fish nutrition by elucidating metabolic pathways and optimising dietary strategies. It will cover the integration of NMR and MS techniques, investigate metabolic networks and pathways, and identify novel biomarkers for assessing fish nutrition, health, and stress responses. The goal is to improve feed efficiency, refine dietary formulations, and advance the overall health of both farmed and wild fish species.

2 Metabolomic Workflow for Samples and Sampling in Fish Nutrition Metabolomics

A traditional metabolomic workflow consists of five key processes: data gathering, processing, statistical analysis, biological interpretation, and sampling, along with analytical sample preparation, as illustrated in Fig. 1 (Brennan 2013; Young and Alfaro 2018). The majority of these procedures are the same for all metabolomic investigations; however, in fish nutrition research, special attention must be paid to collection and analytical sample preparation. The validity of metabolomic data is significantly impacted by the kind of sample used, the collection procedure, and the analytical sample preparation, particularly when biofluids are used. Currently, attention is concentrated on integrative tissues or biofluids, which provide a worldwide perspective on metabolism to evaluate how diets impact fish performance. By using metabolomics, the assessment of fish nutrition has mostly focused on the muscle and, to a lesser degree, the liver and plasma.

2.1 Samples Used for Fish Nutrition Metabolomics

The primary tissue for metabolic studies in fish is muscle, with the dorsal muscle commonly sampled due to its representativeness of overall metabolism. Both muscle tissue and organs like the liver provide valuable insights into metabolic pathways which is influenced mainly by feed. Blood, on the other hand, reflects circulating compounds, resulting from digestion and metabolism. Skeletal muscle is complex, comprising superficial red dawdling oxidative muscle and deep fast white muscle, each with distinct metabolic activities and corresponding metabolite profiles related to anaerobic and glycolytic metabolism (George and Don Stevens 1978; Kiessling et al. 2006).

The skeletal muscle, particularly the deep white muscle, is commonly sampled for fish nutrition and metabolism studies, while the edible portion of the fish is used for quality assessments. Liver sampling often lacks differentiation between its various sections, although variability in metabolite profiles has been observed among different liver sections in Arctic char (*Salvelinus alpinus*) (Cheng et al. 2016; Yadav et al. 2024c). To avoid sampling bias, it is recommended to sample the entire liver or consistently from the same portion of each liver. For example, Abro et al. (2014) systematically sampled the right side of the ventral lobe to ensure uniformity.

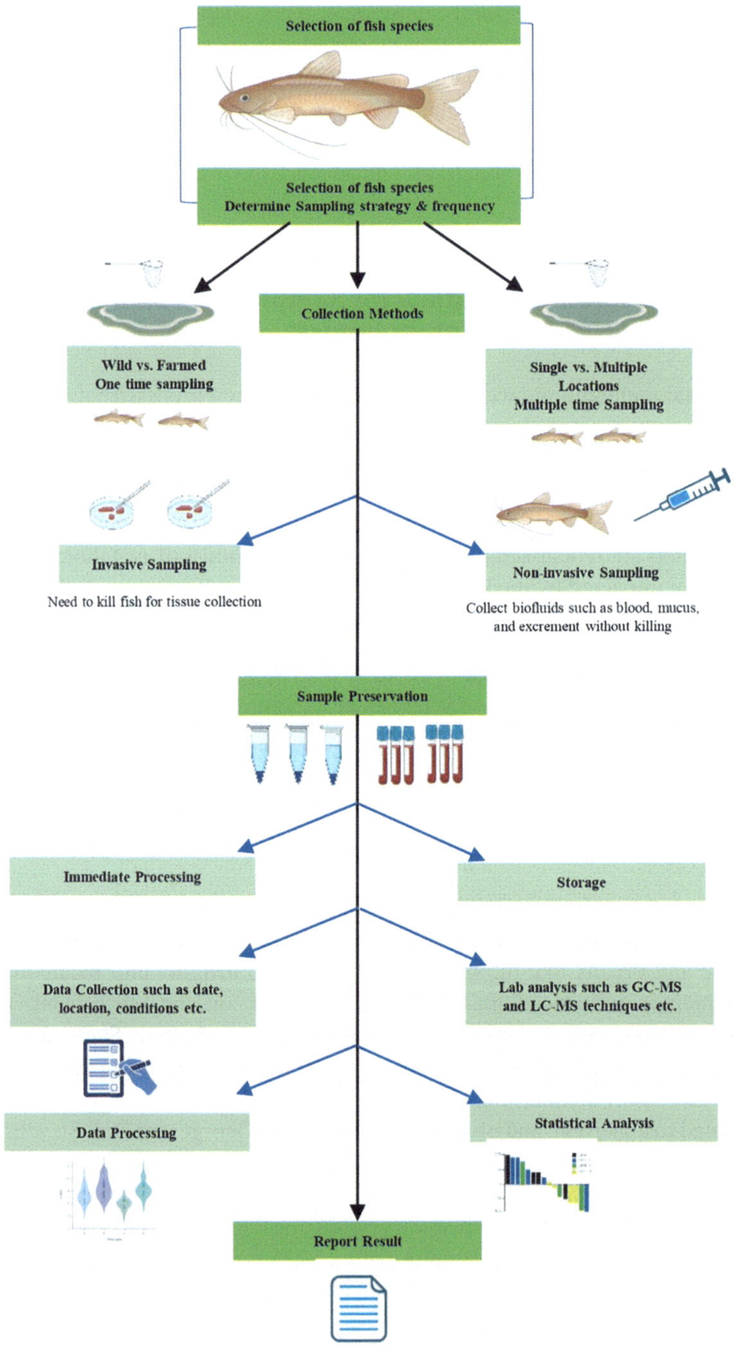

Fig. 1 Flowchart of sampling steps for metabolomics

Urine is commonly analysed in metabolomic studies of humans and some animals (e.g. bovines) as it contains end products of metabolism, allowing for quantitative assessments over a 24-hour period. This analysis provides valuable insights into how food, beverages, and pollutants are processed metabolically. However, in fish, most metabolic waste products are expelled through the gills rather than through urine. Consequently, blood is the preferred biofluid for evaluating fish metabolism. Blood sampling is advantageous as it is less invasive, supports animal welfare, and allows for serial sampling over time. Additionally, faeces and mucus can be used to monitor metabolic end products related to microbiota and immune activities (Asakura et al. 2014; Reverter et al. 2017).

2.2 Sampling Flow in Fish Nutrition Metabolomics

In metabolomics, careful attention must be paid to sampling techniques, as they can significantly impact metabolic patterns and lead to inaccurate results. To ensure robust statistical analysis, it is advisable to collect as many samples as possible. Typically, metabolomics studies in fish nutrition research use a median of 12 fish per condition. This sample size is modest compared to human metabolomic studies, which often involve over 100 samples. However, in controlled livestock studies, fewer than 30 animals are commonly used. This approach helps to minimise the number of experimental animals and reduce costs while still adhering to high standards of animal welfare (Goldansaz et al. 2017). This is over three times greater than what is often done with fish, despite the fact that individual differences in metabolism are typically seen. It's interesting to note that some authors take many samples of the same tissue to make up for the small number of fish per treatment (Mannina et al. 2008). The field of metabolomics for fish nutrition is likely only getting started, with the goal of proving the technology's viability and researching significant feeding impacts. Furthermore, it's possible that the sample sizes utilised in fish studies were only carried over from earlier research on the very stable and repeatable method of controlling the origin and quality of fish based on lipid content.

Furthermore, metabolic processes are dynamics that occur in both living and dead fish, for example, in live fish owing to circadian rhythm and several postmortem modifications. Therefore, it is advised to precisely manage environmental variables like oxygen and temperature that have an impact on the dynamics of metabolites during fishing and raising. Since metabolomics is a time-sensitive approach, we advise taking into account the precise timing of the sample in relation to the most recent eating, since this might have a significant impact on the properties of the sample in the postprandial phase. Longer postprandial sample durations are more suited for researching fish metabolism, but shorter times are better for examining the feed; the timing of the postprandial sampling should be adjusted to account for a number of variables that influence the gut transit time, such as species, gut capacity, temperature, as well as season. Metabolite variability is also influenced by postmortem sampling time. Lactate, biogenic amines, phosphorylated compounds, and other metabolites are altered in concentration by anaerobic reactions that take place in removed tissue (Picone et al. 2011). Consequently, in order to prevent tissue degradation, samples need to be snap-frozen in liquid nitrogen as soon as they are killed. Furthermore, especially when euthanasia is carried out slowly using techniques like hypothermia, tissue metabolite profiles may be significantly impacted. These techniques used on farms put fish under stress and force them to swim for longer periods of time, which raises ATP and causes IMP to break down even more (Özogul and Özogul 2004; Overmyer et al. 2015). Most of the tissue sampling for fish nutrition is done shortly after death, and experimental fish are often quickly killed for animal welfare reasons. However, wild fish that are taken in their native habitat do not follow this pattern. Anaesthetic substances can also alter the quantity of several metabolites in fish under anaesthesia, such as glucose in fish under a benzocaine anaesthesia (Sneddon 2012). In a subsequent metabolic study, it might be challenging to distinguish liver metabolites from blood residual metabolites when liver is obtained without pretreatment. Blood still contains feed nutrients; therefore, this is particularly crucial if sample is done soon after the previous feeding.

In mammals, such as mice, saline solution transfusion can be used to remove residual blood, though this procedure is time-consuming and challenging to implement for metabolomic studies (Ly-Verdú et al. 2014). In fish, washing the liver with saline helps ensure high-quality samples by minimising blood contamination and, in some cases, bile salts from the gall bladder. Noninvasive sampling techniques present additional challenges, particularly concerning the volume of blood that can be collected. Typically, up to 0.5% of the fish's total body weight, or approximately 0.5 ml for a 100 g fish, can be safely collected without causing significant mortality (Iwama et al. 2011; Zang et al. 2015). Metabolomic methods generally require millilitre-sized blood samples. However, depending on the techniques and instrumentation used, microliter samples may suffice, which is particularly important in studies involving small organisms (Dunn 2005). While blood sampling provides serum or plasma, the metabolite profiles of these biofluids can differ significantly from those in human blood. This variability should be considered when studying fish, as their blood metabolite profiles may not directly align with those of other species (Yin et al. 2015). Moreover, the composition of plasma varies with the postprandial sampling period, leading to different concentrations of macromolecules, primarily proteins and lipoproteins, at different times. These plasma lipoproteins can interfere with direct plasma injection in LC-MS (liquid chromatography-mass spectrometry) and hinder the identification of small metabolites in NMR. Consequently, specific sample preparation protocols for LC-MS or specialised protocols for ^{1}H NMR collection must be employed to address these challenges.

For MS, sample preparation involves solvent-based extraction, ultrafiltration, and deproteinisation. Despite these methods, there are significant challenges. For example, lipoproteins can clog ultrafiltration membranes, impeding the passage of small molecules (Tiziani et al. 2008). In NMR, sequences like Carr–Purcell–Meiboom–Gill are employed to reduce signals from macromolecules, but these sequences can also affect small molecule signals differently, complicating the quantification of metabolites. Additionally, the preparation of fish serum is complicated by the presence of nucleated erythrocytes in fish blood, unlike the unnucleated erythrocytes in mammals, which makes controlling clotting more challenging.

When delivering serum for fish, it is important to consider the species-specific clotting mechanism. This is particularly true for cold-water species, as they have less blood cells and are raised in colder climates (Witeska 2013). Fish metabolism may be impacted by other experimental, environmental, and biological variables, which need to be carefully studied and managed. Growth indices in fish have been demonstrated to be impacted by the tank effect, particularly in tanks situated adjacent to the main pathways (Speare et al. 1995). It has been shown that environmental elements like temperature and oxygen have an impact on the fish metabolome (Kullgren et al. 2013; Lardon et al. 2013).

Finally, biological variables such as sexual maturation cause modifications linked to a decrease in muscle fat content (Melis and Anedda 2014; Zhou et al. 2017). Additionally, the liver of zebrafish (*Danio rerio*) varies depending on the gender (Ong et al. 2009). Thus, it is crucial to regulate analytical and experimental parameters to guarantee the accuracy of metabolomics data. Although this is outside the purview of this assessment, such reliability can also be reexamined in terms of analytical accuracy, precision, repeatability, and reproducibility. It may be argued that metabolomics merely offers a cursory understanding of an organism, unable to properly illustrate the changes brought about by dietary changes. Metabolomics, like other omics, is a component of organism-to-organism interactions, although it is more applicable to functional activities, such as those of fish (Goodacre et al. 2004). Consequently, metabolomics may serve as a starting point for understanding metabolic processes, and when combined with other omic techniques, it should increase confidence in the findings. For instance, Mekuchi et al. (2017) evaluated the impact of circadian rhythm on the nutritional status of leopard coral groupers (*Plectropomus leopardus*) by using transcriptomics and metabolomics.

3 Applications of Metabolomics in Fish Nutrition Research

Aquaculture nutritional research seeks to enhance growth trajectories and production yields while also improving the nutritional status of cultivated species. Providing the right amount and quality of nutrition is likely to have a significant impact on other aspects of farming, including improving individual performance; preventing disease; improving the quality of gametes and broodstock; developing high-quality, sustainable feed substitutes; and reducing environmental effects, among other things.

3.1 Biomarkers of Food Intake and Dietary Patterns

Fish metabolic profiles are obviously affected by feeds, indicating that dietary-related metabolic alterations take place. The next section summarises the sorts of research that metabolomics has conducted on experimental diets and then focuses on which major metabolic processes are modified to assess these responses.

In fish nutrition research, there are three types of trials for metabolomic assessment:

(a) Challenge: The entire nutritional influence on fish metabolism is assessed by comparing two distinct feeds or fish circumstances (including disparate eating behaviours). It usually entails evaluating artificially produced fish meals in contrast to the natural meals found in wild fish.
(b) Substitution: This occurs when the goal of the experimental nutrition study is to assess the replacement of one or more feedstuffs, usually FO (fish oil) and FM (fish meal). Typically, the reference diet consists of including either an FO or an FM.
(c) Supplementation: The addition of a single substance, ingredient, or feedstock to enhance or augment the feed composition. Most studies utilise a modern commercial diet that still includes some FM and FO, along with a combination of plant protein and oil sources, as the reference diet; however, others employ a completely plant-based diet.

The National Research Council (U.S.) (2011) states that the diets evaluated in the metabolomics literature for the challenge trials include numerous formulations that clearly fit species needs and eating behaviours (omnivorous, herbivorous, or carnivorous). Fish-derived resources like decontaminated FM or fish protein hydrolysates, as well as microorganism feedstocks like zygomycete biomass or baker's yeast (*Saccharomyces cerevisiae*), are used in substitution experiments in place of FM and/or FO.

The majority of the studies rely on rather brief feeding trials (12–20 weeks). Although they appear to be long-term studies testing diets across the course of the fish life cycle, none of them makes use of metabolomics to evaluate how fish metabolism has adapted to these diets. Furthermore, the evaluation of metabolomics is nearly always carried out at the conclusion of the feeding test; however, only a few numbers of studies have examined the changes over time throughout the feeding test, showcasing adaption trajectories (Schock et al. 2012; Asakura et al. 2014; Casu et al. 2017).

3.2 Disturbances in Muscle Energy Metabolism

Fish muscles are mostly used for motion during rest periods and rapid swimming. To assure these two swimming modes, respectively, the muscle fibres that are red and white fast and slow are specialised. Fish primarily use white muscular fibres, which make up over 80% of their total muscle mass, when they need to burst for a brief period (Beamish 1978). To maintain swimming action, muscles must receive an appropriate amount of energy substrates. Furthermore, the skeletal muscle of ectothermic fish, particularly the white muscle, has a lower metabolic activity than that of endothermic vertebrates. Therefore, it is possible to determine the long-term effects of dietary modifications thanks to the comparatively low turnover of metabolites in fish muscle (Yoshida et al. 2014).

Challenge experiments, which contrast manufactured feeds including plant-based products with natural food sources, consistently demonstrate disruptions in muscle energy metabolism in terms of both the muscle's energy reserve and the metabolites that are involved in energy synthesis. In fact, it was shown that sea bass muscle extracts from fish fed a commercial diet had elevated levels of fumaric and malic acids, two important tricarboxylic acid (TCA) cycle intermediates (Mannina et al. 2008). Additionally, muscle extracts from wild saithe populations near salmon farms that had eaten lost pellets showed greater lactate concentrations (Maruhenda Egea et al. 2015).

The primary byproduct of anaerobic energy metabolism is lactate, which indicates a disruption in muscular energy status caused by nutrition; however, it's also possible that fishing and storing wild fish in various environments have an impact. Numerous variables can contribute to the generation of lactate. Because burst swimming raises the amount of lactate in muscle, lactate is a useful indicator of fish activity (Goolish 1989; Pearson et al. 1990). The way that handling fish increases muscular activity and, thus, lactate generation serves as evidence for this. Higher levels of lactate production are also linked to chronic stress (Karakach et al. 2009). Moreover, lactate is retained in fish muscle, particularly in rainbow trout (Gleeson 1996; Weber et al. 2016).

The quick depletion of glycogen reserves during the postmortem sampling period has an impact on lactate content as well. Lactate, an end product of glycolysis, builds up in the muscle as a result of anaerobic events that occur in the muscle after slaughtering, which favour the utilisation of glycogen reserves to supply glucose for energy (Bankefors et al. 2011). Rich meals high in carbohydrates can also raise the levels of lactate because they enhance glycogen storage and lead to preferred glycogen breakdown for energy needs (Maruhenda Egea et al. 2015). It's fascinating to contrast the metabolic effects of these plant-based diets with fish fasting, which exhibits a pronounced state of energy deficiency.

The preferential utilisation of stored lipids to sustain energy supply was the first metabolic consequence in starved rainbow trout (Kullgren et al. 2010; Baumgarner and Cooper 2012). Second, and perhaps most importantly, NMR analysis showed that the fasting fish had changes in energy metabolites in their muscles other than fatty acid variations. These included an increase in phosphocreatine and alanine and a reduction in lactate and betaine (Kullgren et al. 2010). Depleted levels of oxalate and malonate, which are inhibitors of particular TCA cycle enzymes, also point to a change in the energy status of fasting fish, as demonstrated by GC-MS (Baumgarner and Cooper 2012).

Leucine and isoleucine amino acids increased as a possible signal at the cell level to control energy metabolism as a result of food deprivation and circadian rhythm influence (Mekuchi et al. 2017). These fluctuations in energetic metabolites are also seen in fish given diets that substitute plant products for marine resources, indicating that the fish's energy status has become insufficient. However, compared to feds fish, who receive their energy substrates mostly from feed, the processes in fasted fish, where energy substrates are exclusively obtained through catabolism, may differ. Collectively, these metabolomics findings show that fish fed plant-based diets have an energy-deficient state and that their muscles have adapted to sustain their basic metabolic processes, such swimming (Kullgren et al. 2010).

3.3 To Recognise Impairment of Liver Metabolism

The liver is the primary organ in fish nutrition because it collects and distributes the substantial dietary nutrient supply to the other tissues through the portal vein, which is directly related to the digestive system (Rust 2003). The synthesis of lipoproteins to redistribute lipids to peripheral tissues is part of the delivery function. In addition, the liver performs detoxifying procedures, produces bile for the breakdown of lipids, and stores carbohydrates and fats (Brusle 1996). As a result, the liver is important for the control of metabolism, which has been thoroughly examined in omics research. Metabolomics examination of the liver may thus offer pertinent information on how dietary changes affect metabolism.

Studies on replacement and challenge have shown comparable results for the adaptation of glucose metabolism. Although there are still many unanswered concerns, fish's metabolism of glucose is mostly comparable to that of mammals, with the exception of carnivorous fish, which appear to have poor glucose metabolism. This is particularly true for carnivorous fish, which display a postprandial hyperglycaemia that is occasionally linked to "fatty liver" disease (Polakof et al. 2012). These fish are fed a diet that contains more than 20% carbohydrates. When compared to fish fed diets based on marine resources, plant-based diets in carnivorous species appear to alter glycolysis and gluconeogenesis in the liver. In fact, hepatic glucose content was modulated in red drum fed soybean products, juvenile turbot fed plant-protein-based diets, and Atlantic salmon provided diets enriched with sesame seeds (Wagner et al. 2014).

Lipid metabolism and transport are two key metabolic processes impacted in the liver. Increases in glycerol-3-phosphate and other metabolites have been seen in carnivorous fish fed plant-based diets, indicating improved lipid catabolism (Casu et al. 2017). Conversely, fish given a sesamin-supplemented diet exhibited a drop in hepatic glycerol, which may be the consequence of lipid retention in the liver and therefore a suppression of lipolysis (Wagner et al. 2014). Glycerol-3-phosphate is present at the junction of lipid catabolism and glycolysis in mammals, where it serves as a substrate for mitochondrial glycerol-3-phosphate dehydrogenase (Mráček et al. 2013). According to Casu et al. (2017), the increased level of this metabolite indicates a rise in lipid catabolism to supply a substrate for gluconeogenesis.